Modern Birkhäuser Classics

Many of the original research and survey monographs in pure and applied mathematics published by Birkhäuser in recent decades have been groundbreaking and have come to be regarded as foundational to the subject. Through the MBC Series, a select number of these modern classics, entirely uncorrected, are being re-released in paperback (and as eBooks) to ensure that these treasures remain accessible to new generations of students, scholars, and researchers.

Jan Prüss

Evolutionary Integral Equations and Applications

Reprint of the 1993 Edition

 Birkhäuser

Jan Prüss
Institut für Mathematik
Martin-Luther-Universität
Halle-Wittenberg
Halle
Germany

ISBN 978-3-0348-0498-1   ISBN 978-3-0348-0499-8 (eBook)
DOI 10.1007/978-3-0348-0499-8
Springer Basel Heidelberg New York Dordrecht London

Library of Congress Control Number: 2012944422

Mathematics Subject Classification (2010): 45N05, 34K30, 47D06, 47D09, 47N20, 76A10, 78A25

© Springer Basel 1993
Reprint of the 1st edition 1993 by Birkhäuser Verlag, Switzerland
Originally published as volume 87 in the Monographs in Mathematics series

Cover design: deblik, Berlin

Printed on acid-free paper

Springer Basel AG is part of Springer Science+Business Media (www.birkhauser-science.com)

Contents

Preface

During the last two decades the theory of abstract Volterra equations has undergone rapid development. To a large extent this was due to the applications of this theory to problems in mathematical physics, such as viscoelasticity, heat conduction in materials with memory, electrodynamics with memory, and to the need of tools to tackle the problems arising in these fields. Many interesting phenomena not found with differential equations but observed in specific examples of Volterra type stimulated research and improved our understanding and knowledge. Although this process is still going on, in particular concerning nonlinear problems, the linear theory has reached a state of maturity.

In recent years several good books on Volterra equations have appeared. However, none of them accounts for linear problems in infinite dimensions, and therefore this part of the theory has been available only through the – meanwhile enormous – original literature, so far. The present monograph intends to close this gap. Its aim is a coherent exposition of the state of the art in the linear theory. It brings together and unifies most of the relevant results available at present, and should ease the way through the original literature for anyone intending to work on abstract Volterra equations and its applications. And it exhibits many problems in the linear theory which have not been solved or even not been considered, so far. Even if one is interested in nonlinear problems only, it is useful to know the linear part of the theory; therefore this book offers something to 'nonlinear' people, too.

A word explaining the title of this monograph seems to be necessary. The equations to be studied are termed 'abstract Volterra integral equations' or 'abstract Volterra integrodifferential equations' in the literature, depending on their actual form and on the taste of the authors. These terms are somewhat misleading since they refer also to equations containing an abstract Volterra operator, a different notion. The title of this book, however, refers to the *evolutionary character* of the equations which describe the evolution in time of certain quantities called *state* of the system. And it refers to the *memory* present in the equations of state, expressed by certain *integrals* over a time interval in the past. For these reasons the term 'evolutionary integral equation' seems to be more accurate, although 'Volterra equation' is used also throughout this book.

The 'Introduction' contains the formulation of the problems to be studied, some motivating applications, and an outline of the ideas and methods employed in this monograph. There follows a section on notation and on some background material which has been included for the convenience of the reader. The main text consists of 13 sections which are grouped in three chapters. The last section of each chapter contains applications of the developed results to specific equations which are motivated by mathematical physics. The last paragraph of each section is devoted to bibliographical comments and further discussions.

A word of caution is necessary concerning the list of references. We have not tried to present a complete bibliography for the subject of this book, but instead to have a representative one, restricting it to the more important contributions.

Since it depends very much on personal views and taste what is considered to be essential or important, the selection of the bibliography should by no means expected to be impartial.

The numbering of equations, theorems, etc. is evident; for example Theorem 8.2 means Theorem 2 in Section 8, and (10.5) refers to Equation 5 in Section 10. References are indicated by names followed by numbers in square brackets which are found in the bibliography. The number following an item in the 'Index' refers to the page where it first appears in the text.

Finally, this book would have never been completed without the help of other people. I would like to thank many colleagues for discussions and suggestions, for enduring my smoking habits, and for sending me re- and preprints. Dr. H.J. Warnecke and his crew deserve credit for laser-printing the manuscript several times and for preparing the pictures. Prof. W. Desch, Dr. J. Escher, Dr. R. Schlott read parts of the book, I am very much indebted to them. It is a great pleasure to thank the 'Heisenberg Referat' of the 'Deutsche Forschungsgemeinschaft', Bonn, for continuous financial support. It is not enough to have an idea for a book but there must also be a publisher; in this respect I would like to thank the editors of this series, in particular Prof. H. Amann for his continuous interest and support, but also Dr. T. Hintermann, Birkhäuser Verlag, and collaborators. Most of the credit, however, goes to my family. My wife Bettina TEX-typed the whole manuscript with enthusiasm and carried out the unavoidably many changes and corrections without grumbling. She and also our children Sören and Saskia gave me a lot of encouragement and confidence, and did not complain during the long period of writing this book.

Paderborn, September 1992   Jan Prüss

Introduction

This book is devoted to the study of a class of abstract linear integral equations describing the evolution in time of the state of a system, and to applications of the developed theory to problems in mathematical physics. We begin with the description of the class of equations under consideration.

1. Evolutionary Integral Equations

Let X and Y be Banach spaces such that Y is densely embedded into X, in symbols $Y \overset{d}{\hookrightarrow} X$, and let $\mathcal{B}(Y, X)$ denote the space of bounded linear operators from Y to X, $\mathcal{B}(X) = \mathcal{B}(X, X)$ for short. Suppose $A : \mathbb{R}_+ \to \mathcal{B}(Y, X)$ is measurable and integrable over each compact interval $J \subset \mathbb{R}_+ = [0, \infty)$. Then we consider the equations

$$u(t) = \int_0^t A(\tau) u(t - \tau) d\tau + f(t), \quad t \in \mathbb{R}_+, \tag{1}$$

on the halfline $\mathbb{R}_+$, and

$$v(t) = \int_0^\infty A(\tau) v(t - \tau) d\tau + g(t), \quad t \in \mathbb{R}, \tag{2}$$

on the line $\mathbb{R}$. Here the functions f and g are given, and we are looking for solutions $u(t)$ of (1) and $v(t)$ of (2) in a sense to be specified later.

The family $\{A(t)\}_{t \geq 0}$ of in general unbounded linear operators in the *state space* X with domain $\mathcal{D}(A(t)) \equiv Y$ should be thought of as representing a physical *system*, $u(t) \in X$ resp. $v(t) \in X$ as the *state* of the system at *time* t, $f(t)$ resp. $g(t)$ as given *forces*, and (1) resp. (2) as the equations governing the state of the system. In applications, the space X will typically be a space of functions living on a domain $\Omega \subset \mathbb{R}^n$, $A(t)$ an elliptic differential operator and Y the domain of $A(t)$ which may incorporate certain boundary conditions; see 2. below.

Apparently, the principle of *causality* is valid for (1) and (2), since the present state of the system is determined by its *history* and the present force, but does not depend on the *future*. Therefore (1) and (2) are *evolutionary* in character, which explains the term 'evolutionary integral equations' for (1) and (2).

A second main features of these equations are their *time invariance*. This means that (1) and (2) do not change their form when a time shift is applied to the solutions. In fact, if $u(t)$ is a solution of (1), then $u_s(t) = u(t + s)$, $s \geq 0$, is again a solution of (1), where $f(t)$ is replaced by

$$f_s(t) = f(t + s) + \int_0^s A(t + \tau) u(s - \tau) d\tau, \quad t \geq 0.$$

Equation (2) even has the stronger property of *translation invariance*, i.e. if $v(t)$ is a solution of (2) then the translated function $v_s(t) = v(t+s)$, $s \in \mathbb{R}$, is a solution of (2) with the translate $g_s(t) = g(t + s)$ instead of $g(t)$. These invariance properties mean physically that the system under consideration does not change with time. Mathematically, this makes the powerful tools of transform theory available for the study of (1), and (2); more precisely, Laplace transform techniques can be applied to (1) and (2) is amenable to Fourier transform methods.

In the finite dimensional case dim $X < \infty$ (which implies $Y = X$), (1) and (2) reduce to the well understood 'linear Volterra integral equations' on the halfline resp. on the line. This notion has been employed in the literature also for the general unbounded case $Y \neq X$; we follow this 'tradition' and use the term 'Volterra equation' for (1) and (2) everywhere in this book, for the sake of brevity. We are neither concerned here with dim $X < \infty$ nor with the somewhat more general, but still simple case $Y = X$, dim $X = \infty$. However, since they are needed as a tool, in Sections 0.3 and 0.4 the main results for $Y = X$ are reproduced, namely the unique solvability of (1) on each compact interval $J = [0, T]$ in any of the usual function spaces, and the Paley-Wiener lemma on global solvability of (1) and (2). For an account of the theory of Volterra equations in finite dimensions -linear and also nonlinear- we refer to the recent monograph of Gripenberg, Londen and Staffans [156].

It is time now to present some model problems to explain what kind of concrete equations we have in mind. These problems are described in much greater detail in the sections containing applications, i.e. Sections 5, 9, and 13, where physical interpretations and some background information as well as detailed mathematical treatments can be found.

2. Model Problems
A rich source of problems leading to the Volterra equations (1) and (2) is provided by the theory of viscoelastic material behaviour. We begin with some typical examples from this field.

(i) *Simple Shear*
The following initial-boundary value problem is a typical example of one-dimensional problems in viscoelasticity, like simple shearing motions, torsion of a rod, simple tension; see Section 5.4.

$$\begin{aligned}
u_t(t, x) &= \int_0^t da(\tau) u_{xx}(t - \tau, x) + h(t, x), \quad t \geq 0, \ x \in [0, 1], \\
u(t, 0) &= u(t, 1) = 0, \quad t \geq 0, \\
u(0, x) &= u_0(x), \quad x \in [0, 1].
\end{aligned} \tag{3}$$

Here $a : \mathbb{R}_+ \to \mathbb{R}$ is a function of bounded variation on each compact interval $J = [0, T]$ with $a(0) = 0$, and the subscripts t or x mean partial differentiation w.r.t. the corresponding variable.

To obtain a formulation as an abstract evolutionary integral equation (1), choose a function space, say $X = C_0[0, 1]$, define an operator A_0 by means of $A_0 w(x) = w_{xx}(x)$ with domain $Y = \mathcal{D}(A_0) = \{w \in X : w_{xx} \in X\}$, and let $A(t) = a(t) A_0$. If X and Y are equipped with their natural norms, after an integration w.r.t. time t, (3) becomes (1) with force

$$f(t) = u_0(\cdot) + \int_0^t h(\tau, \cdot)d\tau, \quad t \geq 0,$$

and state $u(t) = u(t, \cdot)$.

This way, boundary conditions other than those of Dirichlet type can be treated, as well, and of course there are other choices for the state space X. $\square$

(ii) *Viscoelastic Fluids*
A model for the dynamic behaviour of the velocity field $u(t, x)$ of a 'linear' homogeneous isotropic incompressible viscoelastic fluid confined to a domain $\Omega \subset \mathbb{R}^3$ and subject to an external force $h(t, x)$ is the boundary value problem (see Section 5.6)

$$
\begin{aligned}
u_t(t, x) &= \int_0^\infty da(\tau)\Delta u(t - \tau, x) - \nabla p(t, x) + h(t, x), \quad t \in \mathbb{R}, \ x \in \Omega, \\
\operatorname{div} u(t, x) &= 0, \quad t \in \mathbb{R}, \ x \in \Omega, \\
u(t, x) &= 0, \quad t \in \mathbb{R}, \ x \in \partial\Omega,
\end{aligned}
\tag{4}
$$

where $p(t, x)$ denotes the also unknown pressure and the scalar function $a(t)$ is of bounded variation on $\mathbb{R}_+$. Observe that for $a(t) \equiv a_0 > 0$, (4) reduces to the linear Navier-Stokes equation for a Newtonian fluid.

Let P_q denote the Helmholtz projection in $L^q(\Omega)^3$ to the subspace $L^q_\sigma(\Omega)$ of divergence-free L^q-vector fields, and $A_q = P_q\Delta$ the Stokes operator; see Section 5.6 for more details. Let $Y = \mathcal{D}(A_q)$ be equipped with the graph norm of A_q and define $A(t) = a(t)A_q$. The application of P_q to (4) yields (2) in differentiated form, i.e.

$$\dot{u}(t) = \int_0^\infty dA(\tau)u(t - \tau) + g(t), \quad t \in \mathbb{R}, \tag{5}$$

where $g(t) = P_q h(t, \cdot)$, and the dot -as always in this book- indicates differentiation w.r.t. time t. $\square$

(iii) *Viscoelastic Beams*
The viscoelastic version of the Timoshenko beam model, where one end of the beam is clamped while the other is free, reads as follows (cp. Section 9.1).

$$
\begin{aligned}
w_{tt}(t, x) &= \int_0^t da(\tau)[w_{txx}(t - \tau, x) + \phi_{tx}(t - \tau, x)] + f_s(t, x), \\
\phi_{tt}(t, x) &= \int_0^t de(\tau)\phi_{txx}(t - \tau, x) \\
&\quad -\gamma \int_0^t da(\tau)[w_{tx}(t - \tau, x) + \phi_t(t - \tau, x)] + f_b(t, x), \\
w(t, 0) &= \phi(t, 0) = \phi_x(t, 1) = w_x(t, 1) + \phi(t, 1) = 0, \\
w(0, x) &= w_0(x), \quad \phi(0, x) = \phi_0(x), \\
w_t(0, x) &= w_1(x), \quad \phi_t(0, x) = \phi_1(x),
\end{aligned}
\tag{6}
$$

where $t \geq 0$, $x \in [0,1]$, and f_s, f_b, w_0, w_1, ϕ_0, ϕ_1 are given. The functions $a(t)$, $e(t)$ are real valued and of bounded variation on each compact interval $J = [0,T]$, and $\gamma > 0$ is a constant; see Section 9.1 for the physical interpretation of the variables w and ϕ, of a and e, as well as of the dynamic equations.

Choose $X = L^2(0,1) \times L^2(0,1)$ and define the operator family $\{A(t)\}_{t \geq 0}$ by

$$A(t) = \begin{pmatrix} a(t)\partial_x^2, & a(t)\partial_x \\ -\gamma a(t)\partial_x, & e(t)\partial_x^2 - \gamma a(t) \end{pmatrix} \tag{7}$$

where $\mathcal{D}(A(t)) \equiv Y = \{(w,\phi) \in W^{2,2}(0,1) : w(0) = \phi(0) = \phi_x(1) = w_x(1) + \phi(1) = 0\}$. After two integrations w.r.t. time, (6) then takes on the form (1). $\quad \square$

The classes of evolutionary integral equations studied in this book typically arise in mathematical physics by some constitutive laws of the memory type when combined with the usual conservation laws. The next examples illustrate this point.

(iv) *Heat Conduction with Memory*
Let $\varepsilon(t,x)$ denote the density of *internal energy*, $\theta(t,x)$ the *temperature*, $q(t,x)$ the *heat flux* vector field in a rigid body $\Omega \subset \mathbb{R}^3$, and let $r(t,x)$ be the external *heat supply*. Balance of energy then reads as

$$\varepsilon_t(t,x) + \operatorname{div} q(t,x) = r(t,x), \quad t \in \mathbb{R},\ x \in \Omega, \tag{8}$$

and boundary conditions are either prescribed temperature or prescribed heat flux through the boundary, i.e.

$$\theta(t,x) = \theta_b(t,x) \text{ on } \overset{\circ}{\Gamma}_b, \quad -q(t,x)n(x) = q_f(x) \text{ on } \overset{\circ}{\Gamma}_f, \tag{9}$$

where $n(x)$ denotes the outer normal of Ω at $x \in \partial\Omega$; Γ_b, $\Gamma_f \subset \partial\Omega$ are closed, $\overset{\circ}{\Gamma}_b \cap \overset{\circ}{\Gamma}_f = \emptyset$, $\overline{\overset{\circ}{\Gamma}_b} = \Gamma_b$, $\overline{\overset{\circ}{\Gamma}_f} = \Gamma_f$, and $\Gamma_b \cup \Gamma_f = \partial\Omega$. For isotropic and homogeneous bodies the general linear time invariant constitutive laws are given by

$$\varepsilon(t,x) = \int_0^\infty dm(\tau)\theta(t-\tau,x) + \varepsilon_\infty, \tag{10}$$

$$q(t,x) = -\int_0^\infty dc(\tau)\nabla\theta(t-\tau,x), \quad t \in \mathbb{R},\ x \in \Omega;$$

the scalar-valued kernels $m(t)$ and $c(t)$ reflect the properties of the material, but are at least of bounded variation on $\mathbb{R}_+$. Note that the case $dm = m_0\delta_0$, $dc = c_0\delta_0$, δ_0 Dirac's distribution, corresponds to the classical Fourier laws of heat conduction. Thus combining (8), (9), and (10) we obtain the boundary value problem

$$\int_0^\infty dm(\tau)\theta_t(t-\tau,x) = \int_0^\infty dc(\tau)\Delta\theta(t-\tau,x) + r(t,x), \quad t \in \mathbb{R},\ x \in \Omega,$$

$$\theta(t,x) = \theta_b(t,x), \quad t \in \mathbb{R},\ x \in \overset{\circ}{\Gamma}_b, \tag{11}$$

$$\int_0^\infty dc(\tau)\frac{\partial\theta}{\partial n}(t-\tau,x) = q_f(t,x), \quad t \in \mathbb{R},\ x \in \overset{\circ}{\Gamma}_f.$$

There are several ways to convert this problem into a Volterra equation (2) on the line. For more background and a thorough mathematical study of (11) see Sections 5.3, 5.5, and 13.2. $\square$

(v) *Electrodynamics with Memory*
The situation is similar in linear electrodynamics with memory. The basic Maxwell equations

$$\mathcal{B}_t + \text{curl } \mathcal{E} = 0, \quad \text{div } \mathcal{B} = 0,$$
$$\mathcal{D}_t - \text{curl } \mathcal{H} + \mathcal{J} = 0, \quad \text{div } \mathcal{D} = \rho, \tag{12}$$

connecting electric field $\mathcal{E}$, magnetic field $\mathcal{H}$, magnetic induction $\mathcal{B}$, electric induction $\mathcal{D}$, free current $\mathcal{J}$, and free charge ρ, have to be supplemented by constitutive equations. For homogeneous and isotropic materials these are given by

$$\mathcal{B}(t, x) = \int_0^\infty d\mu(\tau)\mathcal{H}(t - \tau, x), \quad \mathcal{D}(t, x) = \int_0^\infty d\varepsilon(\tau)\mathcal{E}(t - \tau, x),$$
$$\mathcal{J}(t, x) = \int_0^\infty d\sigma(\tau)\mathcal{E}(t - \tau, x), \quad x \in \Omega, \ t \in \mathbb{R}. \tag{13}$$

The material functions $\mu(t)$, $\varepsilon(t)$, $\sigma(t)$ reflect the properties of the medium under consideration; in the classical case these are simply constant, i.e. $\mathcal{B} = \mu\mathcal{H}$, $\mathcal{D} = \varepsilon\mathcal{E}$, $\mathcal{J} = \sigma\mathcal{E}$. Electrodynamics with memory is discussed in Sections 9.5, 9.6, and 13.3. $\square$

3. Preliminary Discussions
In general, the form in which Volterra equations arise in applications is given by Equation (2) on the line, as shown by the examples just given, and (1) arises from (2) as a *history value problem*; by this we mean the following. Assume $v(t)$ is known for $t < 0$; then $v(t)$ is a solution of (1), where the forcing term $f(t)$ in (1) contains $g(t)$ as well as the forcing effect of the history of v according to

$$f(t) = g(t) + \int_0^\infty A(t + \tau)v(-\tau)d\tau, \quad t \geq 0. \tag{14}$$

In particular (2) reduces to (1) in case the system is *virgin* at $t = 0$, i.e. $g(t) \equiv v(t) \equiv 0$ for $t < 0$; this means that up to time $t = 0$, the system has been in equilibrium and has not been subject to any forces.

It might be questioned whether it makes sense to consider (2) for all times since every man-made material or system has been virgin at some time. However, in this respect the history value problem is even more questionable since it is just impossible to record the complete history of a system and to keep this memory for all time. So in this line of arguments the only reasonable problem is that of systems, virgin at $t = 0$, in which case (1) and (2) coincide. But when considering problems with periodic or almost periodic forcing, the equation to consider is (2), since (1) is only time invariant but not translation invariant, only (2) enjoys the latter property. In this context the important question arises whether the solutions $u(t)$ of (1) and $v(t)$ of (2) are asymptotic to each other, i.e. whether $u(t) - v(t) \to 0$ as $t \to \infty$, whenever $f(t) - g(t) \to 0$ as $t \to \infty$. Under reasonable assumptions this turns out to be the case, and therefore the term *limiting equation* of (1) makes sense for (2); the solution $v(t)$ of (2) describes the limiting behaviour of the solution $u(t)$ of (1) as $t \to \infty$, provided $\dot{f}$ and $\dot{g}$ are asymptotic to each other.

As the above examples show, in many situations the operator family $\{A(t)\}_{t \geq 0}$ turns out to be of the special form $A(t) = a(t)A$, where $a \in L^1_{loc}(\mathbb{R})$ is a scalar-valued kernel, A a closed linear densely defined operator in X, and $Y = \mathcal{D}(A)$ equipped with the graph norm of A. In this case (1) and (2) are said to be of *scalar type*. The theory is considerably simplified for such problems, also much more complete. It contains two important special cases which are obtained as follows.

(i) Choose $a(t) \equiv 1$; differentiation of (1) yields the *abstract Cauchy problem* (of first order)

$$\dot{u}(t) = Au(t) + h(t), \quad t \in \mathbb{R}_+, \ u(0) = u_0, \tag{15}$$

where $u_0 = f(0)$, $h(t) = \dot{f}(t)$, and the 'dot' indicates differentiation w.r.t. time t. Initial value problem (15) has been studied extensively through the last 60 years, and there are many monographs on the subject *semigroup theory* dealing with such abstract Cauchy problems.

(ii) Choose $a(t) \equiv t$; differentiating (1) twice results in the second order problem

$$\ddot{u}(t) = Au(t) + h(t), \quad t \in \mathbb{R}_+, \ u(0) = u_0, \ \dot{u}(0) = u_1, \tag{16}$$

where $u_0 = f(0)$, $u_1 = \dot{f}(0)$, and $h(t) = \ddot{f}(t)$. To some extent it is possible to rewrite (16) as a first order Cauchy problem, i.e. (16) is subsumed by semigroup theory, but there is also an independent theory for (16) which leads to the concept of a *cosine family*.

Similar to the case of second order Cauchy problems (16) attempts have been made to reformulate (1) (and also (2)) as an abstract Cauchy problem (15) in certain spaces of history or forcing functions. In finite dimensions this approach works quite well, although it has the disadvantage of transforming a finite-dimensional problem into an infinite dimensional one. In the infinite dimensional unbounded case, i.e. $Y \neq X$, however, it does not work so well and is problematic in several aspects, some of which have been discussed above. The advantage of this approach is of course that the well established theory of semigroups can be employed; this way Volterra equations become a chapter of semigroup theory.

But this is not the whole story. Already when reformulating (16) as a first order problem there arises the question of the proper *phase space* where $(u, \dot{u})$ should live in. This question can only be resolved *after* (16) has been studied independently: $u(t) \in X_1$, $\dot{u}(t) \in X$, where X_1 denotes the set of all $x \in X$ such that $Co(\cdot)x$ is a C^1-function; here $Co(t)$ means the cosine family corresponding to

(16). For Volterra equations (1) or (2) this problem becomes much worse; even after an independent study of (1) resp. (2) there is no unique or natural choice of a phase space e.g. of history or forcing functions such that (1) resp. (2) reduce to an abstract Cauchy problem (15). Moreover, such reformulations in general change the nature of the original problem; although (1) resp. (2) may be of *parabolic type*, the resulting reformulation will always be of *hyperbolic type*.

For these reasons, we study (1) and (2) as they stand. The theory developed thus forms an extension of semigroup theory, and when specialized to the case $A(t) \equiv A$, reduces to well-known results on semigroups; similarly, for $A(t) = tA$ results on cosine families will be recovered. On the other hand, (1) and (2) are treated very much in the spirit of semigroup theory, and it will be helpful but not neccessary for the reader to be familiar with its basics. Concerning the theory of C_0-semigroups we refer to the monographs of Hille and Phillips [180], Krein [200], Tanabe [319], Davies [77], Fattorini [113], Pazy [267], Goldstein [133], Nagel [255], and Clément, Heijmans, Angenent, van Duijn, de Pagter [47]. Nevertheless, we discuss semigroup methods briefly in Section 13 and give a list of references which is far from being complete. There our point of view is underlined, we show how semigroups in certain phase spaces may be constructed *after* the underlying Volterra equation has been solved.

Nonautonomous problems are also of interest in theory as well as in practice; they occur e.g. as linearizations of nonlinear equations at nonconstant solutions, and arise in the theory of aging of materials with memory. However, due to space considerations and since no ideas other than those presented in this book are known for their treatment (essentially perturbation arguments), the emphasis is put here on autonomous problems for which the theory is much richer. The nonautonomous case is only discussed in some of the comment sections.

For understanding this book, only moderate knowledge in Real and Complex Analysis, Functional Analysis, Differential and Integral Equations, and Harmonic Analysis, especially Transform Theory is needed. For convenience of the reader we have included the preliminary Section 0 which contains several results, scattered in the literature, on the vector-valued Laplace transform, on Volterra equations (1) and (2) in the bounded case, and on spectral theory of vector-valued functions.

4. Equations of Scalar Type

Having said something about the problems to be studied and being motivated by examples from Mathematical Physics, we want to present now some details on the ideas, methods, and contents of this book. Let us begin with equations of the form (1) which are of *scalar type* $A(t) = a(t)A$, i.e. consider

$$u(t) = \int_0^t a(\tau)Au(t - \tau)d\tau + f(t), \quad t \in \mathbb{R}_+, \tag{17}$$

where A is a closed linear, densely defined operator in X, and $a \in L^1_{loc}(\mathbb{R}_+)$ a scalar kernel. The study of this class of Volterra equations is the main theme of Chapter 1. A continuous function $u : \mathbb{R}_+ \to X$ is called a *strong solution* of (17) if

$u(t) \in \mathcal{D}(A)$ for all $t \in \mathbb{R}_+$, $Au(t)$ is continuous, and (17) holds on $\mathbb{R}_+$; u is called a *mild solution* of (17) if the convolution

$$(a * u)(t) = \int_0^t a(\tau)u(t - \tau)d\tau, \quad t \in \mathbb{R}_+,$$

belongs to $\mathcal{D}(A)$ for all $t \in \mathbb{R}_+$, and $u(t) = A(a*u)(t)+f(t)$ holds on $\mathbb{R}_+$. The basic concept concerning (17) is that of *well-posedness* which is the direct extension of the corresponding notion usually employed for the abstract Cauchy problem (15). In Section 1 it is shown that well-posedness is equivalent to the existence of a *resolvent* $\{S(t)\}_{t\geq 0} \subset \mathcal{B}(X)$ for (17), i.e. a strongly continuous family of bounded linear operators in X which commutes with A and satisfies the *resolvent equation*

$$S(t)x = x + \int_0^t a(t - \tau)AS(\tau)x d\tau, \quad t \in \mathbb{R}_+, \ x \in \mathcal{D}(A). \tag{18}$$

The resolvent is the central object to be studied in the theory of Volterra equations; it corresponds to the semigroup in the special case $a(t) \equiv 1$, i.e. for (15), respectively to the cosine family for $a(t) \equiv t$, i.e. for (16). The importance of the resolvent $S(t)$ is shown by the *variation of parameters formula* valid for mild solutions $u(t)$ of (17).

$$u(t) = \frac{d}{dt} \int_0^t S(t - \tau)f(\tau)d\tau, \quad t \in \mathbb{R}_+. \tag{19}$$

This formula is a generalization of the usual variation of parameters formulae for Cauchy problems of first and second order.

Section 1 contains the basic theory of Volterra equations of scalar type. The equivalence between well-posedness of (17) and existence of the resolvent is established, the variation of parameters formula (19) is derived and its elementary properties are studied. It is then used to obtain some simple perturbation results. Elementary spectral theory yields conditions necessary for existence of resolvents which in the case of a normal operator A in a Hilbert space are also sufficient. In particular, we obtain a very useful existence theorem for resolvents.

Due to the time invariance of (17), Laplace transform methods can be employed. Formally the solution of (17) is represented in the frequency domain by

$$\hat{u}(\lambda) = (I - \hat{a}(\lambda)A)^{-1}\hat{f}(\lambda) \tag{20}$$

and the Laplace transform $H(\lambda) = \hat{S}(\lambda)$ of the resolvent by

$$H(\lambda) = \frac{1}{\lambda}(I - \hat{a}(\lambda)A)^{-1}. \tag{21}$$

This formula leads to a characterization of Hille-Yosida type for existence of an at most exponentially growing resolvent for (17) in terms of properties of the holomorphic operator family $H(\lambda)$. This result plays a crucial role in later developments.

In Section 2 we take up the study of *analytic resolvents*, i.e. resolvents $S(t)$ which admit analytic continuation to some sector $\Sigma(0, \theta)$ with vertex 0 and opening angle 2θ. Such resolvents can be easily characterized in terms of boundedness properties of $H(\lambda)$, but require very smooth kernels $a(t)$. It is shown that in analogy to analytic semigroups, $AS(t)$ is bounded for each $t > 0$, and optimal bounds are derived. However, if $A^2S(t)$ is bounded, say on an interval $(t_0-\varepsilon, t_0+\varepsilon)$, $\varepsilon > 0$, then $a(t) \equiv a_0$ and $S(t)$ is already an analytic semigroup. Interpreting A as an elliptic differential operator, this rather striking property of Volterra equations means that there is no infinite spatial smoothing, which is in sharp contrast to partial differential equations of parabolic type.

Equations (17) which admit an analytic resolvent are special cases of *parabolic Volterra equations*, which are studied in Section 3. Parabolicity of (17) is defined in terms of existence and boundedness of $\lambda H(\lambda)$ in the open right halfplane, a condition easy to check in applications. Together with a certain regularity of the involved kernel $a(t)$, called k-regularity, it leads to very simple sufficient conditions for existence and time-regularity of the resolvent. Similar to parabolic partial differential equations, parabolic Volterra equations allow for improved perturbation results and enjoy the property of *maximal regularity* of type C^α, which means that (17) admits a mild solution $u \in C^\alpha(\mathbb{R}_+; X)$, whenever $f \in C^\alpha(\mathbb{R}_+; X)$; here $\alpha \in (0, 1)$.

Section 4 is devoted to a thorough study of the so-called *subordination principle* for Volterra equations. This principle is useful for parabolic as well as for hyperbolic problems and has many applications. To explain it briefly, let $b \in L^1_{loc}(\mathbb{R}_+)$ be Laplace transformable and suppose there is an at most exponentially growing resolvent $S_b(t)$ for (17) with $a(t)$ replaced by $b(t)$. Assume that $c(t)$ is a *completely positive* function, i.e. a function $c \in L^1_{loc}(\mathbb{R}_+)$ such that $k(t)$, defined by

$$\hat{k}(\lambda) = 1/\lambda^2 \hat{c}(\lambda), \quad \lambda > 0, \tag{22}$$

is a *creep function*, i.e. nonnegative, nondecreasing, and concave. Then there is $a \in L^1_{loc}(\mathbb{R}_+)$ such that

$$\hat{a}(\lambda) = \hat{b}(1/\hat{c}(\lambda)), \quad \lambda > \omega, \tag{23}$$

where ω is sufficiently large. By the subordination principle, (17) admits a resolvent $S_a(t)$ as well, which is given by the formula

$$S_a(t) = -\int_0^\infty S_b(\tau) d_\tau w(t; \tau), \quad t > 0. \tag{24}$$

Here $w(t; \tau)$ is a function of two variables defined only in terms of $c(t)$, called the *propagation function*, and it admits for nice physical interpretations in special applications. By the structure of $w(t; \tau)$, (24) yields

$$S_a(t) = S_b(t/\kappa)e^{-\alpha t/\kappa} + \int_0^{t/\kappa} S_b(\tau)v(t; \tau)d\tau, \quad t > 0, \tag{25}$$

where $\kappa = \lim_{t\to 0+} k(t)$ and $\alpha = -\lim_{t\to 0+} \dot{k}(t)$. Note that (25) is a decomposition of the resolvent $S_a(t)$ of the subordinated equation into one part which is the original resolvent S_b propagated with 'speed' $1/\kappa$ and exponentially damped with 'attenuation' α, plus another part which also propagates with speed $1/\kappa$ but due to the presence of the integral is more smooth. Thinking of $S_b(t)$ being a cosine family or a hyperbolic semigroup, (25) with $\kappa > 0$, $\alpha = \infty$ allows for the interesting phenomenon of coexistence of a finite propagation speed and smoothing, i.e. absence of wavefronts. This phenomenon is not observed in linear differential equations, it is a memory effect.

The first three subsections of Section 5 deal with the formulation of the basic equations of viscoelasticity, heat conduction in materials with memory, and thermoviscoelasticity. It is shown that in this connection completely positive functions as well as creep functions occur naturally. The remaining subsections are then devoted to special problems which lead to Volterra equations (17) of scalar type, and the theory developed so far is applied.

5. Nonscalar Problems

In Chapter 2 we study (1) for operator-valued kernels $A \in L^1_{loc}(\mathbb{R}_+; \mathcal{B}(Y, X))$, a situation which is much more complicated than the scalar case. One reason for this is the lack of a 'natural' relation between the spaces Y and X, in contrast to the scalar case where $Y = X_A$, the domain $\mathcal{D}(A)$ of A equipped with the graph norm of A, is the only reasonable choice.

The notion of a *strong solution* of (1) extends to this setting in an obvious way; a continuous function $u : \mathbb{R}_+ \to Y$ is called a strong solution for (1) with $f \in C(\mathbb{R}_+; X)$ if (1) holds on $\mathbb{R}_+$. But already 'mild solution' must be defined differently; $u : \mathbb{R}_+ \to X$ continuous is called a *mild solution* of (1) if it is the limit, uniformly on compact intervals, of strong solutions u_n of (1) with f replaced by f_n, and $f_n \to f$, uniformly on compact intervals. The concepts *well-posedness* and *resolvent* admit natural extensions to the general setting, with the main difference that we have to consider two resolvent equations now, namely

$$S(t)y = y + \int_0^t A(t - \tau)S(\tau)yd\tau, \quad t \geq 0, \ y \in Y, \tag{26}$$

and

$$S(t)y = y + \int_0^t S(t - \tau)A(\tau)yd\tau, \quad t \geq 0, \ y \in Y. \tag{27}$$

Mild solutions are again represented by the variation of parameters formula (19) whenever there is a *weak resolvent* $S(t)$ for (1), i.e. $\{S(t)\}_{t\geq 0} \subset \mathcal{B}(X)$ is strongly continuous and the second resolvent equation (27) holds.

However, in contrast to the case of problems of scalar type, uniqueness of strong solutions does not seem to imply uniqueness of mild solutions, and therefore 'well-posedness' is not equivalent to existence of a resolvent. Also, time regularity of $f(t)$ alone in general does not imply existence of strong solutions, as simple examples show. Another problem arises with the Hille-Yosida type characterization of resolvents of at most exponential growth, which is only valid in spaces with the Radon-Nikodym property. These imperfections require further notions like *pseudo-resolvent* and *a-regularity* of a weak resolvent.

Section 6 deals with the general theory for (1). The basic definitions are introduced, the relations between them are discussed, and the variation of parameters formula is studied. The latter then yields in combination with results on Volterra equations with bounded kernels a fairly general perturbation theorem. The Hille-Yosida type characterization of pseudo-resolvents is proved as well as a convergence theorem for (pseudo-)resolvents of the Trotter-Kato type. This basic theory simplifies considerably if $A(t)$ has a *main part*, i.e. is of the form

$$A(t) = a(t)A + \int_0^t a(t-\tau)dB(\tau), \quad t \geq 0, \tag{28}$$

where A is a closed linear operator in X with dense domain $\mathcal{D}(A)$, $Y = X_A$, $a \in L^1_{loc}(\mathbb{R}_+)$, and $B \in BV_{loc}(\mathbb{R}_+; \mathcal{B}(X_A, X))$, $B(0) = B(0+) = 0$, where $B(0+) = \lim_{t \to 0+} B(t)$.

If X and Y are Hilbert spaces and $A \in BV_{loc}(\mathbb{R}_+; \mathcal{B}(Y, X))$ is such that $-dA$ is of *positive type*, i.e.

$$\text{Re} \int_0^T \int_0^t (dA(\tau)u(t-\tau), u(t))_X dt \leq 0 \quad \text{for each } u \in C(\mathbb{R}_+; Y), \ T > 0, \tag{29}$$

is satisfied, then the *energy inequality*

$$|u(t)|_X \leq |f(0)|_X + \int_0^t |\dot{f}(\tau)|_X d\tau, \quad t > 0, \tag{30}$$

holds for each mild solution $u(t)$ of (1), where $f \in W^{1,1}_{loc}(\mathbb{R}_+; X)$. This inequality then leads to existence of a weak resolvent for (1), even of a resolvent in case $A(t)$ has a main part. Results of this type are important in applications since the abstract energy inequality (30) corresponds to the real energy inequality e.g. in viscoelasticity. It generalizes in a natural way the well-known facts, that m-accretive operators are the negative generators of semigroups of contractions, and that negative semidefinite selfadjoint operators in a Hilbert space generate uniformly bounded cosine families.

The energy inequality (30) can be considerably strengthened if (1) is of *variational form* in the following sense. Let V and H be Hilbert spaces such that $V \overset{d}{\hookrightarrow} H$, and let $((\cdot, \cdot))$ resp. $(\cdot, \cdot)$ denote the inner products in V resp. H, $|| \cdot ||$ resp. $| \cdot |$ the corresponding norms in V resp. H. Identifying the antidual $\overline{H}^*$ of H with H, we obtain $H \overset{d}{\hookrightarrow} \overline{V}^*$ by duality, and the antiduality $< \cdot, \cdot >$ of $V, \overline{V}^*$ is related to the inner product of H via $< v, h > = (v, h)$ for all $v \in V$, $h \in H$. Let $\alpha : \mathbb{R}_+ \times V \times V \to \mathbb{C}$ be such that $\alpha(t, \cdot, \cdot)$ is a bounded sesquilinear form on V for each $t \geq 0$, and consider the problem of variational type

$$(w, u(t)) + \int_0^t \alpha(t-\tau, w, u(\tau))d\tau = < w, f(t) >, \quad t \in \mathbb{R}_+, \ w \in V, \tag{31}$$

where $f \in C(\mathbb{R}_+; \overline{V}^*)$. Representing the forms $\alpha(t, \cdot, \cdot)$ by bounded linear operators $-A(t) \in \mathcal{B}(V, \overline{V}^*)$ it is obvious that (31) can be written as (1) in $X = \overline{V}^*$, with

$Y = V$. Now, if the form α is *coercive* in the sense that there is a constant $\gamma > 0$ with

$$2 \operatorname{Re} \int_0^T \left[\int_0^t d\alpha(\tau, u(t), u(t - \tau)] dt \geq \gamma || \int_0^T u(\tau) d\tau ||^2, \tag{32}$$

for all $u \in C(\mathbb{R}_+; V)$ and $T > 0$, then (30) improves to

$$|u(t)|^2 + \gamma || \int_0^t u(\tau) d\tau ||^2 \leq |f(0)|^2 + 2 \operatorname{Re} \int_0^t < u(t), \dot{f}(\tau) > d\tau. \tag{33}$$

Based on this *strong energy inequality* and some regularity of the form α w.r.t. $t \geq 0$, existence of a resolvent can be proved, along with several uniform estimates and regularity properties. This result is shown to cover nonisotropic hyperbolic viscoelasticity in Section 9, and also applies to thermoviscoelasticity and several other problems.

In Section 7 we extend the results of Sections 2 and 3 on analytic resolvents and parabolic equations to the nonscalar case. This is straightforward and does not lead to further complications; maximal regularity of type C^α carries over easily, too. For equations in variational form, coercive estimates of parabolic type lead to especially nice results, as one should expect. We then consider the perturbed problem

$$u(t) = \int_0^t A(t-\tau)u(\tau) d\tau + \int_0^t a(t-s)[\int_0^s dB(\tau)u(s-\tau)] ds + f(t), \quad t \in \mathbb{R}_+, \tag{34}$$

where the unperturbed equation is parabolic, admits an a-regular resolvent with some additional smoothness properties, and $B \in BV_{loc}(\mathbb{R}_+; \mathcal{B}(Y, X))$, $B(0) = B(0+) = 0$. We show that (34) also enjoys the maximal regularity property of type C^α and of type $B_p^{\alpha,q}$ as well. This result, together with L^∞-estimates for (34) which are quite delicate to prove, yields a resolvent for the perturbed problem (34).

A quite different approach to (1) is taken up in Section 8. In the scalar case $A(t) = a(t)A$, (1) can be reformulated as

$$\mathcal{A}u + \mathcal{B}u = g, \tag{35}$$

in spaces of functions on $\mathbb{R}_+$, where the operators $\mathcal{A}$ and $\mathcal{B}$ are defined by

$$(\mathcal{A}u)(t) = -Au(t), \quad t \in \mathbb{R}_+, \tag{36}$$

and $\mathcal{B}$ formally via Laplace transforms

$$\widehat{(\mathcal{B}u)}(\lambda) = (1/\hat{a}(\lambda))\hat{u}(\lambda), \quad \operatorname{Re} \lambda > 0. \tag{37}$$

One can then apply the sum method to obtain 'solutions' of (35), without going into details here. In case X belongs to the class $\mathcal{HT}$ and the problem under consideration is parabolic, this approach works quite well in L^p-spaces and can even be extended to equations of nonscalar type with main part. Since this approach is still very much under progress, we do not give all of the proofs required, but only an exposition of the general method, as well as some of its implications for (1) which are available at present.

Finally, in Section 9 the discussions of Section 5 are continued. A brief (heuristical) derivation of some common viscoelastic beam models is presented, and the preceding results are applied to study their well-posedness. Models for viscoelastic plates are formulated according to the literature, i.e. the viscoelastic linear Midlin-Timoshenko plate and the linear Kirchhoff model. Concerning three dimensional thermo-viscoelasticity two approaches are persued, one based on results for equations of scalar type plus perturbations, requiring the material relations to be 'almost separable', and the second following the variational approach, which relies on Sections 6.7 and 7.3. The final Sections 9.5 and 9.6 deal with the formulation of linear electrodynamics with memory and applications of several of the results, obtained so far.

6. Equations on the Line

The third chapter is devoted to the study of the equation on the line (2) and its relations to (1). This connection is particularly simple if (1) admits a resolvent $S(t)$ which is *integrable*, i.e.

$$|S(t)| \leq \varphi(t) \text{ for a.a. } t \geq 0, \ \varphi \in L^1(\mathbb{R}_+) \cap C_0(\mathbb{R}_+). \tag{38}$$

In fact, in this case the solution of (2) is given by

$$v(t) = \int_0^\infty S(\tau)\dot{g}(t-\tau)d\tau, \quad t \in \mathbb{R}. \tag{39}$$

From this formula the solvability properties of (2) can easily be read off; moreover, the variation of parameters formula for (1) yields

$$u(t) - v(t) = S(t)g(0) + \int_0^t S(\tau)(\dot{f}(t-\tau) - \dot{g}(t-\tau))d\tau \tag{40}$$

$$- \int_t^\infty S(\tau)\dot{g}(t-\tau)d\tau, \quad t \geq 0,$$

hence $u(t) - v(t) \to 0$ as $t \to \infty$, whenever $\dot{g} \in L^p(\mathbb{R}_-; X)$ for some $p \in [1, \infty]$, and $\dot{f}(t) - \dot{g}(t) \to 0$. This shows that the solutions $u(t)$ of (1) and $v(t)$ of (2) are asymptotic to each other as $t \to \infty$; this justifies the name *limiting equation* of (1) for problem (2).

Thus the solvability behaviour of (1) and (2) is especially nice in the *stable* case, when the resolvent for (1) is integrable. For this reason, but also since stability is very important in applications, Section 10 is devoted to the study of integrability of resolvents. The discussion begins with the derivation of necessary conditions for the important special case of equations of scalar type. These conditions are given in terms of the symbol $H(\lambda)$ of (17) defined by (21), and basically require existence and boundedness of $H(\lambda)$ on the closed right halfplane $\lambda \in \bar{\mathbb{C}}_+$. Unfortunately, these conditions are in general not sufficient. However, for important subclasses of equations (17) it is possible to obtain characterizations of integrability of $S(t)$. These classes contain those equations which admit an analytic resolvent, and parabolic equations with 2-regular kernels, as well as subordinated resolvents.

To mention a result which is particularly striking, suppose A is the generator of a uniformly bounded cosine family $Co(t)$, and let $a(t)$ be of the form

$$a(t) = a_0 + a_\infty t + \int_0^t a_1(\tau)d\tau, \quad t > 0, \tag{41}$$

where $a_0, a_\infty \geq 0$, $a_1 \geq 0$ nonincreasing and $\log a_1(t)$ convex. This situation arises naturally in viscoelasticity; see Section 5.4. Then by the subordination principle the resolvent $S(t)$ exists; it is integrable if and only if A is invertible and $a(t) \not\equiv a_\infty t$. Observing that $a(t) \equiv a_\infty t$ corresponds to purely elastic materials, this result yields the remarkable fact that the presence of a whatsoever small viscoelasticity in the material stabilizes it strongly enough to obtain stability, i.e. integrability of the resolvent; see Section 10.3. A similar phenomenon can also be found in electrodynamics with memory; see Section 13.3.

For problems of nonscalar type which are hyperbolic, integrability of $S(t)$ is in general a too strong concept. However, for equations of variational type (31) we prove *strong integrability*, i.e. $S(\cdot)v^* \in L^1(\mathbb{R}_+; V^*)$ for each $v^* \in V^*$; cp. Section 10.4. This result applies to nonisotropic hyperbolic viscoelasticity. On the other hand, in the stable case resolvents for parabolic problems can even be expected to be *uniformly integrable*, i.e. $S \in L^1(\mathbb{R}_+; \mathcal{B}(X))$ holds; Section 10.5 is devoted to this question.

The general theory of (2) is presented in Sections 11 and 12. To be able to exploit the property of translation invariance which (2) enjoys, we study the solvability behaviour of (2) only in spaces of functions $\mathcal{H}(X)$ on the line which are translation invariant in the sense that the group of translations is strongly continuous and bounded in $\mathcal{H}(X)$. This leads to the concept of *homogeneous spaces*.

For the convolution appearing in (2) to make sense, (2) is actually considered in differentiated form

$$\dot{v}(t) = \int_0^\infty A_0(\tau)\dot{v}(t-\tau)d\tau + \int_0^\infty dA_1(\tau)v(t-\tau) + g(t), \quad t \in \mathbb{R}, \tag{42}$$

where $A(t) = A_0(t) + A_1(t)$, $A_0 \in L^1(\mathbb{R}_+; \mathcal{B}(Y, X))$, $A_1 \in BV(\mathbb{R}_+; \mathcal{B}(Y, X))$. The concept for (2) corresponding to the 'well-posedness' of (1) is that of *b-admissibility* of a homogeneous space $\mathcal{H}(X)$, where $b \in L^1(\mathbb{R})$ is a scalar function with nonvanishing Fourier transform $\tilde{b}(\rho) \neq 0$ on $\mathbb{R}$, i.e. $b \in \mathcal{W}(\mathbb{R})$, the *Wiener class*. This means that for each function $g = b * h$, with $h \in \mathcal{H}(X)$, there is a unique strong solution $u \in \mathcal{H}(X)$ of (42), and the solution operator $G : g \mapsto u$ is bounded in $\mathcal{H}(X)$. Thus the scalar function $b(t)$ measures the amount of time regularity of the forcing function $g(t)$ needed for (42) to admit a strong solution. The main problem then consists in characterizing b-admissibility, and to obtain a representation of the solution operator G as a convolution, i.e.

$$(Gg)(t) = \int_{-\infty}^{\infty} G(\tau)g(t-\tau)d\tau, \quad t \in \mathbb{R}; \tag{43}$$

recall $G(t) = S(t)$ if the resolvent $S(t)$ is integrable.

Formally the Fourier transform $\tilde{u}(\rho)$ of the solution $u(t)$ of (42) is given by

$$\tilde{u}(\rho) = H(i\rho)\tilde{g}(\rho), \quad \rho \in \mathbb{R}, \tag{44}$$

where $H(\lambda)$ again denotes the symbol of (1), (2), i.e.

$$H(\lambda) = \frac{1}{\lambda}(I - \hat{A}(\lambda))^{-1}. \tag{45}$$

Let the (real) spectrum of (42) be defined by

$$\Lambda_0 = \{\rho \in \mathbb{R} : i\rho - i\rho\hat{A}_0(i\rho) - \widehat{dA_1}(i\rho) \in \mathcal{B}(Y, X) \text{ is not invertible}\}; \tag{46}$$

then it is natural to expect that (42) admits a solution if $\tilde{u}(\rho) = 0$ on a neighborhood of Λ_0. This leads to localization in the frequency domain, where the *spectrum* $\sigma(f)$ of functions of subexponential growth becomes important; it is studied in some detail in Sections 0.5 and 0.6. By means of b-admissibility of the spaces

$$\mathcal{H}_\Lambda(X) = \{f \in \mathcal{H}(X) : \sigma(f) \subset \Lambda\}, \tag{47}$$

where $\Lambda \subset \mathbb{R}$ is closed, and $\mathcal{H}(X)$ is homogeneous, the solvability behaviour of (42) can be adequately described.

Necessary conditions for b-admissibility of $C_{ub\Lambda}(X)$ are easily seen to be

$$\Lambda \cap \Lambda_0 = \emptyset, \quad \sup_{\rho \in \Lambda}\{|H(i\rho)|_{\mathcal{B}(X)} + |\tilde{b}(\rho)H(i\rho)|_{\mathcal{B}(X,Y)}\} < \infty. \tag{48}$$

However, these are not always sufficient, but they are if Λ is compact (see Section 11.3), and for several important classes of kernels $A(t)$ for arbitrary Λ. Section 12 is devoted entirely to the sufficiency of (48) for b-admissibility of $\mathcal{H}_\Lambda(X)$ and the construction of Λ-*kernels*, i.e. kernels $G_\Lambda(t)$ representing the solution operator G_Λ for $\mathcal{H}_\Lambda(X)$ by means of (43). The class of equations (42) for which this works, includes parabolic ones, certain hyperbolic problems in Hilbert spaces, and subordinated equations of scalar type.

b-admissibility of $C_{ub\Lambda}(X)$ alone yields nice solvability behaviour of (42) w.r.t. periodic or almost periodic functions. For example, if $\Lambda = (2\pi/\omega)\mathbb{Z}$ then for every ω-periodic function g there is a unique ω-periodic mild solution u of (42) and the Fourier coefficients u_n of u are determined by those of g according to

$$u_n = H(2\pi in/\omega)f_n, \quad n \in \mathbb{Z}. \tag{49}$$

Similar relations are also valid for almost periodic or asymptotically almost periodic or weakly almost periodic functions on the line. If $g \in C_{ub\Lambda}(X)$, $0 \in \Lambda$ and $g(\infty) = \lim_{t\to\infty} g(t)$ exists then $u(t)$ also admits a limit as $t \to \infty$ and

$$u(\infty) = H(0)g(\infty). \tag{50}$$

There is also a satisfactory result on asymptotic equivalence of (42) and its version on $\mathbb{R}_+$, provided $\Lambda_0 = \emptyset$, $\overline{\lim}_{|\lambda|\to\infty} |H(\lambda)| < \infty$ and the complex spectrum of (1),

$$\Sigma_0 = \{\lambda \in \overline{\mathbb{C}}_+ : \lambda - \lambda \hat{A}(\lambda) \in \mathcal{B}(Y, X) \text{ is not invertible}\}, \tag{51}$$

is compact. In fact, if (42) admits a Λ-kernel, with $\Lambda = \mathbb{R}$, $C_{ub}(X)$ is b-admissible and the resolvent $S(t)$ exists, then the solution of (42) is asymptotic to the solution of its local version on $\mathbb{R}_+$, whenever the latter is bounded.

Finally, Section 13 contains several applications of the results of Chapter 3 to problems in viscoelasticity, heat conduction with memory and electrodynamics with memory, which have been introduced and discussed before in Sections 5 and 9. The text is concluded with a discussion and a bibliography of several subjects which are strongly related to the material presented in this book, but due to space considerations have not been treated in detail, here. This discussion includes the semigroup approach, and nonautonomous linear but also some nonlinear equations. Readers with special interest for nonlinear problems in viscoelasticity are refered to the recent monograph by Renardy, Hrusa and Nohel [289], which contains a detailed treatment of nonlinear one-dimensional models.

0 Preliminaries

This section contains some of the notations used throughout this book and collects some material scattered in the literature which is important for the theory to be developed but is also of independent interest. It covers the basic inversion theorems for the vector-valued Laplace transform. The Fourier-Carleman transform and the spectrum of vector-valued functions of subexponential growth are discussed in some detail. The study of Volterra equations in Banach spaces is begun with some elementary but nevertheless important existence results for equations with bounded kernels on finite intervals and is then continued with the Paley-Wiener theorems for the line and the halfline in the vector-valued case.

0.1 Some Notation

Most notations used throughout this book are fairly standard in the modern mathematical literature. So $\mathbb{N}$, $\mathbb{Z}$, $\mathbb{Q}$, $\mathbb{R}$, $\mathbb{C}$ denote the sets of natural numbers, integers, rational numbers, real and complex numbers, respectively, and $\mathbb{N}_0 = \mathbb{N} \cup \{0\}$, $\mathbb{R}_+ = [0, \infty)$, $\mathbb{C}_+ = \{\lambda \in \mathbb{C} : \mathrm{Re}\,\lambda > 0\}$. If (M, d) is a metric space and $N \subset M$, then $\overset{\circ}{N}$, $\bar{N}$, ∂N designate the interior, closure, boundary of N, respectively, and $d(x, N)$ denotes the distance of x to N, while $B_r(x_0)$ and $\bar{B}_r(x_0)$ are the open resp. closed balls with center x_0 and radius r.

X, Y, Z will always be Banach spaces with norms $|\cdot|_X$, $|\cdot|_Y$, $|\cdot|_Z$; the subscripts will be dropped when there is no danger of confusion. $\mathcal{B}(X, Y)$ denotes the space of all bounded linear operators from X to Y, $\mathcal{B}(X) = \mathcal{B}(X, X)$ for short. The dual space of X is $X^* = \mathcal{B}(X, \mathbb{K})$, where $\mathbb{K} = \mathbb{R}$ or $\mathbb{K} = \mathbb{C}$ is the underlying scalar field; $< x, x^* >$ designates the natural pairing between elements $x \in X$ and $x^* \in X^*$. If $(x_n) \subset X$ converges to $x \in X$ we write $x_n \to x$ or $\lim_{n \to \infty} x_n = x$, while $x_n \rightharpoonup x$ or $w - \lim_{n \to \infty} x_n = x$ mean weak convergence; similarly $x_n^* \overset{*}{\rightharpoonup} x^*$ or $w^* - \lim_{n \to \infty} x_n^* = x^*$ stand for weak*-convergence of $(x_n^*) \subset X^*$ to $x^* \in X^*$.

If A is a linear operator in X, $\mathcal{D}(A)$, $\mathcal{R}(A)$, $\mathcal{N}(A)$ denote domain, range, null space of A, respectively, while $\sigma(A)$ and $\rho(A)$ mean spectrum and resolvent set of A. $\sigma(A)$ is further decomposed into $\sigma_p(A)$, $\sigma_c(A)$, $\sigma_r(A)$, the point spectrum, continuous spectrum, and residual spectrum of A. The operator A^* in X^* denotes the dual of A if it exists. If A is closed then $\mathcal{D}(A)$ equipped with the graph norm of A, $|x|_A = |x| + |Ax|$, is a Banach space, for which the symbol X_A is employed.

For the concepts listed above as well as for general reference to operator theory we refer to Dunford and Schwarz [103], Hille and Phillips [180], and Kato [193].

Some frequently used function spaces are the following.

If (M, d) is a metric space, and X a Banach space, then $C(M; X)$ denotes the space of all continuous functions $f : M \to X$. $C_b(M; X)$ resp. $C_{ub}(M; X)$ designate the spaces of all bounded continuous resp. bounded uniformly continuous functions $f : M \to X$; these spaces become Banach spaces when normed by the sup-norm

$$|f|_0 = \sup_{t \in M} |f(t)|. \tag{0.1}$$

1

The space of all functions $f : M \to X$ which are uniformly Lipschitz-continuous is denoted by $Lip(M; X)$, and

$$|f|_{Lip} = \sup_{t \neq s} |f(t) - f(s)|/d(t, s). \qquad (0.2)$$

If (Ω, Σ, μ) is a measure space then $L^p(\Omega, \Sigma, \mu; X)$, $1 \leq p < \infty$, denotes the space of all Bochner-measurable functions $f : \Omega \to X$ such that $|f(\cdot)|^p$ is integrable. This space is also a well-known Banach space when normed by

$$|f|_p = (\int_\Omega |f(t)|^p d\mu(t))^{1/p}, \qquad (0.3)$$

and functions equal a.e. are identified. Similarly, $L^\infty(\Omega, \Sigma, \mu; X)$ denotes the space of (equivalence classes of) Bochner-measurable essentially bounded functions $f : \Omega \to X$, and the norm is defined according to

$$|f|_\infty = \operatorname*{ess\,sup}_{t \in \Omega} |f(t)|. \qquad (0.4)$$

For $\Omega \subset \mathbb{R}^n$ open, Σ the Lebesgue σ−algebra, μ the Lebesgue measure, we abbreviate to $L^p(\Omega; X)$. In this case $W^{m,p}(\Omega; X)$ is the space of all functions $f : \Omega \to X$ having distributional derivatives $D^\alpha f \in L^p(\Omega; X)$ of order $|\alpha| \leq m$; the norm in $W^{m,p}(\Omega; X)$ is

$$|f|_{m,p} = (\sum_{|\alpha| \leq m} |D^\alpha f|_p^p)^{1/p} \quad \text{for } 1 \leq p < \infty, \qquad (0.5)$$

and

$$|f|_{m,\infty} = \max_{|\alpha| \leq m} |D^\alpha f|_\infty \quad \text{for } p = \infty,$$

Also for $\Omega \subset \mathbb{R}^n$ open, $C^m(\bar{\Omega}; X)$ denotes the space of all functions $f : \bar{\Omega} \to X$ which admit continuous partial derivatives $\partial^\alpha f$ in Ω and $\partial^\alpha f$ has continuous extension to $\bar{\Omega}$, for each $|\alpha| \leq m$. The norm in its subspaces $C_b^m(\bar{\Omega}; X)$ and $C_{ub}^m(\bar{\Omega}; X)$ is given by

$$|f|_{m,0} = \sup_{|\alpha| \leq m} |\partial^\alpha f|_0. \qquad (0.6)$$

For $f \in C(\bar{\Omega}; X)$ the *support* of f is defined by

$$\operatorname{supp} f = \overline{\{x \in \Omega : f(x) \neq 0\}}.$$

A usual $C_0^\infty(\Omega; X)$ means the space of all functions $f \in \bigcap_{m \geq 1} C^m(\bar{\Omega}; X)$ such that $\operatorname{supp} f \subset \Omega$ is compact.

Another space of interest is the space $BV(J; X)$ where J is a closed interval. This space consists of all functions $k : J \to X$ of strongly bounded variation, i.e.

$$\operatorname{Var} k|_J = \sup\{\sum_{j=1}^{N} |k(t_j) - k(t_{j-1})| : t_0 < t_1 < \ldots < t_N, \, t_j \in J\} \qquad (0.7)$$

is finite. If $J = [0, T]$ or $J = \mathbb{R}_+$, elements of $BV(J; X)$ are normalized throughout this book by the requirements $k(0) = 0$ and $k(\cdot)$ is left-continuous on J. Sometimes it is convenient to think of $BV(J; X)$ as a subspace of $BV(\mathbb{R}; X)$ by extending functions $k \in BV(J; X)$ by 0 to all of $\mathbb{R}$; the extension will be left-continuous again. By $BV^0(J; X)$, $J = [0, T]$ or $J = \mathbb{R}_+$, we denote the subspace of all functions $k \in BV(J; X)$ which are continuous at 0, i.e. which satisfy $0 = k(0) = k(0+) = \lim_{t \to 0+} k(t)$.

The subscript 'loc' assigned to any of the above function spaces means membership to the corresponding space when restricted to compact subsets of its domain. Usually if $X = \mathbb{K}$ is the underlying scalar field $\mathbb{K} = \mathbb{C}$ or $\mathbb{K} = \mathbb{R}$, the image space in the function space notation introduced above will be dropped. For example $L^1_{loc}(\mathbb{R})$ denotes the space of all measurable scalar-valued functions which are integrable over each compact interval. Other function spaces will be introduced where they are needed for the first time; cp. the index.

The *Fourier transform* of a function $f \in L^1(\mathbb{R}; X)$ is defined by

$$\tilde{f}(\rho) = \int_{-\infty}^{\infty} e^{-i\rho t} f(t) dt, \quad \rho \in \mathbb{R}; \tag{0.8}$$

it is well-known that $\tilde{f} : \mathbb{R} \to X$ is uniformly continuous and tends to 0 as $|\rho| \to \infty$, by the Riemann-Lebesgue lemma. If X is a Hilbert space, by Parseval's theorem the Fourier transform extends to a unitary operator on $L^2(\mathbb{R}; X)$. On the *Schwartz space* $\mathcal{S}(\mathbb{R}; X)$ of all functions $f \in C^\infty(\mathbb{R}; X)$ such that each derivative of f decays faster than any polynomial, the Fourier transform is an isomorphism, and the inversion formula

$$\tilde{\tilde{f}}(t) = 2\pi f(-t), \quad t \in \mathbb{R}, \tag{0.9}$$

holds.

The *Laplace transform* of a function $f \in L^1_{loc}(\mathbb{R}_+; X)$ is denoted by

$$\hat{f}(\lambda) = \int_0^\infty e^{-\lambda t} f(t) dt, \quad \mathrm{Re}\, \lambda > \omega, \tag{0.10}$$

whenever the integral is absolutely convergent for $\mathrm{Re}\, \lambda > \omega$; see Section 0.2 for further discussions. The relation between the Laplace transform of $f \in L^1(\mathbb{R}; X)$, $f(t) \equiv 0$ for $t < 0$, and its Fourier transform is

$$\tilde{f}(\rho) = \hat{f}(i\rho), \quad \rho \in \mathbb{R}. \tag{0.11}$$

Observe that the symbol $\hat{f}(\lambda)$ is also used for the *Fourier-Carleman transform* of $f \in L^1_{loc}(\mathbb{R}; X)$; see Section 0.5 for the definition of the latter.

As usual we employ the star $*$ for the *convolution* of functions defined on the line but also on the halfline

$$(f * g)(t) = \int_{-\infty}^{\infty} f(t - s) g(s) ds, \quad t \in \mathbb{R}, \tag{0.12}$$

e.g. for $f, g \in L^1(\mathbb{R})$, and

$$(f * g)(t) = \int_0^t f(t - s)g(s)ds, \quad t \in \mathbb{R}_+, \tag{0.13}$$

e.g. for $f, g \in L^1(\mathbb{R}_+)$. Observe that (0.12) and (0.13) are equivalent for functions which vanish for $t < 0$; therefore there will be no danger of confusion.

For other symbols and notations the reader should consult the index.

0.2 Laplace Transform

Let $u \in BV_{loc}(\mathbb{R}_+; X)$, where X denotes a complex Banach space. du is said to be of *exponential growth* or *Laplace transformable*, if there is $\omega \in \mathbb{R}$ such that $\int_0^\infty e^{-\omega t}|du(t)| < \infty$. In this case the *Laplace transform* of du,

$$\widehat{du}(\lambda) = \int_0^\infty e^{-\lambda t} du(t), \quad \text{Re } \lambda \geq \omega, \tag{0.14}$$

is well-defined, it is uniformly continuous and bounded in the closed right halfplane Re $\lambda \geq \omega$, and holomorphic in its interior. Similarly, if $v \in L^1_{loc}(\mathbb{R}_+; X)$ is of *exponential growth*, i.e. $\int_0^\infty e^{-\omega t}|v(t)|dt < \infty$ for some $\omega \in \mathbb{R}$, we define

$$\widehat{v}(\lambda) = \int_0^\infty e^{-\lambda t} v(t)dt, \quad \text{Re } \lambda \geq \omega. \tag{0.15}$$

Many of the properties of the scalar-valued Laplace transform are valid in the vector-valued case as well, some of them will be mentioned without poof here. The standard reference for the vector-valued Laplace transform is Chapter VI of Hille and Phillips [180]; for the classical Laplace transform we refer to Widder [339], [340] and Doetsch [100].

The function $u(t)$ can be recovered from its Laplace transform via the *complex inversion formula*

$$\lim_{N \to \infty} \frac{1}{2\pi i} \int_{\gamma - iN}^{\gamma + iN} e^{\lambda t} \widehat{du}(\lambda) \frac{d\lambda}{\lambda} = \frac{1}{2}(u(t) + u(t+)), \quad t \geq 0, \tag{0.16}$$

where $\gamma > \omega$ is arbitrary; here the normalization of $u(t)$ by $u(0) = 0$ and left-continuity is important. The limit in (0.16) is even uniform for t in any finite interval of continuity of $u(t)$. Similarly $v(t)$ can be recovered a.e. from $\hat{v}(\lambda)$ by means of

$$\lim_{N \to \infty} \frac{1}{2\pi} \int_{-N}^N (1 - \frac{|\rho|}{N}) e^{(\sigma + i\rho)t} \hat{v}(\sigma + i\rho)d\rho = v(t), \quad \text{for a.a. } t \geq 0, \tag{0.17}$$

and the limit in (0.17) exists also in $L^1(J; X)$, for every finite interval $J \subset \mathbb{R}_+$; here $\sigma > \omega$ is arbitrary.

By means of Cauchy's theorem, the complex inversion formula (0.16) leads to a simple characterization of functions which are holomorphic and bounded on each sector $|\arg z| \leq \theta, \theta < \theta_0$.

Theorem 0.1 *Let $f : (0, \infty) \to X$ and $\theta_0 \in (0, \pi/2]$. Then the following are equivalent.*
(i) There is a function $v(z)$ holomorphic for $|\arg z| < \theta_0$ and bounded on each sector $|\arg z| \le \theta < \theta_0$ such that $f(\lambda) = \hat{v}(\lambda)$ for each $\lambda > 0$;
(ii) $f(\lambda)$ admits holomorphic extension to the sector $|\arg \lambda| < \frac{\pi}{2} + \theta_0$, and $\lambda f(\lambda)$ is bounded on each sector $|\arg \lambda| \le \frac{\pi}{2} + \theta$, $\theta < \theta_0$.

Proof: ($\Rightarrow$) Let $v(z)$ be holomorphic on the sector $|\arg z| < \theta_0 \le \pi/2$, bounded on each subsector $|\arg z| \le \theta < \theta_0$, and define $f(\lambda) = \hat{v}(\lambda)$, $\lambda > 0$. For $|\phi| < \theta_0$ consider the path $\gamma_R = [0, R]$, $R\exp(i[0, \phi])$, $[R, 0]e^{i\phi}$; by Cauchy's theorem $\int_{\gamma_R} v(z)e^{-\lambda z}dz = 0$, i.e. with $R \to \infty$ we obtain

$$\hat{v}(\lambda) = \int_0^\infty v(t)e^{-\lambda t}dt = \int_0^\infty v(te^{i\phi})e^{-\lambda t e^{i\phi}}e^{i\phi}dt, \quad \lambda > 0.$$

This relation extends $\hat{v}(\lambda)$ holomorphically to the sector $|\arg z| < \theta_0 + \pi/2$, and gives the estimate

$$|\lambda \hat{v}(\lambda)| = |v(\cdot e^{i\phi})|_\infty / \cos(\rho + \phi), \quad \text{for all } \rho = \arg \lambda$$

satisfiying $-\pi/2 - \phi < \rho < \pi/2 - \phi$. Since $\phi \in (-\theta_0, \theta_0)$ can be chosen arbitrarily, there follows boundedness of $\lambda \hat{v}(\lambda)$ on each sector $|\arg \lambda| \le \theta + \pi/2$, where $\theta < \theta_0$.

($\Leftarrow$) Suppose $f(\lambda)$ admits holomorphic extension to $|\arg \lambda| < \pi/2 + \theta_0$, such that $\lambda f(\lambda)$ is bounded on each subsector $|\arg \lambda| \le \phi < \pi/2 + \theta_0$. Consider the path Γ_R consisting of the rays $(\infty, R]e^{-i\phi}$, $[R, \infty)e^{i\phi}$ connected by the part of the circle $|\lambda| = R$ contained in $|\arg \lambda| \le \phi$, and define

$$v(z) = \frac{1}{2\pi i} \int_{\Gamma_R} e^{\lambda z} f(\lambda)d\lambda, \quad |\arg z| < \phi - \pi/2.$$

Since this integral is absolutely convergent for $|\arg z| < \phi - \pi/2$, $\phi < \pi/2 + \theta_0$ is arbitrary and $v(z)$ is independent of ϕ and $R > 0$, by Cauchy's theorem, there follows that $v(z)$ is holomorphic on the sector $|\arg z| < \theta_0$. Choosing $R = 1/|z|$, a simple estimate shows the boundedness of $v(z)$ on each subsector $|\arg z| \le \theta < \theta_0$. Finally, for the Laplace transform of $v(t)$ we obtain

$$\hat{v}(\mu) = \frac{1}{2\pi i} \int_{\Gamma_R} f(\lambda)(\lambda - \mu)^{-1}d\lambda = f(\mu), \quad \mu > R,$$

by Cauchy's theorem again. $\square$

Of interest here are also the real inversion formulae for the Laplace transform. The following one is due to Phillips; cp. Hille and Phillips [180], Chapter VI.

$$\lim_{\sigma \to \infty} e^{-\sigma t} \sum_{n=0}^\infty \frac{(-1)^n(\sigma^2 t)^{n+1}}{n!(n+1)!}\hat{v}^{(n)}(\sigma) = v(t), \quad \text{for a.a. } t > 0, \tag{0.18}$$

and the limit exists also in $L^1(J;X)$, for every finite interval $J \subset \mathbb{R}_+$, it is uniform on any finite interval of continuity of $v(t)$. Similarly, for $u(t)$ we have

$$\lim_{\sigma \to \infty} e^{-\sigma t} \sum_{n=0}^{\infty} \frac{(-1)^n (\sigma^2 t)^n (\sigma t)^2}{n!(n+2)!} \widehat{du}^{(n)}(\sigma) = \frac{1}{2}(u(t) + u(t+)), \quad t > 0, \qquad (0.19)$$

where the limit is uniform on every finite interval of continuity of $u(t)$.

The real inversion formula (0.19) leads to the following extension of Widder's characterizations of scalar-valued L^∞- and BV-functions in terms of their Laplace transform; see Widder [339].

Theorem 0.2 *Let $f : (0, \infty) \to X$. The following are equivalent.*
(i) There is $u \in Lip(\mathbb{R}_+; X)$, $u(0) = 0$, such that $f(\lambda) = \widehat{du}(\lambda)$ for all $\lambda > 0$;
(ii) $f \in C^\infty((0, \infty); X)$ and

$$\sup\{\lambda^{n+1}|f^{(n)}(\lambda)|/n! : \lambda > 0 , \ n \in \mathbb{N}_0\} =: m_\infty(f) < \infty.$$

If this is the case then $|u|_{Lip} = m_\infty(f)$.

Theorem 0.3 *Let $f : (0, \infty) \to X$. The following are equivalent.*
(i) There is $u \in BV(\mathbb{R}_+; X)$ such that $f(\lambda) = \widehat{du}(\lambda)$ for all $\lambda > 0$;
(ii) $f \in C^\infty((0, \infty); X)$ and

$$\sup_{\lambda > 0} \sum_{n=0}^{\infty} \lambda^n |f^{(n)}(\lambda)|/n! =: m_1(f) < \infty.$$

If this is the case then $Var\ u|_0^\infty = m_1(f)$.

For the proof of Theorem 0.2 the following lemma is useful.

Lemma 0.1 *Suppose $f_n \in C^2(J;X)$ satisfies $f_n \to f$ in $C(J;X)$ and $|\ddot{f}_n|_0 \leq M < \infty$, for all $n \in \mathbb{N}$, where $J = [0, a]$.*
Then $f_n \to f$ in $C^1(J;X)$, $\dot{f} \in Lip(J;X)$, and $|\dot{f}|_{Lip} \leq M$.

Proof: Let $g \in C^2(J;X)$; by means of Taylor's formula

$$g(t+h) = g(t) + h\dot{g}(t) + \int_0^h (h-s)\ddot{g}(t+s)ds,$$

we obtain the estimate

$$|h|\,|\dot{g}|_0 \leq 2|g|_0 + (h^2/2)|\ddot{g}|_0, \quad h > 0,$$

which by the choice $|h| = a/2(|g|_0/|g|_{2,0})^{1/2}$ yields the interpolation inequality

$$|\dot{g}|_0 \leq c(a)|g|_0^{1/2}|g|_{2,0}^{1/2}, \quad \text{for all } g \in C_b^2(J;X), \qquad (0.20)$$

with $c(a) = (a/4 + 4/a)$. (0.20) applied to $g = f_n - f_m$ implies $f_n \to f$ in $C^1(J; X)$, and from

$$|\dot{f}_n(t) - \dot{f}_n(s)| \le |\ddot{f}_n|_0 |t - s| \le M(t - s), \quad t, s \in J,$$

with $n \to \infty$ the result follows. $\square$

For the proof of Theorem 0.3 we take advantage of

Lemma 0.2 *Let $J = [0, a]$; suppose $f_n \in C^2(J; X)$ satisfies $f_n \to f$ in $C(J; X)$ and $|\ddot{f}_n|_1 \le M < \infty$.*
Then $f(t)$ is absolutely continuous on J, $\dot{f} \in BV(J; X)$, $\dot{f}_n(t) \to \dot{f}(t)$ for a.a. $t \in J$, $\dot{f}_n \to \dot{f}$ in $L^1(J; X)$, and $\operatorname{Var} \dot{f}_n|_J \le M$.

Proof: Let $k_n(t) = \int_0^t |\ddot{f}_n(s)| ds$, $t \in J$, $n \in \mathbb{N}$; these functions are nondecreasing and uniformly bounded by M. Helly's selection theorem therefore provides a subsequence (n_k) such that $k_{n_k}(t) \to k(t)$ pointwise; the limit function $k(t)$ is nondecreasing and bounded by M as well. Let $t \in J$ be a point of continuity of $k(t)$, and let D_h^+ denote the right difference operator $D_h^+ g(t) = (g(t+h) - g(t))/h$; then

$$
\begin{aligned}
|\dot{f}_n(t) - \dot{f}_m(t)| &\le |\dot{f}_n(t) - D_h^+ f_n(t)| + |D_h^+ (f_n(t) - f_m(t))| \\
&\quad + |\dot{f}_m(t) - D_h^+ f_m(t)| \\
&\le [k_n(t+h) - k_n(t)] + 2h^{-1}|f_n - f_m|_0 + [k_m(t+h) - k_m(t)].
\end{aligned}
$$

Given $\varepsilon > 0$ choose first $h > 0$ small enough such that $k(t+h) - k(t) \le \varepsilon/8$; fix $h > 0$ and choose n, m so large that $|f_n - f_m|_0 \le \varepsilon h/8$, and finally enlarge $n = n_k$, $m = n_l$ such that $|k_n(s) - k(s)| \le \varepsilon/8$, and $|k_m(s) - k(s)| \le \varepsilon/8$, for $s = t$, $s = t + h$. This gives

$$|\dot{f}_n(t) - \dot{f}_m(t)| \le 2(k(t+h) - k(t)) + 4 \cdot \varepsilon/8 + 2h^{-1}\varepsilon h/8 \le \varepsilon,$$

i.e. $\dot{f}_{n_k}(t) \to g(t)$ for each continuity point of $k(t)$. Since for two such points

$$|\dot{f}_n(t) - \dot{f}_n(s)| \le k_n(t) - k_n(s),$$

we obtain in the limit

$$|g(t) - g(s)| \le k(t) - k(s),$$

i.e. $g \in BV(J; X)$. Finally, by Lebesgue's theorem

$$f_{n_k}(t) = f_{n_k}(t_0) + \int_{t_0}^t \dot{f}_{n_k}(s) ds \to f(t_0) + \int_{t_0}^t g(s) ds = f(t),$$

hence $f(t)$ is absolutely continuous on J, and $g(t) = \dot{f}(t)$ a.e. Repeating these arguments with an arbitrary subsequence we may conclude $\dot{f}_n \to \dot{f}$ a.e. and in $L^1(J; X)$. $\square$

The arguments in the proofs of Theorems 0.2 and 0.3 are quite similar, therefore they are carried through simultaneously.

Proofs of Theorems 0.2 and 0.3: (i) $\Rightarrow$ (ii) of Theorem 0.2. Let $u \in Lip(\mathbb{R}_+; X)$, $u(0) = 0$, and define $f(\lambda) = \lambda\hat{u}(\lambda) = \widehat{du}(\lambda)$; then

$$f^{(n)}(\lambda) = \int_0^\infty (-t)^n e^{-\lambda t} du(t), \quad \lambda > 0,$$

hence

$$|f^{(n)}(\lambda)| = \int_0^\infty t^n e^{-\lambda t} |du(t)| \leq |u|_{Lip} \int_0^\infty t^n e^{-\lambda t} dt = |u|_{Lip} n! \lambda^{-(n+1)},$$

i.e. $m_\infty(f) \leq |u|_{Lip} < \infty$.

(i) $\Rightarrow$ (ii) of Theorem 0.3. If $u \in BV(\mathbb{R}_+; X)$, $u(0) = 0$, we obtain similarly

$$\sum_{n=0}^\infty \frac{\lambda^n}{n!} |f^{(n)}(\lambda)| \leq \sum_{n=0}^\infty \frac{\lambda^n}{n!} \int_0^\infty t^n e^{-\lambda t} |du(t)| = \int_0^\infty \left(\sum_{n=0}^\infty \frac{(t\lambda)^n}{n!}\right) e^{-\lambda t} |du(t)|$$

$$= \int_0^\infty |du(t)| = \operatorname{Var} u|_0^\infty,$$

hence $m_1(f) \leq \operatorname{Var} u|_0^\infty < \infty$.

(ii) $\Rightarrow$ (i) in both theorems. Observe first that $f(\lambda)$ admits holomorphic extension to the open right halfplane $\mathbb{C}_+$, as the series expansion

$$f(z) = \sum_{n=0}^\infty (z - \lambda)^n f^{(n)}(\lambda)/n!, \quad \operatorname{Re} z > 0, \ |z - \lambda| < \lambda,$$

and the estimates

$$|f(z)| \leq \sum_{n=0}^\infty |z - \lambda|^n |f^{(n)}(\lambda)|/n! \leq m_1(f),$$

resp.

$$|f(z)| \leq m_\infty(f) \sum_{n=0}^\infty |(z - \lambda)|^n \lambda^{-(n+1)} = \frac{m_\infty(f)}{\lambda - |z - \lambda|} \xrightarrow{\lambda \to \infty} \frac{m_\infty(f)}{\operatorname{Re} z},$$

show. Define functions v_n by means of

$$v_n(t) = e^{-nt} \sum_{k=0}^\infty \frac{(n^2 t)^{k+1}(-1)^k}{(k+1)! k!} f^{(k)}(n), \quad t \geq 0;$$

then

$$|v_n(t)| \leq e^{-nt} \left(\sum_{k=0}^\infty \frac{(nt)^{k+1}}{(k+1)!}\right) m_\infty(f) \leq m_\infty(f),$$

i.e. $|v_n|_0 \leq m_\infty(f)$, and

$$|v_n|_1 \leq \sum_{k=0}^{\infty} (\int_0^\infty e^{-nt} t^{k+1} dt) \frac{n^{2k+2}}{(k+1)! k!} |f^{(k)}(n)| \leq m_1(f).$$

The Laplace transform of $v_n(t)$ is easily computed to the result

$$\hat{v}_n(\lambda) = (\frac{n}{n+\lambda})^2 f(\frac{n\lambda}{n+\lambda}), \quad \text{Re } \lambda > 0,$$

hence $\hat{v}_n(\lambda) \to f(\lambda)$ as $n \to \infty$, uniformly on compact subsets of $\mathbb{C}_+$. Define $w_n(t)$ by

$$w_n(t) = \int_0^t (t-s) v_n(s) ds, \quad t \geq 0, \ n \in \mathbb{N};$$

then $\hat{w}_n(\lambda) = \hat{v}_n(\lambda)/\lambda^2$, Re $\lambda > 0$, hence by boundedness of $f(\lambda)$ on halfplanes Re $\lambda \geq \sigma > 0$, the complex inversion formula for the Laplace transform yields

$$w_n(t) = (2\pi i)^{-1} \int_{\sigma-i\infty}^{\sigma+i\infty} \hat{v}_n(\lambda) e^{\lambda t} d\lambda/\lambda^2, \quad t \geq 0.$$

Next Lebesgue's theorem implies

$$w_n(t) \to_{n\to\infty} w(t) = (2\pi i)^{-1} \int_{\sigma-i\infty}^{\sigma+i\infty} f(\lambda) e^{\lambda t} d\lambda/\lambda^2, \quad t \geq 0,$$

uniformly on bounded intervals $J = [0, a] \subset \mathbb{R}_+$, where the boundedness of $f(\lambda)$ on Re $\lambda \geq \sigma > 0$ was used again.

We are in position now to apply Lemma 0.1 resp. Lemma 0.2. In fact, if $m_\infty(f) < \infty$, Lemma 0.1 yields $w_n \to w$ in $C^1_{loc}(\mathbb{R}_+; X)$, and $\dot{w} \in Lip(\mathbb{R}_+; X)$, as well as $|\dot{w}|_{Lip} \leq m_\infty(f)$; in case $m_1(f) < \infty$, Lemma 0.2 implies $w_n \to w$ in $W^{1,1}_{loc}(\mathbb{R}_+; X)$ and $\dot{w} \in BV(\mathbb{R}_+; X)$, Var $\dot{w}|_0^\infty \leq m_1(f)$. Setting $u(t) = \dot{w}(t)$, and observing that $w(0) = \dot{w}(0) = 0$, the assertions of Theorems 0.2 and 0.3 follow. $\square$

In case X has the Radon-Nikodym property, then a Lipschitz function has an L^∞-derivative a.e. In this case Theorem 0.2 becomes

Corollary 0.1 *Suppose X has the Radon-Nikodym property, and let $f : (0, \infty) \to X$. Then the following are equivalent.*
(i) There is $v \in L^\infty(\mathbb{R}_+; X)$ such that $f(\lambda) = \hat{v}(\lambda)$ for all $\lambda > 0$;
(ii) $f \in C^\infty((0, \infty); X)$ and $m_\infty(f) < \infty$.
If this is the case then $|v|_\infty = m_\infty(f)$.

Theorem 0.2 as well as Corollary 0.1 are due to Arendt [10].

In practice $m_1(f)$ as well as $m_\infty(f)$ are difficult to estimate, and so Theorems 0.2 and 0.3 are not easy to apply. However, quite often one encounters functions

$g(\lambda)$, holomorphic in the right halfplane Re $\lambda > 0$ and such that $\lambda g(\lambda)$ is bounded there. In general it is not true that this already implies $m_\infty(f) < \infty$ as one might expect; see Desch and Prüss [91]. However, the following result holds.

Propositon 0.1 *Suppose $g : \mathbb{C}_+ \to X$ is holomorphic and satisfies*

$$|\lambda g(\lambda)| + |\lambda^2 g'(\lambda)| \leq M \quad \text{for all} \quad Re\ \lambda > 0, \tag{0.21}$$

Then $m_\infty(g) < \infty$.

Proof: To estimate $g^{(n)}(\lambda)$ for $n > 1$, we use the Cauchy integral representation

$$g'(\lambda) = \frac{1}{2\pi i} \int_{\varepsilon-i\infty}^{\varepsilon+i\infty} g'(z)(\lambda - z)^{-1}dz, \quad \lambda > \varepsilon.$$

Differentiating this formula $(n - 1)$-times yields

$$g^{(n)}(\lambda) = (-1)^{(n-1)}(n - 1)!(2\pi i)^{-1} \int_{\varepsilon-i\infty}^{\varepsilon+i\infty} g'(z)(\lambda - z)^{-n}dz, \quad \lambda > \varepsilon.$$

By means of (0.21) we obtain

$$
\begin{aligned}
|g^{(n)}(\lambda)| &\leq (n - 1)!(M/2\pi) \int_{-\infty}^{\infty} |\varepsilon + i\rho|^{-2}|\lambda - \varepsilon - i\rho|^{-n}d\rho \\
&\leq (n - 1)!(M/2\pi)(\lambda - \varepsilon)^{-n} \int_{-\infty}^{\infty} (\varepsilon^2 + \rho^2)^{-1}d\rho \\
&\leq (n - 1)!(M/2\pi)(\lambda - \varepsilon)^{-n}(\pi/\varepsilon).
\end{aligned}
$$

Choosing $\varepsilon = \lambda/n$ the latter gives

$$|g^{(n)}(\lambda)| \leq n!(M/2)\lambda^{-(n+1)}(1 - 1/n)^{-n} \leq n!(Me/2)\lambda^{-(n+1)},$$

thereby proving $m_\infty(g) < \infty$. □

If only $\lambda g(\lambda)$ is bounded we can prove the following weaker result which nevertheless is useful as well.

Proposition 0.2 *Suppose $g : \mathbb{C}_+ \to X$ is holomorphic and satisfies*

$$|\lambda g(\lambda)| \leq M, \quad Re\ \lambda > 0. \tag{0.22}$$

Then there is $u \in C(\mathbb{R}_+; X)$ with $u(0) = 0$ such that $g(\lambda) = \lambda \hat{u}(\lambda)$, $Re\ \lambda > 0$; moreover, there is a constant $L > 0$ such that

$$|u(t) - u(s)| \leq L(t - s)[1 + \log(t/(t - s))] \quad \text{for all } t > s \geq 0. \tag{0.23}$$

Proof: We define

$$u(t) = (2\pi i)^{-1} \int_{\Gamma_{\varepsilon,r}} g(\lambda) e^{\lambda t} d\lambda/\lambda, \quad t \geq 0, \tag{0.24}$$

where $\Gamma_{\varepsilon,r}$ denotes the contour $\varepsilon + i(-\infty, -r)$, $\varepsilon + re^{i[-\pi/2,\pi/2]}$, $\varepsilon + i(r, \infty)$, with $\varepsilon, r > 0$. Observe that the integral is absolutely convergent, by virtue of (0.22). Clearly, the definition of $u(t)$ is independent of $\varepsilon, r > 0$, and contracting the contour in the right halfplane it follows that $u(0) = 0$, by Cauchy's theorem. Fix $t > s \geq 0$ and estimate by means of (0.22) to the result

$$|u(t) - u(s)| \leq (M/2\pi) \int_{\Gamma_{\varepsilon,r}} |e^{\lambda t} - e^{\lambda s}| |d\lambda|/|\lambda|^2.$$

Letting $\varepsilon \to 0$ this yields

$$
\begin{aligned}
|u(t) - u(s)| \quad &\leq \quad (M/\pi)[\int_r^\infty |e^{i\rho t} - e^{i\rho s}| d\rho/\rho^2 + \int_0^{\pi/2} |e^{rte^{i\phi}} - e^{rse^{i\phi}}| d\phi/r] \\
&\leq \quad (M/\pi)[2\int_r^\infty |\sin(\rho(t-s)/2)| d\rho/\rho^2 + e^{rt}(t-s)\pi/2] \\
&\leq \quad M(t-s)[\pi^{-1} \int_{r(t-s)/2}^\infty |\sin\tau| d\tau/\tau^2 + e^{rt}/2].
\end{aligned}
$$

Choosing $r = 2/t$ we obtain

$$
\begin{aligned}
|u(t) - u(s)| \quad &\leq \quad M(t-s)[\pi^{-1} \int_{1-s/t}^\infty |\sin\tau| d\tau/\tau^2 + e^2/2] \\
&\leq \quad L(t-s)[1 + \log(t/(t-s))],
\end{aligned}
$$

what was to be proved.

Finally, we have by means of Cauchy's theorem and Fubini's theorem

$$
\begin{aligned}
\lambda \hat{u}(\lambda) \quad &= \quad \lambda \int_0^\infty e^{-\lambda t} u(t) dt = (2\pi i)^{-1} \int_{\Gamma_{\varepsilon,r}} g(\mu) \lambda(\lambda - \mu)^{-1} d\mu/\mu \\
&= \quad (2\pi i)^{-1} \int_{\Gamma_{\varepsilon,r}} g(\mu)(1 + \mu/(\lambda - \mu)) d\mu/\mu \\
&= \quad (2\pi i)^{-1} \int_{\Gamma_{\varepsilon,r}} g(\mu)(\lambda - \mu)^{-1} d\mu \\
&= \quad g(\lambda) \quad \text{for all } \operatorname{Re} \lambda > r + \varepsilon. \qquad \square
\end{aligned}
$$

Combining Propositions 0.1 and 0.2 with Theorem 0.2 we obtain the following result on inversion of the vector-valued Laplace transform.

Theorem 0.4 *Suppose $g : \mathbb{C}_+ \to X$ is holomorphic and satisfies*

$$|\lambda^{n+1} g^{(n)}(\lambda)| \leq M, \quad \text{for} \quad \operatorname{Re} \lambda > 0, \, 0 \leq n \leq k+1, \tag{0.25}$$

*where $k \geq 0$. Then there is a function $u \in C^k((0, \infty); X)$ such that $g(\lambda) = \hat{u}(\lambda)$
for Re $\lambda > 0$. Moreover, there is a constant $C > 0$ such that*

$$|t^n u^{(n)}(t)| \leq C, \quad \text{for } t > 0, \ 0 \leq n \leq k, \tag{0.26}$$

and

$$|t^{k+1} u^{(k)}(t) - s^{k+1} u^{(k)}(s)| \leq C|t - s|[1 + \log(t/(t - s))], \quad 0 \leq s < t < \infty. \tag{0.27}$$

Proof: For $n \leq k + 1$ we define $g_n(\lambda) = \lambda^n g^{(n)}(\lambda)$; then for $n \leq k$, $g_n(\lambda)$ satisfies
the assumptions of Proposition 0.1, hence there are functions $u_n \in C(\mathbb{R}_+; X)$ with
$u_n(0) = 0$ such that $g_n(\lambda) = \lambda \hat{u}_n(\lambda)$ and there is $L > 0$ such that

$$|u_n(t) - u_n(s)| \leq L|t - s|, \quad \text{for all } t, s \in \mathbb{R}_+, \ 0 \leq n \leq k.$$

Proposition 0.2 yields $u_{k+1} \in C(\mathbb{R}_+; X)$ with $u_{k+1}(0) = 0$ such that $g_{k+1}(\lambda) = \lambda \hat{u}_{k+1}(\lambda)$ and

$$|u_{k+1}(t) - u_{k+1}(s)| \leq L(t - s)[1 + \log(t/(t - s))], \quad 0 \leq s < t < \infty.$$

Since

$$g'_n(\lambda) = n\lambda^{n-1} g^{(n)}(\lambda) + \lambda^n g^{(n+1)}(\lambda) = (ng_n(\lambda) + g_{n+1}(\lambda))/\lambda$$

for all Re $\lambda > 0$ and $0 \leq n \leq k$, we obtain

$$\hat{u}'_n(\lambda) = ((n - 1)\hat{u}_n(\lambda) + \hat{u}_{n+1}(\lambda))/\lambda,$$

which implies

$$-tu_n(t) = (n - 1) \int_0^t u_n(\tau)d\tau + \int_0^t u_{n+1}(\tau)d\tau, \quad t > 0, \ 0 \leq n \leq k,$$

by uniqueness of the Laplace transform. This identity shows $u_n \in C^1((0, \infty); X)$
and

$$-tu'_n(t) = nu_n(t) + u_{n+1}(t), \quad t > 0, \ 0 \leq n \leq k.$$

Let $u(t) = u'_0(t)$; then $|u(t)| = |u_1(t)|/t \leq L$, hence

$$\hat{u}(\lambda) = \lambda \hat{u}_0(\lambda) = g_0(\lambda) = g(\lambda), \quad \text{Re } \lambda > 0,$$

and $u_n(t) = (-1)^n (t^n u(t))^{(n-1)}$ as is easily seen by induction. The assertion now
follows from the properties of $u_n(t)$. □

Propositions 0.1, 0.2 and Theorem 0.4 are taken from Prüss [278].

0.3 Volterra Equations with Bounded Kernels
Let X be a Banach space and consider the linear Volterra equation

$$u(t) = f(t) + \int_0^t dK(\tau)u(t - \tau), \quad t \in J, \tag{0.28}$$

where $K \in BV^0(J; \mathcal{B}(X))$, $f \in C(J; X)$, and $J = [0, T]$. Along with (0.28) we also consider the *resolvent equations* for (0.28) in $\mathcal{B}(X)$

$$L(t) = K(t) + \int_0^t dK(\tau)L(t - \tau), \quad t \in J, \tag{0.29}$$

and

$$L(t) = K(t) + \int_0^t L(t - \tau)dK(\tau), \quad t \in J. \tag{0.30}$$

The following result is basic.

Theorem 0.5 *Suppose $K \in BV^0(J; \mathcal{B}(X))$. Then there is a unique function $L \in BV^0(J; \mathcal{B}(X))$ such that (0.29) and (0.30) are satisfied. $L(t)$ can be obtained via the Neumann series*

$$L(t) = \sum_{n=0}^{\infty} K_n(t), \quad t \in J, \tag{0.31}$$

where

$$K_0(t) = K(t), \ K_{n+1}(t) = \int_0^t dK(\tau)K_n(t - \tau), \quad t \in J, \ n \in \mathbb{N}_0. \tag{0.32}$$

Proof: (Uniqueness) Let $L_1, L_2 \in L^1(J; \mathcal{B}(X))$ be solutions of (0.29), (0.30), respectively. Then

$$(L_2 - K) * L_1 = (L_2 * dK) * L_1 = L_2 * (dK * L_1) = L_2 * (L_1 - K),$$

hence $K * L_1 = L_2 * K$, and consequently $1 * L_1 = 1 * L_2$. This implies $L_1 = L_2$ by differentiation, and so the solution L of (0.29), (0.30) is unique if it exists.

(Existence) Let $k(t) = \text{Var } K|_0^t$; then $k \in BV^0(J)$ and by induction we obtain that K_n defined by (0.32) belongs to $BV^0(J; \mathcal{B}(X))$, and $\text{Var } K_n|_s^t \le k_n(t) - k_n(s)$, where $k_0(t) = k(t)$, and $k_{n+1}(t) = (dk * k_n)(t)$. Let $\omega \ge 0$ be so large that $\eta = \int_0^T e^{-\omega t} dk(t) < 1$; such ω certainly exists since $k(0) = k(0+) = 0$. Then by induction $\int_0^T e^{-\omega t} dk_n(t) \le \eta^{n+1}$, $n \in \mathbb{N}_0$, and therefore

$$\text{Var } K_n|_0^T \le k_n(t) \le e^{\omega T} \int_0^T e^{-\omega t} dk_n(t) \le \eta^{n+1} e^{\omega T}, \quad n \in \mathbb{N}_0.$$

This shows that the series defined by (0.31) converges absolutely in the norm of $BV^0(J; \mathcal{B}(X))$, and so $L \in BV^0(J; \mathcal{B}(X))$ exists. Since

$$L - K = L - K_0 = \sum_{n=1}^{\infty} K_n = dK * \sum_{n=1}^{\infty} K_{n-1} = dK * L$$

and $dK * K_n = K_n * dK$ for all $n \in \mathbb{N}_0$, $L(t)$ satisfies the resolvent equations (0.29) and (0.30). The proof is complete. $\square$

The kernel $L(t)$ is called the *resolvent kernel* associated with $K(t)$. If $u(t)$ is a continuous solution of (0.28), convolving (0.28) with $dL(t)$ and adding the result to (0.28) we obtain the *variation of parameters formula*

$$u(t) = f(t) + \int_0^t dL(\tau)f(t-\tau), \quad t \in J. \tag{0.33}$$

Thus as a consequence of Theorem 0.5 we have

Corollary 0.2 *Suppose $K \in BV^0(J; \mathcal{B}(X))$ and let $L \in BV^0(J; \mathcal{B}(X))$ denote the resolvent kernel associated with K. Then for any $f \in C(J; X)$ with $f(0) = 0$, there is a unique solution $u \in C(J; X)$ of (0.28) which is given by the variation of parameters formula (0.33). Moreover, the solution map $f \mapsto u$ preserves the spaces $L^p(J; X)$, $1 \le p \le \infty$, and $BV^0(J; X)$ and the estimates*

$$|u|_p \le (1 + l(T))|f|_p \,, \quad f \in L^p(J; X), \quad 1 \le p \le \infty, \tag{0.34}$$
$$\mathrm{Var}\, u|_0^T \le (1 + l(T))\mathrm{Var}\, f|_0^T, \quad f \in BV^0(J; X), \tag{0.35}$$

hold, where $l(t)$ denotes the resolvent kernel of $k(t)$, i.e. the solution of the scalar equation $l = k + dk * l$.

Observe that the solution of (0.28) with $f(t) \equiv x$ is $x + L(t)x$, which is in general not continuous, but left-continuous only. (0.34) and (0.35) follow directly from (0.33), observing that $l(T) \ge \mathrm{Var}\, L|_0^T$.

The resolvent kernel $L(t)$ associated with $K(t)$ is also useful in connection with the operator-valued Volterra equations

$$U(t) = F(t) + \int_0^t dK(\tau)U(t-\tau), \quad t \in J, \tag{0.36}$$

and

$$V(t) = G(t) + \int_0^t V(t-\tau)dK(\tau), \quad t \in J. \tag{0.37}$$

In fact, the solution of (0.36) is given by

$$U(t) = F(t) + \int_0^t dL(\tau)F(t-\tau), \quad t \in J, \tag{0.38}$$

while that of (0.37) is obtained via

$$V(t) = G(t) + \int_0^t G(t-\tau)dL(\tau), \quad t \in J. \tag{0.39}$$

Observe that by application to a vector $x \in X$, (0.36) reduces to (0.28) and (0.38) to (0.33); however this is not the case for (0.37). This leads to a slightly different behaviour of (0.36) and (0.37).

Corollary 0.3 *Let Z denote another Banach space, $K \in BV^0(J; \mathcal{B}(X))$ and let $L \in BV^0(J; \mathcal{B}(X))$ be the resolvent kernel associated with K. Then*

(i) for every $F \in C_s(J; \mathcal{B}(Z; X))$, $F \in C(J; \mathcal{B}(Z; X))$, $F \in BV(J; \mathcal{B}(Z; X))$, with $F(0) = 0$, there is a unique solution $U \in C_s(J; \mathcal{B}(Z; X))$, $U \in C(J; \mathcal{B}(Z; X))$, $U \in BV^0(J; \mathcal{B}(Z; X))$, of (0.36), respectively, and it is given by (0.38);

(ii) for every $G \in C(J; \mathcal{B}(X; Z))$ or $G \in BV(J; \mathcal{B}(X; Z))$, with $G(0) = 0$, there is a unique solution $V \in C(J; \mathcal{B}(X; Z))$ or $V \in BV^0(J; \mathcal{B}(X; Z))$, of (0.37), respectively, and it is given by (0.39).

Here, by $C_s(J; \mathcal{B}(Z; X))$ we mean the space of all *strongly continuous families* $\{F(t)\}_{t \in J} \subset \mathcal{B}(Z; X)$, i.e. the functions $F(\cdot)z$ are continuous in X for each $z \in Z$.

0.4 Convolution Equations with Bounded Kernels on the Line
Let X denote again a Banach space, and $K \in L^1(\mathbb{R}; \mathcal{B}(X))$. In this subsection, we consider the convolution equation on the line

$$u(t) = f(t) + \int_{-\infty}^{\infty} K(\tau)u(t - \tau)d\tau, \quad t \in \mathbb{R}, \tag{0.40}$$

where $f \in L^p(\mathbb{R}; X)$, $1 \le p \le \infty$.

Suppose (0.40) admits a unique solution $u \in C_{ub}(\mathbb{R}; X)$ for each given $f \in C_{ub}(\mathbb{R}; X)$. Then by the translation invariance of (0.40), the solution of (0.40) for any τ-periodic continuous f must be τ-periodic. In particular, it must be constant whenever f is constant. Now choose $f(t) = e^{i\rho t}y$, $\rho \in \mathbb{R}$, $y \in X$; then $v(t) = u(t)e^{-i\rho t}$ satisfies $v = y + K_\rho * v$, where $K_\rho(t) = e^{-i\rho t}K(t)$; hence $v(t) \equiv x$ and so $u(t) \equiv e^{i\rho t}x$, for some $x \in X$; (0.40) then implies $x = y + \tilde{K}(\rho)x$. This shows that $I - \tilde{K}(\rho)$ is invertible for every $\rho \in \mathbb{R}$. We state this necessary condition as

Proposition 0.3 *Let $K \in L^1(\mathbb{R}; \mathcal{B}(X))$, and suppose (0.40) admits a unique solution $u \in C_{ub}(\mathbb{R}; X)$, for every $f \in C_{ub}(\mathbb{R}; X)$. Then $I - \tilde{K}(\rho)$ is invertible for each $\rho \in \mathbb{R}$.*

Next consider the whole-line resolvent equations

$$R(t) = K(t) + \int_{-\infty}^{\infty} K(\tau)R(t - \tau), \quad t \in \mathbb{R}, \tag{0.41}$$

and

$$R(t) = K(t) + \int_{-\infty}^{\infty} R(\tau)K(t - \tau), \quad t \in \mathbb{R}. \tag{0.42}$$

Suppose (0.41) and (0.42) admit a solution $R \in L^1(\mathbb{R}; \mathcal{B}(X))$. Then R is necessarily unique and the unique solution u of (0.40) is given by the variation of parameters formula

$$u(t) = f(t) + \int_{-\infty}^{\infty} R(\tau)f(t - \tau)d\tau, \quad t \in \mathbb{R}. \tag{0.43}$$

It is clear that (0.43) leaves invariant $L^p(\mathbb{R}; X)$, $1 \leq p \leq \infty$, as well as $C_{ub}(\mathbb{R}; X)$. It turns out that the necessary condition obtained in Proposition 0.3 is also sufficient for existence of $R \in L^1(\mathbb{R}; \mathcal{B}(X))$. This is the main result of this subsection; in the case $X = \mathbb{C}$ it is due to Paley and Wiener [266], while the vector-valued case was proved in Prüss [271] and Gripenberg [154]; for more general versions see Gripenberg [155], and Gripenberg, Londen and Staffans [156].

Theorem 0.6 *Let $K \in L^1(\mathbb{R}; \mathcal{B}(X))$. Then (0.41), (0.42) admit a unique solution $R \in L^1(\mathbb{R}; \mathcal{B}(X))$ if and only if $I - \tilde{K}(\rho)$ is invertible for each $\rho \in \mathbb{R}$.*

Proof: ($\Rightarrow$) Let $R \in L^1(\mathbb{R}; \mathcal{B}(X))$ be a solution of (0.41), (0.42); then Fourier transform yields

$$(I + \tilde{R}(\rho))(I - \tilde{K}(\rho)) = (I - \tilde{K}(\rho))(I + \tilde{R}(\rho)) = I, \quad \rho \in \mathbb{R},$$

and so $I - \tilde{K}(\rho)$ is invertible for each $\rho \in \mathbb{R}$.

($\Leftarrow$) Let $F \in C_{ub}(\mathbb{R}; \mathcal{B}(X))$ be defined by

$$F(\rho) = \tilde{K}(\rho)(I - \tilde{K}(\rho))^{-1}, \quad \rho \in \mathbb{R}. \tag{0.44}$$

We are going to show that $F(\rho) = \tilde{R}(\rho)$, $\rho \in \mathbb{R}$, for some $R \in L^1(\mathbb{R}; \mathcal{B}(X))$; once this is known to be true it then follows easily that R solves (0.41), (0.42), by uniqueness of the Fourier transform.

(i) Let $\varphi \in L^1(\mathbb{R})$ be defined by $\varphi(t) = 2(\pi t^2)^{-1} \sin(3t/2) \sin(t/2)$, $t \in \mathbb{R}$; a simple computation yields

$$\tilde{\varphi}(\rho) = \begin{cases} 1 & \text{for } |\rho| \leq 1 \\ 2 - |\rho| & \text{for } 1 \leq |\rho| \leq 2 \\ 0 & \text{for } |\rho| \geq 2. \end{cases}$$

Define $\varphi_\varepsilon(t) = \varepsilon\varphi(\varepsilon t)$, $t \in \mathbb{R}$; then $\tilde{\varphi}_\varepsilon(\rho) = \tilde{\varphi}(\rho/\varepsilon)$, $\rho \in \mathbb{R}$, and with $\rho_n = 3n\varepsilon$, $|n| \leq N$, we obtain the partition of unity

$$\sum_{-N}^{N} \tilde{\varphi}_\varepsilon(\rho - \rho_n) + \psi_N(\rho) = 1, \quad \rho \in \mathbb{R}, \tag{0.45}$$

where

$$\psi_N(\rho) = \begin{cases} 0 & \text{for } |\rho| \leq (3N + 1)\varepsilon, \\ |\rho|/\varepsilon - 3N - 1 & \text{for } (3N + 1)\varepsilon \leq |\rho| \leq (3N + 2)\varepsilon, \\ 1 & \text{for } |\rho| \geq (3N + 2)\varepsilon. \end{cases}$$

By means of this partition of unity we obtain the decomposition

$$F(\rho) = \sum_{-N}^{N} \tilde{\varphi}_{\varepsilon}(\rho - \rho_n)F(\rho) + \psi_N(\rho)F(\rho), \quad \rho \in \mathbb{R},$$

and it is sufficient to show that each term of the right hand side is the Fourier-transform of an $L^1(\mathbb{R}; \mathcal{B}(X))$-function.

(ii) Define $K_n(t)$ by

$$K_n(t) = [((\varphi_{2\varepsilon}e^{i\rho_n\cdot}) * K)(t) - \varphi_{2\varepsilon}(t)e^{i\rho_n t}\tilde{K}(\rho_n)](I - \tilde{K}(\rho_n))^{-1}, \quad t \in \mathbb{R};$$

then

$$
\begin{aligned}
|K_n|_1 &\leq M \int_{-\infty}^{\infty} |\int_{-\infty}^{\infty} K(\tau)(\varphi_{2\varepsilon}(t - \tau) - \varphi_{2\varepsilon}(t))e^{i\rho_n(t-\tau)}d\tau|dt \\
&\leq M \int_{-\infty}^{\infty} |K(\tau)|(\int_{-\infty}^{\infty} |\varphi_{2\varepsilon}(t - \tau) - \varphi_{2\varepsilon}(t)|dt)d\tau \\
&= M \int_{-\infty}^{\infty} |K(\tau)|(\int_{-\infty}^{\infty} |\varphi(t - 2\varepsilon\tau) - \varphi(t)|dt)d\tau \\
&\leq 2M|\varphi|_1(\int_{|\tau| \geq 1/\sqrt{\varepsilon}} |K(\tau)|d\tau) + M|K|_1 \cdot \sup_{|h| \leq 2\sqrt{\varepsilon}} |\varphi(\cdot - h) - \varphi|_1,
\end{aligned}
$$

where $M = \sup\{|(I - \tilde{K}(\rho))^{-1}| : \rho \in \mathbb{R}\} < \infty$ by assumption and since $|\tilde{K}(\rho)| \to 0$ as $|\rho| \to \infty$, by the Riemann-Lebesgue lemma. Choosing $\varepsilon > 0$ small enough we obtain $|K_n|_1 \leq 1/2$ for each $|n| \leq N$. The Fourier transform of $K_n(t)$ is easily evaluated to the result

$$\tilde{K}_n(\rho) = \tilde{\varphi}_{2\varepsilon}(\rho - \rho_n)(\tilde{K}(\rho) - \tilde{K}(\rho_n))(I - \tilde{K}(\rho_n))^{-1}, \quad \rho \in \mathbb{R},$$

hence

$$F_n(\rho) = \tilde{\varphi}_{\varepsilon}(\rho - \rho_n)\tilde{K}(\rho)(I - \tilde{K}(\rho))^{-1} = \tilde{\varphi}_{\varepsilon}(\rho - \rho_n)\tilde{K}(\rho)(I - \tilde{K}(\rho_n))^{-1}(I - \tilde{K}_n(\rho))^{-1},$$

with $\tilde{\varphi}_{2\varepsilon}(\rho) \equiv 1$ on supp $\tilde{\varphi}_{\varepsilon}$. Since $|K_n|_1 \leq 1/2$, $F_n(\rho)$ is the Fourier transform of $R_n \in L^1(\mathbb{R}; \mathcal{B}(X))$ given by

$$R_n = \varphi_{\varepsilon}e^{i\rho_n} * K * (I - \tilde{K}(\rho_n))^{-1} \sum_{l=0}^{\infty} K_n^{*l}.$$

(iii) Let $\chi(t) = 2(\pi t^2)^{-1} \sin^2(t/2)$ denote Fejer's kernel, define dilations of them by $\chi_N(t) = 3N\varepsilon\chi(3N\varepsilon t)$, and let $K_N = \chi_N * K$. Then $K_N \to K$ in $L^1(\mathbb{R}; \mathcal{B}(X))$ as $N \to \infty$, $\varepsilon > 0$ fixed, and

$$\tilde{K}_N(\rho) = \begin{cases} (1 - |\rho|/3N\varepsilon)\tilde{K}(\rho) & \text{for } |\rho| \leq 3N\varepsilon, \\ 0 & \text{otherwise.} \end{cases}$$

In particular $\tilde{K}_N(\rho) = 0$ on supp ψ_N, and therefore

$$
\begin{aligned}
F_N(\rho) &= \psi_N(\rho)\tilde{K}(\rho)(I - \tilde{K}(\rho))^{-1} \\
&= (\tilde{K}(\rho) - \sum_{-N}^{N} \tilde{\varphi}_\varepsilon(\rho - \rho_n)\tilde{K}(\rho))(I - (\tilde{K}(\rho) - \tilde{K}_N(\rho)))^{-1}
\end{aligned}
$$

is the Fourier transform of a function $R_N \in L^1(\mathbb{R}; \mathcal{B}(X))$ as at the end of step (ii), provided N is chosen so large that $|K - K_N|_1 < 1$ holds. The proof is now complete. $\square$

As indicated in front of Theorem 0.6, we obtain the following result on solvability properties of (0.40) as a consequence of this result.

Corollary 0.4 *Let $K \in L^1(\mathbb{R}; \mathcal{B}(X))$ be such that $I - \tilde{K}(\rho)$ is invertible for $\rho \in \mathbb{R}$. Then for every $f \in L^p(\mathbb{R}; X)$ there is a unique solution $u \in L^p(\mathbb{R}; X)$ of (0.40), where $1 \le p \le \infty$. If $R \in L^1(\mathbb{R}; \mathcal{B}(X))$ denotes the whole-line resolvent of (0.40) the solution u of (0.40) is given by (0.43). Moreover, the spaces $C_{ub}(\mathbb{R}; X)$ and $C_b(\mathbb{R}; X)$ are also invariant under the solution map $f \mapsto u$.*

In the Volterra case we have supp $K \subset \mathbb{R}_+$; this does not necessarily imply supp $R \subset \mathbb{R}_+$, in case $R \in L^1(\mathbb{R}; \mathcal{B}(X))$ exists. In fact, $R \in L^1(\mathbb{R}_+; \mathcal{B}(X))$ yields

$$
(I - \hat{K}(\lambda))(I - \hat{R}(\lambda)) = (I - \hat{R}(\lambda))(I - \hat{K}(\lambda) = I, \quad \text{Re } \lambda \ge 0,
$$

i.e. $I - \hat{K}(\lambda)$ must necessarily be invertible for all Re $\lambda \ge 0$, not only on the imaginary axis. It turns out that this condition is also sufficient; this is the Paley-Wiener lemma on the halfline. The argument used in the proof is a straightforward adaption of the original idea of Paley and Wiener [266]; see also Gripenberg, Londen, and Staffans [156].

Theorem 0.7 *Let $K \in L^1(\mathbb{R}_+; \mathcal{B}(X))$, and extend $K(t)$ to all of $\mathbb{R}$ by 0. Then (0.41), (0.42) admit a unique solution $R \in L^1(\mathbb{R}_+; \mathcal{B}(X))$ ($R(t) \equiv 0$ for $t < 0$) if and only if $I - \hat{K}(\lambda)$ is invertible for each Re $\lambda \ge 0$.*

Proof: Extend $K(t)$ by 0 to all of $\mathbb{R}$. Since $I - \hat{K}(i\rho) = I - \tilde{K}(\rho)$ is invertible for all $\rho \in \mathbb{R}$, by Theorem 0.6 there is a solution $R \in L^1(\mathbb{R}; \mathcal{B}(X))$ of (0.41), (0.42). Let $R_+(t) = R(t)e_0(t)$, $R_-(t) = R(-t) - R_+(-t)$, $t \in \mathbb{R}$, where $e_0(t)$ denotes the Heaviside function. Then R_+ and R_- vanish for $t < 0$, hence their Laplace transforms are well-defined, analytic for Re $\lambda > 0$, bounded and continuous on $\bar{\mathbb{C}}_+$, and tend to zero as $|\lambda| \to \infty$. Define

$$
G(\lambda) = \begin{cases} \hat{K}(\lambda)(I - \hat{K}(\lambda))^{-1} - \hat{R}_+(\lambda), & \text{Re } \lambda > 0 \\ \hat{R}_-(-\lambda), & \text{Re } \lambda < 0; \end{cases}
$$

then $G(i\rho + 0) = \tilde{R}(\rho) - \tilde{R}_+(\rho) = \tilde{R}_-(\rho) = G(i\rho - 0)$, $\rho \in \mathbb{R}$, i.e. $G(\lambda)$ is entire by Morera's theorem, (cp. eg. Conway [59]), bounded and $G(\lambda) \to 0$ as $|\lambda| \to \infty$.

Liouville's theorem yields $G(\lambda) \equiv 0$, hence $\hat{R}_+(\lambda) = \hat{K}(\lambda)(I - \hat{K}(\lambda))^{-1}$, $\mathrm{Re}\,\lambda \geq 0$, i.e. $R = R_+$ by uniqueness of the Fourier transform. $\quad\square$

0.5 The Spectrum of Functions of Subexponential Growth

Let $f \in L^1_{loc}(\mathbb{R}; X)$ be of subexponential growth, where X denotes a complex Banach space; by this we mean

$$\int_{-\infty}^{\infty} e^{-\varepsilon|t|} |f(t)| dt < \infty, \quad \text{for each } \varepsilon > 0.$$

The *Fourier-Carleman transform* $\hat{f}$ of f is defined by

$$\hat{f}(\lambda) = \begin{cases} \int_0^\infty e^{-\lambda t} f(t) dt, & \mathrm{Re}\,\lambda > 0 \\ -\int_{-\infty}^0 e^{-\lambda t} f(t) dt, & \mathrm{Re}\,\lambda < 0. \end{cases}$$

Obviously, $\hat{f}(\lambda)$ is holomorphic in $\mathbb{C} \setminus i\mathbb{R}$. Define

$$\rho(f) = \{\rho \in \mathbb{R} : \hat{f}(\lambda) \text{ is analytically extendable to some } B_\varepsilon(i\rho)\}; \qquad (0.46)$$

then $\sigma(f) = \mathbb{R} \setminus \rho(f)$ is called the *spectrum* of f. The Fourier-Carleman transform and this approach of defining the spectrum of functions in the scalar case are due to Carleman [37]; see also Katznelson [194].

Example 0.1 (i) Suppose $f \in L^1_{loc}(\mathbb{R}; X)$ is τ-periodic. Then an easy computation yields

$$\hat{f}(\lambda) = (1 - e^{-\lambda\tau})^{-1} \int_0^\tau e^{-\lambda t} f(t) dt, \quad \mathrm{Re}\,\lambda \neq 0, \qquad (0.47)$$

i.e. $\hat{f}(\lambda)$ extends to a meromorphic function with simple poles at most at $\lambda_n = 2\pi i n/\tau$, $n \in \mathbb{Z}$. The residuum at $\lambda = \lambda_n$ is given by

$$\mathrm{Res}\hat{f}(\lambda)|_{\lambda=\lambda_n} = f_n = \frac{1}{\tau} \int_0^\tau e^{-2\pi i n t/\tau} f(t) dt, \quad n \in \mathbb{Z},$$

the nth Fourier coefficient of f. Therefore we have

$$\sigma(f) = \{2\pi n/\tau : n \in \mathbb{Z}, \; f_n \neq 0\} \subset (2\pi/\tau)\mathbb{Z}. \qquad (0.48)$$

(ii) Let $f \in L^1(\mathbb{R}; X)$ and let $\tilde{f}(\rho)$ denote its Fourier transform

$$\tilde{f}(\rho) = \int_{-\infty}^{\infty} f(t) e^{-i\rho t} dt, \quad \rho \in \mathbb{R}. \qquad (0.49)$$

Then

$$\lim_{\sigma\to 0+} (\hat{f}(i\rho + \sigma) - \hat{f}(i\rho - \sigma)) = \lim_{\sigma\to 0+} \int_{-\infty}^{\infty} f(t) e^{-i\rho t} e^{-\sigma|t|} dt = \tilde{f}(\rho),$$

for every $\rho \in \mathbb{R}$; therefore $\rho_0 \notin \sigma(f)$ implies $\tilde{f}(\rho) \equiv 0$ in a neighborhood of ρ_0, and so $\rho_0 \notin \operatorname{supp} \tilde{f}$. Conversely, if $\rho_0 \notin \operatorname{supp} \tilde{f}$ then $\hat{f}(\lambda)$ is continuous on a ball $B_\varepsilon(i\rho_0)$, hence by Morera's theorem (cp. Conway [59]), $\hat{f}(\lambda)$ is holomorphic there, i.e. $\rho_0 \notin \sigma(f)$. This implies the relation

$$\sigma(f) = \operatorname{supp} \tilde{f}, \quad \text{for all } f \in L^1(\mathbb{R}; X). \qquad \square \qquad (0.50)$$

Below we will show that (0.50) holds for all $f \in L^1_{loc}(\mathbb{R}; X)$ of polynomial growth provided the Fourier transform of f is understood in the sense of vector-valued distributions.

   Next we summarize some of the properties of the spectrum of functions of subexponential growth.

Proposition 0.4 *Let* $f, g \in L^1_{loc}(\mathbb{R}; X)$ *be of subexponential growth,* $\alpha \in \mathbb{C} \setminus \{0\}$, $K \in BV(\mathbb{R}; \mathcal{B}(X; Z))$ *with* $\operatorname{supp} dK$ *compact. Then*
(i) $\sigma(f)$ *is closed;* $\qquad\qquad\qquad\qquad$ *(ii)* $\sigma(\alpha f) = \sigma(f)$;
(iii) $\sigma(f(\cdot + h)) = \sigma(f)$; $\qquad\qquad\qquad$ *(iv)* $\sigma(\dot{f}) \subset \sigma(f)$;
(v) $\sigma(f + g) \subset \sigma(f) \cup \sigma(g)$; $\qquad\qquad$ *(vi)* $\sigma(dK * f) \subset \sigma(f)$;
(vii) $\sigma(\widetilde{dK}f) \subset \sigma(f) - \operatorname{supp} dK$.

Proof: (i) and (ii) are trivial by the definition of $\sigma(f)$.
   (iii) follows from the identity

$$\hat{f}(\cdot + h)(\lambda) = e^{\lambda h}\hat{f}(\lambda) - \int_0^h e^{\lambda(h-t)}f(t)dt, \quad \operatorname{Re}\lambda \neq 0, \qquad (0.51)$$

since the second term on the right is an entire function.
   (iv) If $\dot{f} \in L^1_{loc}(\mathbb{R}; X)$ is of subexponential growth, then

$$\widehat{\dot{f}}(\lambda) = \lambda\hat{f}(\lambda) - f(0), \quad \operatorname{Re}\lambda \neq 0;$$

this implies (iv).
   (v) If $\rho \notin \sigma(f) \cup \sigma(g)$ then $\hat{f}(\lambda)$ and $\hat{g}(\lambda)$ admit analytic extension to a ball $B_\varepsilon(i\rho_0)$, hence $\rho \notin \sigma(f + g)$.
   (vi) Since $\operatorname{supp} dK$ is compact, i.e. $K(t)$ is constant for $t \geq N$, $t \leq -N$, say, $dK * f$ is well-defined and

$$\int_{-\infty}^\infty e^{-\varepsilon|t|}|dK * f|dt \leq \int_{-N}^N |dK(\tau)|(\int_{-\infty}^\infty e^{-\varepsilon|t+\tau|}|f(t)|dt)$$

$$\leq (\int_{-N}^N |dK(\tau)|e^{\varepsilon|\tau|})\int_{-\infty}^\infty e^{-\varepsilon|t|}|f(t)|dt,$$

hence $dK * f$ is again of subexponential growth. A simple computation yields for $\operatorname{Re}\lambda \neq 0$

$$(dK * f)\widehat{}(\lambda) = (\int_{-\infty}^\infty dK(\tau)e^{-\lambda\tau})\hat{f}(\lambda) + \int_{-\infty}^\infty dK(\tau)\int_0^\tau f(t - \tau)e^{-\lambda t}dt, \qquad (0.52)$$

and so (vi) follows since the coefficient of $\hat{f}(\lambda)$ and the second term are entire.

(vii) We have for Re $\lambda > 0$

$$
\begin{aligned}
(\widetilde{dK} \cdot f)\widehat{\,}(\lambda) &= \int_0^\infty \widetilde{dK}(t)f(t)e^{-\lambda t}dt = \int_{-\infty}^\infty dK(\rho)\int_0^\infty f(t)e^{-(\lambda+i\rho)t}dtds \\
&= \int_{-\infty}^\infty dK(\rho)\hat{f}(\lambda+i\rho)d\rho, \tag{0.53}
\end{aligned}
$$

and similarly for Re $\lambda < 0$. This formula defines an analytic extension to all $\lambda = i\rho_0$, such that $\rho_0 + \rho \notin \sigma(f)$, for $\rho \in \operatorname{supp} dK$ i.e. (vii) holds. $\square$

The spectrum of a function also enjoys a certain continuity property which turns out to be very useful.

Theorem 0.8 *Let $f_n, f \in L^1_{loc}(\mathbb{R}; X)$ be of subexponential growth and such that*
(i) $\int_{-\infty}^\infty e^{-\varepsilon|t|}|f_n(t) - f(t)|dt \to 0$ as $n \to \infty$, for each $\varepsilon > 0$;
(ii) there are $c > 0$ and $k \in \mathbb{N}_0$ such that

$$
|\widehat{f_n}(\lambda)| \leq c|Re\,\lambda|^{-k}, \quad 0 < |Re\,\lambda| < \delta, \; n \in \mathbb{N}.
$$

Then $\sigma(f_n) \subset \Lambda$ for all $n \in \mathbb{N}$ implies $\sigma(f) \subset \bar{\Lambda}$; equivalently

$$
\sigma(f) \subset \cap_{m \geq 1}\overline{\cup_{n \geq m}\sigma(f_n)}. \tag{0.54}
$$

Proof: (i) implies $\widehat{f_n}(\lambda) \to \hat{f}(\lambda)$ uniformly on compact subsets of $\mathbb{C} \setminus i\,\mathbb{R}$. Let $\Lambda \subset \mathbb{R}$ be closed and such that $\sigma(f_n) \subset \Lambda$ for all $n \in \mathbb{N}$. Choose $\rho_0 \notin \Lambda$, and let $r \leq \operatorname{dist}(\rho_0, \Lambda)/4$, $r \leq \delta/2$. We want to prove that $\hat{f}(\lambda)$ admits analytic extension to $B_\varepsilon(i\rho_0)$; for this it sufficient to prove

$$
|\widehat{f_n}(\lambda)| \leq M \quad \text{for all } n \in \mathbb{N}, \; |\lambda - i\rho_0| \leq r, \tag{0.55}
$$

by Montel's theorem (cp. Conway [59]). For this purpose observe that $\widehat{f_n}(\lambda)$ is holomorphic on $B_\varepsilon(i\rho_0)$, hence Cauchy's theorem yields with

$$
h(\lambda) = [(\lambda - i\rho_0)(1 + \frac{(\lambda - i\rho_0)^2}{4r^2})]^k
$$

the relation

$$
h(\lambda)\widehat{f_n}(\lambda) = \frac{1}{2\pi i}\int_{|z-i\rho_0|=2r} h(z)\widehat{f_n}(z)\frac{dz}{z - \lambda}, \quad \lambda \in B_r(i\rho_0), \; n \in \mathbb{N}.
$$

Since for $|z - i\rho_0| = 2r$ we have $|h(z)| = 2^k|\operatorname{Re}\,z|^k$, (ii) yields

$$
|h(\lambda)\widehat{f_n}(\lambda)| \leq \frac{c}{2\pi}\int_{|z-i\rho_0|=2r} 2^k|\operatorname{Re}\,z|^k \cdot |\operatorname{Re}\,z|^{-k}\frac{|dz|}{|z - \lambda|} \leq 2^{k+1}c,
$$

for all $\lambda \in B_r(i\rho_0)$, $n \in \mathbb{N}$, and so

$$\sup_{\lambda \in B_r(i\rho_0)} |\widehat{f_n}(\lambda)| = \sup_{|\lambda - i\rho_0| = r} |\widehat{f_n}(\lambda)| \le c2^{k+1} \sup_{|\lambda - i\rho_0| = r} |\frac{1}{h(\lambda)}|$$

$$\le 2^{k+1} c(\frac{4}{3})^k r^{-k} = M,$$

independently of $n \in \mathbb{N}$, i.e. (0.55) follows and the proof is complete. $\square$

Observe that (i) of Theorem 0.8 is implied by

$$|f_n| \le g \quad \text{a.e.,} \quad f_n \to f \text{ a.e.} \tag{0.56}$$

where $g \in L^1_{loc}(\mathbb{R})$ is of subexponential growth, while (ii) follows if g is *growing polynomially*, i.e.

$$\int_{-\infty}^{\infty} |g(t)|(1 + |t|)^{-m} dt < \infty \quad \text{for some } m \ge 0.$$

In fact, then

$$|\widehat{f_n}(\lambda)| \le \int_0^{\infty} g(t)e^{-\operatorname{Re}\lambda t} dt \le (\int_0^{\infty} g(t)(1+t)^{-m} dt) \cdot \sup_{t>0}[(1+t)^m e^{-\operatorname{Re}\lambda t}]$$

$$\le (\int_0^{\infty} g(t)(1+t)^{-m} dt)(\frac{m}{\operatorname{Re}\lambda})^m e^{1-m}, \quad 0 < \operatorname{Re}\lambda < 1,$$

and similarly for $0 > \operatorname{Re}\lambda > -1$.

0.6 The Spectrum of Functions Growing Polynomially

We have seen in Example 0.1, (ii) that $\sigma(f) = \operatorname{supp}\tilde{f}$ holds for $f \in L^1(\mathbb{R})$. This characterization of the spectrum allows for a considerable extension if the Fourier transform is understood in the sense of distributions. For this purpose let $f \in L^1_{loc}(\mathbb{R}; X)$ be of polynomial growth; the tempered distribution D_f corresponding to f is then defined according to

$$[D_f, \varphi] = \int_{-\infty}^{\infty} f(t)\varphi(t) dt, \quad \varphi \in \mathcal{S}, \tag{0.57}$$

where $\mathcal{S}$ denotes the Schwartz space of all C^{∞}-functions on $\mathbb{R}$ with each of its derivatives decaying faster than any polynomial. The Fourier transform of D_f is defined by

$$[\widetilde{D_f}, \varphi] = [D_f, \tilde{\varphi}], \quad \varphi \in \mathcal{S}. \tag{0.58}$$

Recall also the definition of $\operatorname{supp} D_f$. A number $\rho \in \mathbb{R}$ belongs to $\operatorname{supp} D_f$ if for every $\varepsilon > 0$ there is $\varphi \in \mathcal{S}$ such that $\operatorname{supp} \varphi \subset B_\varepsilon(\rho)$ and $[D_f, \varphi] \ne 0$.

Proposition 0.5 *Let $f \in L^1_{loc}(\mathbb{R}; X)$ be of polynomial growth. Then*
(i) $\sigma(f) = \operatorname{supp}\widetilde{D_f}$;

(ii) $\sigma(f) = \emptyset \iff f \equiv 0$;

(iii) $\sigma(f) = \{\rho_1, \ldots, \rho_n\} \iff f(t) = \sum_{k=1}^{n} p_k(t) e^{i\rho_k t}$, *for some polynomials* $p_k(t)$.

(iv) $\sigma(f)$ *compact* $\iff$ f *admits extension to an entire function of exponential growth.*

Proof: (i) If $\rho_0 \notin \sigma(f)$ then $\hat{f}(\lambda)$ admits holomorphic extension to some ball $B_\varepsilon(i\rho_0)$. Let $\varphi \in \mathcal{S}$ be such that supp $\varphi \subset B_\varepsilon(\rho_0)$; then

$$\int_{-\infty}^{\infty} (\hat{f}(i\rho + \sigma) - \hat{f}(i\rho - \sigma))\varphi(\rho)d\rho = \int_{-\infty}^{\infty} (\int_{-\infty}^{\infty} f(t)e^{-i\rho t}e^{-\sigma|t|}dt)\varphi(\rho)d\rho$$

$$= \int_{-\infty}^{\infty} f(t)e^{-\sigma|t|}\tilde{\varphi}(t)dt$$

implies with $\sigma \to 0+$

$$[\widetilde{D_f}, \varphi] = \int_{-\infty}^{\infty} f(t)\tilde{\varphi}(t)dt = \lim_{\sigma \to 0+} \int_{-\infty}^{\infty} f(t)e^{-\sigma|t|}\tilde{\varphi}(t)dt$$

$$= \lim_{\sigma \to 0+} \int_{-\infty}^{\infty} (\hat{f}(i\rho + \sigma) - \hat{f}(i\rho - \sigma))\varphi(\rho)d\rho = 0,$$

i.e. $\rho_0 \notin \text{supp}\widetilde{D_f}$. This shows the inclusion $\text{supp}\widetilde{D_f} \subset \sigma(f)$.

To prove the converse inclusion we use Theorem 0.8. For this purpose let $\psi \in C_0^\infty(\mathbb{R})$ be such that supp $\psi \subset (-1,1)$, $\psi \geq 0$, and $\int_{-\infty}^{\infty} \psi(\rho)d\rho = 1$; define $\psi_n(\rho) = n\psi(n\rho)$. Then $\widetilde{\psi_n} \to 1$ as $n \to \infty$, uniformly for bounded t, and $\widetilde{\psi_n} \in \mathcal{S}$. Since f is polynomially growing, $f_n = \widetilde{\psi_n}f$ belongs to $L^1(\mathbb{R}; X)$, $f_n \to f$ uniformly for bounded t, and $|f_n| \leq |f| = g$. Thus (0.56) holds, which implies (i) and (ii) of Theorem 0.8 since f is growing polynomially. Example 0.1, (ii) yields

$$\sigma(f_n) = \text{supp}\widetilde{f_n} \subset B_\varepsilon(\text{supp}\widetilde{D_f}) \quad \text{for } n \geq n(\varepsilon).$$

In fact, if $\varphi \in \mathcal{S}$ is such that supp $\varphi \subset B_{\varepsilon/2}(\rho_0)$, then

$$\int_{-\infty}^{\infty} \widetilde{f_n}(\rho)\varphi(\rho)d\rho = \int_{-\infty}^{\infty} f_n(t)\tilde{\varphi}(t)dt = \int_{-\infty}^{\infty} f(t)\widetilde{\psi_n}(t)\widetilde{\varphi_n}(t)dt$$

$$= \int_{-\infty}^{\infty} f(t)\widetilde{\psi_n * \varphi}(t)dt = [\widetilde{D_f}, \psi_n * \varphi];$$

with supp $(\psi_n * \varphi) \subset$ supp $\varphi +$ supp $\psi_n \subset B_{\varepsilon/2}(\rho_0) + (-1/n, 1/n)$ we conclude $[\widetilde{D_f}, \psi_n * \varphi] = 0$ if dist $(\rho_0, \text{supp}\,\widetilde{D_f}) \geq \varepsilon$ and $\varepsilon/2 > 1/n$. Therefore, by Theorem 0.8 we obtain $\sigma(f) \subset B_\varepsilon(\text{supp}\,\widetilde{D_f})$ for every $\varepsilon > 0$, i.e. $\sigma(f) \subset \text{supp}\,\widetilde{D_f}$.

(ii) $\sigma(f) = \emptyset$ implies $\widetilde{D_f} = 0$ by (i), hence $D_f = 0$, and so $f = 0$.

(iii) Consider first the special case $\sigma(f) = \{0\}$; we then have to show that f is a polynomial. For this purpose consider the Laurent expansion of $\hat{f}(\lambda)$ at $\lambda = 0$

$$\hat{f}(\lambda) = \sum_{-\infty}^{\infty} a_n \lambda^n, \quad \lambda \in \mathbb{C} \setminus \{0\}.$$

The coefficients a_n are given by

$$a_n = \frac{1}{2\pi i} \int_{|z|=r} \hat{f}(z) z^{-n-1} dz, \quad n \in \mathbb{Z},$$

where $r > 0$ is arbitrary. Since f is of polynomial growth by assumption, there follows the estimate

$$|\hat{f}(\lambda)| \le c |\mathrm{Re}\,\lambda|^{-k}, \quad 0 < |\mathrm{Re}\,\lambda| < 1, \tag{0.59}$$

for some $k \in \mathbb{N}_0$. This implies

$$\left| \frac{1}{2\pi i} \int_{|z|=r} z^{n+k-1} \left(1 + \frac{z^2}{r^2}\right)^k \hat{f}(z) dz \right| \le c 2^k r^n, \quad \text{for all } n \in \mathbb{Z},\ r > 0,$$

hence

$$c 2^k r^n \ge \left| \sum_{l=0}^{k} \binom{k}{l} r^{-2l} \cdot \frac{1}{2\pi i} \int_{|z|=r} z^{n+k-1+2l} \hat{f}(z) dz \right| = \left| \sum_{l=0}^{k} \binom{k}{l} r^{-2l} a_{-n-k-2l} \right|.$$

Multiplying with r^{2k} and letting $r \to 0$ then implies $a_{-n-k} = 0$ for all $n \ge 1$, i.e. $\lambda = 0$ is a pole of order at most k of $\hat{f}(\lambda)$. Let $f_0(t) = \sum_{l=0}^{k} a_{-l-1} t^l / l!$; obviously $\hat{f}_0(\lambda) = \sum_{-k}^{-1} a_l \lambda^{-l}$ and so $\hat{f}(\lambda) - \hat{f}_0(\lambda)$ is entire, i.e. $\sigma(f - f_0) = \emptyset$, hence $f = f_0$. The general case can now easily be reduced to this special case.

(iv) Suppose $\sigma(f)$ is compact; we then define

$$f_0(z) = \frac{-1}{2\pi i} \int_{\Gamma} \hat{f}(\lambda) e^{\lambda z} d\lambda, \quad z \in \mathbb{C},$$

where Γ denotes any closed rectifiable simple contour surrounding $\sigma(f)$ clockwise. It is easy to verify that for any $|\mathrm{Re}\,\lambda| > \max_{\mu \in \Gamma} |\mathrm{Re}\,\mu|$ the identity

$$\hat{f}_0(\lambda) = \frac{1}{2\pi i} \int_{\Gamma} \hat{f}(\mu) \frac{d\mu}{\mu - \lambda}$$

holds. Thus it remains to show that the right hand side of this equation equals $\hat{f}(\lambda)$. For this purpose let $R > 0$ be sufficiently large, i.e. $R > |\lambda|$, and $R > \max\{|\rho| : \rho \in \sigma(f)\}$. Then Cauchy's theorem yields the identity

$$\left(1 + \frac{\lambda^2}{R^2}\right)^k \hat{f}(\lambda) = \frac{1}{2\pi i} \int_{\Gamma \cup \Gamma_R} \left(1 + \frac{\mu^2}{R^2}\right)^k \hat{f}(\mu) \frac{d\mu}{\mu - \lambda} \tag{0.60}$$

where $\Gamma = \Gamma^- \cup \Gamma^+$, $\Gamma_R = \Gamma_R^- \cup \Gamma_R^+$ (cp. Fig. 0.1) and k from (0.59). Similarly to the estimate in step (iii) of this proof, we obtain

$$\left| \frac{1}{2\pi i} \int_{\Gamma_R} \left(1 + \frac{\mu^2}{R^2}\right)^k \hat{f}(\mu) \frac{d\mu}{\mu - \lambda} \right| \le c 2^k R^{-k} \sup_{|\lambda|=R} \frac{1}{|\mu - \lambda|} \to 0 \quad \text{as } R \to \infty,$$

hence with $R \to \infty$, (0.60) implies $\hat{f}_0(\lambda) = \hat{f}(\lambda)$ for all $\mathrm{Re}\,\lambda \ne 0$. By uniqueness of the Laplace transform this yields $f_0 \equiv f$. The converse follows from the Taylor series of entire functions. $\square$

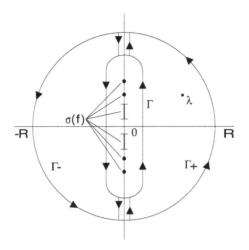

Figure 0.1: Integration path for (0.60)

The characterization $\sigma(f) = \text{supp } \widetilde{D}_f$ allows for some further useful properties.

Proposition 0.6 *Let $f \in L^1_{loc}(\mathbb{R}; X)$ be of polynomial growth, and let $\psi \in \mathcal{S}$, the Schwartz space. Then*
(i) $\sigma(\psi * f) \subset \sigma(f) \cap \text{supp } \widetilde{\psi}$;
(ii) $\sigma(f - \psi * f) \subset \sigma(f) \cap \text{supp}(1 - \widetilde{\psi})$;
(iii) $\psi * f = f$ if $\widetilde{\psi} = 1$ on a neighbourhood of $\sigma(f)$;
(iv) $\sigma(f) \cap \text{supp } \widetilde{\psi} = \emptyset$ implies $\psi * f = 0$.

Proof: Observe first that $\psi * f$ is well-defined and of polynomial growth again when f is of this class and ψ belongs to $\mathcal{S}$. The principal property which leads to Proposition 0.6 is the identity

$$[\widetilde{D}_{\psi*f}, \varphi] = [\widetilde{D}_f, \tilde{\psi}\varphi], \quad \varphi \in \mathcal{S}. \tag{0.61}$$

To prove (i), suppose $\rho \notin \sigma(f)$; then $\sigma(f) \cap \overline{B}_\varepsilon(\rho) = \emptyset$ for some $\varepsilon > 0$. Let $\varphi \in C_0^\infty(\mathbb{R})$ such that $\text{supp } \varphi \subset \overline{B}_\varepsilon(\rho)$; with (0.61) we obtain $[\widetilde{D}_{\psi*f}, \varphi] = [\widetilde{D}_f, \tilde{\psi}\varphi] = 0$, since $\text{supp } \tilde{\psi}\varphi \subset \overline{B}_\varepsilon(\rho)$, by the definition of $\text{supp } \widetilde{D}_f$. Therefore $\sigma(\psi * f) \subset \sigma(f)$.

On the other hand, suppose $\rho \notin \text{supp } \tilde{\psi}$; then $\text{supp } \tilde{\psi} \cap \overline{B}_\varepsilon(\rho) = \emptyset$ for some $\varepsilon > 0$. Let $\varphi \in C_0^\infty(\mathbb{R})$, $\text{supp } \varphi \subset \overline{B}_\varepsilon(\rho)$; (0.61) yields this time $[\widetilde{D}_{\psi*f}, \varphi] = [\widetilde{D}_f, \tilde{\psi}\varphi] = 0$, since $\text{supp } \tilde{\psi}\varphi = \text{supp } \tilde{\psi} \cap \text{supp } \varphi = \emptyset$, i.e. $\tilde{\psi}\varphi = 0$. Therefore $\sigma(\psi * f) \subset \text{supp } \tilde{\psi}$ follows.

To prove (ii) observe that (i) and Proposition 0.4 already imply $\sigma(f - \psi * f) \subset \sigma(f) \cup \sigma(\psi * f) \subset \sigma(f)$. So let $\rho \notin \text{supp } (1 - \tilde{\psi})$, and choose $\varepsilon > 0$ such that $\text{supp } (1 - \tilde{\psi}) \cap \overline{B}_\varepsilon(\rho) = \emptyset$. Then for $\varphi \in C_0^\infty(\mathbb{R})$, with $\text{supp } \varphi \subset \overline{B}_\varepsilon(\rho)$; we have $[\widetilde{D}_{f-\psi*f}, \varphi] = [\widetilde{D}_f, (1 - \tilde{\psi})\varphi] = 0$, since $(1 - \tilde{\psi})\varphi = 0$. Therefore $\sigma(f - \psi * f) \subset \text{supp } (1 - \tilde{\psi})$, and (ii) follows.

The remaining assertions are direct consequences of (i) and (ii). $\square$

The last result of this subsection is useful in detecting the spectrum of bounded functions, in particular of almost periodic and asymptotically almost periodic functions.

Proposition 0.7 *Let $f \in L^\infty(\mathbb{R}; X)$ and $\rho \notin \sigma(f)$. Then there is a constant $M(\rho) > 0$ such that*

$$\left| \int_0^t e^{-i\rho s} f(s) ds \right| \leq M(\rho), \quad \text{for all } t > 0.$$

In particular

$$\alpha(\rho, f) := \lim_{N \to \infty} N^{-1} \int_0^N e^{-i\rho s} f(s) ds = 0 \quad \text{for all } \rho \notin \sigma(f).$$

Proof: Replacing $f(t)$ by $e^{-i\rho t} f(t)$ if necessary, we may assume $\rho = 0$. If $0 \notin \sigma(f)$ then $\hat{f}(\lambda)$ is holomorphic and bounded on some ball $B_{2r}(0)$. Let Γ_r denote the circle $|\lambda| = r$ oriented counterclockwise and Γ_r^+, Γ_r^- its parts located in the right resp. left halfplane. Define $g_t(\lambda) = \int_0^t e^{-\lambda s} f(s) ds$, where $t > 0$ is fixed but arbitrary otherwise; $g_t(\lambda)$ is an entire function, hence Cauchy's theorem yields

$$g_t(0) = \frac{1}{2\pi i} \int_{\Gamma_r} g_t(\lambda) e^{\lambda t} (1 + \frac{\lambda^2}{r^2}) \frac{d\lambda}{\lambda}.$$

Similarly

$$\hat{f}(0) = \frac{1}{2\pi i} \int_{\Gamma_r} \hat{f}(\lambda) e^{\lambda t} (1 + \frac{\lambda^2}{r^2}) \frac{d\lambda}{\lambda},$$

by holomorphy of $\hat{f}(\lambda)$ in $B_{2r}(0)$. Subtracting and collecting terms we obtain

$$\begin{aligned}
\int_0^t f(s) ds &= g_t(0) = \hat{f}(0) - \int_{\Gamma_r^-} \hat{f}(\lambda) h_t(\lambda) d\lambda \\
&\quad + \int_{\Gamma_r^+} (g_t(\lambda) - \hat{f}(\lambda)) h_t(\lambda) d\lambda + \int_{\Gamma_r^-} g_t(\lambda) h_t(\lambda) d\lambda,
\end{aligned}$$

where

$$h_t(\lambda) = (2\pi i \lambda)^{-1} e^{\lambda t} (1 + \lambda^2/r^2).$$

Since $\hat{f}(\lambda)$ is bounded on Γ_r, say by M, and $h_t(\lambda)$ is so on Γ_r^-, the first integral is bounded by M. For the second one we have

$$|g_t(\lambda) - \hat{f}(\lambda)| \leq \int_t^\infty |f(\tau)| \exp(-\text{Re } \lambda \tau) d\tau \leq |f|_\infty \cdot \exp(-\text{Re } \lambda t)/ \text{Re } \lambda,$$

for all $\text{Re } \lambda > 0$, and

$$|h_t(\lambda)| \leq (2\pi r)^{-1} \exp(\text{Re } \lambda t)(2\text{Re } \lambda)/r,$$

therfore the second integral is estimated by $|f|_\infty/r$. For the last one we obtain similarly

$$\begin{aligned} |g_t(\lambda)h_t(\lambda)| &\le (2\pi r)^{-1}\exp(\operatorname{Re}\lambda t)(2|\operatorname{Re}\lambda|)/r \cdot \int_0^t |f(\tau)|\exp(-\operatorname{Re}\lambda\tau)d\tau \\ &\le |f|_\infty/\pi r^2, \end{aligned}$$

hence the integral is dominated also by $|f|_\infty/r$. Summing up there follows

$$|\int_0^t f(\tau)d\tau| \le 2(M + |f|_\infty/r), \quad t > 0,$$

hence the result is proved. □

Chapter I

Equations of Scalar Type

1   Resolvents

The concept of the resolvent which is central for the theory of linear Volterra equations is introduced and discussed. It is applied to the inhomogeneous equation to derive various variation of parameters formulas. The main tools for proving existence theorems for the resolvent are described in detail; these methods are the operational calculus in Hilbert spaces, perturbation arguments, and the Laplace-transform method. The generation theorem, the analog of the Hille-Yosida theorem of semigroup theory for Volterra equations, proved in Section 1.5, is of fundamental importance in later chapters. The theory is completed with several counterexamples, and with a discussion of the integral resolvent.

1.1 Well-posedness and Resolvents

Let X be a complex Banach space, A a closed linear unbounded operator in X with dense domain $\mathcal{D}(A)$, and $a \in L^1_{loc}(\mathbb{R}_+)$ a scalar kernel $\not\equiv 0$. We consider the Volterra equation

$$u(t) = f(t) + \int_0^t a(t-s)Au(s)ds, \quad t \in J, \tag{1.1}$$

where $f \in C(J; X)$, $J = [0, T]$. In the sequel we denote by X_A the domain of A equipped with the graph norm $|\cdot|_A$ of A, i.e. $|x|_A = |x| + |Ax|$; X_A is a Banach space since A is closed, and it is continuously and densely embedded into X. We shall use the abbreviation

$$(a * f)(t) = \int_0^t a(t-s)f(s)ds, \quad t \in J,$$

for the convolution.

The following notions of solutions of (1.1) are natural.

Definition 1.1 *A function $u \in C(J; X)$ is called*
(a) **strong solution** *of (1.1) on J if $u \in C(J; X_A)$ and (1.1) holds on J;*
(b) **mild solution** *of (1.1) on J if $a * u \in C(J; X_A)$ and $u(t) = f(t) + A(a * u)(t)$ on J;*
(c) **weak solution** *of (1.1) on J if*

$$< u(t), x^* > = < f(t), x^* > + < (a * u)(t), A^* x^* >$$

on J, for each $x^* \in D(A^*)$.

Obviously, every strong solution of (1.1) is a mild solution, and each mild solution is a weak one. If in addition, the resolvent set $\rho(A)$ of A is nonempty, then every weak solution is also a mild one; see Proposition 1.4 in Section 1.2. However, in general not every mild solution is a strong solution.

Note that in case $a(t) \equiv 1$ and $f \in C^1(J; X)$, (1.1) is equivalent to the Cauchy problem

$$\dot{u}(t) = \dot{f}(t) + Au(t), \quad u(0) = f(0). \tag{1.2}$$

A moment of reflection shows that our solution concepts coincide with those which are usually employed with (1.2). Similarly, if $a(t) \equiv t$ and $f \in C^2(J; X)$, (1.1) is equivalent to

$$\ddot{u}(t) = \ddot{f}(t) + Au(t), \quad u(0) = f(0), \quad \dot{u}(0) = \dot{f}(0), \tag{1.3}$$

and our solution concepts for (1.1) become those for second order problems (1.3).

There are two natural transformations associated with (1.1). The first one consists of an exponential shift which results from multiplication of (1.1) with $e^{-\omega t}$. Then $u_\omega(t) = u(t)e^{-\omega t}$ will be again a solution of (1.1) with $f(t)$ replaced by $f_\omega(t) = f(t)e^{-\omega t}$, and $a(t)$ by $a_\omega(t) = a(t)e^{-\omega t}$. This way the kernel can be made integrable whenever $a(t)$ is of exponential growth.

The second transformation concerns the operator A. Adding $\omega a * u$ to both sides of (1.1) and defining $v = u + \omega a * u$, (1.1) transforms into an equation of the same form, but with A replaced by $A + \omega$ and u by v, and a by the resolvent kernel r_ω, i.e. the solution of $r + \omega a * r = a$. This way one can assume $0 \in \rho(A)$ whenever $\rho(A) \neq \emptyset$.

These transformations can always be carried out when considering (1.1) locally on $\mathbb{R}_+$, however, they change asymptotic behaviour.

The following definition of well-posedness is a direct extension of well-posed Cauchy problems (1.2).

Definition 1.2 *Equation (1.1) is called* **well-posed** *if for each $x \in \mathcal{D}(A)$ there is a unique strong solution $u(t; x)$ on $\mathbb{R}_+$ of*

$$u(t) = x + (a * Au)(t), \quad t \geq 0, \tag{1.4}$$

and $(x_n) \subset \mathcal{D}(A), x_n \to 0$ imply $u(t; x_n) \to 0$ in X, uniformly on compact intervals.

Suppose (1.1) is well-posed; we then may introduce the solution operator $S(t)$ for (1.1).

$$S(t)x = u(t; x), \quad x \in \mathcal{D}(A), t \geq 0. \tag{1.5}$$

By uniqueness, $S(t)$ is unambigously defined and linear for each $t \geq 0$, $S(0)x = x$ holds on $\mathcal{D}(A)$, and $S(t)x$ is continuous on $\mathbb{R}_+$, for each $x \in \mathcal{D}(A)$. We show that $S(t)$ is also uniformly bounded on compact intervals. To see this, assume on the contrary there are sequences $(t_n) \subset [0, T]$ and $(y_n) \subset \mathcal{D}(A), |y_n| = 1$, such that $|S(t_n)y_n| \geq n$ for each $n \in \mathbb{N}$; then $x_n = y_n/n \in \mathcal{D}(A)$, $x_n \to 0$, hence from Definition 1.2 we obtain the contradiction

$$1 \leq |S(t_n)x_n| = |u(t_n; x_n)| \to 0 \quad as \quad n \to \infty.$$

Thus $S(t)$ is bounded and therefore admits extension to all of X, $S(t)x$ is continuous for each $x \in X$. By definition of strong solutions it is further clear that the solution operator maps $\mathcal{D}(A)$ into itself, $AS(t)x$ is continuous on $\mathbb{R}_+$ for each $x \in \mathcal{D}(A)$ and

$$S(t)x = x + a * AS(t)x = x + Aa * S(t)x$$

holds on $\mathbb{R}_+$. Since $S(t)$ is bounded we obtain further $\mathcal{R}(a * S(t)) \subset \mathcal{D}(A)$ and $Aa * S(t) = S(t) - I$ is strongly continuous on $\mathbb{R}_+$; in other words $u(t; x) = S(t)x$ is a mild solution of (1.4) for each $x \in X$. Next we observe that also the mild solutions of (1.4) are unique. In fact, if $u = Aa * u$ then $v = a * u$ is a strong solution of (1.4) with $x = 0$, hence $v = 0$ by uniqueness of the strong solutions; Titchmarsh's theorem (cp. e.g. Yosida [347], p. 166) then implies $u = 0$. Finally, we see that A commutes with $S(t)$; in fact, for $x \in \mathcal{D}(A)$, both $u(t; Ax)$ and $Au(t; x)$ are mild solutions of (1.4) with x replaced by Ax, hence

$$S(t)Ax = u(t; Ax) = Au(t; x) = AS(t)x \quad \text{for all } x \in \mathcal{D}(A), \ t \geq 0.$$

These considerations give rise to the following

Definition 1.3 *A family $\{S(t)\}_{t \geq 0} \subset \mathcal{B}(X)$ of bounded linear operators in X is called a* **resolvent** *for (1.1) [or* **solution operator** *for (1.1)] if the following conditions are satisfied.*
(S1) *$S(t)$ is strongly continuous on $\mathbb{R}_+$ and $S(0) = I$;*
(S2) *$S(t)$ commutes with A, which means that $S(t)\mathcal{D}(A) \subset \mathcal{D}(A)$ and $AS(t)x = S(t)Ax$ for all $x \in \mathcal{D}(A)$ and $t \geq 0$;*
(S3) *the* **resolvent equation** *holds*

$$S(t)x = x + \int_0^t a(t-s)AS(s)x \, ds \quad \text{for all } x \in \mathcal{D}(A), \ t \geq 0. \tag{1.6}$$

We have proved so far that a well-posed problem (1.1) admits a resolvent; the converse is also true.

Proposition 1.1 *(1.1) is well-posed iff (1.1) admits a resolvent $S(t)$. If this is the case we have in addition $\mathcal{R}(a * S(t)) \subset \mathcal{D}(A)$ for all $t \geq 0$ and*

$$S(t)x = x + A \int_0^t a(t-s)S(s)x \, ds \quad \text{for all } x \in X, \ t \geq 0. \tag{1.7}$$

In particular, $Aa * S$ is strongly continuous in X.

Proof: The only thing left to prove is that existence of a resolvent $S(t)$ implies well-posedness of (1.1). So suppose $S(t)$ is a resolvent for (1.1). Then, by the properties of $S(t)$, for each $x \in \mathcal{D}(A)$ the function $u(t) = S(t)x$ is a strong solution of (1.4). To prove its uniqueness let $v(t)$ be another strong solution of (1.4); by (S2) and (S3) as well as by (1.4) we obtain

$$\begin{aligned} 1 * v &= (S - a * AS) * v = S * v - S * a * Av \\ &= S * (v - a * Av) = S * x = 1 * Sx, \end{aligned}$$

hence after differentiation $v(t) = S(t)x$. Continuous dependence of the solutions on x follows directly from uniform boundedness of $S(t)$ on compact intervals which in turn is a consequence of (S1), by the uniform boundedness principle.   □

Corollary 1.1 *(1.1) admits at most one resolvent $S(t)$.*

Proof: If $S_1(t)$ and $S_2(t)$ are both resolvents for (1.1), then for $x \in \mathcal{D}(A)$ we obtain

$$\begin{aligned} 1 * S_1 x &= (S_2 - a * AS_2) * S_1 x = S_2 * S_1 x - a * S_2 * AS_1 x \\ &= S_2 * (S_1 x - a * AS_1 x) = S_2 * x = 1 * S_2 x, \end{aligned}$$

hence $S_1(t)x = S_2(t)x$ for each $x \in \mathcal{D}(A)$, $t \geq 0$, and so $S_1(t) = S_2(t)$ by density of $\mathcal{D}(A)$.   □

For the special cases $a(t) = 1$ and $a(t) = t$ mentioned above the resolvent $S(t)$ becomes the C_0-semigroup e^{At} generated by A, resp. the cosine family $Co(t)$ generated by A.

1.2 Inhomogeneous Equations

Suppose $S(t)$ is a resolvent for (1.1) and let $u(t)$ be a mild solution of (1.1). Then from (S1) $\sim$ (S3) we obtain

$$\begin{aligned} 1 * u &= (S - A(a * S)) * u = S * u - AS * (a * u) \\ &= S * u - S * (Aa * u) = S * (u - Aa * u) = S * f, \end{aligned}$$

i.e. $S * f$ is continuously differentiable and

$$u(t) = \frac{d}{dt} \int_0^t S(t - s)f(s)ds, \quad t \in J. \tag{1.8}$$

This is the variation of parameters formula for Volterra equations (1.1).

Proposition 1.2 *Suppose (1.1) admits a resolvent $S(t)$ and let $f \in C(J; X)$. Then*
(i) if $u \in C(J; X)$ is a mild solution of (1.1), then $S * f$ is continuously differentiable on J and

$$u(t) = \frac{d}{dt} \int_0^t S(t - s)f(s)ds, \quad t \in J; \tag{1.9}$$

in particular, mild solutions of (1.1) are unique,
(ii) if $f \in W^{1,1}(J; X)$ then

$$u(t) = S(t)f(0) + \int_0^t S(t - s)\dot{f}(s)ds, \quad t \in J, \tag{1.10}$$

is a mild solution of (1.1),

(iii) if $f = x + a * g$, with $g \in W^{1,1}(J; X)$ and $x \in \mathcal{D}(A)$, then

$$u(t) = S(t)x + a * S(t)g(0) + a * S * \dot{g}(t), \quad t \in J, \qquad (1.11)$$

is a strong solution of (1.1);
(iv) if $f \in W^{1,1}(J; X_A)$ then $u(t)$ given by (1.10) is a strong solution of (1.1).

Proof: (ii) If $f \in W^{1,1}(J; X)$ then $u(t)$ defined by (1.10) is continuous on J and we obtain from Proposition 1.1

$$a * u = a * S f(0) + a * S * \dot{f} \in \mathcal{D}(A) \quad \text{for each} \quad t \in J,$$

and

$$
\begin{aligned}
Aa * u &= Aa * Sf(0) + (Aa * S) * \dot{f} \\
&= (S - I)f(0) + S * \dot{f} - 1 * \dot{f} \\
&= Sf(0) + S * \dot{f} - f(0) - 1 * \dot{f} = u - f,
\end{aligned}
$$

i.e. $u(t)$ is a mild solution of (1.1).

(iii) Let $v = Sg(0) + S * \dot{g}$; by (ii) v is a mild solution of (1.1) with f replaced by g, in particular we have $a * v \in C(J; X_A)$. Now, $u(t)$ given by (1.11) becomes $u = Sx + a * v$, hence $u \in C(J; X_A)$ since $x \in \mathcal{D}(A)$ and so u is a strong solution of (1.1).

(iv) This is a direct consequence of (ii) since $S(t)$ commutes with A. $\qquad \square$

Suppose for a moment $a \in BV_{loc}(\mathbb{R}_+)$; then, given $x \in \mathcal{D}(A)$, we obtain from the resolvent equation (1.6) that $S(\cdot)x$ is of class $W^{1,\infty}_{loc}(\mathbb{R}_+; X)$ and

$$\dot{S}(t)x = \int_0^t da(s)S(t - s)Ax, \quad t \geq 0. \qquad (1.12)$$

In this case (1.8) can be rewritten as

$$u(t) = f(t) + \int_0^t \dot{S}(t - s)f(s)ds, \quad t \in J, \qquad (1.13)$$

and the right hand side of this equation makes sense if we only know $f \in C(J; X_A)$ rather than $f \in W^{1,1}(J; X_A)$. In general, $S(t)$ will not be differentiable on $\mathcal{D}(A)$, this leads to a restricted class of resolvents.

Definition 1.4 *A resolvent $S(t)$ for (1.1) is called **differentiable**, if $S(\cdot)x \in W^{1,1}_{loc}(\mathbb{R}_+; X)$ for each $x \in \mathcal{D}(A)$ and there is $\varphi \in L^1_{loc}(\mathbb{R}_+)$ such that*

$$|\dot{S}(t)x| \leq \varphi(t)|x|_A \quad a.e. \text{ on } \mathbb{R}_+, \text{ for each } x \in \mathcal{D}(A). \qquad (1.14)$$

We have shown above that $S(t)$ is differentiable in case $a(t)$ belongs to $BV_{loc}(\mathbb{R}_+)$. Whenever $S(t)$ is differentiable then (1.13) makes sense for each $f \in C(J; X_A)$ and from this we obtain.

Proposition 1.3 *Suppose (1.1) admits a differentiable resolvent $S(t)$. Then*
(i) if $f \in C(J; X_A)$ then $u(t)$ defined by (1.13) is a mild solution of (1.1);
(ii) if $f = x + a * g$, with $g \in C(J; X_A)$ and $x \in \mathcal{D}(A)$, then

$$u(t) = S(t)x + a * g(t) + a * \dot{S} * g(t), \quad t \in J, \tag{1.15}$$

is a strong solution of (1.1).

Proof: If $S(t)$ is differentiable, then $\dot{S}(t)$ satisfies

$$\dot{S}(t)x = a(t)Ax + Aa * \dot{S}(t)x, \quad \text{a.a. } t \geq 0, \ x \in \mathcal{D}(A); \tag{1.16}$$

in particular $a * \dot{S}(t)$ maps $\mathcal{D}(A)$ into $\mathcal{D}(A)$ for a.a. $t \geq 0$ and

$$|Aa * \dot{S}(t)x| \leq |\dot{S}(t)x| + |a(t)Ax| \leq (\varphi(t) + |a(t)|)|x|_A.$$

This implies for $f \in C(J; X_A)$ that $a * u = a * f + a * \dot{S} * f$ belongs to $C(J; X_A)$, hence u is a mild solution of (1.1) by (1.16).

If $f = x + a * g$ with $g \in C(J; X_A)$ then $v = g + \dot{S} * g$ is a mild solution of (1.1) with f replaced by g, hence $a * v \in C(J; X_A)$. But (1.15) reads $u = Sx + a * v$ and so $u \in C(J; X_A)$ since $x \in \mathcal{D}(A)$ and u is a strong solution of (1.1) by (1.16). □

We conclude this section demonstrating that the concepts mild solution and weak solution for well-posed equations of the form (1.1) coincide since then $\rho(A) \neq \emptyset$, as we shall see in the next section.

Proposition 1.4 *Let $f \in C(J; X)$ and suppose $\rho(A) \neq \emptyset$. Then each weak solution u on J of (1.1) is a mild solution of (1.1) on J.*

Proof: Suppose $\mu \in \rho(A)$ and let $u \in C(J; X)$ be a weak solution of (1.1); then for each $x^* \in \mathcal{D}(A^*)$ we have

$$< a * u, A^* x^* > = < u - f, x^* > \quad \text{on } J,$$

hence

$$< a * u, (\mu - A^*)x^* > = < f - u + \mu a * u, x^* > \quad \text{on } J.$$

Since $\mu - A^*$ is invertible, we obtain with $y^* = (\mu - A^*)x^*$

$$\begin{aligned} < a * u, y^* > &= < f - u + \mu a * u, (\mu - A^*)^{-1} y^* > \\ &= < (\mu - A)^{-1}(f - u + \mu a * u), y^* > \end{aligned}$$

for all $y^* \in X^*$, hence

$$a * u = (\mu - A)^{-1}(f - u + \mu a * u) \quad \text{on } J.$$

But this implies $a * u(t) \in \mathcal{D}(A)$ for each $t \in J$ and

$$Aa * u(t) = u(t) - f(t) \quad \text{on } J,$$

i.e. $u(t)$ is a mild solution of (1.1). □

1.3 Necessary Conditions for Well-posedness

Suppose $S(t)$ is a resolvent for (1.1) and let $-\mu \in \sigma(A)$ be an eigenvalue of A with eigenvector $x \neq 0$. Then

$$S(t)x = s(t; \mu)x, \quad t \geq 0, \tag{1.17}$$

where $s(t; \mu)$ is the solution of the one-dimensional Volterra equation

$$s(t; \mu) + \mu \int_0^t a(t - \tau)s(\tau; \mu)d\tau = 1, \quad t \geq 0. \tag{1.18}$$

In fact,

$$Aa * s(t; \mu)x = a * s(t; \mu)Ax = -\mu a * s(t; \mu)x = s(t; \mu)x - x,$$

hence (1.17) holds, by uniqueness of the solutions of (1.4). Similarly, if $S(t)$ is differentiable then

$$\dot{S}(t)x = \mu r(t; \mu)x, \quad t \geq 0, \tag{1.19}$$

where $r(t; \mu)$ is the solution of the one-dimensional equation

$$r(t; \mu) + \mu \int_0^t a(t - \tau)r(\tau; \mu)d\tau = a(t), \quad t \geq 0. \tag{1.20}$$

These observations lead to necessary conditions for well-posedness of (1.1).

Proposition 1.5 *Suppose (1.1) admits a resolvent $S(t)$, and let $s(t; \mu)$ and $r(t; \mu)$ be defined as the solutions of the scalar Volterra equations (1.18) and (1.20). Then*
(i) there is a locally bounded lower semicontinuous function $\psi(t)$ such that

$$|s(t; \mu)| \leq \psi(t) \quad \text{for all } -\mu \in \sigma(A) \text{ and } t \geq 0; \tag{1.21}$$

(ii) if $S(t)$ is differentiable, then there is $\varphi \in L_{loc}^1(\mathbb{R}_+)$ such that

$$|r(t; \mu)| \leq \varphi(t) \quad \text{for all } -\mu \in \sigma(A) \text{ and a.a. } t \geq 0. \tag{1.22}$$

In particular, there is a sequence $(\mu_n) \subset \rho(A)$ such that $|\mu_n| \to \infty$.

Proof: (i) Suppose $-\mu \in \sigma(A) \backslash \sigma_r(A)$; then there are sequences $(x_n) \subset D(A)$, $|x_n| = 1$ and $(y_n) \subset X$, $y_n \to 0$ such that $(\mu + A)x_n = y_n$. Set $u_n(t) = S(t)x_n - s(t; \mu)x_n$; then $u_n(t)$ satisfies the equation

$$u_n + \mu a * u_n = a * Sy_n,$$

hence

$$u_n = r(\cdot; \mu) * Sy_n \to 0 \quad \text{as } n \to \infty,$$

uniformly on compact intervals. Therefore

$$|s(t;\mu)| = |s(t;\mu)x_n| \le |u_n(t)| + |S(t)x_n| \le |u_n(t)| + |S(t)|,$$

i.e. with $n \to \infty$

$$|s(t;\mu)| \le |S(t)| \quad \text{for all } t \ge 0.$$

On the other hand, if $-\mu \in \sigma_r(A)$ then there is $x^* \in D(A^*)$, $|x^*| = 1$, such that $A^*x^* = -\mu x^*$; given $x \in X$, this implies by (1.7)

$$< Sx, x^* > = < x, x^* > -\mu a* < Sx, x^* >,$$

i.e.

$$< Sx, x^* > = s(\cdot;\mu) < x, x^* > \quad \text{on } \mathbb{R}_+,$$

by uniqueness. Thus we obtain

$$|s(t;\mu)| = \sup_{|x|\le 1} |s(t;\mu) < x, x^* >| = \sup_{|x|\le 1} | < S(t)x, x^* > | \le |S(t)|;$$

this means that $\psi(t) = |S(t)|$ is the function we are looking for. It follows from the strong continuity of $S(t)$ that $\psi(t)$ is locally bounded and lower semicontinuous.

(ii) is proved by a similar argument, the function $\varphi(t)$ comes from the definition of a differentiable resolvent.

Finally, we show there is a sequence $(\mu_n) \subset \rho(A)$ such that $|\mu_n| \to \infty$. Assume on the contrary $\rho(A) \cap \{\mu \in \mathbb{C} : |\mu| \ge R\} = \emptyset$; then we have by (i)

$$|s(t;\mu)| \le \psi(t), \quad \text{for all } t \ge 0 \quad \text{and all} \quad \mu \in \mathbb{C}, \ |\mu| \ge R.$$

Since $s(t;\cdot)$ is an entire function for each fixed $t \ge 0$, the latter implies $s(t;\mu) \equiv s(t;0) = 1$ by Liouville's theorem and so (1.18) shows $a(t) = 0$ a.e., in contradiction to the standing assumption $a(t) \not\equiv 0$. $\square$

If X is a Hilbert space and A is a normal operator then the operational calculus for such operators can be used to construct the resolvent $S(t)$.

Theorem 1.1 *Let A be a normal operator in the Hilbert space X. Then (1.1) is well-posed iff there is a locally bounded lower semicontinuous function $\psi(t)$ such that*

$$|s(t;\mu)| \le \psi(t) \quad \text{for all} \ -\mu \in \sigma(A) \ \text{and } t \ge 0. \tag{1.23}$$

The resolvent $S(t)$ is then given by

$$S(t) = \int_{\sigma(A)} s(t;-\mu)dE(\mu), \quad t \ge 0, \tag{1.24}$$

where E denotes the spectral measure of A. $S(t)$ is differentiable iff there is $\varphi \in L^1_{loc}(\mathbb{R}_+)$ such that

$$|r(t;\mu)| \le \varphi(t) \quad \text{for all} \ -\mu \in \sigma(A) \ \text{and a.a.} \ t \ge 0. \tag{1.25}$$

For the concept of the spectral measure of a normal operator and to its operational calculus we refer to Dunford and Schwartz [103]. The proof of Theorem 1.1 is a simple exercise to this operational calculus and is therefore omitted.

An easy computation shows that for $a(t) = 1$ we have

$$r(t;\mu) = s(t;\mu) = e^{-\mu t}, \quad t \geq 0, \ \mu \in \mathbb{C},$$

i.e. (1.23) means in this case $\mathrm{Re}\ \sigma(A) \leq \omega < \infty$, the spectrum of A must necessarily be contained in a left halfplane.

On the other hand, for $a(t) = t$ we obtain

$$s(t;\mu) = \cos(\sqrt{\mu}t), \quad r(t;\mu) = \sin(\sqrt{\mu}t)/\sqrt{\mu}, \quad t \geq 0, \ \mu \in \mathbb{C}.$$

In this case (1.23) and (1.25) reduce to $\mathrm{Re}\ \sqrt{\sigma(A)} \leq \omega$, i.e. $\sigma(A)$ must necessarily be contained inside of the parabola $p(y) = \omega^2 - y^2 + 2i\omega y$, $y \in \mathbb{R}$.

In general, it turns out to be quite difficult to obtain estimates like (1.23) and (1.25) and although Theorem 1.1 deals only with really nice operators it is in general only of limited use. However, two cases are worthy of mention since they appear naturally in applications. Recall that a measure $da(t)$ is said to be of *positive type* if

$$\mathrm{Re}\ \int_0^T (da * \varphi)(t)\overline{\varphi}(t)dt \geq 0, \quad \text{for each}\ \varphi \in C(\mathbb{R}_+, \mathbb{C})\ \text{and}\ T > 0.$$

If $da(t)$ is of subexponential growth, it is of positive type iff $\mathrm{Re}\ \widehat{da}(\lambda) \geq 0$ for all $\mathrm{Re}\ \lambda > 0$, cp. Gripenberg, Londen and Staffans [156] or the original paper of Nohel and Shea [258].

Corollary 1.2 *Suppose A is selfadjoint and negative semidefinite in the Hilbert space X. Then the following assertions are valid.*

(i) Let $a \in BV_{loc}(\mathbb{R}_+)$ be such that the measure da is of positive type. Then (1.1) admits a differentiable resolvent $S(t)$ such that

$$|S(t)| \leq 1, \quad \text{and} \quad |\dot{S}(t)x| \leq \mathrm{Var}\ a|_0^t |Ax|, \quad \text{for all}\ t \geq 0, \ x \in \mathcal{D}(A).$$

(ii) Let $a \in L^1_{loc}(\mathbb{R}_+)$ be nonnegative and nonincreasing. Then (1.1) admits a resolvent $S(t)$ which is selfadjoint and such that $0 \leq (S(t)x, x) \leq 1$ for all $x \in X$.

Proof: (i) Since A is selfadjoint and negative semidefinite we have $\sigma(-A) \subset \mathbb{R}_+$; let $\mu \in \mathbb{R}_+$ be fixed. Differentiate (1.18) and multiply with $\overline{s(t;\mu)}$; this yields

$$\frac{\partial}{\partial t}|s(t;\mu)|^2 = 2\ \mathrm{Re}\ (\frac{\partial}{\partial t}s(t;\mu)\overline{s(t;\mu)}) = -2\mu\ \mathrm{Re}\ [da * s(t;\mu)\overline{s(t;\mu)}],$$

hence

$$|s(t;\mu)|^2 = 1 - 2\mu\ \mathrm{Re}\ \int_0^t da * s(\tau;\mu)\overline{s(\tau;\mu)}d\tau \leq 1,$$

since $\mu \geq 0$ and da is of positive type. Hence (1.23) holds with $\psi(t) \equiv 1$, and therefore (1.25) with $\varphi(t) = \text{Var } a|_0^t$ since $r = da * s$.

(ii) If $a(t)$ is nonnegative and nonincreasing then $0 \leq s(t; \mu) \leq 1$, for all $t, \mu \geq 0$; this is a special case of a result due to Friedman [120]. The assertion follows then from Theorem 1.1. $\Box$

Observe that if $a(t)$ is nonnegative, nonincreasing, then $a(0+) < \infty$ implies $a \in BV(\mathbb{R}_+)$ and da of positive type; this exhibits a connection between (i) and (ii) of Corollary 1.2. In the sections containing applications we shall refer frequently to this corollary.

1.4 Perturbed Equations
Consider the perturbed equation

$$u(t) = f(t) + (a + a * k) * Au(t) + b * u(t), \quad t \in J \qquad (1.26)$$

where $a(t)$, A, $f(t)$ are as before and $k, b \in L^1_{loc}(\mathbb{R}_+)$. The concepts of solutions, well-posedness and resolvents have obvious extension to the perturbed problem (1.26); note that $a * u \in C(J; X_A)$ if and only if $(a + a * k) * u \in C(J; X_A)$. By means of the variation of parameters formula obtained in Section 1.2 it is possible to show that (1.26) is well-posed iff (1.1) has this property, provided both, b and k, are locally of bounded variation.

Theorem 1.2 *Suppose $k, b \in BV_{loc}(\mathbb{R}_+)$. Then (1.26) is well-posed iff (1.1) is well-posed. Also, the resolvent $S_1(t)$ of (1.26) is differentiable iff the resolvent $S(t)$ of (1.1) has this property.*

Proof: Suppose (1.1) admits a resolvent $S(t)$. The resolvent $S_1(t)$ must satisfy the equation

$$S_1 = I + Aa * S_1 + k * Aa * S_1 + b * S_1, \qquad (1.27)$$

hence by the variation of parameters formula (1.8) it must be a solution of

$$
\begin{aligned}
S_1 &= \frac{d}{dt}(S * (I + b * S_1 + k * Aa * S_1)) \\
&= S + db * S * S_1 + dk * (Aa * S) * S_1 \\
&= S + [(db + dk) * S - k] * S_1,
\end{aligned}
$$

since b, k are of locally bounded variation. Thus S_1 must be a solution of

$$S_1 = S + K * S_1, \qquad (1.28)$$

where $1 * K = (b + k) * S - 1 * k$ belongs to $BV^0_{loc}(\mathbb{R}_+; \mathcal{B}(X))$. Therefore, by Theorem 0.5, (1.28) has a unique strongly continuous solution $S_1(t)$ on $\mathbb{R}_+$. Convolving (1.28) with $a(t)$ we see that $a * S_1(t)$ maps into $\mathcal{D}(A)$, and applying A we obtain strong continuity of $Aa * S_1(t)$, since $Aa * S(t)$ has this property; note that

K commutes with A. For the same reason $S_1(t)$ commutes with A and it satisfies (1.27) by construction and by Proposition 1.2. If $S(t)$ is also differentiable, differentiation of (1.28) yields

$$\dot{S}_1 x = \dot{S}x + K * \dot{S}_1 x + Kx,$$

hence S_1 has also this property. The converse implications follow reversing the roles of $S(t)$ and $S_1(t)$. $\quad\square$

Corollary 1.3 *Suppose $k, b \in W_{loc}^{1,1}(\mathbb{R}_+)$ and let $S(t)$ be a resolvent for (1.1). Then the resolvent $S_1(t)$ for (1.26) admits the decomposition*

$$S_1(t) = S(t) + S_2(t), \quad t \geq 0,$$

where $S_2(t)$ is continuous in $\mathcal{B}(X)$ for all $t \geq 0$. If $S(t)$ is also differentiable then $\dot{S}_2 \in L_{loc}^1(\mathbb{R}_+, \mathcal{B}(X_A, X))$.

This corollary is simply due to the fact that for a strongly continuous (resp. strongly locally integrable) family of operators $T(t)$ and a scalar kernel $c \in L_{loc}^1(\mathbb{R}_+)$ the convolution $c * T$ is continuous in $\mathcal{B}(X)$ (resp. locally integrable in $\mathcal{B}(X)$).

For the special case $a(t) = 1$, (1.26) is equivalent to

$$\begin{aligned}
\dot{u}(t) &= \dot{f}(t) + Au(t) + k * Au(t) + db * u(t), \quad t \geq 0, \\
u(0) &= f(0);
\end{aligned} \tag{1.29}$$

Theorem 1.2 then states that (1.29) is well-posed iff A generates a C_0-semigroup in X. This observation yields

Corollary 1.4 *Suppose $k,\ b \in BV_{loc}(\mathbb{R}_+)$. Then (1.29) is well-posed iff A generates a C_0-semigroup in X. If this in the case, the resolvent for (1.29) is differentiable.*

On the other hand, for $a(t) = t$, (1.26) is equivalent to

$$\begin{aligned}
\ddot{u} &= \dot{f} + Au + k * Au + \dot{b}(t)f(0) + db * \dot{u}, \quad t \geq 0, \\
u(0) &= f(0), \quad \dot{u}(0) = \dot{f}(0) + b(0+)f(0);
\end{aligned} \tag{1.30}$$

in this case Theorem 1.2 states that (1.30) is well-posed iff A generates a cosine family in X, provided $k \in BV_{loc}(\mathbb{R}_+)$ and $b \in W_{loc}^{1,1}(\mathbb{R}_+)$.

Corollary 1.5 *Suppose $b \in W_{loc}^{1,1}(\mathbb{R}_+)$ and $k \in BV_{loc}(\mathbb{R}_+)$. Then (1.30) is well-posed iff A generates a cosine family in X. If this is the case, the resolvent for (1.30) is differentiable.*

We conclude this subsection with examples showing that the assumptions of Theorem 1.2 cannot be relaxed considerably.

Example 1.1 Consider the problem

$$
\begin{aligned}
\dot{u}(t) &= Au(t), \quad t \le 1, \\
\dot{u}(t) &= Au(t) + Au(t-1), \quad t > 1, \\
u(0) &= x;
\end{aligned}
\tag{1.31}
$$

(1.31) is of the form (1.26) with $b(t) \equiv 0$ and $k(t) = \delta_0(t-1)$, $a(t) \equiv 1$, i.e. it is of the form (1.29). We show that (1.31) admits a resolvent $S(t)$ iff A generates an analytic C_0-semigroup in X. In fact, suppose (1.31) admits a resolvent $S(t)$; (1.31) for $t \le 1$ then implies that A must generate a C_0-semigroup e^{At} in X and $S(t) = e^{At}$ holds for $t \le 1$. The formula of variation of parameters yields an explicit representation of $S(t)$ for all $t \ge 0$, namely

$$
S(t) = \sum_{n=0}^{\infty} \frac{1}{n!}(t-n)_+^n A^n e^{A(t-n)_+}, \quad t \ge 0,
\tag{1.32}
$$

where $(\tau)_+ = \max(\tau, 0)$ for each $\tau \in \mathbb{R}$. Let $t = 1 + \tau, \tau \in (0,1)$; (1.32) yields the identity

$$
\tau A e^{A\tau} = S(1+\tau) - e^{A(1+\tau)}, \quad \tau \in (0,1),
$$

hence $|\tau A e^{A\tau}| \le M$ for $\tau \le 1$. But this estimate is well-known to be equivalent to analyticity of e^{At}. Conversely, if e^{At} is analytic it is easy to verify that formula (1.32) defines the resolvent for (1.31). □

Example 1.2 Consider the initial-boundary value problem

$$
\begin{aligned}
u_t(t,x) &= u_x(t,x) + \int_0^t k(t-\tau)u_x(\tau,x)d\tau, \quad t \ge 0, \ x \in [0,2\pi], \\
u(0,x) &= u_0(x), \quad x \in [0,2\pi], \\
u(t,0) &= u(t,2\pi), \quad t \ge 0.
\end{aligned}
\tag{1.33}
$$

Let $X = L^2(0,2\pi)$, $Au = u_x$ with $D(A) = \{u \in W^{1,2}(0,2\pi), u(0) = u(1)\}$; it is well-known that A is a normal, even skewadjoint operator in X. We will show that (1.33) does not admit a resolvent in X in case the kernel $k(t)$ is chosen as

$$
k(t) = t^{\alpha-1}/\Gamma(\alpha), \quad t > 0, \ \alpha \in (0,1);
\tag{1.34}
$$

note that $k(t)$ is even analytic for $t > 0$ but has a singularity at $t = 0$, it is not of bounded variation near $t = 0$. Since $\sigma(A) = \{in : n \in \mathbb{Z}\}$, by Theorem 1.1 we have to study the functions $s_n(t) = s(t; -in)$, i.e. the solutions of

$$
\dot{s}_n = ins_n + ink * s_n, \quad s_n(0) = 1, \quad t \ge 0, \ n \in \mathbb{Z}.
\tag{1.35}
$$

It suffices to show that $\{s_n(t) : n \in \mathbb{Z}\}$ is not locally uniformly bounded, by Theorem 1.1.

(i) The Laplace transform $\widehat{s_n}(\lambda)$ of s_n is easily seen to be

$$
\widehat{s_n} = (\lambda - in - in/\lambda^\alpha)^{-1}, \quad \lambda \in \mathbb{C}_+;
$$

$\widehat{s_n}(\lambda)$ clearly has essential singularities for $\lambda \leq 0$, but it also has poles in $\mathbb{C}\backslash\mathbb{R}_-$. To obtain the asymptotic behaviour of these poles we study the zeros of

$$h_n(\lambda) = \lambda - in - in/\lambda^\alpha$$

as $n \to \pm\infty$. Rescaling by $z = \lambda/|n|$, these are the solutions of the equation

$$z^{1+\alpha} = (iz^\alpha + i|n|^{-\alpha})\text{sgn } n. \tag{1.36}$$

As $n \to \pm\infty$ they are given by $z_\infty^1 = \pm i$ and $z_\infty^2 = 0$, hence for $|n|$ large, there is always a solution z_n of (1.36) close to $\pm i$. But this means $|n|^\alpha(z_n \mp i) = \pm i/z_n^\alpha \to (\pm i)^{1-\alpha}$ as $n \to \pm\infty$, i.e.

$$\lambda_n = |n|z_n \sim in + (in)^{1-\alpha} \text{ as } n \to \pm\infty;$$

note that $\text{Re } \lambda_n = \text{Re } (in)^{1-\alpha} = |n|^{1-\alpha}\cos(\frac{\pi}{2}(1-\alpha)) \to \infty$ as $|n| \to \infty$.

(ii) The complex inversion formula for the Laplace transform yields

$$s_n(t) = (2\pi i)^{-1} \int_{\omega_n-i\infty}^{\omega_n+i\infty} e^{\lambda t}(\lambda - in - in\lambda^{-\alpha})^{-1}d\lambda,$$

where ω_n has to be chosen sufficiently large, in particular $\omega_n > \text{Re } \lambda_n$. Since for large $|n|$ the functions h_n have no zeros strictly away from the set $\mathbb{R}_-$ other than λ_n we may deform the integration path into Γ, which consists of the ray $(\infty, R]e^{-i\varphi}$, the arc $Re^{i[-\varphi,\varphi]}$ and the ray $[R, \infty)e^{i\varphi}$. By Cauchy's theorem this yields

$$s_n(t) = a_n e^{\lambda_n t} + (2\pi i)^{-1} \int_\Gamma e^{\lambda t}(\lambda - in - in\lambda^{-\alpha})^{-1}d\lambda, \tag{1.37}$$

where by residue calculus

$$a_n = \text{ Res } (\lambda - in - in\lambda^{-\alpha})^{-1}|_{\lambda=\lambda_n} = (1 + in\alpha/\lambda_n^{1+\alpha}) \to 1, \quad \text{as } n \to \infty.$$

The integral can be estimated by $ce^{Rt}/|n|$, provided $\varphi \in (\frac{\pi}{2}, \frac{\pi}{2\alpha})$ and R is large. Thus

$$|s_n(t)| \sim |a_n|e^{\text{Re}\lambda_n t} \sim e^{\text{Re}\lambda_n t} \to \infty \text{ as } |n| \to \infty,$$

i.e. (1.23) is violated and so (1.33) does not admit a resolvent. □

1.5 The Generation Theorem

Since (1.1) is a convolution equation on the halfline it is natural to employ the Laplace transform for its study. Besides the standing assumptions on $a(t)$ and A we therefore suppose that $a(t)$ is *Laplace transformable*, i.e. there is $\omega \in \mathbb{R}$ such that $\int_0^\infty e^{-\omega t}|a(t)|dt < \infty$. But we also have to restrict the class of resolvents, they must be Laplace transformable as well.

Definition 1.5 *Suppose $S(t)$ is a resolvent for (1.1).*
(i) $S(t)$ is called **exponentially bounded** *if there are constants $M \geq 1$ and $\omega \in \mathbb{R}$ such that*

$$|S(t)| \leq Me^{\omega t}, \quad \text{for all } t \geq 0; \tag{1.38}$$

ω or more precisely (M,ω) is called a **type** *of $S(t)$.*

*(ii) The **growth bound** $\omega_0(S)$ of a resolvent $S(t)$ for (1.1) is defined by*

$$\omega_0(S) = \overline{\lim}_{t \to \infty} t^{-1} \log |S(t)|. \tag{1.39}$$

Obviously, we have the relation $\omega_0(S) = \inf\{\omega : \exists M > 0 \text{ such that (1.38) holds }\}$.

Now, suppose $S(t)$ is an exponentially bounded resolvent for (1.1); its Laplace transform

$$H(\lambda) = \widehat{S}(\lambda) = \int_0^\infty e^{-\lambda t} S(t) dt, \quad \text{Re } \lambda > \omega,$$

is then well-defined and holomorphic for Re $\lambda > \omega$, and satisfies the estimates

$$|H^{(n)}(\lambda)| \leq Mn! \, (\text{Re } \lambda - \omega)^{-(n+1)}, \quad \text{Re } \lambda > \omega, \, n \in \mathbb{N}_0. \tag{1.40}$$

To compute $H(\lambda)$ we use the resolvent equations (1.6) and (1.7); in fact, we obtain from the convolution theorem for Re $\lambda > \omega$ the relation

$$H(\lambda)x = x/\lambda + \widehat{a}(\lambda)H(\lambda)Ax,$$

valid for each $x \in D(A)$, and

$$H(\lambda)x = x/\lambda + A\widehat{a}(\lambda)H(\lambda)x,$$

for each $x \in X$. Thus the operators $I - \widehat{a}(\lambda)A$ are bijective, i.e. invertible, and we obtain

$$H(\lambda) = (I - \widehat{a}(\lambda)A)^{-1}/\lambda \quad \text{for Re } \lambda > \omega.$$

In particular, $1/\widehat{a}(\lambda) \in \rho(A)$ for all such λ, provided $\widehat{a}(\lambda) \neq 0$. Assume $\widehat{a}(\lambda_0) = 0$ for some λ_0 with Re $\lambda_0 > \omega$; since $\widehat{a}(\lambda)$ is holomorphic, λ_0 is an isolated zero of finite multiplicity, and $H(\lambda_0) = I/\lambda_0$. Choose a small circle Γ around λ_0 which is entirely contained in the halfplane Re $\lambda > \omega$ such that $\widehat{a}(\lambda) \neq 0$ on Γ; then $AH(\lambda) = (H(\lambda) - I/\lambda)/\widehat{a}(\lambda)$ is well-defined and holomorphic on Γ, hence by Cauchy's formula we obtain

$$A = A\lambda_0 H(\lambda_0) = A(2\pi i)^{-1} \int_\Gamma \frac{\lambda H(\lambda)}{\lambda - \lambda_0} d\lambda = (2\pi i)^{-1} \int_\Gamma \frac{\lambda H(\lambda) - I}{(\lambda - \lambda_0)\widehat{a}(\lambda)} d\lambda,$$

and so A is bounded, a contradiction to our standing hypotheses. Thus $\widehat{a}(\lambda) \neq 0$ for Re $\lambda > \omega$.

It turns out that these properties of $\widehat{a}(\lambda)$ and A in combination with (1.40) characterize resolvents of type (M, ω).

Theorem 1.3 *Let A be a closed linear unbounded operator in X with dense domain $\mathcal{D}(A)$ and let $a \in L^1_{loc}(\mathbb{R}_+)$ satisfy $\int_0^\infty e^{-\omega t} |a(t)| dt < \infty$. Then (1.1) admits a resolvent $S(t)$ of type (M, ω) iff the following conditions hold.*

(H1) $\widehat{a}(\lambda) \neq 0$ *and* $1/\widehat{a}(\lambda) \in \rho(A)$ *for all* $\lambda > \omega$;
(H2) $H(\lambda) = (I - \widehat{a}(\lambda)A)^{-1}/\lambda$ *satisfies the estimates*

$$|H^{(n)}(\lambda)| \leq Mn!(\lambda - \omega)^{-(n+1)}, \quad \lambda > \omega, \, n \in \mathbb{N}_0. \tag{1.41}$$

Proof: It remains to prove sufficiency of (H1) and (H2). By Theorem 0.2 there is a Lipschitz family $\{U_\omega(t)\}_{t\geq 0} \subset \mathcal{B}(X)$ with $\widehat{U}_\omega(\lambda) = H(\lambda + \omega/\lambda)$, $\lambda > 0$. Define $\{U(t)\}_{t\geq 0} \subset \mathcal{B}(X)$ by

$$U(t) = e^{\omega t}U_\omega(t) - \int_0^t e^{\omega s}U_\omega(s)ds, \, t \geq 0;$$

then $U(t)$ is locally Lipschitz and

$$\widehat{U}(\lambda) = H(\lambda)/\lambda, \quad \lambda > \omega.$$

The definition of $H(\lambda)$ shows that $U(t)$ commutes with A and yields the identity

$$\widehat{U}(\lambda) = \lambda^{-2} + \widehat{a}(\lambda)\widehat{U}(\lambda)A,$$

i.e.

$$U(t)x = tx + a * U(t)Ax \quad \text{for all } x \in \mathcal{D}(A), \, t \geq 0.$$

Since for each $x \in \mathcal{D}(A)$, $f(t) = U(t)Ax$ is locally lipschitz, i.e. $f \in BV_{loc}(\mathbb{R}_+; X)$, we see that $U(t)x$ belongs to $W^{1,\infty}_{loc}(\mathbb{R}_+; X)$ and

$$(U(t)x)\dot{} = x + (a * f(t))\dot{} = x + a * df(t), \quad \text{for a.a. } t \geq 0.$$

But this implies $a * df(t) = a * (U(t)Ax)\dot{}$ for all $x \in \mathcal{D}(A^2)$ and so $a * df(t)$ is even continuous, from which in turn we see that $U(t)x$ is continuously differentiable on $\mathbb{R}_+$ for each $x \in \mathcal{D}(A^2)$. $\mathcal{D}(A^2)$ is dense in X and the operators $(U(t + h) - U(t))/h$ are uniformly bounded for $0 < h \leq 1$, t bounded, hence by the Banach-Steinhaus Theorem $U(t)x$ is continuously differentiable on $\mathbb{R}_+$ for each $x \in X$. Define $S(t)x = \dot{U}(t)x$, $t \geq 0, x \in X$; it is clear that $S(t)x$ is strongly continuous, of type (M, ω) and $S(t)$ commutes with A since $U(t)$ does and A is closed. Finally, $\widehat{S}(\lambda) = H(\lambda)$ for $\text{Re } \lambda > \omega$ and by (H2) we obtain

$$\widehat{S}(\lambda) = \lambda^{-1} + \widehat{a}(\lambda)\widehat{S}(\lambda)A,$$

i.e. the resolvent equation (1.6) is satisfied. The proof is complete. □

For the special case $a(t) = 1$, Theorem 1.3 becomes the celebrated generation theorem for C_0-semigroups, while for $a(t) = t$ it reduces to the generation theorem for cosine families due to Da Prato and Giusti [63].

In general, (H2) is quite difficult to check, however there are several important special classes of operators A and kernels $a(t)$ such that this is possible. Many of the results in subsequent sections rely on Theorem 1.3.

By means of Theorem 1.3 several important conclusions can be derived from existence of an exponentially bounded resolvent.

Corollary 1.6 *Suppose (1.1) admits a resolvent* $S(t)$ *of type* (M, ω). *Then*
(i) $\widehat{a}(\lambda) \neq 0$ *for* $Re \; \lambda > \omega$ *and* $\Omega = \{1/\widehat{a}(\lambda) : \; Re \; \lambda > \omega\} \subset \rho(A)$;
(ii) A *is Abel-ergodic at* 0, *i.e.* $\mu(\mu - A)^{-1} \to I$ *strongly as* $|\mu| \to \infty$, $\mu \in \Omega_\theta = \{1/\widehat{a}(\lambda) : \; Re \; \lambda > \omega, |\arg \; \lambda| \leq \theta\}$, *for each* $\theta < \pi/2$;
(iii) If $a(t)$ *is real, then w.l.o.g.* $\widehat{a}(\lambda) > 0$ *for* $\lambda > \omega$, *hence* $(\mu_0, \infty) \subset \rho(A)$ *for some* $\mu_0 > 0$ *and* $\mu(\mu - A)^{-1} \to I$ *strongly as* $\mu \to \infty$.

Proof: (i) has already been proved in front of Theorem 1.3.
(ii) Equation (1.40) with $n = 0$ yields $|\mu(\mu - A)^{-1}| \leq M \frac{|\lambda|}{Re \; \lambda - \omega} \leq M_\theta$, where $\mu = 1/\widehat{a}(\lambda)$, $Re \; \lambda \geq \omega + 1, |\arg \; \lambda| \leq \theta < \pi/2$, thus $\mu(\mu - A)^{-1}$ is bounded for such μ. For $x \in \mathcal{D}(A)$ we have

$$\mu(\mu - A)^{-1}x = x + (\mu - A)^{-1}Ax \to x \quad \text{as } |\mu| \to \infty,$$

hence $\mu(\mu - A)^{-1} \to I$ by the Banach-Steinhaus theorem.
(iii) is obvious from (ii) and (i). $\quad \square$

In contrary to the case of semigroups and cosine families, resolvents if they exist need not to be exponentially bounded even if the kernel $a(t)$ involved is of class $C^\infty(\mathbb{R}_+)$ and belongs to $L^1(\mathbb{R}_+)$. The reason for this phenomenon is the fact that there exist kernels $a \in C^\infty(\mathbb{R}_+) \cap L^1(\mathbb{R}_+)$ with $a(0) = 1$ such that $\widehat{a}(\lambda)$ admits zeros with arbitrarily large real part. By Corollary 1.6, (1.1) cannot admit an exponentially bounded resolvent. On the other hand, if A merely generates a C_0-semigroup then (1.1) for such kernel $a(t)$ admits a resolvent by Corollary 1.4 with $k(t) = \dot{a}(t)$ and $b(t) = 0$. For the construction of a kernel with the described properties we refer to Desch and Prüss [91]. One should note, however, that the derivative $\dot{a}(t)$ of such kernels behaves very badly, i.e. is not Laplace transformable, since otherwise $\widehat{a}(\lambda) = (1 + \hat{\dot{a}}(\lambda))/\lambda$ would be nonzero for large real parts of λ.

1.6 Integral Resolvents
Suppose (1.1) admits a resolvent $S(t)$ and let $a \in BV_{loc}(\mathbb{R}_+)$. We have seen in Section 1.2 that $S(t)$ then is differentiable and for $x \in \mathcal{D}(A)$, $\dot{S}(t)x$ is a solution of

$$\dot{S}(t)x = a(t)Ax + A \int_0^t a(t - \tau)\dot{S}(\tau)xd\tau;$$

Proposition 1.3 proved that $\dot{S}(t)$ is useful for variation of parameters formulas as well.

Returning to the general case, we consider the integral equation

$$R(t)x = a(t)x + A \int_0^t a(t - \tau)R(\tau)xd\tau, \quad t \geq 0, \tag{1.42}$$

instead of (1.7). A solution of (1.42) is called integral resolvent for (1.1); the precise definition is as follows.

Definition 1.6 *A family* $\{R(t)\}_{t\geq 0} \subset \mathcal{B}(X)$ *is called* **integral resolvent** *for (1.1) if the following conditions are satisfied.*
(R1) $R(\cdot)x \in L^1_{loc}(\mathbb{R}_+; X)$ *for each* $x \in X$, *and* $|R(t)| \leq \varphi(t)$ *a.e. on* $\mathbb{R}_+$, *for some* $\varphi \in L^1_{loc}(\mathbb{R}_+)$;
(R2) $R(t)$ *commutes with* A *for each* $t \geq 0$;
(R3) *the* **integral resolvent equation** *holds*

$$R(t)x = a(t)x + \int_0^t a(t-\tau)AR(\tau)x\,d\tau \quad \text{for all } x \in \mathcal{D}(A), \text{ a.a. } t \geq 0. \quad (1.43)$$

Suppose $R(t)$ is an integral resolvent for (1.1), let $f \in C(J; X)$ and $u \in C(J; X)$ be a mild solution for (1.1). Then $R * u$ is well-defined and continuous and we obtain from (1.43) and (1.1)

$$a * u = (R - Aa * R) * u = R * u - R * Aa * u = R * f,$$

i.e. $R * f \in C(J; X_A)$ and from (1.1) we obtain

$$u(t) = f(t) + A \int_0^t R(t-s)f(s)\,ds, \quad t \in J; \quad (1.44)$$

compare (1.44) with (1.13) above. (1.44) easily yields the uniqueness of the integral resolvent for (1.1).

If both, $S(t)$ and $R(t)$ exist for (1.1), the relations between S and R are given by

$$R(t)Ax = \dot{S}(t)x \quad \text{for } x \in \mathcal{D}(A), \quad t \geq 0, \quad (1.45)$$

and

$$R(t)x = (a * S)^{\cdot}(t)x \quad \text{for } x \in X, \ t \geq 0. \quad (1.46)$$

For the case of the Cauchy problem (1.2), i.e. $a(t) \equiv 1$, we have $R(t) \equiv S(t)$ while for the second order equation (1.3) $S(t) = \mathrm{Co}(t)$, the cosine family generated by A and $R(t) = \mathrm{Si}(t)$, the sine family generated by A. It is possible that $R(t)$ exists but not $S(t)$, and vice versa.

If $R(t)$ is an integral resolvent for (1.1) then (1.44) yields a mild solution to (1.1) in case $f \in C(J; X_A)$, and even a strong solution in case $f = a * g$ with $g \in C(J; X_A)$; this can be seen as in the proof of Proposition 1.3.

There is a result similar to Theorem 1.3 of the last section for integral resolvents.

Theorem 1.4 *Let A be a closed linear unbounded operator in X with dense domain $D(A)$ and let $a \in L^1_{loc}(\mathbb{R}_+)$ satisfy $\int_0^\infty e^{-\omega t}|a(t)|dt < \infty$. Then (1.1) admits an integral resolvent $R(t)$ such that*

$$|R(t)| \leq e^{\omega t}\varphi(t), \quad \text{for a.a. } t > 0, \quad (1.47)$$

holds for some $\varphi \in L^1(\mathbb{R}_+)$ iff the following conditions are satisfied.

(K1) $\widehat{a}(\lambda) \neq 0$ and $1/\widehat{a}(\lambda) \in \rho(A)$ for all $\lambda > \omega$;

(K2) $K(\lambda) = \widehat{a}(\lambda)(I - \widehat{a}(\lambda)A)^{-1}$ satisfies the estimates

$$\sum_{n=0}^{\infty} (\lambda - \omega)^n |K^{(n)}(\lambda)|/n! \leq M \quad \text{for all } \lambda > \omega. \tag{1.48}$$

The proof of this result is similar to that of Theorem 1.3, based on Theorem 0.3. Since it is not needed later on, we therefore omit it here. Corollary 1.6 remains true with $S(t)$ replaced by $R(t)$.

1.7 Comments
a) For the finite-dimensional case the solution of

$$T = a(t)A + a * AT$$

is usually called the resolvent of (1.1). In our terminology this corresponds to the integral resolvent, i.e. we have $T = RA$. For equations with unbounded operators in infinite dimensions our concept of the resolvent $S(t)$ seems to be more appropriate since it is a direct generalization of C_0-semigroups and cosine families. $S(t)$ is sometimes called *differential resolvent* to distinguish it from the *integral resolvent* $R(t)$.

b) Corollary 1.2(i) is essentially due to Carr and Hannsgen [38], while the second part of this corollary is taken from Friedman [121]; see also Sforza [301] where mainly the second order case is treated. For L^∞-estimates of $t^\beta \mu^\gamma s(t; \mu)$ which imply estimates of $(-A)^\gamma S(t)$ and $S^{(n)}(t)$ in this setting, see Engler [112].

c) The perturbation approach for (1.1) like in Section 1.4 was one of the first methods available to solve linear Volterra equations involving unbounded operators and has been used by many authors. It applies to much more general, even nonconvolution equations and will be taken up in Chapter II again. Therefore we postpone a discussion of the relevant literature.

   Perturbations of the operator A have been considered by Rhandi [291]. In this paper it is proved that in case $a \in BV_{loc}(\mathbb{R}_+)$ and $B \in \mathcal{B}(X)$ or $B \in \mathcal{B}(X_A)$ then well-posedness of (1.1) and of (1.1) with $A + B$ instead of A are equivariant.

d) Examples 1.1 and 1.2 are taken from Prüss [271]; see also Grimmer and Prüss [145] for an earlier version of Example 1.2. These papers contain another interesting example which we briefly describe now.

Example 1.3 Consider the initial boundary value problem

$$\begin{aligned}
u_t(t, x) &= Du(t, x) - \int_0^t b(t - \tau)Du(\tau, x)d\tau, \quad t \geq 0, x \in [0, 2\pi], \\
u(t, 0) &= u(t, 2\pi), \quad t \geq 0, \\
u_x(t, 0) &= u_x(t, 2\pi), \quad t \geq 0, \\
u(0, x) &= u_0(x), \quad x \in [0, 2\pi],
\end{aligned} \tag{1.49}$$

where $b(t) = 1/\sqrt{\pi t}$ and $D = -2i(d/dx)^2 + (1 - i)(d/dx)$. Obviously, (1.49) is of the form (1.1) with $a(t) = 1 - 2\sqrt{t/\pi}$ and A defined in $X = L^2(0, 2\pi)$ by $(Au)(x) = Du(x)$, $\mathcal{D}(A) = \{u \in W^{2,2}(0, 2\pi) : u(0) = u(2\pi), u'(0) = u'(2\pi)\}$. It is not difficult to show that A does not generate a C_0-semigroup since $\sigma(A) = \{n(1 + i) + 2in^2 : n \in \mathbb{Z}\}$ is not contained in any left halfplane. On the other hand, in Prüss [271] or Grimmer and Prüss [145] it is shown that (1.49) admits a resolvent, i.e. is well-posed.

e) The generation theorem in Section 1.5 is essentially due to Da Prato and Iannelli [66] who proved it for the case $a \in BV_{loc}(\mathbb{R}_+)$. Nonscalar versions of this result will be discussed in Chapter II.

f) Theorem 1.3 admits a slight extension involving only the growth bound of $S(t)$.

Theorem 1.3': *Let A be a closed linear unbounded operator in X with dense domain $\mathcal{D}(A)$ and let $a \in L^1_{loc}(\mathbb{R}_+)$ satisfy $\int_0^\infty e^{-\omega_a t}|a(t)|dt < \infty$. Then (1.1) admits a resolvent $S(t)$ with growth bound $\omega_0 \in \mathbb{R}$ iff the following conditions are satisfied.*
(H3) $\hat{a}(\lambda) \neq 0$ *for* $Re\ \lambda > max(\omega_0, \omega_a)$; *if* $\omega_0 < \omega_a$ *then* $1/\hat{a}(\lambda)$ *admits analytic continuation to the halfplane* $Re\ \lambda > \omega_0$, *and* $1/\hat{a}(\lambda) \in \rho(A)$ *for all* $Re\ \lambda > \omega_0$;
(H4) *for each* $\varepsilon > 0$ *there is a constant* $M_\varepsilon \geq 1$ *such that* $H(\lambda) = (1/\hat{a}(\lambda) - A)^{-1}/(\lambda\hat{a}(\lambda))$ *satisfies*

$$|H^{(n)}(\lambda)| \leq M_\varepsilon n!(\lambda - \omega_0 - \varepsilon)^{-(n+1)}, \quad Re\ \lambda > \omega_0 + \varepsilon, \ n \in \mathbb{N}_0.$$

Except for the analytic continuation of $1/\hat{a}(\lambda)$ in case $\omega_0 < \omega_a$, which will be obtained in Section 2.1, Theorem 1.3' is a simple consequence of Theorem 1.3. Observe that $\omega_0 < 0$ in Theorem 1.3' forces $a(t)$ to be nonintegrable, i.e. $\omega_0 < \omega_a$.

g) As in the case of the abstract Cauchy problem $\dot{u} = Au + f$ which corresponds to $a(t) \equiv 1$, a theory of *integrated resolvent* and *distribution resolvents* even for non-densely defined operators A has been developed; see Arendt and Kellermann [13] for integrated resolvents and Da Prato and Iannelli [67] for distribution resolvents. The *duality theory* of resolvents has not been studied so far.

i) An important practical problem is the identification problem, which can be roughly stated as follows. Suppose the space X and the operator A are given and let some (or just one) solution $u(t)$ for different (only one) forcing functions $f(t)$ be known; determine the kernel $a(t)$! This so-called 'inverse problem' is not considered in this book, and there are not many papers dealing with it; so far, the only ones we know of are the papers Lorenzi [221], and Lorenzi and Sinestrari [222], [223]. However it will become clear in Section 2.1 that the kernel $a(t)$ is already uniquely determined from the knowledge of just one observation.

2 Analytic Resolvents

This section is devoted to the theory of analytic resolvents, the analog of analytic semigroups for Volterra equations of scalar type. A complete characterization of such resolvents in terms of Laplace transforms is given. In contrast to the general generation theorem of Section 1, the main result of this section, Theorem 2.1, requires conditions which are much simpler to check; this is done in several illustrating examples. The spatial regularity of analytic resolvents is studied and a characterization of analytic semigroups in these terms is derived. It is shown that analytic resolvents lead to improved perturbation results and stronger properties of the variation of parameter formulas.

2.1 Definition and First Properties

In this section we consider again the Volterra equation

$$u(t) = f(t) + \int_0^t a(t-s)Au(s)ds, \quad t \in J, \tag{2.1}$$

where $a(t)$ and A as well as f are as in Section 1, and X is a complex Banach space. In the following, we denote by $\Sigma(\omega, \theta)$ the open sector with vertex $\omega \in \mathbb{R}$ and opening angle 2θ in the complex plane which is symmetric w.r.t. the real positive axis, i.e.

$$\Sigma(\omega, \theta) = \{\lambda \in \mathbb{C} : |\arg(\lambda - \omega)| < \theta\}.$$

We consider in this section a special class of equations of the form (2.1), namely which are such that the resolvent $S(t)$ of (2.1) is analytic in the sense of the following definition.

Definition 2.1 *A resolvent $S(t)$ for (2.1) is called **analytic**, if the function $S(\cdot)$: $\mathbb{R}_+ \to \mathcal{B}(X)$ admits analytic extension to a sector $\Sigma(0, \theta_0)$ for some $0 < \theta_0 \le \pi/2$. An analytic resolvent $S(t)$ is said to be of **analyticity type** (ω_0, θ_0) if for each $\theta < \theta_0$ and $\omega > \omega_0$ there is $M = M(\omega, \theta)$ such that*

$$|S(z)| \le Me^{\omega Rez}, \quad z \in \Sigma(0, \theta). \tag{2.2}$$

As a simple consequence of the definition of analytic resolvents we obtain estimates on the derivatives of $S(t)$ by means of Cauchy's integral formula.

Corollary 2.1 *Suppose $S(t)$ is an analytic resolvent for (2.1) of analyticity type (ω_0, θ_0). Then for each $\omega > \omega_0$, $\theta < \theta_0$ there is $M = M(\omega, \theta)$ such that*

$$|S^{(n)}(t)| \le Mn!e^{\omega t(1+\alpha)}(\alpha t)^{-n}, \quad t > 0, \ n \in \mathbb{N}, \tag{2.3}$$

where $\alpha = \sin\theta$.

Proof: Let $\omega > \omega_0$ and $\theta < \theta_0$ be fixed and choose $M = M(\omega, \theta)$ according to Definition 2.1. Cauchy's formula then yields

$$S^{(n)}(t) = (2\pi i)^{-1} n! \int_{|z-t|=r} S(z)(z-t)^{-n-1} dz, \tag{2.4}$$

where $r = t \sin \theta$. Estimate (2.2) gives

$$|S^{(n)}(t)| \leq (2\pi)^{-1} n! M \int_{-\pi}^{\pi} e^{\omega(t+r\cos\varphi)} r^{-n} d\varphi$$
$$\leq n! M e^{\omega(t+r)} r^{-n}, \quad n \in \mathbb{N}, \; t > 0,$$

and this is precisely (2.3). $\quad\square$

Estimate (2.3) shows that for an analytic resolvent $S(t)$ for (2.1), $|\dot{S}(t)| \sim 1/t$, and so $|\dot{S}(\cdot)|$ is almost integrable at $t = 0$. However, unless A is bounded, $|\dot{S}(\cdot)|$ will not be integrable near $t = 0$. In fact, if the latter would be the case then $F(t) = \dot{S}(t)e^{-\omega t} \in L^1(\mathbb{R}_+, \mathcal{B}(X))$, hence $|\widehat{F}(\lambda)| \to 0$ as $|\lambda| \to \infty$; with $\widehat{F}(\lambda) = (\lambda + \omega)H(\lambda + \omega) - I = (I - \widehat{a}(\lambda + \omega)A)^{-1} - I$, $(I - \mu A)^{-1}$ is surjective for some $\mu \in \mathbb{C}$, hence $\mathcal{D}(A) = X$, and so A must be bounded. The case $a(t) \equiv 1$, i.e. the case of analytic semigroups, shows that (2.3) cannot be improved, in general.

In analogy to the theory of analytic semigroups, a complete characterization of analytic resolvents of analyticity type (ω_0, θ_0) is possible in terms of the spectrum of A and the Laplace-transform of $a(t)$. In the sequel it will be assumed that $a(t)$ is of exponential growth.

(i) To derive these conditions assume $S(t)$ is an analytic resolvent of type (ω_0, θ_0) and define $H(\lambda) = \widehat{S}(\lambda)$, Re $\lambda > \omega_0$. Since $S(z)$ is holomorphic in $\Sigma(0, \theta_0)$, $H(\lambda)$ admits analytic extension to the sector $\Sigma(\omega_0, \pi/2 + \theta_0)$, by Theorem 0.1, and for each $\omega > \omega_0$, $\theta < \theta_0$ there is a constant $C(\omega, \theta)$ such that the estimate

$$|H(\lambda)| \leq C(\omega, \theta)/|\lambda - \omega|, \quad \lambda \in \Sigma(\omega, \theta + \pi/2), \tag{2.5}$$

holds.

(ii) Next we show that the holomorphic function $\widehat{a}(\lambda)$ admits meromorphic extension to the same sector $\Sigma(\omega_0, \theta_0 + \pi/2)$. In fact, from Theorem 1.3 we know already

$$H(\lambda) = (I - \widehat{a}(\lambda)A)^{-1}/\lambda, \quad \text{Re } \lambda > \omega_1, \tag{2.6}$$

where $\omega_1 > \omega_0$ is chosen large enough so that $\int_0^\infty |a(t)|e^{-\omega_1 t} dt < \infty$. Let $x \in \mathcal{D}(A)$ and $x^* \in X^*$ be such that $\varphi(\lambda) = \lambda < H(\lambda)x, x^* > \neq < x, x^* >$; such x, x^* always exist since otherwise $\lambda H(\lambda) \equiv I$ which implies $\widehat{a}(\lambda) \equiv 0$ or $A = 0$, in contradiction to our standing hypotheses $a(t) \not\equiv 0$ and A unbounded. $\varphi(\lambda)$ is holomorphic on $\Sigma(\omega_0, \theta_0 + \pi/2)$, hence its derivative $\varphi'(\lambda)$ is so as well. For Re $\lambda > \omega_1$ we obtain

$$\varphi'(\lambda) = \widehat{a}'(\lambda) < (I - \widehat{a}(\lambda)A)^{-2} Ax, x^* >$$
$$= \widehat{a}'(\lambda)\lambda^2 < H(\lambda)^2 Ax, x^* >$$
$$= \widehat{a}'(\lambda)\psi(\lambda),$$

where $\psi(\lambda) = \lambda^2 < H(\lambda)^2 Ax, x^* >$ is holomorphic on $\Sigma(\omega_0, \theta_0 + \pi/2)$, and $\psi(\lambda) \not\equiv 0$, since otherwise $\varphi(\lambda) \equiv \lim_{\lambda \to \infty} \varphi(\lambda) = < \lim_{\lambda \to \infty} \lambda H(\lambda)x, x^* > = < x, x^* >$ by Corollary 1.6, in contradiction to the choice of x and x^*. This shows that $\widehat{a}'(\lambda) = \varphi(\lambda)/\psi(\lambda)$ is meromorphically extendible to this sector. We also have

$$
\begin{aligned}
\varphi'(\lambda) &= \widehat{a}'(\lambda) < A(I - \widehat{a}(\lambda)A)^{-2}x, x^* > \\
&= [\widehat{a}'(\lambda)/\widehat{a}(\lambda)] \cdot < \lambda^2 H(\lambda)^2 x - \lambda H(\lambda)x, x^* > \\
&= [\widehat{a}'(\lambda)/\widehat{a}(\lambda)] \cdot \chi(\lambda),
\end{aligned}
$$

where $\chi(\lambda)$ is holomorphic on $\Sigma(\omega_0, \theta_0 + \pi/2)$; this implies

$$
\widehat{a}(\lambda) = \widehat{a}'(\lambda)\chi(\lambda)/\varphi'(\lambda) = \chi(\lambda)/\psi(\lambda), \qquad \text{Re } \lambda > \omega_1,
$$

i.e. $\widehat{a}(\lambda)$ admits a meromorphic continuation to $\Sigma(\omega_0, \theta_0 + \pi/2)$.

(iii) Relation (2.6) can now be extended to $\Sigma(\omega_0, \theta_0 + \pi/2)$ as follows. Let $\Omega = \{\lambda \in \Sigma(\omega_0, \theta_0 + \pi/2) : \lambda \neq 0, \, \widehat{a}(\lambda) \neq 0, \infty\}$; for $x \in \mathcal{D}(A)$ we have

$$
\lambda H(\lambda)(I - \widehat{a}(\lambda)A)x = x \quad \text{for all } \lambda \in \Omega, \tag{2.7}
$$

since this relation holds for Re $\lambda > \omega_1$, i.e. $I - \widehat{a}(\lambda)A$ is injective for all $\lambda \in \Omega$. On the other hand, for Re $\lambda > \omega_1$, $H(\lambda)$ commutes with A, i.e. for each $\mu \in \rho(A)$ we have

$$
(\mu - A)^{-1}H(\lambda) = H(\lambda)(\mu - A)^{-1}.
$$

By analytic continuation, this relation holds on all of Ω and therefore $H(\lambda)\mathcal{D}(A) \subset \mathcal{D}(A)$ for all $\lambda \in \Omega$, and $AH(\lambda)x$ is analytic on Ω. But this implies also the identity

$$
\lambda(I - \widehat{a}(\lambda)A)H(\lambda)x = x, \quad \lambda \in \Omega, \tag{2.8}
$$

for all $x \in \mathcal{D}(A)$, since this is valid for Re $\lambda > \omega_1$. Therefore, the operators $\lambda(I - \widehat{a}(\lambda)A)$ have dense ranges in X, and so they are invertible for all $\lambda \in \Omega$, and (2.6) holds on Ω as well.

(iv) If $\lambda_0 \in \Sigma(\omega_0, \theta_0 + \pi/2)$ is such that $\widehat{a}(\lambda_0) = 0$, then - since λ_0 is isolated - we may choose a small circle Γ around λ_0 contained in $\Sigma(\omega_0, \theta_0 + \pi/2)$, and apply the Cauchy integral formula to the operator-valued function $\lambda H(\lambda)$ to obtain

$$
A = A\lambda_0 H(\lambda_0) = A(2\pi i)^{-1} \int_\Gamma \frac{\lambda H(\lambda)}{\lambda - \lambda_0} d\lambda = (2\pi i)^{-1} \int_\Gamma \frac{\lambda H(\lambda) - I}{(\lambda - \lambda_0)\widehat{a}(\lambda)} d\lambda,
$$

i.e. A is bounded, in contradiction to the standing hypotheses on A. Thus $\widehat{a}(\lambda)$ does not have zeros in $\Sigma(\omega_0, \theta_0 + \pi/2)$.

(v) If $\lambda_0 \in \Sigma(\omega_0, \theta_0 + \pi/2)$ is a pole of $\widehat{a}(\lambda)$, say of order n, then the Laurent-expansion

$$
\widehat{a}(\lambda) = a_{-n}(\lambda - \lambda_0)^{-n} + (\lambda - \lambda_0)^{-n+1}\varphi(\lambda)
$$

with a holomorphic function $\varphi(\lambda)$ near λ_0 and Taylor-expansion of $\lambda H(\lambda)$ show that $\lambda H(\lambda)$ has a zero of order n; passing to the limit $\lambda \to \lambda_0$ in (2.7) and (2.8) we obtain with

$$
H_n = \lim_{\lambda \to \lambda_0} \lambda H(\lambda)(\lambda - \lambda_0)^{-n}
$$

the identity

$$-a_{-n}H_nAx = -a_{-n}AH_nx = x \quad \text{for all } x \in D(A).$$

Since $-a_{-n} \neq 0$ this yields $0 \in \rho(A)$ and $H_n = -\frac{1}{a_{-n}}A^{-1}$. Thus $\hat{a}(\lambda)$ may only have poles in $\Sigma(\omega_0, \theta_0 + \pi/2)$ if $0 \in \rho(A)$; note that in case $\omega_0 < 0$ holds, $\lambda = 0$ must be a pole for $\hat{a}(\lambda)$.

2.2 Generation of Analytic Resolvents

It turns out that the conditions derived so far are characterizing analytic resolvents of type (ω_0, θ_0).

Theorem 2.1 *Let A be a closed linear unbounded operator in X with dense domain $\mathcal{D}(A)$ and let $a \in L^1_{loc}(\mathbb{R}_+)$ satisfy $\int_0^\infty |a(t)|e^{-\omega_a t}dt < \infty$ for some $\omega_a \in \mathbb{R}$. Then (2.1) admits an analytic resolvent $S(t)$ of analyticity type (ω_0, θ_0) iff the following conditions hold.*

(A1) $\hat{a}(\lambda)$ *admits meromorphic extension to $\Sigma(\omega_0, \theta_0 + \pi/2)$;*

(A2) $\hat{a}(\lambda) \neq 0$, *and $1/\hat{a}(\lambda) \in \rho(A)$ for all $\lambda \in \Sigma(\omega_0, \theta_0 + \pi/2)$;*

(A3) *For each $\omega > \omega_0$ and $\theta < \theta_0$ there is a constant $C = C(\omega, \theta)$ such that $H(\lambda) = (1/\hat{a}(\lambda) - A)^{-1}/(\lambda\hat{a}(\lambda))$ satisfies the estimate*

$$|H(\lambda)| \leq C/|\lambda - \omega| \quad \text{for all } \lambda \in \Sigma(\omega, \theta + \pi/2). \tag{2.9}$$

Proof: It remains to prove the sufficiency part. Let $\omega > \omega_0$ and $\theta < \theta_0$ be given and choose $\theta' \in (\theta, \theta_0)$ as well as $C = C(\omega, \theta')$ from (A3). Define $S(t)$ by

$$S(z) = (2\pi i)^{-1} \int_{\Gamma_R} e^{\lambda z} H(\lambda) d\lambda, \quad z \in \Sigma(0, \theta), \tag{2.10}$$

where Γ_R denotes the contour consisting of the two rays $\omega + ire^{i\theta'}$ and $\omega - ire^{-i\theta'}$ with $r \geq R$ and the larger part of the circle $|\lambda - \omega| = R$ connecting these rays. Let $z = te^{i\varphi}$, $R = 1/t$, and $\alpha = \sin(\theta' - \theta)$; estimating (2.10) we obtain

$$
\begin{aligned}
|S(z)| &\leq (C/2\pi) \int_{\Gamma_R} e^{Re(\lambda z)} |\lambda - \omega|^{-1} |d\lambda| \\
&\leq (C/\pi)e^{\omega Rez} \{ \int_1^\infty e^{-\alpha r} dr/r + \int_0^\pi e^{\cos\varphi} d\varphi \} \\
&\leq Me^{\omega Rez}.
\end{aligned}
$$

This estimate shows that the integral in (2.10) is absolutely convergent for $z \in \Sigma(0, \theta)$, hence $S(z)$ is holomorphic in this region and (2.2) of Definition 2.1 is satisfied. Moreover, for the Laplace-transform of $S(t)$ we obtain for $\lambda > \omega$

$$\hat{S}(\lambda) = \int_0^\infty e^{-\lambda t} S(t)dt = (2\pi i)^{-1} \int_0^\infty \int_{\Gamma_0} e^{-\lambda t} e^{\mu t} H(\mu) d\mu dt$$

$$= (2\pi i)^{-1} \int_{\Gamma_0} (\int_0^\infty e^{-(\lambda - \mu)t} dt) H(\mu) d\mu = (2\pi i)^{-1} \int_{\Gamma_0} H(\mu)(\lambda - \mu)^{-1} d\mu$$

$$= H(\lambda),$$

where we used Fubini's theorem and Cauchy's integral formula. Thus $H(\lambda)$ is the Laplace-transform of $S(t)$ and therefore by Theorem 1.3, $S(t)$ is a resolvent for (2.2). Finally, since $\omega > \omega_0$ and $\theta < \theta_0$ are arbitrary, $S(t)$ is an analytic resolvent of type (ω_0, θ_0). $\square$

For an analytic resolvent $S(t)$ for (2.1) we so far only know that $S(z) \to I$ strongly as $z \to 0+$ for real z, and so a natural question is whether this convergence is true in every sector $\Sigma(0, \theta)$ with $\theta < \theta_0$. The answer is in the affirmative.

Corollary 2.2 *Suppose $S(t)$ is an analytic resolvent for (2.1) of type (ω_0, θ_0). Then $S(z)x \to x$ for $z \in \Sigma(0, \theta)$ as $z \to 0$, for each $x \in X$ and $\theta < \theta_0$.*

Proof: Let $x \in X$ and $x^* \in X^*$ and consider $\varphi(z) = < S(z)x, x^* >$. Since $\varphi(t) \to \varphi(0) = < x, x^* >$ as $t \to 0+$ and φ is holomorphic and bounded on $\Sigma(0, \theta_0)$, a well-known result about the boundary behaviour of analytic functions (cp. Duren [106], Theorem 1.3) implies $\varphi(z) \to \varphi(0)$ as $z \to 0+$, uniformly for $z \in \Sigma(0, \theta)$, for all $\theta < \theta_0$. Therefore, the Laplace-transform $\varphi(\lambda)$ has the convergence property

$$\lambda \widehat{\varphi}(\lambda) = < \lambda H(\lambda)x, x^* > \to < x, x^* > = \varphi(0) \quad \text{as } |\lambda| \to \infty,$$

where $\lambda \in \Sigma(\omega, \theta + \pi/2)$, for any $\omega > \omega_0$, $\theta < \theta_0$; similarly,

$$\phi(\lambda) = < \lambda^2 H(\lambda)^2 x, x^* > \to < x, x^* > \quad \text{as } |\lambda| \to \infty, \ \lambda \in \Sigma(\omega, \theta + \pi/2),$$

where x and x^* are arbitrary. Therefore

$$\widehat{a}(\lambda) = \chi(\lambda)/\psi(\lambda) \to 0 \quad \text{as } |\lambda| \to \infty, \quad \text{uniformly in } \lambda \in \Sigma(\omega, \theta + \pi/2),$$

for each $\omega > \omega_0, \theta < \theta_0$, where $\chi(\lambda)$ and $\psi(\lambda)$ are defined as in (ii) of Section 2.1, i.e.

$$\psi(\lambda) = \lambda^2 < H(\lambda)^2 Ax, x^* >, \quad \chi(\lambda) = < \lambda H(\lambda)(\lambda H(\lambda)x - x), x^* >,$$

with $x \in \mathcal{D}(A)$ and $x^* \in X^*$ such that $< Ax, x^* > \neq 0$. (2.10) and (2.7) then imply for $x \in \mathcal{D}(A)$

$$S(z)x - x = (2\pi i)^{-1} \int_{\Gamma_R} e^{\lambda z} H(\lambda)\widehat{a}(\lambda) Ax d\lambda,$$

hence with $z = te^{i\varphi}, R = 1/t$, as in the proof of Theorem 2.1

$$|S(z)x - x| \leq M e^{\omega Rez} |Ax| \sup\{|\widehat{a}(\lambda)| : |\lambda| \geq 1/t, \lambda \in \Sigma(\omega, \theta' + \pi/2)\} \to 0$$

as $|z| = t \to 0$. Thus $S(z)x \to x$ as $z \to 0$, $z \in \Sigma(0,\theta)$ for each $x \in \mathcal{D}(A)$, and since $S(z)$ is uniformly bounded for such z, by the Banach-Steinhaus theorem $S(z) \to I$ strongly as $z \to 0$, $z \in \Sigma(0,\theta)$. $\square$

Another natural question which arises in connection with analytic resolvent is what regularity is imposed on $a(t)$ by conditions $(A1) \sim (A3)$. It turns out that $a(t)$ itself admits analytic continuation to the same sector $\Sigma(0,\theta_0)$ of analyticity of the resolvent $S(t)$.

Corollary 2.3 *Suppose $S(t)$ is an analytic resolvent for (2.1) of analyticity type (ω_0, θ_0). Then $a(t)$ admits analytic extension to $\Sigma(0,\theta_0)$. Furthermore, on each subsector $\Sigma(0,\theta)$, $\theta < \theta_0$, there is a decomposition of the form*

$$a(z) = \sum_j p_j(z)e^{\lambda_j z} + a_1(z), \quad z \in \Sigma(0,\theta), \tag{2.11}$$

where the λ_j denote the finitely many poles of $\widehat{a}(\lambda)$ contained in $\overline{\Sigma(\omega, \theta + \pi/2)}$, the $p_j(z)$ are polynomials, and $a_1(z)$ is analytic in $\Sigma(0,\theta)$ and satisfies

$$|a_1(z)| \le Ce^{\omega Re z}/|z|, \quad z \in \Sigma(0,\theta), \tag{2.12}$$

$$z a_1(z) \to 0 \quad \text{as } z \to 0, \, z \in \Sigma(0,\theta_0), \tag{2.13}$$

and $\widehat{a}(\lambda)$ may only have poles in case $0 \in \rho(A)$.

Proof: Let $\omega > \omega_0$ and $\theta < \theta_0$ be fixed but arbitrary otherwise, and choose $\omega' \in (\omega_0, \omega)$, $\theta' \in (\theta, \theta_0)$ such that no poles of $\widehat{a}(\lambda)$ are located on the rays $\omega' \pm ire^{\pm i\theta'}$. From the proof of Corollary 2.2 we already know $\widehat{a}(\lambda) \to 0$ as $|\lambda| \to \infty, \lambda \in \Sigma(\omega', \theta' + \pi/2)$, hence there are only finitely many poles $\lambda_1, \ldots, \lambda_n \in \Sigma(\omega', \theta' + \pi/2)$ of $\widehat{a}(\lambda)$. Deforming the contour Γ in the complex inversion formula for the Laplace-transform of $a(t)$ into Γ_0, by the theorem of residues we obtain

$$
\begin{aligned}
a(z) &= \sum_{j=1}^{n} \text{Res}(e^{\lambda z}\widehat{a}(\lambda))_{\lambda=\lambda_j} + (2\pi i)^{-1} \int_{\Gamma_0} e^{\lambda z}\widehat{a}(\lambda)d\lambda \\
&= \sum_{j=1}^{n} p_j(z)e^{\lambda_j z} + a_1(z),
\end{aligned}
$$

i.e. decomposition (2.11) holds. $a_1(z)$ is estimated as follows

$$
\begin{aligned}
|a_1(z)| &\le \sup\{|\widehat{a}(\lambda)| : \lambda \in \Gamma_0\}\pi^{-1} \int_0^\infty e^{Re(\lambda z)}|d\lambda| \\
&\le Me^{\omega' Re z}/|z\cos(\pi/2 + \theta' - \theta)| \le Ce^{\omega Re z}/|z|, \quad z \in \Sigma(\omega, \theta + \pi/2),
\end{aligned}
$$

i.e. (2.12) holds. (2.13) follows similarly by choosing Γ_R as the integration contour, and the last statement is already clear from (v) in Section 2.1. $\square$

2.3 Examples

It is worthwhile to mention that Hille's generation theorem for analytic C_0-semi-groups cp. Hille-Phillips [180], Theorem 12.8.1, is a special case of Theorem 2.1. In fact, this corresponds to the case $a(t) \equiv 1$; (A1) is then trivially satisfied since $\widehat{a}(\lambda) = 1/\lambda$ and (A2) and (A3) reduce to

$$\Sigma(\omega_0, \theta_0 + \pi/2) \subset \rho(A) \quad \text{and} \quad |(\lambda - A)^{-1}| \leq C/|\lambda - \omega|, \quad \lambda \in \Sigma(\omega, \theta + \pi/2),$$

which are precisely Hille's conditions.

Typical examples for kernels $a(t)$ and operators A satisfying (A1) $\sim$ (A3) are given in

Example 2.1 Consider the kernels

$$a(t) = t^{\beta-1}/\Gamma(\beta), \quad t > 0$$

where $\beta \in (0, 2)$ and Γ denotes the gamma-function. Let us examine for which operators (2.1) admits a bounded analytic resolvent $S(t)$. For the Laplace-transform of $a(t)$ we obtain

$$\widehat{a}(\lambda) = \lambda^{-\beta}, \quad \text{Re } \lambda > 0,$$

hence $\widehat{a}(\lambda)$ admits analytic extension to the complex plane sliced along the negative real axis; thus (A1) holds for $\Sigma(0, \pi)$. The function $1/\widehat{a}(\lambda)$ maps the sector $\Sigma(0, \theta_0 + \pi/2)$ onto the sector $\Sigma(0, \theta_1)$, $\theta_1 = \beta(\theta_0 + \pi/2)$, hence (A2) with $\omega_0 = 0$ is equivalent to $\rho(A) \supset \Sigma(0, \theta_1)$, for some $\theta_1 > \beta\pi/2$. Finally, $S(t)$ will be bounded in some sector $\Sigma(0, \theta)$ if (A3) holds uniformly in $\omega > 0$, which is equivalent to

$$|\mu(\mu - A)^{-1}| \leq M \quad \text{for all } \mu \in \Sigma(0, \beta\pi/2).$$

Thus the pair $(t^{\beta-1}/\Gamma(\beta), A)$ generates a bounded analytic resolvent iff

$$\rho(A) \supset \Sigma(0, \beta\pi/2) \quad \text{and} \quad |\mu(\mu - A)^{-1}| \leq M \quad \text{for all } \mu \in \Sigma(0, \beta\pi/2).$$

Example 2.2 An important class of kernels $a(t)$ which satisfy (A1) is the class of completely monotonic kernels, i.e. kernels which are represented as Laplace transforms of positive measures

$$a(t) = \int_0^\infty e^{-st} d\alpha(s), \quad t > 0,$$

with $\alpha(t)$ nondecreasing, and such that $\int_1^\infty d\alpha(s)/s < \infty$ (the latter is equivalent to $a \in L^1_{loc}(\mathbb{R}_+)$). $\widehat{a}(\lambda)$ is then given by the integral representation

$$\widehat{a}(\lambda) = \int_0^\infty \frac{1}{\lambda + s} d\alpha(s), \quad \text{Re } \lambda > 0,$$

hence $\widehat{a}(\lambda)$ admits analytic extension to $\Sigma(0, \pi)$. The kernels of Example 1 with $\beta \leq 1$ are of this kind. Moreover, the decomposition

$$\widehat{a}(\sigma + i\rho) = \int_0^\infty \frac{\sigma + s}{(\sigma + s)^2 + \rho^2} d\alpha(s) - i\rho \int_0^\infty \frac{1}{(\sigma + s)^2 + \rho^2} d\alpha(s)$$

shows $\widehat{a}(\lambda) \neq 0$ on $\mathbb{C} \setminus \mathbb{R}_-$ and

$$|\arg \widehat{a}(\lambda)| \leq |\arg \lambda| \quad \text{for all } \lambda \in \Sigma(0, \pi).$$

Thus if A generates an analytic C_0-semigroup bounded in some sector $\Sigma(0, \theta)$, then (A2) and (A3) are satisfied since then

$$|\mu(\mu - A)^{-1}| \leq M, \quad \text{for all } \mu \in \Sigma(0, \theta + \pi/2).$$

Therefore by Theorem 2.1, (2.1) admits an analytic resolvent $S(t)$ of type $(0, \theta)$. Let us summarize this in

Corollary 2.4 *Let $a \in C(0, \infty) \cap L^1(0, 1)$ be completely monotonic and let A generate an analytic semigroup $T(t)$ such that $|T(t)| \leq M$ on $\Sigma(0, \theta)$. Then (2.1) admits an analytic resolvent $S(t)$ of type $(0, \theta)$.*

Example 2.3 Consider the kernel

$$a(t) = \int_0^\infty t^{\rho-1}/\Gamma(\rho)d\rho, \quad t > 0;$$

its Laplace transform is given by

$$\widehat{a}(\lambda) = 1/\log \lambda.$$

The function $\log \lambda$ maps the sector $\Sigma(0, \theta)$ onto the strip $\mathcal{S}_\theta = \{z \in \mathbb{C} : |\operatorname{Im} z| < \theta\}$. Hence if A is such that $\overline{\mathcal{S}}_{\pi/2} \subset \rho(A)$ and $|(\mu - A)^{-1}| \leq C/|\mu|$ holds on $\overline{\mathcal{S}}_{\pi/2}$ then $|H(\lambda)| \leq C/|\lambda|$ for $\lambda \in \Sigma(0, \pi/2 + \theta), \theta$ sufficiently small and Theorem 2.1 yields a bounded analytic resolvent $S(t)$.

On the other hand, if A is such that $\Sigma(0, \phi) \subset \rho(A)$ for some $\phi > 0$ and $|(\mu - A)^{-1}| \leq C/|\mu|$, then $1/\widehat{a}(\lambda)$ maps $\Sigma(\omega, \pi/2 + \theta)$ into the halfstrip $\mathcal{S}_{\pi/2+\theta} \cap \{\mu \in \mathbb{C} : \operatorname{Re} \mu \geq R\}$ where $R = \log(\omega \cos \theta)$. Choosing ω large enough this halfstrip will be contained in $\Sigma(0, \phi)$, hence Theorem 2.1 yields a resolvent of type (ω, θ), where $\omega \geq \exp[(\pi/2 + \theta)/\sin \phi]/\cos \theta$.

This example shows that there are kernels such that (2.1) admits an analytic resolvent for every operator which is such that $(\alpha, \infty) \subset \rho(A)$ for some $\alpha \geq 0$ and $|\mu(\mu - A)^{-1}| \leq C$ as $\mu \to \infty$, in particular for every *sectorial* operator, cp. Section 8. Note that the singularity of $a(t)$ at $t = 0$ is quite strong, although $a(t)$ is still integrable at $t = 0$; in fact, $a(t) \sim 1/(t \ln^2 t)$ as $t \to 0$ can be shown.

2.4 Spatial Regularity

For the special case $a(t) \equiv 1$, an analytic resolvent for (2.1) becomes the analytic semigroup $S(t) = e^{At}$ generated by A. The semigroup property then yields the relations

$$S^{(n)}(t) = A^n S(t), \quad t > 0, \ n \in \mathbb{N},$$

and therefore we have $\mathcal{R}(S(t)) \subset \bigcap_{n=1}^\infty \mathcal{D}(A^n)$ for $t > 0$, as well as the estimates

$$|A^n S(t)| = |S^{(n)}(t)| \leq Mn! e^{\omega t(1+\alpha)}(\alpha t)^{-n}, \quad t > 0, \ n \in \mathbb{N},$$

by Corollary 2.1. For the general case, we still obtain $\mathcal{R}(S(t)) \subset \mathcal{D}(A)$, but nothing more, in general, as the next theorem shows.

Theorem 2.2 *Suppose $S(t)$ is an analytic resolvent for (2.1) of type (ω_0, θ_0). Then (i) $\mathcal{R}(S(t)) \subset \mathcal{D}(A)$ for all $t > 0$, and for each $\omega > \omega_0$, $\theta < \theta_0$, there are constants c, $C > 0$ and $\beta \in (\frac{\pi}{2\theta_0}, \frac{\pi}{2\theta})$ such that*

$$|AS(t)| \leq Ce^{\omega t}(t^{\beta} + e^{c/t^{\beta}}), \quad t > 0. \tag{2.14}$$

(ii) If in addition, there are $\omega > \omega_0$, $0 < \theta < \theta_0$, $c > 0$, and $\alpha > 0$ such that

$$|\widehat{a}(\lambda)| \geq c(|\lambda - \omega|^{\alpha} + 1)^{-1} \text{ for all } \lambda \in \Sigma(\omega, \pi/2 + \theta), \tag{2.15}$$

then there is a constant $C > 0$ such that

$$|AS(t)| \leq Ce^{\omega t}(1 + t^{-\alpha}), \quad t > 0. \tag{2.16}$$

(iii) If (2.15) holds and if there is a set $E \subset \mathbb{R}_+$ which is not discrete and such that $A^2 S(t)$ is bounded for each $t \in E$, then $a(t) \equiv a_0$, and $a_0 A$ generates an analytic semigroup $T(t)$.

The proof of (i) is based on the following Lemma on holomorphic functions on the unit disc $\mathbb{D} = \{z \in \mathbb{C} : |z| < 1\}$; although the proof is standard it is included here for the sake of completeness.

Lemma 2.1 *Suppose $\phi_1(z), \phi_2(z)$ are bounded holomorphic functions on the open unit disk $\mathbb{D} \subset \mathbb{C}$ such that $\phi_2(z) \neq 0$ on $\mathbb{D}$. Then there is a constant $c > 0$ such that $\varphi(z) = \phi_1(z)/\phi_2(z)$ satisfies*

$$|\varphi(z)| \leq \exp(c \cdot \frac{1 + |z|}{1 - |z|}) \text{ for all } z \in \mathbb{D}. \tag{2.17}$$

Proof: By the canonial factorization theorem for bounded analytic functions (cp. Duren [106], Theorem 2.8), there are real numbers γ_j, functions $\mu_j(t) \geq 0$ with $\log \mu_j(\cdot) \in L^1(0, 2\pi)$, singular nonnegative measures ν_j on $[0, 2\pi]$, and Blaschke-products $b_j(z)$ such that

$$\phi_j(z) = e^{i\gamma_j} \cdot b_j(z) \cdot \exp(\frac{1}{2\pi} \int_0^{2\pi} \frac{e^{it} + z}{e^{it} - z} \log \mu_j(t) dt) \cdot \exp(\int_0^{2\pi} \frac{e^{it} + z}{e^{it} - z} d\nu_j(t))$$

holds for $j = 1, 2$. Since $\phi_2(z) \neq 0$ on $\mathbb{D}$ we have $b_2(z) \equiv 1$; on the other hand, $|b_1(z)| \leq 1$ and with $z = re^{i\varphi}$

$$0 \leq \text{Re} \frac{e^{it} + z}{e^{it} - z} = \frac{1 - r^2}{1 - 2r\cos(\varphi - t) + r^2} \leq \frac{1 - r^2}{(1 - r)^2} = \frac{1 + r}{1 - r} = \frac{1 + |z|}{1 - |z|}.$$

From this we obtain (2.17) with

$$c = \ \text{Var} \ \nu_1|_0^{2\pi} + \ \text{Var} \ \nu_2|_0^{2\pi} + |\log \mu_1|_1 + |\log \mu_2|_1. \quad \square$$

Proof of Theorem 2.2: (i) By means of Lemma 2.1 we first derive an estimate for $1/\widehat{a}(\lambda)$ on a sector $\Sigma(\omega, \frac{\pi}{2} + \theta)$, where $\omega > \omega_0$, $\theta < \theta_0$, are fixed but arbitrary otherwise. Choose $\beta = \frac{\pi}{2\theta'}$, with $\theta < \theta' < \theta_0$, $\omega > \omega' > \omega_0$, and consider the function

$$\lambda(z) = \omega' + (\frac{1+z}{1-z})^p, \quad z \in \mathbb{D}, \tag{2.18}$$

where $p = 1 + 1/\beta$. λ maps the unit disk $\mathbb{D}$ conformally onto the sector $\Sigma(\omega', \frac{\pi}{2} + \theta')$. $\widehat{a}(\lambda)$ has only finitely many poles within this sector, since $\widehat{a}(\lambda)$ is bounded as $|\lambda| \to \infty$, $\lambda \in \Sigma(\omega', \frac{\pi}{2} + \theta')$, and therefore $\widehat{a}(\lambda)$ can be factored according to

$$\widehat{a}(\lambda) = \frac{(\lambda - \omega_0)^m}{\prod_{j=1}^n (\lambda - \lambda_j)^{p_j}} \cdot \chi_2(\lambda) = \chi_2(\lambda)/\chi_1(\lambda),$$

where the λ_j denote the poles of $\widehat{a}(\lambda)$ within $\Sigma(\omega', \frac{\pi}{2} + \theta')$, p_j their orders, and $m = \sum_j p_j$. Obviously, $\chi_1(\lambda)$ and $\chi_2(\lambda)$ are both holomorphic on $\Sigma(\omega', \pi/2 + \theta')$, $\chi_1(\lambda)$ is bounded and has the zeros λ_j, whereas $\chi_2(\lambda)$ is nonvanishing and bounded, too. Let $\psi_j(z) = \chi_j(\lambda(z))$; then Lemma 2.1 yields the estimate

$$\left|\frac{1}{\widehat{a}(\lambda)}\right| = |\varphi(z)| = \left|\frac{\psi_1(z)}{\psi_2(z)}\right| \leq \exp(c \cdot \frac{1+|z|}{1-|z|}) \quad \text{on } \mathbb{D},$$

for some $c > 0$. Defining $\mu = \frac{1+z}{1-z}$ we obtain from (2.18) $\mu = (\lambda - \omega')^{1/p}$; on the other hand

$$\frac{1+|z|}{1-|z|} = \frac{|\mu+1| + |\mu-1|}{|\mu+1| - |\mu-1|} = \frac{(|\mu+1| + |\mu-1|)^2}{|\mu+1|^2 - |\mu-1|^2} \leq \frac{(1+|\mu|)^2}{\text{Re}\,\mu},$$

and so with $\alpha = \frac{1}{p}$

$$|\mu| = |\lambda - \omega'|^\alpha, \quad \text{Re}\,\mu = \text{Re}(\lambda - \omega')^\alpha = |\lambda - \omega'|^\alpha \cos(\arg(\lambda - \omega')\alpha).$$

Since for $\lambda \in \Sigma(\omega', \frac{\pi}{2} + \theta)$

$$|\arg(\lambda - \omega')|\alpha \leq (\frac{\pi}{2} + \theta)/p < \frac{\pi}{2}(1 + 1/\beta)/(1 + 1/\beta) = \frac{\pi}{2},$$

these estimates yield

$$\left|\frac{1}{\widehat{a}(\lambda)}\right| \ \leq \ \exp(c \cdot \frac{(1 + |\lambda - \omega'|^\alpha)^2}{|\lambda - \omega'|^\alpha})$$

$$\leq \ \exp(c(|\lambda - \omega|^\alpha + 1)), \quad \lambda \in \Sigma(\omega, \frac{\pi}{2} + \theta). \tag{2.19}$$

Next, we use representation (2.10) with $R = t^{-(1-\alpha)^{-1}}$ to show (2.14). In fact

$$
\begin{aligned}
AS(t) &= (2\pi i)^{-1} \int_{\Gamma_R} e^{\lambda t} AH(\lambda) d\lambda \\
&= (2\pi i)^{-1} \int_{\Gamma_R} e^{\lambda t} [H(\lambda) - 1/\lambda] \frac{1}{\widehat{a}(\lambda)} d\lambda,
\end{aligned}
\tag{2.20}
$$

hence

$$
\begin{aligned}
|AS(t)| &\leq C \int_{\Gamma_R} e^{\operatorname{Re}\lambda t} e^{c|\lambda-\omega|^\alpha} \frac{|d\lambda|}{|\lambda-\omega|} \\
&\leq Ce^{\omega t} \int_R^\infty e^{-rt\sin\theta} e^{r^\alpha c} \frac{dr}{r} + Ce^{\omega t} \int_0^{\frac{\pi}{2}+\theta} e^{Rt\cos\varphi} e^{cR^\alpha} d\varphi \\
&\leq Ce^{\omega t}(t^\beta + e^{ct^\beta});
\end{aligned}
$$

in particular $\mathcal{R}(S(t)) \subset \mathcal{D}(A)$ for each $t > 0$ and (2.14) holds.

(ii) If (2.15) holds, then we obtain from (2.20) with $R = 1/t$

$$
\begin{aligned}
|AS(t)| &\leq C \int_{\Gamma_R} e^{\operatorname{Re}\lambda t} (1 + |\lambda-\omega|^\alpha) \frac{|d\lambda|}{|\lambda-\omega|} \\
&\leq Ce^{\omega t} \int_R^\infty e^{-rt\sin\theta} (1 + r^\alpha) \frac{dr}{r} + Ce^{\omega t} \int_0^{\pi/2+\theta} e^{Rt\cos\varphi} (1 + R^\alpha) d\varphi \\
&\leq Ce^{\omega t}(1 + t^{-\alpha}), \quad t > 0,
\end{aligned}
$$

and so (2.16) follows.

(iii) Suppose $A^2 S(t)$ is bounded for $t \in E$, where $E \subset \mathbb{R}_+$ is nondiscrete. Define the function $c(z)$ by means of

$$
c(z) = \frac{1}{2\pi i} \int_{\Gamma_0} e^{\lambda z} [\lambda \widehat{a}(\lambda)]^{-1} d\lambda, \quad |\arg z| \leq \theta < \theta_0;
$$

$c(z)$ is holomorphic in the sector $\Sigma(0, \theta)$ and from (2.20) we obtain $c(t) = 0$ for each $t \in E$, since A is unbounded by assumption. Since E is nondiscrete this implies $c(z) \equiv 0$, by uniqueness of holomorphic functions.

Similarly, for some $n > \alpha - 2$, we define $b(z)$ by means of

$$
b(z) = \frac{1}{2\pi i} \int_{\Gamma_0} e^{\lambda z} [\lambda^{n+2} \widehat{a}(\lambda)]^{-1} d\lambda, \quad |\arg z| \leq \theta < \theta_0;
$$

$b(z)$ is holomorphic on $\Sigma(0, \theta)$, and we have $b^{(n+1)}(z) = c(z) = 0$ on this sector, i.e. $b(z)$ is a polynomial of order at most n. By the choice of n, the Laplace transform of $b(t)$ becomes

$$
\widehat{b}(\lambda) = [\lambda^{(n+2)} \widehat{a}(\lambda)]^{-1}, \quad \lambda > \omega;
$$

use Fubini's theorem and the Cauchy integral formula. On the other hand, since $b(t)$ is a polynomial of order at most n, we have

$$
\widehat{b}(\lambda) = \sum_0^n b_{n-j} \lambda^{-j-1},
$$

for some constants $b_k \in \mathbb{C}$. Therefore, we obtain the representation

$$\frac{1}{\widehat{a}(\lambda)} = \sum_0^n b_j \lambda^{j+1},$$

i.e. the function $1/\widehat{a}(\lambda)$ is a polynomial, too. If the degree of this polynomial is greater than one, then it maps the sets $\Sigma(\omega, \pi/2 + \theta) \cap B_R^c(0)$ for large R onto sets containing the complement of a disk. By Theorem 2.1 this implies that $\rho(A)$ contains the complement of a disk, and $|\mu(\mu - A)^{-1}| \leq C$ as $|\mu| \to \infty$; thus A is bounded, thereby contradicting our standing assumption. Therefore $1/\widehat{a}(\lambda) = b_0 \lambda$, or in other words $a(t) \equiv a_0$. But then $S(t) = e^{a_0 At}$, i.e. $a_0 A$ generates an analytic semigroup. The proof is complete. □

2.5 Perturbed Equations

Suppose that (2.1) admits an analytic resolvent $S_0(t)$ and consider the perturbed equation

$$u(t) = f(t) + (a + a * dk) * Au(t) + b * u(t), \quad t \in J, \tag{2.21}$$

where a, A, f are as before, $b \in L^1_{loc}(\mathbb{R}_+)$ and $k \in BV^0_{loc}(\mathbb{R}_+)$. In contrast to Theorem 1.2, no further assumptions are needed to obtain a resolvent for (2.2).

Theorem 2.3 *Suppose (2.1) admits an analytic resolvent $S_0(t)$ of analyticity type (ω_0, θ_0) and let $b \in L^1_{loc}(\mathbb{R}_+)$, $k \in BV^0_{loc}(\mathbb{R}_+)$. Then (2.21) admits an exponentially bounded resolvent $S(t)$.*

Proof: Without loss of generality we may assume that b and dk are Laplace transformable, i.e.

$$\int_0^\infty |dk(t)| e^{-\omega t} < \infty \quad \text{and} \quad \int_0^\infty |b(t)| e^{-\omega t} < \infty;$$

even more, since $k(t)$ does not have a jump at zero, choosing ω large enough we can achieve

$$\int_0^\infty |b(t)| e^{-\omega t} + \int_0^\infty |dk(t)| e^{-\omega t} \leq \eta, \tag{2.22}$$

where $\eta > 0$ is any prescribed small number. The resolvent $S(t)$ for (2.21) is the solution of

$$S(t) = I + Aa * S(t) + dk * Aa * S(t) + b * S(t), \quad t \in \mathbb{R}_+, \tag{2.23}$$

and so its Laplace-transform is given by

$$\widehat{S}(\lambda) = \frac{1}{\lambda}(1 - \widehat{b}(\lambda) - (1 + \widehat{dk}(\lambda))\widehat{a}(\lambda)A)^{-1}.$$

Denote the resolvent kernel of $b(t)$ by $r(t)$, i.e. $r(t)$ is the solution of $r = b + b * r$, and let $l(t)$ be defined by $l = k + 1 * r + k * r$. Rewrite $\widehat{S}(\lambda)$ as follows.

$$
\begin{aligned}
\widehat{S}(\lambda) &= (1 + \widehat{r}(\lambda)) \frac{1}{\lambda} (1 - \widehat{a}(\lambda)A - \widehat{dl}(\lambda)\widehat{a}(\lambda)A)^{-1} \\
&= (1 + \widehat{r}(\lambda))(1 - \widehat{dl}(\lambda)\widehat{a}(\lambda)A(1 - \widehat{a}(\lambda)A)^{-1})^{-1} \frac{1}{\lambda}(1 - \widehat{a}(\lambda)A)^{-1} \\
&= (1 + \widehat{r}(\lambda)) \sum_{n=0}^{\infty} [\widehat{dl}(\lambda)]^n [\widehat{a}(\lambda)A(1 - \widehat{a}(\lambda)A)^{-1}]^n \frac{1}{\lambda}(1 - \widehat{a}(\lambda)A)^{-1} \\
&= (1 + \widehat{r}(\lambda)) \sum_{n=0}^{\infty} [\widehat{dl}(\lambda)]^n H_n(\lambda), \quad \lambda > \omega, \qquad (2.24)
\end{aligned}
$$

where

$$
H_n(\lambda) = \frac{\widehat{a}(\lambda)^n}{\lambda} A^n (1 - \widehat{a}(\lambda)A)^{-(n+1)}, \quad \lambda > \omega, \ n \in \mathbb{N}_0. \qquad (2.25)
$$

We now recover the operator-valued functions $H_n(\lambda)$ as Laplace transforms of functions $S_n(t)$. For this purpose note that from Theorem 2.1 we obtain constants $\omega_1 > \max(0, \omega_0)$ and $\theta \in (0, \theta_0)$ such that

$$
|(1 - \widehat{a}(\lambda)A)^{-1}| \le C \cdot \frac{|\lambda|}{|\lambda - \omega_1|}, \quad \lambda \in \Sigma(\omega_1, \theta)
$$

holds; for $\lambda \in \Sigma(2\omega_1, \theta)$ we therefore have, with a probably different constant C,

$$
|H_0(\lambda)| \le \frac{C}{|\lambda - 2\omega_1|}, \quad |\widehat{a}(\lambda)A(1 - \widehat{a}(\lambda)A)^{-1}| \le C.
$$

Let $\omega > 2\omega_1$ be sufficiently large, and consider the integrals

$$
S_n(t) = (2\pi i)^{-1} \int_{\Gamma_R} e^{\lambda t} H_n(\lambda) d\lambda, \quad t > 0, \qquad (2.26)
$$

where Γ_R denotes again the contour as in Section 2.2. It is easily seen that $S_n(t)$ admits analytic extension to a sector $\Sigma(0, \theta_1)$ and that

$$
|S_n(t)| \le C^n \cdot M e^{\omega t}, \quad t > 0, \ n \in \mathbb{N}_0, \qquad (2.27)
$$

is satisfied. Finally, from (2.24) we obtain the resolvent $S(t)$ for (2.21).

$$
S(t) = (\delta_0 + r) * \sum_{n=0}^{\infty} (dl)^{*n} * S_n(t), \quad t > 0, \qquad (2.28)
$$

and the series converges absolutely and uniformly on bounded intervals. In fact, from (2.22) and by the definition of r and l, we obtain

$$
\begin{aligned}
|dl^{*n} * S_n(t)| &\le M e^{\omega t} \cdot C^n \Big(\int_0^{\infty} |dl(s)| e^{-\omega s} \Big)^n \\
&\le M e^{\omega t} (2\eta C/(1 - \eta))^n \le M 2^{-n} e^{\omega t},
\end{aligned}
$$

provided η is chosen small enough, i.e. ω is large enough. Since $S_n(t)$ is even uniformly continuous for $t > 0$ and commutes with A it follows from uniqueness of the Laplace-transform that $S(t)$ defined by (2.28) is the resolvent for (2.21). $\square$

Note that in the contrast to the functions $S_n(t)$, the resolvent $S(t)$ of (2.21) is not analytic, in general, it even need not be continuous in $\mathcal{B}(X)$, as Example 1.1 shows.

2.6 Maximal Regularity

Suppose $S(t)$ is an analytic resolvent for (2.1), and let $f \in C(J;X)$, where $J = [0,T]$. Since $S(t)$ belongs to $C^1((0,\infty); \mathcal{B}(X))$, a much better behaviour of the variation of parameters formula (1.8) can be expected, which we now employ in the form

$$u(t) = f(t) + \int_0^t \dot{S}(t-s)f(s)ds, \quad t \in J. \tag{2.29}$$

For (2.29) to make sense, continuity of f will in general not be enough, however, if f is Hölder-continuous with exponent $\alpha \in (0,1)$, i.e. $f \in C^\alpha(J;X)$, then $u(t)$ given by (2.29) is well-defined and a mild solution of (2.1).

We denote by $C_0^\alpha(J;X)$ the spaces

$$C_0^\alpha(J;X) = \{f : J \to X : f(0) = 0, \ |f|_\alpha < \infty\}$$

normed by

$$|f|_\alpha = \sup\{|f(t) - f(s)|(t-s)^{-\alpha} : 0 \le s < t \le T\};$$

obviously $C_0^\alpha(J;X)$ is a Banach space. It turns out that the solution operator $f \mapsto u$ is leaving these spaces invariant, (2.1) has the property of *maximal regularity of type C^α*.

Theorem 2.4 *Suppose (2.1) admits an analytic resolvent $S(t)$, let $f \in C(J;X)$, and $u(t)$ be defined by (2.29), where $\alpha \in (0,1)$. Then*
(i) if $f \in C_0^\alpha(J;X)$ then $u \in C_0^\alpha(J;X)$ is a mild solution of (2.1);
(ii) if $f \in C_0^\alpha(J;X_A)$ then $u \in C_0^\alpha(J;X_A)$ is a strong solution of (2.1);
(iii) if $f = a * g$, $g \in C_0^\alpha(J;X)$, then $u \in C_0^\alpha(J;X_A)$ is a strong solution of (2.1).

Proof: (a) Let $f \in C_0^\alpha(J;X)$; rewrite (2.29) as

$$u(t) = S(t)f(t) + \int_0^t S'(t-s)(f(s) - f(t))ds, \quad t \in J, \tag{2.30}$$

and estimate according to

$$\begin{aligned}
|u(t)| &\le |S(t)||f(t)| + \int_0^t |S'(t-s)||f(s) - f(t)|ds \\
&\le t^\alpha M_T |f|_\alpha + \int_0^t M_T(t-s)^{-1} \cdot |f|_\alpha(t-s)^\alpha ds \\
&\le t^\alpha M_T |f|_\alpha + M_T |f|_\alpha \cdot t^\alpha/\alpha,
\end{aligned}$$

where $M_T = \max\{|S(t)|, t|S'(t)| : t \in J\}$. This estimate shows that $u(t)$ defined by (2.29), or equivalently by (2.30) exists, is bounded on J, and satisfies $u(0) = 0$.

(b) To see that $u(t)$ is again Hölder-continuous let $t, \bar{t} \in J, h = t - \bar{t} > 0$, and decompose as follows.

$$
\begin{aligned}
u(t) - u(\bar{t}) &= (S(t) - S(\bar{t}))f(\bar{t}) + S(h)(f(t) - f(\bar{t})) \\
&\quad + \int_{\bar{t}}^{t} S'(t - \tau)(f(\tau) - f(t))d\tau \\
&\quad + \int_{0}^{\bar{t}} (S'(h + \tau) - S'(\tau))(f(\bar{t} - \tau) - f(\bar{t}))d\tau \\
&= I_1 + I_2 + I_3 + I_4.
\end{aligned}
$$

The integrals I_j are estimated as follows.

$$
\begin{aligned}
|I_1| &\leq |S(t) - S(\bar{t})||f(\bar{t})| \leq M_T \log(t/\bar{t}) \cdot |f|_\alpha \cdot \bar{t}^\alpha \\
&\leq M_T |f|_\alpha h^\alpha / \alpha,
\end{aligned}
$$

where we used the elementary inequality

$$
\log(1 + \rho) \leq \rho^\alpha / \alpha \quad \text{for all } \rho > 0, \ \alpha \in (0, 1),
$$

and $f(0) = 0$.

$$
|I_2| \leq |S(h)||f(t) - f(\bar{t})| \leq M_T |f|_\alpha h^\alpha,
$$

since $f \in C_0^\alpha(J; X)$.

$$
|I_3| \leq M_T |f|_\alpha \int_{\bar{t}}^{t} (t - \tau)^{\alpha - 1} d\tau = M_T |f|_\alpha h^\alpha / \alpha.
$$

To estimate I_4 we observe that from Corollary 2.1 we have

$$
|S''(t)| \leq M_T t^{-2}, \quad t \in J,
$$

where M_T is large enough, hence

$$
|S'(t + h) - S'(t)| \leq \int_{t}^{t+h} |S''(\tau)|d\tau \leq M_T \cdot \frac{h}{t(t + h)}
$$

for all $t, t + h \in J$. This yields

$$
\begin{aligned}
|I_4| &\leq M_T |f|_\alpha \int_{0}^{\bar{t}} h \tau^{\alpha - 1} (\tau + h)^{-1} d\tau = M_T |f|_\alpha h^\alpha \int_{0}^{\bar{t}/h} \tau^{\alpha - 1} (\tau + 1)^{-1} d\tau \\
&\leq M_T |f|_\alpha h^\alpha \int_{0}^{\infty} \tau^{\alpha - 1} (\tau + 1)^{-1} d\tau = M_T |f|_\alpha h^\alpha \pi / \sin(\alpha \pi).
\end{aligned}
$$

Thus $u \in C_0^\alpha(J; X)$ and for some constant depending only on J we have the estimate

$$
|u|_\alpha \leq C |f|_\alpha / \sin(\alpha \pi). \tag{2.31}
$$

(c) We next show that $u(t)$ is a mild solution of (2.1); this follows from a simple approximation argument. Let $\varphi_\varepsilon \in C_0^\infty(0, \infty)$ be such that supp $\varphi_\varepsilon \subset (0, \varepsilon), \varphi_\varepsilon(t) \geq 0$ on $\mathbb{R}_+$, $\int_0^\infty \varphi_\varepsilon(t)dt = 1$, and define $f_\varepsilon = \varphi_\varepsilon * f$. Then $f_\varepsilon \in C^1(J; X)$ and so $u_\varepsilon = \varphi_\varepsilon * u$ is the mild solution of (2.1) with f replaced by f_ε. Since $f_\varepsilon \to f$ and $u_\varepsilon \to u$, hence also $a * u_\varepsilon \to a * u$ in $C(J; X)$ we obtain $Aa * u_\varepsilon = u_\varepsilon - f_\varepsilon \to u - f$, and from closedness of A there follows $a * u \in C(J; X_A)$ and $Aa * u = u - f$, i.e. u is a mild solution (2.1).

(d) Part (ii) follows directly from (i) since A commutes with the resolvent $S(t)$.

(e) Let $v \in C_0^\alpha(J; X)$ defined by (2.29) with f replaced by g denote the mild solution of

$$v = g + Aa * v; \tag{2.32}$$

and let u be defined by (2.29); then $u = a * v$ belongs to $C^\alpha(J; X_A)$ and convolving (2.32) with a we see that u is a strong solution of (2.1). $\square$

Corollary 2.5 *Under the assumptions of Theorem 2.4, there is a constant $C > 0$ depending only on T, A and $a(t)$ such that $u(t)$ defined by (2.29) satisfies the following estimates.*
(i) $|u|_\alpha \leq C|f|_\alpha / \sin(\pi\alpha)$, $f \in C_0^\alpha(J; X)$;
(ii) $|Au|_\alpha \leq C|Af|_\alpha / \sin(\pi\alpha)$, $f \in C_0^\alpha(J; X_A)$;
(iii) $|Au|_\alpha \leq C|g|_\alpha / \sin(\pi\alpha)$, $f = a * g, g \in C_0^\alpha(J; X)$.

Note that the case $f(0) = x \neq 0$ can be reduced to $f(0) = 0$; rewrite (2.29) as

$$u(t) = (f(t) - x) + \int_0^t S'(t - \tau)(f(\tau) - x)d\tau + S(t)x. \tag{2.33}$$

Thus if $f \in C^\alpha(J; X)$ with $f(0) = x \neq 0$ it is not always true that $u \in C^\alpha(J; X)$ holds. However, $u(t)$ given by (2.33) still is a mild solution of (2.1) for each $x \in X$. From Theorem 2.4 and the properties of $S(t)$ we have

Corollary 2.6 *Let $S(t)$ be an analytic resolvent for (2.1), let $u(t)$ be defined by (2.33) and $\alpha \in (0, 1)$. Then*
(i) if $f \in C^\alpha(J; X)$ then u is a mild solution of (2.1);
(ii) if $f \in C^\alpha(J; X_A)$ then u is a strong solution of (2.1);
(iii) if $f = x + a * g$ with $x \in \mathcal{D}(A)$ and $g \in C^\alpha(J; X)$ then u is a strong solution of (2.1).

In Section 3.5 it will be shown that Theorem 2.4 and Corollaries 2.5, 2.6 remain valid for the perturbed equation (2.21). This is at first sight somewhat surprising since the resolvent of (2.21) does *not* satisfy estimates (2.3) for $n = 1, 2$ which were used in the proof of Theorem 2.4.

2.7 Comments
a) Analytic resolvents have been introduced by Da Prato and Iannelli [65] who

also proved a weaker version of the sufficiency part of Theorem 2.1; see also Grimmer and Pritchard [144], and Da Prato and Iannelli [68]. The neccessity part of Theorem 2.1 is new as is Corollary 2.3 showing the restriction on the kernel $a(t)$. Examples 2.1, 2.2, and 2.3 are standard, although Example 2.3 is probably not so well-known.

b) The second part of Theorem 2.2 in a somewhat weaker form is due to Da Prato, Iannelli and Sinestrari [70]; parts (i) and (iii) of Theorem 2.2 seem to be new. A counterexample of Desch and Prüss [91] shows that, in the setting of Theorem 2.2, in general (2.15) does not hold, and that the estimate (2.19) leading to (2.14) is optimal. On the other hand, if $\sigma(A)$ contains a halfray $[r, \infty)e^{i\varphi}$ or maybe $\arg \hat{a}(\lambda)$ is bounded by some $\theta \in (0, \infty)$ then (2.15) is valid, as Proposition 3.2 and Lemma 8.1 show.

c) The perturbation result in Section 2.5 seems to be new; a much more general version is presented in Chapter II. It is also not difficult to show that perturbations $A + B$ of the involved operator A are possible as in the case of analytic semigroups. If B satisfies $\mathcal{D}(B) \supset \mathcal{D}(A)$, and for each $\varepsilon > 0$ there is $C(\varepsilon) > 0$ such that

$$|Bx| \leq \varepsilon |Ax| + C(\varepsilon)|x|, \quad x \in \mathcal{D}(A),$$

holds, then the $(a, A + B)$ generates an analytic resolvent whenever (a, A) does so, and the angle θ_0 is the same, only the growth bound ω_0 possibly changes. This follows from simple estimates in frequency domain, which employ Theorem 2.1 and $\hat{a}(\lambda) \to 0$ as $|\lambda| \to \infty$, in $\Sigma(\omega, \theta)$, $\theta < \pi/2 + \theta_0$, $\omega > \omega_0$.

d) It is not clear in general whether an analytic resolvent $S(t)$ for (2.1) is differentiable in the sense of Definition 1.4, i.e. whether the integral resolvent $R(t)$ for (2.1) also exists. This is somewhat surprising since $S(t)$ is of course continuously differentiable on $(0, \infty)$; however, it does not seem to be possible to show $R \in L^1_{loc}(\mathbb{R}_+, \mathcal{B}(X))$, unless more is assumed on the behaviour of $a(z)$ near zero, or of $\hat{a}(\lambda)$ near infinity. For example, a condition like

$$\int_1^\infty |\hat{a}(\omega + re^{\pm i\theta})| \frac{dr}{r} < \infty, \quad \text{for some } \theta \in (\pi/2, \pi), \ \omega > \omega_0,$$

is sufficient for integrability of $R(t)$ near zero, $-\int_0^1 (\log t)a(t)dt < \infty$ is another one. These are, however, not necessary, as Example 2.3 shows.

e) The results in Section 2.6 have been first obtained by Da Prato, Iannelli, and Sinestrari [69] for the case $a \in BV_{loc}(\mathbb{R}_+)$. There are many other function classes for which maximal regularity results like Theorem 2.4 hold. This subject will be taken up in Sections 7 and 8 again.

f) It is instructive to have a look at the fundamental solution $\phi_\alpha(t, x)$ of the problem

$$u(t, x) = f(t, x) + \Gamma(\alpha)^{-1} \int_0^t (t - s)^{\alpha - 1} u_{xx}(s, x)ds, \quad t \geq 0, \ x \in \mathbb{R}, \quad (2.34)$$

where $\alpha \in (0, 2)$; cp. Example 2.1 in Section 2.3. The fundamental solution represents the resolvent $S(t)$ according to

$$(S(t)v)(x) = \int_{-\infty}^{\infty} \phi_\alpha(t, x - y)v(y)dy, \quad t \geq 0, \ x \in \mathbb{R},$$

and is well-known for $\alpha = 1$, and for the limiting cases $\alpha = 0$ and $\alpha = 2$.

$$\phi_0(t, x) = e^{-|x|}/2, \quad t > 0, \ x \in \mathbb{R};$$
$$\phi_1(t, x) = e^{-x^2/4t}/\sqrt{4\pi t}, \quad t > 0, \ x \in \mathbb{R};$$
$$\phi_2(t, x) = (\delta_0(t - x) + \delta_0(t + x))/2, \quad t > 0, \ x \in \mathbb{R}.$$

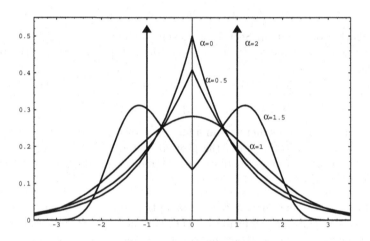

Figure 2.1: Fundamental solutions $\phi_\alpha(1, x)$ of (2.34)

For general $\alpha \in (0, 2)$ it is possible to obtain a representation of ϕ_α in terms of the Mittag-Leffler functions $ML_\beta(z)$ defined by

$$ML_\beta(z) = \sum_{k=0}^{\infty} \frac{z^k}{\Gamma(k\beta + 1)}, \quad z \in \mathbb{C}.$$

This was first observed in Friedman [121] and recently studied in detail by Fujita [124]. The identity

$$\widetilde{\widehat{\phi_\alpha}}(\lambda, \xi) = \widehat{s}(\lambda, \xi^2) = (1 + \lambda^{-\alpha}\xi^2)^{-1}/\lambda$$

implies

$$s(t, \xi^2) = \widetilde{\phi_\alpha}(t, \xi) = \sum_{k=0}^{\infty} (-\xi)^{2k} \frac{t^{k\alpha}}{\Gamma(k\alpha + 1)} = ML_\alpha(-\xi^2 t^\alpha), \quad t > 0, \ x \in \mathbb{R};$$

thus $\phi_\alpha(t, x)$ can be obtained as the inverse Fourier transform of $s(t, \xi^2)$. Note that with $\varphi_\alpha(x) = \phi_\alpha(1, x)$ we have

$$\phi_\alpha(t, x) = t^{-\alpha/2}\varphi_\alpha(xt^{-\alpha/2}), \quad t > 0, \ x \in \mathbb{R},$$

and

$$\varphi_\alpha(x) = (2\pi)^{-1} \int_{-\infty}^{\infty} e^{ix\xi} ML_\alpha(-\xi^2)d\xi, \quad x \in \mathbb{R}.$$

Figure 2.1 shows the graphs of $\phi_\alpha(1, x)$ for different values of α. Note that except for $\alpha = 2$, the support of $\phi_\alpha(t, \cdot)$ is the whole line; therefore the propagation speed is always infinite, except for $\alpha = 2$. For $\alpha \in (0, 1)$ the maximum stays at $x = 0$, but as α is increasing beyond 1, this maximum splits into two which are sharpening and moving in different direction to ± 1 as α approaches 2. Thus for $\alpha \in (1, 2)$ there is a sort of 'main speed', this way resembling some properties of the wave equation.

3 Parabolic Equations

As a continuation of Section 2, the concept of parabolicity for Volterra equations of scalar type is introduced and resolvents for such equations are discussed in detail. If the kernel $a(t)$ has some extra regularity property, like convexity, then the resolvent exists, and exhibits the same stability under perturbations as analytic resolvents. Again the maximal regularity property of type C^α is valid, even for the perturbed equation. In Section 3.6 we derive a representation formula for the resolvent in case A is the generator of a C_0-semigroup.

3.1 Parabolicity

Let X be a complex Banach space, A a closed linear operator in X with dense domain $\mathcal{D}(A)$, and let $a \in L^1_{loc}(\mathbb{R}_+)$ be of subexponential growth, which means $\int_0^\infty e^{-\varepsilon t}|a(t)|dt < \infty$ for all $\varepsilon > 0$. We consider the Volterra equation of scalar type

$$u(t) = \int_0^t a(t - \tau)Au(\tau)d\tau + f(t), \quad t \geq 0. \tag{3.1}$$

If (3.1) admits a bounded resolvent $S(t)$, then by Corollary 1.6 we have $\hat{a}(\lambda) \in \rho(A)$ for all $\lambda \in \mathbb{C}$ with $\operatorname{Re}\lambda > 0$, and $H(\lambda) = (I - \hat{a}(\lambda)A)^{-1}/\lambda$ satisfies

$$|H(\lambda)| \leq \frac{M}{\operatorname{Re}\lambda}, \quad \operatorname{Re}\lambda > 0, \tag{3.2}$$

for some constant $M \geq 1$. If $S(t)$ is even analytic then by Theorem 2.1 we even have

$$|H(\lambda)| \leq \frac{M}{|\lambda|}, \quad |\arg\lambda| \leq \frac{\pi}{2} + \theta, \tag{3.3}$$

for some $\theta > 0$. The following concept is between these two extremes.

Definition 3.1 *Equation (3.1) is called* **parabolic**, *if the following conditions hold.*
(P1) $\hat{a}(\lambda) \neq 0$, $1/\hat{a}(\lambda) \in \rho(A)$ *for all* $\operatorname{Re}\lambda > 0$.
(P2) *There is a constant* $M \geq 1$ *such that* $H(\lambda) = (I - \hat{a}(\lambda)A)^{-1}/\lambda$ *satisfies*

$$|H(\lambda)| \leq \frac{M}{|\lambda|} \quad \text{for all } \operatorname{Re}\lambda > 0. \tag{3.4}$$

Thus if (3.1) admits an analytic resolvent $S(t)$ which is bounded on some sector $\Sigma(0, \theta)$ then (3.1) is parabolic; the converse is of course not true. The main difference between (3.3) and parabolicity of (3.1) is that the existence of an analytic continuation of $\hat{a}(\lambda)$ to some sector $\Sigma(0, \theta + \pi/2)$ is no longer required; recall that (3.3) implies analyticity of $a(z)$ on $\Sigma(0, \theta)$ as we have seen in Corollary 2.3.

Definition 3.2 *Let* $a \in L^1_{loc}(\mathbb{R}_+)$ *be of subexponential growth and suppose* $\hat{a}(\lambda) \neq 0$ *for all* $\operatorname{Re}\lambda > 0$. *$a$ is called* **sectorial with angle** $\theta > 0$ *(or merely θ-sectorial) if*

$$|\arg\hat{a}(\lambda)| \leq \theta \quad \text{for all } \operatorname{Re}\lambda > 0. \tag{3.5}$$

Here $\arg \hat{a}(\lambda)$ is defined as the imaginary part of a fixed branch of $\log \hat{a}(\lambda)$, and θ in (3.5) need not be less than π. If a is sectorial, we always choose that branch of $\log \hat{a}(\lambda)$ which gives the smallest angle θ; in particular, in case $\hat{a}(\lambda)$ is real for real λ we choose the principal branch.

A standard situation leading to parabolic equations is described in

Proposition 3.1 *Let $a \in L^1_{loc}(\mathbb{R}_+)$ be θ-sectorial for some $\theta < \pi$, suppose A is closed linear densely defined, such that $\rho(A) \supset \Sigma(0, \theta)$, and*

$$|(\mu - A)^{-1}| \leq \frac{M}{|\mu|} \quad \text{for all } \mu \in \Sigma(0, \theta). \tag{3.6}$$

Then (3.1) is parabolic.

Proof: Since $1/\hat{a}(\lambda) \in \overline{\Sigma}(0, \theta)$ for all $\mathrm{Re}\, \lambda > 0, H(\lambda) = (I - \hat{a}(\lambda)A)^{-1}/\lambda$ is well-defined, in particular (P1) holds; (3.6) yields (P2). $\quad \square$

Since $a(t)$ is $\frac{\pi}{2}$-sectorial if and only if $a(t)$ is of positive type, we obtain the following nice class of parabolic equations.

Corollary 3.1 *Let $a \in L^1_{loc}(\mathbb{R}_+)$ be of subexponential growth and of positive type, and let A generate a bounded analytic C_0-semigroup in X. Then (3.1) is parabolic.*

In the next subsections we show under some additional regularity assumption on the kernel $a(t)$ that parabolic equations admit bounded resolvents which enjoy some further smoothness properties. It will be shown that the maximal regularity result Theorem 2.4 remains true for parabolic equations and that strong perturbation results similar to Theorem 2.3 also hold. For these purposes we first need to discuss the kind of regularity of the kernels which is appropriate for parabolic problems.

3.2 Regular Kernels

The notion of regularity needed below is introduced in

Definition 3.3 *Let $a \in L^1_{loc}(\mathbb{R}_+)$ be of subexponential growth and $k \in \mathbb{N}$. $a(t)$ is called k-regular if there is a constant $c > 0$ such that*

$$|\lambda^n \hat{a}^{(n)}(\lambda)| \leq c|\hat{a}(\lambda)| \quad \text{for all } \mathrm{Re}\, \lambda > 0, \ 0 \leq n \leq k. \tag{3.7}$$

Observe that any k-regular kernel $a(t)$, $k \geq 1$ has the property that $\hat{a}(\lambda)$ has no zeros in the open right halfplane. In fact, if $\hat{a}(\lambda_0) = 0$ then $\hat{a}(\lambda) = (\lambda - \lambda_0)^m \varphi(\lambda)$ for some holomorphic function $\varphi(\lambda)$ with $\varphi(\lambda_0) \neq 0$; differentiation yields

$$m(\lambda - \lambda_0)^{m-1} \varphi(\lambda) = \hat{a}'(\lambda) - (\lambda - \lambda_0)^m \varphi'(\lambda) = [(\hat{a}'(\lambda)/\hat{a}(\lambda))\varphi(\lambda) - \varphi'(\lambda)](\lambda - \lambda_0)^m,$$

hence (3.7) and $\varphi(\lambda_0) \neq 0$ yield a contradiction.

Convolutions of k-regular kernels are again k-regular; this follows easily from the product rule of differentiation. Moreover, integration and differentation are operations which preserve k-regularity as well. On the other hand, sums and differences of k-regular kernels need not be k-regular; consider $a(t) = 1$ and $b(t) = t^2$ to see this. However, if $a(t)$ and $b(t)$ are k-regular and

$$|\arg \hat{a}(\lambda) - \arg \hat{b}(\lambda)| \le \theta < \pi, \quad \mathrm{Re}\,\lambda > 0$$

then $a(t) + b(t)$ is k-regular as well.

It is not difficult to verify that (3.7) is equivalent to

$$|(\lambda^n \hat{a}(\lambda))^{(n)}| \le c'|\hat{a}(\lambda)|, \quad \mathrm{Re}\,\lambda > 0, \ 0 \le n \le k. \tag{3.8}$$

and to

$$|\lambda^n (\log \hat{a}(\lambda))^{(n)}| \le c'', \quad \mathrm{Re}\,\lambda > 0, \ 0 \le n \le k. \tag{3.9}$$

where $c', c'' > 0$ are also constants.

If $a(t)$ is real-valued and 1-regular then $a(t)$ is sectorial. Indeed,

$$\arg \hat{a}(re^{i\phi}) = \mathrm{Im} \log \hat{a}(re^{i\phi}) = \mathrm{Im} \int_0^\phi \frac{d}{dt} \log \hat{a}(re^{it}) dt = \mathrm{Im} \int_0^\phi \frac{\hat{a}'(re^{it}) i r e^{it}}{\hat{a}(re^{it})} dt,$$

hence

$$|\arg \hat{a}(re^{i\phi})| \le c\phi \le \frac{c\pi}{2},$$

i.e. $a(t)$ is $c\frac{\pi}{2}$-sectorial with c from (3.7). The converse of this is not true as the example $a(t) = 1$ for $t \in [0,1], a(t) = 0$ for $t > 1$ shows. In fact, then $\hat{a}(\lambda) = (1 - e^{-\lambda})/\lambda$, hence a is π-sectorial but

$$\frac{-\lambda \hat{a}'(\lambda)}{\hat{a}(\lambda)} = 1 - \frac{\lambda}{e^\lambda - 1}, \quad \mathrm{Re}\,\lambda > 0,$$

shows that this function is not bounded in the right halfplane.

We shall make use of

Lemma 3.1 *Suppose $g : \mathbb{C}_+ \to \mathbb{C}$ is holomorphic and satisfies $g(\lambda) \ne 0$, and $|\arg g(\lambda)| \le \theta$ for $\mathrm{Re}\,\lambda > 0$. Then for each $n \in \mathbb{N}$ there is a constant $c_n > 0$ depending only on n such that*

$$(\mathrm{Re}\,\lambda)^n |g^{(n)}(\lambda)| \le c_n \theta |g(\lambda)|, \quad \mathrm{Re}\,\lambda > 0. \tag{3.10}$$

Proof: Let $u(\lambda) = \arg g(\lambda)$. The Poisson formula for the halfplane and its analytic completion yield with some constant $\gamma \in \mathbb{R}$

$$\log g(\lambda) = \gamma + \frac{i}{\pi} \int_{-\infty}^\infty [\frac{1 - i\rho\lambda}{\lambda - i\rho}] u(i\rho) \frac{d\rho}{1 + \rho^2},$$

hence by differentiation

$$(-1)^n (\log g(\lambda))^{(n)} = i\frac{n!}{\pi} \int_{-\infty}^{\infty} \frac{1}{(\lambda - i\rho)^{n+1}} u(i\rho) d\rho, \quad \text{Re } \lambda > 0.$$

This identity implies for $\lambda = \sigma + i\tau$

$$\sigma^n |(\log g(\lambda))^{(n)}| \leq \frac{n!}{\pi} \int_{-\infty}^{\infty} \frac{\sigma^n}{(\sigma^2 + (\tau - \rho)^2)^{\frac{n+1}{2}}} |u(i\rho)| d\rho$$

$$\leq \theta \frac{n!}{\pi} \int_{-\infty}^{\infty} (1 + \rho^2)^{-\frac{(n+1)}{2}} d\rho,$$

and from this (3.10) follows. $\quad\square$

By means of Lemma 3.1 we can now show that kernels which are sectorial on a sector larger than the right halfplane are k-regular, for every $k \in \mathbb{N}$.

Proposition 3.2 *Suppose $a \in L^1_{loc}(\mathbb{R}_+)$ is such that $\hat{a}(\lambda)$ admits analytic extension to $\Sigma(0, \phi)$, where $\phi > \frac{\pi}{2}$, and there is $\theta \in (0, \infty)$ such that*

$$|\arg \hat{a}(\lambda)| \leq \theta \quad \text{for all} \quad \lambda \in \Sigma(0, \phi). \tag{3.11}$$

Then $a(t)$ is k-regular for every $k \in \mathbb{N}$.

Proof: Let $g : \mathbb{C}_+ \to \mathbb{C}$ be defined by $g(z) = \hat{a}(z^p)$, where $p = \frac{2\phi}{\pi}$. Then $g(z)$ fulfills the assumptions of Lemma 3.1, hence (3.10) holds. This implies with $\hat{a}(\lambda) = g(\lambda^\alpha), \alpha = \frac{1}{p}$,

$$\lambda^n \hat{a}^{(n)}(\lambda) = \sum_{l=1}^{n} b_l^n g^{(l)}(\lambda^\alpha) \lambda^{\alpha l}, \quad \text{Re } \lambda > 0, \; n \in \mathbb{N},$$

for some constants b_l^n. Therefore, by Lemma 3.1 we obtain

$$|\lambda^n \hat{a}^{(n)}(\lambda)| \leq \sum_{l=1}^{n} |b_l^n| |g^{(l)}(\lambda^\alpha)| |\lambda^\alpha|^l$$

$$\leq c|\hat{a}(\lambda)| \sum_{l=1}^{n} |b_l^n| (\frac{|\lambda|^\alpha}{\text{Re } \lambda^\alpha})^l \leq c(n)|\hat{a}(\lambda)|, \quad \text{Re } \lambda > 0,$$

since with $\alpha = \frac{1}{p} < 1$

$$\text{Re } \lambda^\alpha = |\lambda|^\alpha \cos(\alpha \arg \lambda) \geq |\lambda|^\alpha \cos \frac{\alpha\pi}{2} = c(\alpha)|\lambda|^\alpha. \quad\square$$

We have seen above that nonnegative, nonincreasing kernels are in general not 1-regular. However, if the kernel is also convex, then we can show that it is 1-regular. More generally, let us introduce

Definition 3.4 *Let* $a \in L^1_{loc}(\mathbb{R}_+)$ *and* $k \geq 2$. $a(t)$ *is called* k-**monotone** *if* $a \in C^{k-2}(0, \infty)$, $(-1)^n a^{(n)}(t) \geq 0$ *for all* $t > 0$, $0 \leq n \leq k - 2$, *and* $(-1)^{k-2} a^{(k-2)}(t)$ *is nonincreasing and convex.*

Thus by definition a 2-monotone kernel $a(t)$ is nonnegative, nonincreasing and convex, and $a(t)$ is completely monotonic if and only if $a(t)$ is k-monotone for all $k \geq 2$.

Proposition 3.3 *Suppose* $a \in L^1_{loc}(\mathbb{R}_+)$ *is* $(k + 1)$-*monotone,* $k \geq 1$. *Then* $a(t)$ *is* k-*regular and of positive type.*

Proof: Suppose $a \in L^1_{loc}(\mathbb{R}_+)$ is $(k + 1)$-monotone, for some $k \geq 1$. Since 2-monotone kernels are of positive type (see below), Lemma 3.1 yields with $\lambda = \sigma + i\rho, \sigma > 0, \rho \in \mathbb{R}$

$$\sigma^n |\hat{a}^{(n)}(\lambda)| \leq c_n |\hat{a}(\lambda)|, \quad \text{Re } \lambda > 0, \ n \in \mathbb{N}.$$

Therefore it is sufficient to prove

$$|\rho|^n |\hat{a}^{(n)}(\lambda)| \leq c_n |\hat{a}(\lambda)|, \quad \text{Re } \lambda > 0, \ n \leq k.$$

Let $a_\sigma(t) = a(t) e^{-\sigma t}, t, \sigma > 0$; then $a_\sigma(t)$ is $(k + 1)$-monotone again, and so by Proposition 3.9 below we have

$$|\hat{a}_\sigma(i\rho)| \geq (\sqrt{2})^{-3} \int_0^{\frac{1}{|\rho|}} a_\sigma(t) dt, \quad \rho \in \mathbb{R}, \ \sigma > 0,$$

as well as

$$|\hat{a}_\sigma^{(n)}(i\rho)| \leq c_k \int_0^{\frac{1}{|\rho|}} t^n a_\sigma(t) dt, \quad \rho \in \mathbb{R}, \ \sigma > 0, \ n \leq k,$$

where c_k only depends on k, by Proposition 3.8. These inequalities combined yield

$$|\rho^n| |\hat{a}^{(n)}(\lambda)| \leq c_k |\rho|^n \int_0^{\frac{1}{|\rho|}} t^n a_\sigma(t) dt \leq c_k \int_0^{\frac{1}{|\rho|}} a_\sigma(t) dt \leq 2^{\frac{3}{2}} c_k |\hat{a}(\lambda)|$$

for all $\sigma = \text{Re } \lambda > 0$, $\rho \in \mathbb{R}$, and $n \leq k$. Therefore a is k-regular.

To prove that $a(t)$ is of positive type we use (3.46) for $k = 2$ and $\lambda = i\rho$

$$\hat{a}(i\rho) = a_\infty / i\rho + \frac{1}{\rho^2} \int_0^\infty (1 - e^{-i\rho t} - i\rho t) d\dot{a}(t),$$

which gives

$$\text{Re } \hat{a}(i\rho) = \rho^{-2} \int_0^\infty (1 - \cos \rho t) d\dot{a}(t), \quad \rho \in \mathbb{R} \backslash \{0\}.$$

Since $a(t)$ is convex, $d\dot{a}(t) \geq 0$, and so $\text{Re } \hat{a}(i\rho) \geq 0, \rho \in \mathbb{R}$. $\text{Re } \hat{a}(\lambda)$ is harmonic in the right halfplane and tends to zero as $|\lambda| \to \infty$; therefore by the minimum principle we may conclude Re $\hat{a}(\lambda) \geq 0$ for all Re $\lambda \geq 0$, i.e. $a(t)$ is of positive type.   □

Another sufficient condition for k-regularity is contained in

Proposition 3.4 *Suppose* $a \in W_{loc}^{k+m,1}(0,\infty)$ *is such that* $\int_0^\infty t^l |a^{(m+l)}(t)| dt < \infty$ *for all* $0 \leq l \leq k$ *and* $a^{(j)}(0) = 0$ *for all* $0 \leq j < m-1$; *assume there is* $b \in BV(\mathbb{R}_+)$ *with* $\hat{db}(\lambda) = [\lambda^m \hat{a}(\lambda)]^{-1}$ *for all* $\lambda > 0$. *Then* a *is* k-*regular.*

Proof: For Re $\lambda > 0$ we have the identity

$$\lambda^n \frac{\hat{a}^{(n)}(\lambda)}{\hat{a}(\lambda)} = \hat{db}(\lambda) \lambda^{n+m} \hat{a}^{(n)}(\lambda)$$

$$= \hat{db}(\lambda) \cdot \{[[(-t)^n a]^{(n+m)}]^\wedge(\lambda) + (t^n a(t))^{(n+m-1)}|_{t=0}\}$$

$$= (-1)^n \hat{db}(\lambda) \cdot \sum_{l=0}^n \binom{n}{l} \frac{(n+m)!}{(m+l)!} \{(t^l a^{(m+l)}(t))^\wedge(\lambda) + \frac{m+l}{n+m} t^l a^{(l+m-1)}(t)|_{t=0}\},$$

hence $\lambda^n \hat{a}^{(n)}(\lambda)/\hat{a}(\lambda)$ is bounded for Re $\lambda > 0$, $n \leq k$.   □

The cases $m = 1$ and $m = 2$ are the most important in applications of Proposition 3.4.

3.3 Resolvents for Parabolic Equations
The first main result on resolvents for parabolic equations is the following

Theorem 3.1 *Let* X *be a Banach space,* A *a closed linear operator in* X *with dense domain* $\mathcal{D}(A)$, $a \in L_{loc}^1(\mathbb{R}_+)$. *Assume (3.1) is parabolic, and* $a(t)$ *is* k-*regular, for some* $k \geq 1$.

Then there is a resolvent $S \in C^{k-1}((0,\infty); \mathcal{B}(X))$ *for (3.1), and there is a constant* $M \geq 1$ *such that the estimates*

$$|t^n S^{(n)}(t)| \leq M, \quad \text{for all } t \geq 0, \ n \leq k-1, \tag{3.12}$$

$$|t^k S^{(k-1)}(t) - s^k S^{(k-1)}(s)| \leq M|t - s|[1 + \log \frac{t}{t-s}], \quad 0 \leq s < t < \infty, \tag{3.13}$$

are valid.

Observe that (3.13) implies

$$|t^k S^{(k-1)}(t) - s^k S^{(k-1)}(s)| \leq M(t-s)(1 + \frac{1}{\varepsilon}(\frac{s}{t-s})^\varepsilon \leq M_T^\varepsilon (t-s)^{1-\varepsilon}$$

for all $0 \leq s < t < T$, i.e. $S^{(k-1)} \in C_{loc}^{1-\varepsilon}((0,\infty); \mathcal{B}(X))$ for each $\varepsilon \in (0,1)$.

Proof: The proof of Theorem 3.1 is based on a result on the inversion of the Laplace transform, proved in Section 0.2. Since (3.1) is parabolic by assumption, we have

$$|H(\lambda)| \leq \frac{M}{|\lambda|} \quad \text{for} \quad \text{Re } \lambda > 0.$$

For $H'(\lambda)$ one obtains

$$H'(\lambda) = -H(\lambda)/\lambda + \hat{a}'(\lambda)A(I - \hat{a}(\lambda)A)^{-1}H(\lambda),$$

and since $a(t)$ is 1-regular this formula gives

$$|H'(\lambda)| \leq \frac{M}{|\lambda|^2} + |\frac{\hat{a}'(\lambda)}{\hat{a}(\lambda)}| \frac{M(M+1)}{|\lambda|} \leq \frac{M_1}{|\lambda|^2}, \quad \text{Re } \lambda > 0.$$

Proceeding this way, inductively there follows

$$|H^{(n)}(\lambda)| \leq \frac{M_n}{|\lambda|^{n+1}}, \quad \text{Re } \lambda > 0, \ n \leq k,$$

with some constants M_n. Thus by Proposition 0.1, $H(\lambda)$ satisfies (H1) and (H2) of Theorem 1.3, and so there is a resolvent $S(t)$ for (3.1). Theorem 0.4 shows moreover that $S(t)$ belongs to $C^{k-1}((0, \infty); \mathcal{B}(X))$ and satisfies (3.12) and (3.13). The proof is complete. □

Combination of Corollary 3.1, Proposition 3.3 and Theorem 3.1 yields

Corollary 3.2 *Suppose A generates an analytic C_0-semigroup in X, bounded on some sector $\Sigma(0, \theta)$, and let $a \in L^1_{loc}(\mathbb{R}_+)$ be 2-monotone. Then (3.1) admits a bounded resolvent $S(t)$ in X which moreover belongs to $C^\alpha_{loc}((0, \infty); \mathcal{B}(X))$ for each $\alpha < 1$.*

Proof: By Proposition 3.3, $a(t)$ is 1-regular and of positive type, and therefore (3.1) is parabolic by Corollary 3.1. Theorem 3.1 with $k = 1$ yields the assertion. □

In Section 1.3, Corollary 1.2, we have seen that (3.1) is well-posed if A is negative semidefinite in the Hilbert space X, and $a \in BV_{loc}(\mathbb{R}_+)$ is such that the measure da is of positive type. This is in general a hyperbolic situation; however, we can prove now the following parabolic extension of this result.

Corollary 3.3 *Suppose A generates a C_0-semigroup $T(z)$ which is analytic in $\Sigma(0, \frac{\pi}{2})$, and uniformly bounded in each subsector $\Sigma(0, \theta)$. Let $a \in BV_{loc}(\mathbb{R}_+)$ be of the form*

$$a(t) = a_0 + \int_0^t a_1(s)ds, \quad t > 0, \tag{3.14}$$

where $a_0 \geq 0, a_1 \in L^1(\mathbb{R}_+)$ is 3-monotone, and

$$\overline{\lim}_{t \to 0} \frac{t^{-1} \int_0^t \tau a_1(\tau) d\tau}{a_0 + \int_0^t -\tau \dot{a}_1(\tau) d\tau} < \infty. \tag{3.15}$$

Then there exists a bounded resolvent $S(t)$ for (3.1) in X, which moreover belongs to $C_{loc}^{1+\alpha}((0,\infty); \mathcal{B}(X))$ for each $\alpha < 1$, and $tS'(t)$ is bounded on $(0,\infty)$.

Observe that (3.15) holds if $a_0 > 0$ or if $\underline{\lim}_{t \to 0} - t\dot{a}_1(t)/a_1(t) > 0$; on the other hand (3.15) implies $a_0 > 0$ or $a_1(0+) = \infty$, i.e. $da(t)$ must have a singularity at $t = 0$.

Proof: The assumption on A implies $\sigma(A) \subset (-\infty, 0]$ and (3.6) for every $\theta < \pi$. For parabolicity of (3.1) it is therefore sufficient to show that $a(t)$ is θ-sectorial for some $\theta < \pi$. By Proposition 3.1, $\hat{a}(\lambda)$ is of the form

$$\hat{a}(\lambda) = \lambda^{-1}(a_0 + \hat{a}_1(\lambda)), \qquad \text{Re } \lambda > 0.$$

and so we only need to show

$$|\arg(a_0 + \hat{a}_1(\lambda))| \leq \theta_0 < \frac{\pi}{2}, \qquad \text{Re } \lambda > 0. \tag{3.16}$$

Since $a_1(t)$ is 3-monotone, it is 2-regular and of positive type, by Proposition 3.3; the remarks following Definition 3.3 then show that $a(t)$ is 2-regular again. Since $a_1(t)$ is 3-monotone, Proposition 3.10 yields

$$\frac{3}{5}|\rho| \int_0^{\frac{1}{|\rho|}} t a_1(t) dt \leq |\text{Im } \hat{a}_1(i\rho)| \leq 12|\rho| \int_0^{\frac{1}{|\rho|}} t a_1(t) dt, \qquad \rho \in \mathbb{R},$$

as well as

$$\frac{3}{5} \int_0^{\frac{1}{|\rho|}} -t\dot{a}_1(t) dt \leq \text{Re } \hat{a}_1(i\rho) \leq 4 \int_0^{\frac{1}{|\rho|}} -t\dot{a}_1(t) dt, \qquad \rho \in \mathbb{R},$$

and therefore (3.16) is equivalent to

$$\phi(\rho) = |\rho| \int_0^{\frac{1}{|\rho|}} t a_1(t) dt / (a_0 + \int_0^{\frac{1}{|\rho|}} -t\dot{a}_1(t) dt) \leq C, \qquad \rho \in \mathbb{R}; \tag{3.17}$$

apply the maximum principle to the harmonic function $\arg(a_0 + \hat{a}_1(\lambda))$ in the right halfplane $\mathbb{C}_+$ to see this. $\phi(\rho)$ is a continuous function on $\mathbb{R}\backslash\{0\}$, therefore (3.17) needs only to be checked at $\rho = 0$ and $\rho = \infty$. Condition (3.15) shows that (3.17) holds at $\rho = \infty$; the estimate

$$\phi(\rho) \leq \int_0^\infty a_1(t) dt / (a_0 + \int_0^1 -\tau\dot{a}_1(\tau) dt) < \infty \qquad \text{for } |\rho| \leq 1$$

yields (3.17) near $\rho = 0$.

Thus Proposition 3.1 shows that (3.1) is parabolic. As observed already, $a(t)$ is 2-regular, and so Theorem 3.1 yields the assertion.   □

It should be observed that in Corollary 3.3 the condition "$a_1 \in L^1(\mathbb{R}_+)$" can be replaced by "$0 \in \rho(A)$" since $\hat{a}(\lambda)^{-1} \to 0$ as $\lambda \to 0$, and so (3.17) then is not needed at $\rho = 0$.

3.4 Perturbations

Consider as in Section 2.5 the perturbed equation

$$u(t) = f(t) + (a + a * dk) * Au(t) + b * u(t), \quad t \geq 0, \tag{3.18}$$

where $a(t)$ is such that (3.1) is parabolic, $b \in L^1_{loc}(\mathbb{R}_+)$, and $k \in BV_{loc}(\mathbb{R}_+)$ is continuous at zero. If $a(t)$ is in addition 1-regular then by Theorem 3.1, Equation (3.1) admits a resolvent. It turns out that in this situation (3.18) is well-posed as well. This shows that for parabolic equations a perturbation theorem like in the case of analytic resolvents is valid, provided $a(t)$ is in addition 1-regular.

Theorem 3.2 *Suppose (3.1) is parabolic, $a \in L^1_{loc}(\mathbb{R}_+)$ is 1-regular, and let $b \in L^1_{loc}(\mathbb{R}_+)$, $k \in BV_{loc}(\mathbb{R}_+)$ such that $k(0) = k(0+)$. Then (3.18) admits a resolvent $S(t)$.*

Proof: Without loss of generality we may assume that $b(t)$ and $dk(t)$ are Laplace transformable, and since $k(t)$ is continuous at zero, we may even assume

$$|\hat{b}(\lambda)| + |\hat{dk}(\lambda)| \leq \eta \quad \text{for Re } \lambda \geq 0, \tag{3.19}$$

where $\eta > 0$ is any prescribed number; multiply (3.18) with $e^{-\omega t}$ with some sufficiently large ω, to see this. As in the proof of Theorem 2.4 we let $r(t)$ and $l(t)$ be defined by

$$r = b + b * r, \quad l = k + 1 * r + k * r. \tag{3.20}$$

The resolvent of (3.18) then must satisfy the equation

$$S = I + Aa * S + dk * Aa * S + b * S, \tag{3.21}$$

hence as in the proof of Theorem 2.4

$$\hat{S}(\lambda) = \frac{1}{\lambda}(1 - \hat{b}(\lambda) - (1 + \hat{dk}(\lambda))\hat{a}(\lambda)A)^{-1} = (1 + \hat{r}(\lambda)) \sum_{n=0}^{\infty} [\hat{dl}(\lambda)]^n H_n(\lambda),$$

for Re $\lambda > 0$, where

$$H_n(\lambda) = \frac{1}{\lambda}\hat{a}(\lambda)^n A^n (1 - \hat{a}(\lambda)A)^{-(n+1)}, \quad \text{Re } \lambda > 0. \tag{3.22}$$

Since (3.1) is parabolic we obtain

$$|H_n(\lambda)| \le |H_0(\lambda)||\hat{a}(\lambda)A(I - \hat{a}(\lambda)A)^{-1}|^n \le \frac{M_0}{|\lambda|} \cdot M^n, \quad \text{Re } \lambda > 0,$$

for some constants $M_0, M \ge 1$. 1-regularity of $a(t)$ and

$$H_n'(\lambda) = -\frac{1}{\lambda}H_n(\lambda) + n\frac{\hat{a}'(\lambda)}{\hat{a}(\lambda)}H_n(\lambda) + (n+1)\frac{\hat{a}'(\lambda)}{\hat{a}(\lambda)}H_{n+1}(\lambda),$$

yields the estimate

$$|H_n'(\lambda)| \le \frac{M_0 M^n}{|\lambda|^2} + nC\frac{M_0 M^n}{|\lambda|^2} + (n+1)C\frac{M_0 M^{n+1}}{|\lambda|^2} \le \frac{M_1 M^n (n+1)}{|\lambda|^2}.$$

Therefore, there are functions $S_n \in L^\infty(\mathbb{R}_+; \mathcal{B}(X)) \cap C_{loc}^\alpha((0,\infty); \mathcal{B}(X))$, $\alpha \in (0,1)$, such that

$$\hat{S}_n(\lambda) = H_n(\lambda) \quad \text{for Re } \lambda > 0,$$

and

$$|S_n(t)| \le M_2 \cdot M^n(n+1), \quad t > 0,$$

for some constant M_2 which is independent of n, by Theorem 0.4. Define

$$S(t) = (\delta + r) * \sum_{n=0}^{\infty} [dl]^{*n} * S_n(t), \quad t > 0; \tag{3.23}$$

this series converges absolutely if $\eta > 0$ is sufficiently small and it is clear from the construction that (3.21) holds. Each function $S_n(t)$ commutes with A, and so (3.21) evaluated at $x \in \mathcal{D}(A)$ shows that $S(t)x$ is continuous on $\mathbb{R}_+$, hence $S(t)$ is strongly continuous on $\mathbb{R}_+$, by the Banach-Steinhaus Theorem. Thus $S(t)$ is the resolvent for (3.18) and the proof is complete. □

It should be noted that in case $a(t)$ is k-regular, $k > 1$, then $S_n(t)$ defined in the proof of Theorem 3.2 belongs to $C^{k-1}((0,\infty); \mathcal{B}(X))$; however, $S(t)$ in general will only be continuous, since the discontinuities in $l(t)$ will be reproduced in $S(t)$. No matter how smooth $a(t)$ is, the resolvent $S(t)$ can be only as smooth as $l(t)$.

3.5 Maximal Regularity
Consider again the perturbed equation (3.18) where $a(t)$ is such that (3.1) is parabolic, $b \in L_{loc}^1(\mathbb{R}_+)$, and $k \in BV_{loc}(\mathbb{R}_+)$ is continuous at zero. It is a surprising fact that (3.18) has the maximal regularity property of type C^α in case $a(t)$ is 2-regular, although the resolvent $S(t)$ of (3.18) is merely continuous.

Theorem 3.3 *Suppose $a \in L_{loc}^1(\mathbb{R}_+)$ is 2-regular and such that (3.1) is parabolic, let $b \in BV_{loc}(\mathbb{R}_+)$, and $k \in BV_{loc}(\mathbb{R}_+)$ with $k(0) = k(0+)$, let $J = [0,T]$ and $\alpha \in (0,1)$. Then*
(i) for each $f \in C_0^\alpha(J;X)$ there is a mild solution $u \in C_0^\alpha(J;X)$ of (3.18);

(ii) for each $f \in C_0^\alpha(J; X_A)$ there is a strong solution $u \in C_0^\alpha(J; X_A)$ of (3.18);
(iii) for each $f = a * g, g \in C_0^\alpha(J; X)$ there is a strong solution $u \in C_0^\alpha(J; X_A)$ of (3.18).

Proof: It is sufficient to prove (i); (ii) and (iii) follow from (i) as in the proof of Theorem 2.4. According to the proof of Theorem 3.2, the resolvent $S(t)$ for (3.18) is of the form

$$S(t) = (\delta + r) * \sum_{n=0}^{\infty} [dl]^{*n} * S_n(t), \quad t > 0, \tag{3.24}$$

where $r(t)$ and $l(t)$ are defined as in (3.20), and w.l.o.g.

$$\int_0^\infty |r(t)| + \int_0^\infty |dl(t)| \leq \eta, \tag{3.25}$$

$\eta > 0$ being any prescribed number. $S_n(t)$ satisfies

$$\hat{S}_n(\lambda) = H_n(\lambda) = \frac{1}{\lambda} \hat{a}(\lambda)^n A^n (1 - \hat{a}(\lambda)A)^{-(n+1)}, \quad \mathrm{Re}\,\lambda > 0.$$

Since $a(t)$ is by assumption 2-regular, as in the proof of Theorem 3.2 there are constants M_0 and M such that

$$|H_n(\lambda)| + |\lambda H_n'(\lambda)| + |\lambda^2 H_n''(\lambda)| \leq \frac{M_0}{|\lambda|} M^n, \quad \mathrm{Re}\,\lambda > 0, \ n \in \mathbb{N}. \tag{3.26}$$

Theorem 0.4 then shows $S_n \in C^1((0, \infty); \mathcal{B}(X))$ with

$$|S_n(t)| + |t S_n'(t)| \leq M_1 M^n, \quad t > 0 \tag{3.27}$$

and

$$|t^2 S_n'(t) - s^2 S_n'(s)| \leq M_1 M^n (t - s)(1 + \log \frac{t}{t - s}), \quad 0 \leq s < t < \infty. \tag{3.28}$$

The latter yields the estimate

$$|S_n'(t + h) - S_n'(t)| \leq M_2 M^n \frac{h}{t(t + h)}(1 + \log(1 + \frac{t}{h})) \quad \text{for all } t, h > 0. \tag{3.29}$$

Given $f \in C_0^\alpha(J; X)$ we set

$$u_n(t) = \frac{d}{dt}(S_n * f)(t) = S_n(t)f(t) + \int_0^t S_n'(t - \tau)(f(\tau) - f(t))d\tau, \quad t \in J, \tag{3.30}$$

and follow the proof of Theorem 2.4 to obtain $u_n \in C_0^\alpha(J; X)$ as well as the estimate

$$|u_n|_\alpha \leq C M^n |f|_\alpha, \quad n \in \mathbb{N}_0, \tag{3.31}$$

where C is a constant independent of n; observe that the estimate of I_4 there has to be modified slightly using (3.29). Then set

$$u(t) = (\delta + r) * \sum_{n=0}^{\infty} [dl]^{*n} * u_n(t), \quad t \in J;$$

the series converges absolutely in $C_0^\alpha(J; X)$ provided $\eta M < 1$, and we obtain

$$|u|_\alpha \le (1 + \eta) C \sum_{n=0}^{\infty} (\eta M)^n |f|_\alpha = C_1 |f|_\alpha.$$

If $f \in C^1(J; X), f(0) = 0$ then $u_n = S_n * f'$ implies that $u = S * f'$ is the unique mild solution of (3.18). The result now follows from an approximation argument like in proof of Theorem 2.4. □

Corollaries analogous to Corollaries 2.5 and 2.6 can be obtained as well; we leave this to the reader.

3.6 A Representation Formula
Let $a \in L^1_{loc}(\mathbb{R}_+)$ be 1-regular and sectorial with angle $\theta < \frac{\pi}{2}$ and suppose A generates a bounded C_0-semigroup $T(\tau)$ in the Banach space X. Then $\rho(A) \supset \mathbb{C}_+$ and there is a constant $M_1 \ge 1$ such that

$$|(\mu - A)^{-1}| \le \frac{M_1}{\operatorname{Re} \mu} \quad \text{for } \operatorname{Re} \mu > 0,$$

in particular, (3.6) holds on each sector $\Sigma(0, \theta)$ with $\theta < \frac{\pi}{2}$. By Proposition 3.1 this implies that (3.1) is parabolic, and Theorem 3.1 shows the existence of a bounded resolvent $S(t)$ for (3.1). In this section we want to derive a representation formula for $S(t)$ in terms of the semigroup $T(\tau)$ and certain functions $w^t \in L^1(\mathbb{R}_+)$. To see how this becomes possible recall that

$$(I - \mu A)^{-1} = \mu^{-1} \int_0^\infty e^{-\tau/\mu} T(\tau) d\tau, \quad \operatorname{Re} \mu > 0; \tag{3.32}$$

thus we obtain

$$\hat{S}(\lambda) = H(\lambda) = \lambda^{-1}(I - \hat{a}(\lambda)A)^{-1} = \frac{1}{\lambda \hat{a}(\lambda)} \int_0^\infty e^{-\tau/\hat{a}(\lambda)} T(\tau) d\tau,$$

i.e.

$$\hat{S}(\lambda) = \int_0^\infty h(\lambda, \tau) T(\tau) d\tau, \quad \operatorname{Re} \lambda > 0, \tag{3.33}$$

where

$$h(\lambda; \tau) = \frac{e^{-\tau/\hat{a}(\lambda)}}{\lambda \hat{a}(\lambda)}, \quad \operatorname{Re} \lambda > 0, \tau \ge 0. \tag{3.34}$$

The Laplace transform of $h(\lambda; \tau)$ w.r.t. τ is then given by

$$\sigma(\lambda; \mu) = \frac{1}{\lambda(1 + \mu\hat{a}(\lambda))}, \quad \text{Re } \lambda > 0, \text{ Re } \mu > 0, \tag{3.35}$$

which on the other hand is the Laplace transform of the function $s(t; \mu)$ introduced in Section 1.3 w.r.t. t. Next we show that $s(t; \mu)$ is the Laplace transform w.r.t. τ of an L^1-function $w^t(\tau)$.

Proposition 3.5 *Suppose the kernel* $a \in L^1_{loc}(\mathbb{R}_+)$ *is 1-regular and sectorial with angle* $\theta < \frac{\pi}{2}$, *and let* $s(t; \mu)$ *denote the solution of*

$$s(t) + \mu(a * s)(t) = 1, \quad t \geq 0, \ \mu \in \mathbb{C}.$$

Then there is a uniformly bounded family $\{w^t\}_{t>0} \subset BV(\mathbb{R}_+) \cap W^{1,1}_{loc}(\mathbb{R}_+)$ *such that*

$$s(t; \mu) = -\int_0^\infty e^{-\mu\tau} \dot{w}^t(\tau) d\tau, \quad t > 0, \text{ Re } \mu > 0. \tag{3.36}$$

The function $\dot{w} : (0, \infty) \to L^1(\mathbb{R}_+)$ *is continuous and* $w^t \overset{*}{\rightharpoonup} -e_0$ *in* $BV(\mathbb{R}_+)$ *as* $t \to 0$, *where* $e_0(t)$ *denotes the Heaviside function.*

Proof: Fix $t > 0$ and $\varepsilon > 0$; then $s(t; \mu)$ is represented as

$$s(t; \mu) = \frac{e^{\varepsilon t}}{2\pi} \int_{-\infty}^\infty e^{i\rho t} \sigma(\varepsilon + i\rho, \mu) d\rho, \quad t > 0, \text{ Re } \mu \geq 0, \tag{3.37}$$

by inversion of the Laplace transform. Integration by parts and a differentiation w.r.t. μ yields

$$-\frac{\partial}{\partial\mu} s(t; \mu) = \frac{e^{\varepsilon t}}{2\pi t} \int_{-\infty}^\infty e^{i\rho t} \frac{\partial^2}{\partial\lambda\partial\mu} \sigma(\varepsilon + i\rho; \mu) d\rho. \tag{3.38}$$

By a simple computation we obtain

$$\frac{\partial^2}{\partial\lambda\partial\mu} \sigma(\lambda; \mu) = \frac{1}{\lambda^2} \Big[\frac{\hat{a}(\lambda)}{(1 + \mu\hat{a}(\lambda))^2} \Big(1 - \frac{\lambda\hat{a}'(\lambda)}{\hat{a}(\lambda)}\Big) + 2\frac{\lambda\hat{a}'(\lambda)}{\hat{a}(\lambda)} \frac{\mu\hat{a}(\lambda)^2}{(1 + \mu\hat{a}(\lambda))^3} \Big],$$

and a direct estimate which employs 1-regularity and $\theta < \frac{\pi}{2}$ gives

$$\Big| \frac{\partial^2}{\partial\lambda\partial\mu} \sigma(\lambda; \mu) \Big| \leq \frac{C}{|\lambda|^2} \cdot \frac{|\hat{a}(\lambda)|}{(1 + |\mu||\hat{a}(\lambda)|)^2} \quad \text{for Re } \lambda > 0, \text{ Re } \mu \geq 0, \tag{3.39}$$

where $C > 0$ denotes a constant independent of μ and λ. Combining (3.38) and (3.39) yields

$$\int_{-\infty}^\infty \Big| \frac{\partial}{\partial\mu} s(t; i\tau) \Big| d\tau \leq 2C \frac{e^{\varepsilon t}}{2\pi t} \int_0^\infty \int_{-\infty}^\infty \frac{|\hat{a}(\varepsilon + i\rho)|}{(1 + \tau|\hat{a}(\varepsilon + i\rho)|)^2} \cdot \frac{d\rho}{\varepsilon^2 + \rho^2} d\tau$$

$$= 2C \frac{e^{\varepsilon t}}{2\pi t} \int_{-\infty}^\infty \Big(\int_0^\infty \frac{dr}{(1 + r)^2} \Big) \frac{d\rho}{\varepsilon^2 + \rho^2} = \frac{C}{\pi} \frac{e^{\varepsilon t}}{\varepsilon t} = C \frac{e}{\pi}$$

with the choice $\varepsilon = \frac{1}{t}$. Thus for fixed $t > 0$, $s(t; \cdot)$ belongs to the Hardy space $H^\infty(\mathbb{C}_+)$ and $\frac{\partial}{\partial \mu} s(t; \cdot) \in H^1(\mathbb{C}_+)$; by Hardy's inequality (see Duren [106]), there are functions $v^t \in L^1(\mathbb{R}_+)$ such that $\widehat{v^t}(\mu) = s(t; \mu)$ for Re $\mu > 0$, and $|v^t|_1 \leq \frac{Ce}{2\pi}$, for all $t > 0$. Define

$$w^t(\tau) = \int_\tau^\infty v^t(s)ds, \quad t > 0, \ \tau \geq 0;$$

then $\dot{w}^t(\tau) = -v^t(\tau)$, hence (3.36) is valid, and the family $\{w^t\}_{t>0} \subset BV(\mathbb{R}_+)$ is uniformly bounded.

A similar argument involving Hardy's theorem shows that the function $\dot{w}$: $(0, \infty) \to L^1(\mathbb{R}_+)$ is continuous (even $C^\alpha_{loc}((0; \infty); L^1(\mathbb{R}_+))$ for any $\alpha \in (0, 1)$) and $w^t \overset{*}{\to} -e_0$ follows from $s(t; \mu) \to 1$ as $t \to 0$ for every $\mu \in \mathbb{C}$, since span$\{e^{\mu\tau} : \mu < 0\}$ is dense in $C_0(\mathbb{R}_+)$. $\square$

Since $\dot{w} : (0, \infty) \to L^1(\mathbb{R}_+)$ is continuous and bounded, the vector valued Laplace transform $\widehat{\dot{w}}(\lambda)$ exists for Re $\lambda > 0$. The ordinary Laplace transform $\mathcal{L} : L^1(\mathbb{R}_+) \to H^\infty(\mathbb{C}_+)$ is linear and bounded, hence

$$\mathcal{L}\widehat{\dot{w}}(\lambda) = \widehat{\mathcal{L}\dot{w}}(\lambda) = -\hat{s}(\lambda; \mu) = -\sigma(\lambda; \mu)$$

holds for Re $\lambda > 0$ and Re $\mu \geq 0$. Since on the other hand

$$\sigma(\lambda; \cdot) = \mathcal{L}h(\lambda, \cdot) \quad \text{for Re } \lambda > 0,$$

we obtain from uniqueness of the ordinary Laplace transform

$$\widehat{\dot{w}}(\lambda) = -h(\lambda; \cdot), \quad \text{Re } \lambda > 0. \tag{3.40}$$

Via (3.33) this yields

$$\begin{aligned}
\hat{S}(\lambda) &= \int_0^\infty h(\lambda; \tau)T(\tau)d\tau = -\int_0^\infty \widehat{\dot{w}}(\lambda; \tau)T(\tau)d\tau \\
&= -\left(\int_0^\infty \dot{w}^t(\tau)T(\tau)d\tau\right)^{\wedge}(\lambda),
\end{aligned}$$

and so by uniqueness of the vector-valued Laplace transform we arrive at the representation formula

$$S(t) = -\int_0^\infty \dot{w}^t(\tau)T(\tau)d\tau, \quad t > 0, \tag{3.41}$$

we have been looking for.

3.7 Comments

a) The concept of parabolicity employed in this section has been introduced in Prüss [278] as well as the notion of k-regular kernels, and the results of Sections 3.2

and 3.3 appear also there. The perturbation theorem, Theorem 3.2, and Theorem 3.3 on maximal regularity are new, as is the representation formula in Section 3.6 for the case considered here.

b) In the situation of Proposition 3.1 one can show without further assumptions that there is an L_{loc}^p-resolvent $S(t)$ for (3.1), i.e. a function $S \in L_{loc}^p(\mathbb{R}_+; \mathcal{B}(X))$, for all $1 \leq p < \infty$, such that $\hat{S}(\lambda) = H(\lambda)$, Re $\lambda > 0$, and the resolvent equation holds a.e. The proof of this result relies on the Dunford integral representation

$$S(t) = \frac{-1}{2\pi i} \int_\Gamma (\mu + A)^{-1} s(t; \mu) d\mu, \quad t > 0,$$

where Γ is an appropriate contour in the complex plane, and on L^p-estimates for the scalar function $s(\cdot; \mu)$. The latter in turn can be obtained from standard estimates in the theory of H^p-spaces, and from lower bounds for the Laplace transform of sectorial kernels. This is similar to the method used by Friedman and Shinbrot [123] and Friedman [121] for the case $a(t) = a_0 + \int_0^t a_1(s)ds$ with $a_1 \in W_{loc}^{1,1}(\mathbb{R}_+)$. However, the latter result is a special case of Theorem 3.2, even $a_1 \in L_{loc}^1(\mathbb{R}_+)$ is sufficent.

c) 2- and 3-monotone kernels have been used extensively in the theory of scalar and vector-valued Volterra equations; see e.g. Levin [213], Shea and Wainger [302], Carr and Hannsgen [38], [39], Hannsgen [163], [162], Kiffe and Stecher [196], and the recent monograph by Gripenberg, Londen, and Staffans [156], as well as the references given there.

d) The estimates on $\hat{a}^{(n)}$ in the appendix are generalizations of the results of Shea and Wainger [302], Hannsgen [161], Carr and Hannsgen [38] for 2-monotone kernels; cp. also Gripenberg, Londen, and Staffans [156].

e) For parabolic equations perturbations $A + B$, where $D(B) \supset D(A)$, and for each $\varepsilon > 0$ there is $C(\varepsilon) > 0$ such that

$$|Bx| \leq \varepsilon|Ax| + C(\varepsilon)|x|, \quad x \in D(A),$$

are again possible, if one applies an exponential shift. Parabolicity is in this sense invariant under such perturbations.

f) It is also possible to obtain spatial regularity of the resolvent $S(t)$ for equations of parabolic type, provided $a(t)$ is k-regular, for some large enough k. In fact, after an exponential shift we obtain by the arguments in the proof of Theorem 3.1 and by Lemma 8.1 the estimate

$$|AH^{(k)}(\lambda)| \leq M/(|\lambda|^2 + 1), \quad \text{Re } \lambda \geq 0,$$

for some sufficiently large k. This yields

$$|t^k AS(t)| \leq Ce^{\omega t}, \quad t \geq 0,$$

i.e. $\mathcal{R}(S(t)) \subset D(A)$ for all $t > 0$; cp. Section 2.4.

g) Observe that Proposition 3.5 yields an estimate of the form

$$|s(t;\mu)| \leq M \quad \text{for all } t > 0, \ \operatorname{Re} \mu > 0,$$

provided $a(t)$ is 1-regular and θ-sectorial with $\theta < \pi/2$. The family $w(t;\tau) = w^t(\tau)$ constructed there satisfies

$$\operatorname{Var} w(t;\cdot)|_0^\infty \leq M, \quad \text{for all } t > 0.$$

It seems to be unknown whether in this situation also

$$\operatorname{Var} w(\cdot;\tau)|_0^\infty \leq M, \quad \text{for all } \tau > 0$$

holds; the latter would imply

$$\int_0^\infty |r(t;\mu)| dt \leq \frac{M'}{|\mu|} \quad \text{for all } t > 0, \ \operatorname{Re} \mu > 0.$$

However, this in turn can be proved at least if $a(t)$ is in addition 2-monotone. It will be seen that the family $w(t;\tau)$ is of central importance for the theory presented in Section 4.

Appendix: k-monotone Kernels

A kernel $a \in L^1_{loc}(\mathbb{R}_+)$ is called *1-monotone* if $a(t)$ is nonnegative and nonincreasing; a is called *k-monotone* $(k \geq 2)$ if $a \in C^{k-2}(0,\infty)$ satisfies $(-1)^n a^{(n)}(t) \geq 0$ for all $t > 0$, $n \leq k - 2$, and $(-1)^{k-2} a^{(k-2)}(t)$ is nonincreasing and convex. To keep the presentation self-contained we now prove some estimates for k-monotone kernels which have been used above.

The behaviour of such kernels at $t = 0$ is described in

Proposition 3.6 *Let $a \in L^1_{loc}(\mathbb{R}_+)$ be k-monotone for some $k \geq 1$. Then*
(i) $a^{(n)}(t)t^{n+1} \to 0$ as $t \to 0$, $n \leq k - 1$;
(ii) $a^{(n)}(t)t^n \in L^1(0,1)$ for all $n \leq k - 1$, and $\int_0^1 (-t)^k da^{(k-1)}(t) < \infty$;
(iii) For $t > 0, l \in \mathbb{N}_0$ we have the identity

$$\int_0^t \frac{\tau^l}{l!} a(\tau) d\tau = \sum_{j=0}^{k-1} (-1)^j a^{(j)}(t) \frac{t^{j+l+1}}{(j+l+1)!} + \int_0^t \frac{\tau^{k+l}(-1)^k}{(k+l)!} da^{(k-1)}(\tau). \quad (3.42)$$

Proof: Since $a(t)$ is nonincreasing and nonnegative and belongs to $L^1(0,1)$, we have

$$ta(t) \leq \int_0^t a(s)ds \to 0 \quad \text{as } t \to 0,$$

as well as

$$0 \leq \int_\varepsilon^t -\tau \dot{a}(\tau) d\tau = -ta(t) + \varepsilon a(\varepsilon) + \int_\varepsilon^t a(\tau)d\tau \to -ta(t) + \int_0^t a(\tau)d\tau$$

as $\varepsilon \to 0$; this proves (i) and (ii) for $n = 0, n = 1$ respectively. Suppose (ii) holds for some $n \le k - 1$; then

$$0 \le \frac{t^{n+1}}{(n+1)!}(-1)^n a^{(n)}(t) = (\int_0^t \frac{\tau^n}{n!}d\tau)(-1)^n a^{(n)}(t) \le \int_0^t \frac{\tau^n}{n!}(-1)^n a^{(n)}(\tau)d\tau \to 0$$

as $t \to 0$, since $(-1)^n a^{(n)}(t)$ is nonincreasing and $t^n a^{(n)}(t) \in L^1(0,1)$. Thus (i) holds for n as well and the relation

$$\begin{aligned}0 &\le \int_\varepsilon^t (-1)^{n+1} \frac{\tau^{n+1}}{(n+1)!} a^{(n+1)}(\tau) d\tau \\ &= (-1)^{n+1} \frac{t^{n+1}}{(n+1)!} a^{(n)}(t) + (-1)^n \frac{\varepsilon^{n+1}}{(n+1)!} a^{(n)}(\varepsilon) + \int_\varepsilon^t (-1)^n \frac{\tau^n}{n!} a^{(n)}(\tau)\end{aligned}$$

shows that (ii) also holds for $n + 1$. Thus (i) and (ii) follow by induction. (iii) is obtained by integration by parts. $\quad\square$

The next proposition contains the behaviour of $a(t)$ at $t = \infty$.

Proposition 3.7 *Let $a \in L_{loc}^1(\mathbb{R}_+)$ be k-monotone for some $k \ge 1$, and let $a_\infty = \lim_{t\to\infty} a(t)$. Then*
(i) $a^{(n)}(t)t^n \to 0$ as $t \to \infty$, $1 \le n \le k - 1$;
(ii) $a^{(n)}(t)t^{n-1} \in L^1(1,\infty)$, $1 \le n \le k - 1$, and $\int_1^\infty t^{k-1}(-1)^k da^{(k-1)}(t) < \infty$;
(iii) For $t > 0$, $l \le k - 1$ we have the identity

$$\int_t^\infty (-1)^k \frac{\tau^l}{l!} da^{(k-1)}(\tau) = \sum_{j=1}^{l+1} (-1)^{k-j} a^{(k-j)}(t) \frac{t^{l-j+1}}{(l-j+1)!} - \delta_{l,k-1} a_\infty. \quad (3.43)$$

Proof: Since $a(t)$ is nonincreasing and nonnegative, $a_\infty = \lim_{t\to\infty} a(t)$ exists and is nonnegative; from

$$0 \le \int_t^R -\dot{a}(\tau)d\tau = a(t) - a(R) \to a(t) - a_\infty, \quad \text{as } R \to \infty$$

there follows $\dot{a} \in L^1(1,\infty)$. Suppose $a^{(n)}(t)t^{n-1} \in L^1(1,\infty)$ for some $1 \le n \le k-1$; then

$$\begin{aligned}0 &\le (-1)^n a^{(n)}(t) \frac{t^n}{n!}(1 - 2^{-n}) = (-1)^n a^{(n)}(t) \int_{\frac{t}{2}}^t \frac{\tau^{n-1}}{(n-1)!} d\tau \\ &\le \int_{\frac{t}{2}}^t \frac{\tau^{n-1}}{(n-1)!}(-1)^n a^{(n)}(\tau)d\tau \to 0 \quad \text{as } t \to \infty,\end{aligned}$$

since $(-1)^n a^{(n)}(t)$ is nonincreasing, hence $a^{(n)}(t)t^n \to 0$ as $t \to \infty$. Furthermore, the identity

$$\begin{aligned}0 &\le \int_t^R (-1)^{n+1} \frac{\tau^n}{n!} a^{(n+1)}(\tau) d\tau \\ &= (-1)^{n+1} \frac{R^n}{n!} a^{(n)}(R) + (-1)^n \frac{t^n}{n!} a^{(n)}(t) + \int_t^R (-1)^n \frac{\tau^{n-1}}{(n-1)!} a^{(n)}(\tau)d\tau\end{aligned}$$

shows $t^n a^{(n+1)}(t) \in L^1(1, \infty)$. Thus (i) and (ii) follow by induction. (iii) is obtained by integration by parts. $\square$

Next we consider the Laplace-transform of a k-monotone kernel $a(t)$. An integration by parts shows

$$\lambda \hat{a}(\lambda) = a_\infty + \int_0^\infty -\dot{a}(t)(1 - e^{-\lambda t})dt, \quad \text{Re } \lambda > 0; \tag{3.44}$$

thus by Propositions 3.6 and 3.7 we conclude that $\lambda \hat{a}(\lambda)$ admits continuous extension to $\overline{\mathbb{C}}_+$ if $a(t)$ is merely 1-monotone, and (3.44) holds for all Re $\lambda \geq 0$. Furthermore,

$$\begin{aligned}
|\lambda \hat{a}(\lambda)| &\leq a_\infty + \int_0^\varepsilon -\dot{a}(t) t \frac{1 - e^{-\lambda t}}{t} |dt + \int_\varepsilon^\infty -\dot{a}(t)|1 - e^{-\lambda t}|dt \\
&\leq a_\infty + |\lambda| \int_0^\varepsilon -\dot{a}(t) t dt + 2 \int_\varepsilon^\infty -\dot{a}(t) dt \\
&= a_\infty + |\lambda|(-\varepsilon a(\varepsilon) + \int_0^\varepsilon a(t) dt) + 2(a(\varepsilon) - a_\infty),
\end{aligned}$$

since $|\frac{1-e^{-z}}{z}| \leq 1$ for Re $z \geq 0$. Hence with $\varepsilon = 1/|\lambda|$,

$$|\lambda \hat{a}(\lambda)| \leq |\lambda| \int_0^{\frac{1}{|\lambda|}} a(t) dt + a(1/|\lambda|) - a_\infty \leq 2|\lambda| \int_0^{\frac{1}{|\lambda|}} a(t) dt,$$

i.e. we have the upper estimate

$$|\hat{a}(\lambda)| \leq 2 \int_0^{\frac{1}{|\lambda|}} a(t) dt, \quad \text{Re } \lambda \geq 0. \tag{3.45}$$

Integrating $(k-1)$-times by parts in (3.44), when $a(t)$ is k-monotone we obtain by means of Propositions 3.6 and 3.7

$$\lambda^k \hat{a}(\lambda) = \lambda^{k-1} a_\infty - \int_0^\infty \left(\sum_{j=0}^{k-1} \frac{(-\lambda t)^j}{j!} - e^{-\lambda t} \right) da^{(k-1)}(t), \quad \text{Re } \lambda \geq 0. \tag{3.46}$$

With

$$f_k(z) = \frac{(k-1)!}{z^k} \left(\sum_{j=0}^{k-1} \frac{(-z)^j}{j!} - e^{-z} \right) = (-1)^{k-1} \sum_{j=0}^\infty \frac{(k-1)! j!}{(j+k)!} \cdot \frac{(-z)^j}{j!}, \tag{3.47}$$

(3.46) can be rewritten as

$$\hat{a}(\lambda) = \frac{a_\infty}{\lambda} - \int_0^\infty f_k(\lambda t) \frac{t^k}{(k-1)!} da^{(k-1)}(t), \quad \text{Re } \lambda \geq 0, \tag{3.48}$$

and this gives the following representation of $\hat{a}^{(n)}(\lambda)$.

$$\hat{a}^{(n)}(\lambda) = \frac{(-1)^n n! a_\infty}{\lambda^{n+1}} - \int_0^\infty f_k^{(n)}(\lambda t) t^{n+k} \frac{da^{(k-1)}(t)}{(k-1)!}, \quad \text{Re } \lambda \geq 0. \quad (3.49)$$

(3.49) can be used to obtain estimates for $\hat{a}^{(n)}(\lambda)$, $n \leq k-1$; for this purpose we have to study the functions $f_k(z)$ in detail.

Lemma 3.2 *Let $k \geq 1$ and $f_k(z)$ be defined by (3.47). Then*
(i) $f_k(z) = \int_0^1 (t-1)^{k-1} e^{-tz} dt$, *for all $k \geq 1, z \in \mathbb{C}$;*
(ii) $|f_k^{(n)}(z)| \leq \int_0^1 (1-t)^{k-1} t^n dt = \frac{n!(k-1)!}{(n+k)!}$, $k \geq 1, n \geq 0$, Re $z \geq 0$;
(iii) $|f_k^{(n)}(z)| \leq \frac{n!}{|z|^{n+1}} 2^{n+1}$, $0 \leq n \leq k-1$, Re $z \geq 0$.

Proof: Let $g_k(z) = \int_0^1 (t-1)^{k-1} e^{-tz} dt$; we want to show $f_k = g_k$. For $k = 1$ we have

$$g_1(z) = \int_0^1 e^{-tz} dt = \frac{1}{z}(1 - e^{-z}) = f_1(z).$$

Next we observe that

$$g_{k+1}(z) = \int_0^1 (t-1)^k e^{-tz} dt = (t-1)^k \frac{e^{-tz}}{-z}\Big|_0^1 + \frac{k}{z} \int_0^1 (t-1)^{k-1} e^{-tz} dt$$

$$= \frac{(-1)^k}{z} + \frac{k}{z} g_k(z) \quad \text{for all } k \geq 1, \; z \in \mathbb{C}.$$

On the other hand, we also have

$$f_{k+1}(z) = \frac{k}{z} f_k(z) + \frac{(-1)^k}{z} \quad \text{for all } k \geq 1, \; z \in \mathbb{C},$$

hence g_k and f_k satisfy the same recursion, and so $f_k = g_k$, by $f_1 = g_1$; this proves (i).

To prove (ii) we simply estimate $f_k^{(n)}(z)$, which is given by

$$f_k^{(n)}(z) = \int_0^1 (t-1)^{k-1}(-t)^n e^{-tz} dt, \quad (3.50)$$

to the result

$$|f_k^{(n)}(z)| \leq \int_0^1 (1-t)^{k-1} t^n dt = \frac{n!(k-1)!}{(n+k)!}.$$

To prove (iii), integrate $(n+1)$-times by parts in (3.50); this yields

$$f_k^{(n)}(z) = \frac{(-1)^n}{z^{(n+1)}} \int_0^1 \{(t-1)^{k-1} t^n\}^{n+1} e^{-tz} dt - \frac{(-1)^n}{z^{n+1}} \{(t-1)^{k-1} t^n\}^{(n)} e^{-tz}\Big|_0^1,$$

since the integrand vanishes at $t = 0, 1$ to the order n. We have

$$(\frac{d}{dt})^{n+1}\{(t-1)^{k-1}t^n\} = \sum_{l=1}^{n+1} \binom{n+1}{l} \frac{t^{l-1}}{(l-1)!} \frac{(t-1)^{k-1-l}}{(k-1-l)!} \cdot n!(k-1)!,$$

and the term with $l = n + 1$ is absent in case $n = k - 1$. This yields

$$|f_k^{(n)}(z)| \leq \frac{1}{|z|^{n+1}}\{n! + n!(k-1)! \sum_{l=1}^{n+1} \binom{n+1}{l} \int_0^1 \frac{t^{l-1}}{(l-1)!} \frac{(1-t)^{k-1-l}}{(k-1-l)!} dt\}$$

$$= \frac{n!}{|z|^{n+1}}\{1 + \sum_{l=1}^{n+1} \binom{n+1}{l}\} = \frac{2^{n+1}n!}{|z|^{n+1}} \quad \text{for } n < k - 1,$$

and similarly also for $n = k - 1$. $\square$

Lemma 3.2 now yields the desired estimates for $\hat{a}^{(n)}(\lambda)$. From (3.49) we obtain

$$|\hat{a}^{(n)}(\lambda)| \leq \frac{n!a_\infty}{|\lambda|^{n+1}} + \int_0^\infty |f_k^{(n)}(\lambda t)| \frac{t^{n+k}}{(k-1)!}(-1)^k da^{(k-1)}(t)$$

$$\leq \frac{n!a_\infty}{|\lambda|^{n+1}} + n! \int_0^{\frac{1}{|\lambda|}} \frac{t^{n+k}}{(n+k)!}(-1)^k da^{(k-1)}(t) + \frac{n!2^{n+1}}{|\lambda|^{n+1}} \int_{\frac{1}{|\lambda|}}^\infty \frac{t^{k-1}}{(k-1)!}(-1)^k da^{(k-1)}(t)$$

$$\leq \frac{n!a_\infty}{|\lambda|^{n+1}} + \int_0^{\frac{1}{|\lambda|}} t^n a(t)dt - n! \sum_{j=0}^{k-1} \frac{(-1)^j}{(j+n+1)!} a^{(j)}(1/|\lambda|)(|\lambda|)^{-(j+n+1)}$$

$$+ \frac{n!2^{n+1}}{|\lambda|^{n+1}} \sum_{j=0}^{k-1} \frac{(-1)^j}{j!} a^{(j)}(1/|\lambda|)|\lambda|^{-j} - \frac{n!2^{n+1}}{|\lambda|^{n+1}} a_\infty$$

by Lemma 3.2 and Propositions 3.6 and 3.7. Since $\frac{(j+n+1)!}{j!} \leq \frac{(k+n)!}{(k-1)!}$ another application of Proposition 3.6 finally yields

$$|\hat{a}^{(n)}(\lambda)| \leq \int_0^{\frac{1}{|\lambda|}} t^n a(t)dt + n!(2^{n+1}\frac{(k+n)!}{(k-1)!} - 1) \int_0^{\frac{1}{|\lambda|}} \frac{t^n}{n!}a(t)dt$$

$$= 2^{n+1}\frac{(k+n)!}{(k-1)!} \int_0^{\frac{1}{|\lambda|}} t^n a(t)dt, \quad \text{Re } \lambda \geq 0, \ n \leq k - 1.$$

Summarizing we have

Proposition 3.8 *Let* $a \in L^1_{loc}(\mathbb{R}_+)$ *be k-monotone for some* $k \geq 1$. *Then*

$$|\hat{a}^{(n)}(\lambda)| \leq 2^{n+1}\frac{(2n+1)!}{n!} \int_0^{\frac{1}{|\lambda|}} t^n a(t)dt, \quad \text{Re } \lambda \geq 0, \ 0 \leq n \leq k - 1. \quad (3.51)$$

To derive lower bounds for $\hat{a}(\lambda)$ we write $\lambda = \sigma + i\rho$; then

$$\sqrt{2}|\hat{a}(\lambda)| \geq \operatorname{Re} \hat{a}(\lambda) - \operatorname{Im} \hat{a}(\lambda) = \int_0^\infty a_\sigma(t) \cos \rho t\, dt + \int_0^\infty a_\sigma(t) \sin \rho t\, dt$$

$$= \int_0^{\frac{\pi}{2|\rho|}} a_\sigma(t) \cos \rho t\, dt + \int_0^\infty (a_\sigma(t) - a_\sigma(t + \frac{\pi}{\rho})) \sin \rho t\, dt$$

$$\geq (\cos 1) \int_0^{\frac{1}{|\rho|}} a_\sigma(t)\, dt \geq \frac{1}{2} e^{-\frac{\sigma}{|\rho|}} \int_0^{\frac{1}{|\rho|}} a(t)\, dt.$$

if $a(t)$ is nonnegative and convex; observe that $a(t)$ is convex if and only if $a(t) - a(t+h)$ is nonincreasing in t, for every $h > 0$. This yields

Proposition 3.9 *Let $a \in L^1_{loc}(\mathbb{R}_+)$ be 2-monotone. Then*

$$2\int_0^{\frac{1}{|\rho|}} a(t)\, dt \geq |\hat{a}(i\rho)| \geq (2\sqrt{2})^{-1} \int_0^{\frac{1}{|\rho|}} a(t)\, dt, \quad \text{for all } \rho \neq 0. \tag{3.52}$$

Observe that the constants appearing in Propositions 3.8 and 3.9 do not depend on the particular kernel in question, they only depend on the property of k-monotonicity.

Next we derive estimates for $\operatorname{Re} \hat{a}(i\rho)$ and $-\operatorname{Im} \hat{a}(i\rho)$ from above and below.

Proposition 3.10 *Let $a \in L^1_{loc}(\mathbb{R}_+)$ be 2-monotone. Then*
(i) $\frac{3}{5}\rho \int_0^{\frac{1}{\rho}} ta(t)\, dt \leq -\operatorname{Im} \hat{a}(i\rho) \leq 12\rho \int_0^{\frac{1}{\rho}} ta(t)\, dt$, $\rho > 0$;
(ii) $\operatorname{Re} \hat{a}(i\rho) \leq 4\int_0^{\frac{1}{\rho}} -t\dot{a}(t)\, dt$, $\rho > 0$.
(iii) *If a is also 3-monotone then* $\operatorname{Re} \hat{a}(i\rho) \geq \frac{3}{5} \int_0^{\frac{1}{\rho}} -t\dot{a}(t)\, dt$, $\rho > 0$.

Proof: By (3.46) with $k = 2$ we have

$$\hat{a}(i\rho) = \frac{a_\infty}{i\rho} - \frac{1}{(i\rho)^2} \int_0^\infty (1 - i\rho t - e^{-i\rho t})\, d\dot{a}(t),$$

hence with $0 \leq t - \sin t \leq \min(1+t, t^3/3!)$

$$-\operatorname{Im} \hat{a}(i\rho) = \frac{a_\infty}{\rho} + \frac{1}{\rho^2} \int_0^\infty (\rho t - \sin \rho t)\, d\dot{a}(t)$$

$$\leq \frac{a_\infty}{\rho} + \frac{1}{\rho^2} \int_{\frac{1}{\rho}}^\infty (\rho t + 1)\, d\dot{a}(t) + \frac{1}{\rho^2} \int_0^{\frac{1}{\rho}} \frac{(\rho t)^3}{3!}\, d\dot{a}(t) \tag{3.53}$$

$$= \frac{a_\infty}{\rho} - \frac{2}{\rho^2}\dot{a}(\frac{1}{\rho^2}) + \frac{1}{\rho}(a(\frac{1}{\rho}) - a_\infty) + \rho \int_0^{\frac{1}{\rho}} \frac{t^3}{3!}\, d\dot{a}(t) \leq 12\rho \int_0^{\frac{1}{\rho}} ta(t)\, dt$$

by Proposition 3.6. Similarly, with $t - \sin t \geq \frac{t^3}{3!}(1 - \frac{t^2}{20}) \geq \frac{t^3}{10}$ for $t \in [0,1]$, we obtain

$$-\operatorname{Im} \hat{a}(i\rho) \geq \frac{a_\infty}{\rho} + \frac{6\rho}{10}\int_0^{\frac{1}{\rho}} \frac{t^3}{3!}d\dot{a}(t) + \frac{1}{\rho^2}\int_{\frac{1}{\rho}}^\infty (\rho t - \sin \rho t)d\dot{a}(t)$$

$$\geq \frac{7a_\infty}{10\rho} + \frac{6\rho}{10}[\frac{1}{6\rho^3}\dot{a}(\frac{1}{\rho}) - \frac{1}{2\rho^2}a(\frac{1}{\rho}) + \int_0^{\frac{1}{\rho}} ta(t)dt] + \frac{3}{10\rho}a(\frac{1}{\rho}) - \frac{7}{10\rho^2}(1 - \sin 1)\dot{a}(\frac{1}{\rho})$$

$$\geq \frac{7a_\infty}{10\rho} + \frac{6}{10}\rho\int_0^{\frac{1}{\rho}} ta(t)dt \geq \frac{3}{5}\rho\int_0^{\frac{1}{\rho}} ta(t)dt,$$

since $7(1 - \sin 1) \geq 1$; this proves (i).

On the other hand, with $1 - \cos t \leq \min(t^2/2, 2)$

$$\operatorname{Re} \hat{a}(i\rho) = \frac{1}{\rho^2}\int_0^\infty (1 - \cos \rho t)d\dot{a}(t) \leq \frac{2}{\rho^2}\int_{\frac{1}{\rho}}^\infty d\dot{a}(t) + \int_0^{\frac{1}{\rho}} \frac{t^2}{2}d\dot{a}(t)$$

$$\leq -\frac{2}{\rho^2}\dot{a}(1/\rho) + \frac{1}{2\rho^2}\dot{a}(1/\rho) - \int_0^{\frac{1}{\rho}} t\dot{a}(t)dt \leq 4\int_0^{\frac{1}{\rho}} -t\dot{a}(t)dt, \tag{3.54}$$

since by convexity

$$-\frac{1}{2\rho^2}\dot{a}(1/\rho) = (\int_0^{\frac{1}{\rho}} t\,dt)(-\dot{a}(1/\rho)) \leq \int_0^{\frac{1}{\rho}} -t\dot{a}(t)dt.$$

This proves (ii).

To derive the bound from below for $\operatorname{Re} \hat{a}(\lambda)$ we use (3.48) with $k = 3$.

$$\hat{a}(i\rho) = \frac{a_\infty}{i\rho} - \frac{1}{(i\rho)^3}\int_0^\infty (1 - i\rho t + \frac{(-i\rho t)^2}{2} - e^{-i\rho t})d\ddot{a}(t),$$

which implies

$$\operatorname{Re} \hat{a}(i\rho) = -\frac{1}{\rho^3}\int_0^\infty (\rho t - \sin \rho t)d\ddot{a}(t).$$

Estimating $t - \sin t$ as above we obtain

$$-\frac{1}{\rho^3}\int_0^{\frac{1}{\rho}}(\rho t - \sin \rho t)d\ddot{a}(t) \geq -\frac{6}{10}\int_0^{\frac{1}{\rho}} \frac{t^3}{3!}d\ddot{a}(t)$$

$$= -\frac{6}{10}[\frac{1}{6\rho^3}\ddot{a}(1/\rho) - \frac{1}{2\rho^2}\dot{a}(1/\rho) + \int_0^{\frac{1}{\rho}} t\dot{a}(t)dt].$$

On the other hand,

$$-\rho^3\int_{\frac{1}{\rho}}^\infty (\rho t - \sin \rho t)d\ddot{a}(t) \geq -\frac{3}{10\rho^3}\int_{\frac{1}{\rho}}^\infty (\rho t - 1)d\ddot{a}(t) - \frac{7}{10\rho^3}\int_{\frac{1}{\rho}}^\infty (1 - \sin 1)d\ddot{a}(t)$$

$$\geq \frac{3}{10\rho^2}\int_{\frac{1}{\rho}}^\infty \ddot{a}(t)dt + \frac{1}{10\rho^3}\ddot{a}(1/\rho) \geq -\frac{3}{10\rho^2}\dot{a}(1/\rho) + \frac{1}{10\rho^3}\ddot{a}(1/\rho).$$

Combining these inequalities yields (iii). $\quad\square$

4 Subordination

The class of completely positive kernels plays a prominent role in the theory of vector-valued Volterra equations, and appears in applications quite naturally. This class of kernels, its properties and associated creep functions are discussed thoroughly in this section. By means of the principle of subordination it is possible to construct new resolvents from a given one, e.g. from a C_0-semigroup or from a cosine family. The new resolvent can be explicitly represented in terms of the given one, and of the propagation function associated with a completely positive kernel. This representation is particularly useful for the understanding of the regularity and the asymptotic behaviour of the resolvent.

4.1 Bernstein Functions

The theorem of Bernstein on completely monotonic functions will play a fundamental role in this section. Before we state this result let us recall the definition of a completely monotonic function on $(0, \infty)$.

Definition 4.1 *A C^∞-function $f : (0, \infty) \to \mathbb{R}$ is called* **completely monotonic** *if*

$$(-1)^n f^{(n)}(\lambda) \geq 0 \quad \text{for all } \lambda > 0, \ n \in \mathbb{N}_0. \tag{4.1}$$

The class of completely monotonic functions will be denoted by $\mathcal{CM}$.

Bernstein's theorem gives the characterization of completely monotonic functions as Laplace transforms of positive measures supported on $\mathbb{R}_+$.

Bernstein's Theorem *A C^∞-function $f : (0, \infty) \to \mathbb{R}$ is completely monotonic iff there is a nondecreasing function $b : \mathbb{R}_+ \to \mathbb{R}$, such that*

$$f(\lambda) = \int_0^\infty e^{-\lambda t} db(t), \quad \lambda > 0. \tag{4.2}$$

Normalizing $b(t)$ by '$b(0) = 0$ and $b(t)$ left-continuous', $b(t)$ is uniquely determined by f. Moreover,

$$(-1)^n f^{(n)}(\lambda) = \int_0^\infty e^{-\lambda t} t^n db(t), \quad \lambda > 0, \ n \in \mathbb{N}_0, \tag{4.3}$$

and

$$(-1)^n f^{(n)}(0^+) = \int_0^\infty t^n db(t), \quad n \in \mathbb{N}_0.$$

Bernstein's Theorem shows in particular, that every $f \in \mathcal{CM}$ admits a holomorphic extension to $\mathbb{C}_+$. The class $\mathcal{CM}$ is easily seen to be closed under pointwise addition, multiplication and convergence. However, the composition of completely monotonic functions is in general not completely monotonic, for this another class of functions is needed.

Definition 4.2 *A C^∞-function $h : (\alpha, \beta) \to \mathbb{R}$ is called* **absolutely monotonic** *on (α, β) if $h^{(n)}(\mu) \geq 0$ for all $\mu \in (\alpha, \beta), n \in \mathbb{N}_0$. Let $\mathcal{AM}(\alpha, \beta)$ denote this class of functions.*

Frequently, the following result on composition of functions will be used.

Proposition 4.1 *Let $h \in \mathcal{AM}(\alpha, \beta)$ and $f \in \mathcal{CM}$ such that*

$$f(0+) \leq \beta \quad \text{and} \quad f(\infty) \geq \alpha. \tag{4.4}$$

Then $h \circ f$ defined by $(h \circ f)(\lambda) = h(f(\lambda))$, $\lambda > 0$, is completely monotonic.

The proof follows by an application of the chain rule and the product rule for differentiation of composite functions. Observe that $(h \circ f)(\lambda) = h(f(\lambda))$ is well-defined on $(0, \infty)$ by (4.4).

Definition 4.3 *A C^∞-function $\varphi : (0, \infty) \to \mathbb{R}$ is called a* **Bernstein function** *if $\varphi(\lambda) \geq 0$ for $\lambda > 0$ and φ' is completely monotonic. The class of Bernstein functions will be denoted by $\mathcal{BF}$.*

The class $\mathcal{BF}$ is easily seen to be closed under pointwise addition, multiplication with positive numbers and convergence, however, in general not under multiplication. Concerning composition we have

Proposition 4.2 *Let $f \in \mathcal{CM}$ and $\varphi, \psi \in \mathcal{BF}$. Then*
(i) $f \circ \varphi$ is completely monotonic;
(ii) $\psi \circ \varphi$ is a Bernstein function.

The proof again follows by application of the chain and the product rule of differentiation. The following respresentation of Bernstein functions will be of central importance in the sequel; it is a consequence of Bernstein's Theorem.

Proposition 4.3 *A C^∞-function $\varphi : (0, \infty) \to \mathbb{R}$ is a Bernstein function iff there exist unique constants $k_0, k_\infty \geq 0$, a unique function $k_1 \in L^1_{loc}(\mathbb{R}_+)$, nonnegative and nonincreasing with $\lim_{t \to \infty} k_1(t) = 0$ such that*

$$\varphi(\lambda) = \lambda\left(k_0 + \frac{k_\infty}{\lambda} + \hat{k}_1(\lambda)\right), \quad \lambda > 0. \tag{4.5}$$

Moreover,

$$k_0 = \lim_{\lambda \to \infty} \frac{\varphi(\lambda)}{\lambda} \quad \text{and} \quad k_\infty = \varphi(0+). \tag{4.6}$$

Proof: ($\Leftarrow$) Suppose $\varphi(\lambda)$ is of the form (4.5) with $k_0, k_\infty \geq 0$, and $k_1 \in L^1_{loc}(\mathbb{R}_+)$ nonnegative, nonincreasing. Then φ is of class C^∞ on $(0, \infty)$, $\varphi(\lambda) \geq 0$ for all $\lambda > 0$; therefore it is sufficient to show that $\varphi' \in \mathcal{CM}$, i.e. that $f(\lambda) = (d/d\lambda)(\lambda\hat{k}_1(\lambda))$

is completely monotonic. To verify this, assume first $k_1(0+) < \infty$. Then $k_1 \in BV(\mathbb{R}_+)$ and $\lambda \hat{k}_1(\lambda) = \widehat{dk_1}(\lambda)$, hence

$$l(t) = -\int_0^t s\,dk_1(s) = -\int_{0+}^t s\,dk_1(s)$$

is nondecreasing since k_1 is nonincreasing. Therefore

$$f(\lambda) = (\widehat{dk_1})'(\lambda) = -\widehat{tdk_1}(\lambda) = \widehat{dl}(\lambda)$$

is completely monotonic by Bernstein's Theorem. If $k_1(0+) = \infty$, we approximate $k_1(t)$ by $k_1^\varepsilon(t) = k_1(t+\varepsilon)$, $\varepsilon, t > 0$, and let

$$\varphi_\varepsilon(\lambda) = \lambda(k_0 + \frac{k_\infty}{\lambda} + \widehat{k_1^\varepsilon}(\lambda)), \quad \lambda > 0.$$

Then $\varphi_\varepsilon(\lambda)$ is a Bernstein function and $\varphi_\varepsilon(\lambda) \to \varphi(\lambda)$ pointwise as $\varepsilon \to 0+$, since

$$\widehat{k_1^\varepsilon}(\lambda) = e^{\lambda\varepsilon}\hat{k}_1(\lambda) - \int_0^\varepsilon e^{\lambda(\varepsilon-t)}k_1(t)dt \to \hat{k}_1(\lambda)$$

as $\varepsilon \to 0+$. Hence $\varphi(\lambda)$ is a Bernstein function.

($\Rightarrow$) Assume that $\varphi(\lambda)$ is a Bernstein function. Then $k_\infty = \varphi(0+)$ exists and is nonnegative; subtracting k_∞ from $\varphi(\lambda)$ we may assume w.l.o.g. $k_\infty = 0$. Since $\varphi(\lambda)$ is positive and concave,

$$k_0 = \lim_{\lambda\to\infty} \frac{\varphi(\lambda)}{\lambda} = \inf_{\lambda>0} \frac{\varphi(\lambda)}{\lambda}$$

exists and is nonnegative; subtracting $k_0\lambda$ from $\varphi(\lambda)$ we may assume w.l.o.g. $k_0 = 0$.

Next observe that

$$\varphi(\lambda) = \lambda \int_0^1 \varphi'(t\lambda)dt, \quad \lambda > 0;$$

therefore the function $\psi(\lambda) = \varphi(\lambda)/\lambda$ is completely monotonic. Bernstein's Theorem provides nondecreasing functions k, l, normalized by $k(0) = l(0) = 0$ and left-continuity, such that $\varphi(\lambda) = \lambda\widehat{dk}(\lambda)$ and $\varphi'(\lambda) = \widehat{dl}(\lambda)$ for $\lambda > 0$. Then

$$\frac{\hat{l}(\lambda)}{\lambda} = \frac{\varphi'(\lambda)}{\lambda^2} = \frac{(\lambda^2\hat{k}(\lambda))'}{\lambda^2} = 2\frac{\hat{k}(\lambda)}{\lambda} + \hat{k}'(\lambda) = 2\frac{\hat{k}(\lambda)}{\lambda} - \widehat{tk}(\lambda),$$

i.e.

$$tk(t) = 2\int_0^t k(\tau)d\tau - \int_0^t l(\tau)d\tau, \quad t > 0.$$

From this identity and $k(0^+) = k(0) = 0$ (which follows from $k_0 = 0$) we obtain $k \in W_{loc}^{1,1}(\mathbb{R}_+)$, and with $k_1(t) = \dot{k}(t)$

$$tk_1(t) = k(t) - l(t), \quad \text{for a.a. } t > 0.$$

Extend k_1 to all of $(0, \infty)$ by means of this relation; then $k_1(t)$ is nonnegative, left-continuous and $tk_1 \in BV_{loc}(\mathbb{R}_+)$, hence $k_1 \in BV[\varepsilon, 1/\varepsilon]$, for every $\varepsilon > 0$.

We now show that $k_1(t)$ is nonincreasing. For this purpose, let $p \in C^1(0, \infty)$ be nonnegative and with compact support; then

$$0 \le \int_0^\infty p(t)dl(t) = -\int_0^\infty p'(t)l(t)dt = \int_0^\infty (tp(t))'k_1(t)dt = -\int_0^\infty tp(t)dk_1(t),$$

hence $\int_0^\infty q(\tau)dk_1(\tau) \le 0$ for all nonnegative $q \in C^1(0, \infty)$ with compact support, and so also for all such $q \in C(0, \infty)$. This implies $k_1(t)$ nonincreasing. Finally, we have

$$\lim_{t \to \infty} k_1(t) = \lim_{\lambda \to 0} \lambda \hat{k}_1(\lambda) = \lim_{\lambda \to 0} \varphi(\lambda) = \varphi(0+) = 0,$$

and from this also (4.6) follows. □

Decomposing $k_1(t) = k_2(t) + k_3(t)$, $t > 0$, where

$$k_2(t) = \max(k_1(t) - k_1(1), 0), \quad t > 0, \tag{4.7}$$

and

$$k_3(t) = \min(k_1(t), k_1(1)), \quad t > 0, \tag{4.8}$$

(4.5) becomes

$$\varphi(\lambda) = \lambda k_0 + \lambda \hat{k}_2(\lambda) + k_\infty + \widehat{dk_3}(\lambda), \quad \lambda > 0, \tag{4.9}$$

since $k_3 \in BV(\mathbb{R}_+)$. As a consequence of (4.9) we obtain

Corollary 4.1 *Let $\varphi(\lambda)$ be a Bernstein function. Then $\varphi(\lambda)$ admits a continuous extension to $\overline{\mathbb{C}}_+$ - again denoted by $\varphi(\lambda)$ - which is holomorphic in $\mathbb{C}_+$, and satisfies $\mathrm{Re}\, \varphi(\lambda) > 0$ for all $\mathrm{Re}\, \lambda > 0$.*

Proof: Since $k_2 \in L^1(\mathbb{R}_+)$ and $k_3 \in BV(\mathbb{R}_+)$, (4.9) shows that $\varphi(\lambda)$ can be continuously extended to $\overline{\mathbb{C}}_+$, such that the extension which is again denoted by $\varphi(\lambda)$ is holomorphic in $\mathbb{C}_+$. To see that $\mathrm{Re}\, \varphi(\lambda) > 0$ for $\mathrm{Re}\, \lambda > 0$, we assume first $k_1(0+) < \infty$. Then $\varphi(\lambda) = \lambda k_0 + k_\infty + \widehat{dk_1}(\lambda)$, $\mathrm{Re}\, \lambda > 0$, and since $k_1(t)$ is nonincreasing

$$\mathrm{Re}\, \widehat{dk_1}(\lambda) = k_1(0+) + \mathrm{Re} \int_{0+}^\infty e^{-\lambda t}dk_1(t) \ge k_1(0+) + \int_{0+}^\infty dk_1(t) = 0,$$

hence $\mathrm{Re}\, \varphi(\lambda) \ge 0$ for all $\mathrm{Re}\, \lambda \ge 0$. If $k_1(0+) = \infty$, approximate $k_1(t)$ by $k_1^\varepsilon(t) = k_1(t + \varepsilon)$, $t, \varepsilon > 0$, as in the proof of Proposition 4.3; then $\varphi_\varepsilon(\lambda) = \lambda k_0 + k_\infty + \lambda \hat{k}_1^\varepsilon(\lambda) \to \varphi(\lambda)$ on $\overline{\mathbb{C}}_+$ as $\varepsilon \to 0+$, and therefore $\mathrm{Re}\, \varphi(\lambda) \ge 0$ on $\overline{\mathbb{C}}_+$. But we cannot have $\mathrm{Re}\, \varphi(\lambda_0) = 0$ for some $\lambda_0 \in \mathbb{C}_+$ since $u(\lambda) = \mathrm{Re}\, \varphi(\lambda)$ is harmonic in $\mathbb{C}_+$. □

In virtue of the representation of Bernstein functions obtained in Proposition 4.3, it is convenient to introduce the following

Definition 4.4 *A function* $k : (0, \infty) \to \mathbb{R}$ *is called a* **creep function** *if* $k(t)$ *is nonnegative, nondecreasing, and concave. A creep function* $k(t)$ *has the* **standard form**

$$k(t) = k_0 + k_\infty t + \int_0^t k_1(\tau) d\tau, \quad t > 0, \tag{4.10}$$

where $k_0 = k(0+) \geq 0$, $k_\infty = \lim_{t \to \infty} k(t)/t = \inf_{t>0} = k(t)/t \geq 0$, *and* $k_1(t) = \dot{k}(t) - k_\infty$ *is nonnegative, nonincreasing,* $\lim_{t \to \infty} k_1(t) = 0$. *The class of creep functions will be denoted by* $\mathcal{CF}$.

Thus Proposition 4.3 states that $\varphi \in \mathcal{BF}$ iff $\varphi(\lambda) = \lambda \widehat{dk}(\lambda)$ for some $k \in \mathcal{CF}$. Observe that $\mathcal{BF} \subset \mathcal{CF}$. The class of Bernstein functions φ which are represented as $\varphi(\lambda) = \lambda \widehat{dk}(\lambda)$ with some $k \in \mathcal{BF}$ will be called $\mathcal{CBF}$, the class of *complete Bernstein functions*. We have the inclusions $\mathcal{CBF} \subset \mathcal{BF} \subset \mathcal{CF}$.

4.2 Completely Positive Kernels

Let $\varphi(\lambda)$ be a Bernstein function. Then by Proposition 4.2, the function $1/\varphi(\lambda) = (\frac{1}{z} \circ \varphi)(\lambda)$ is completely monotonic, and so by Bernstein's theorem there exists a unique nondecreasing $c \in BV_{loc}(\mathbb{R}_+)$ such that $\widehat{dc} = 1/\varphi(\lambda)$ for $\lambda > 0$. On the other hand, by Proposition 4.3, $\varphi(\lambda) = \lambda \widehat{dk}(\lambda)$, $\lambda > 0$, for some creep function $k(t)$. Thus we have the relation

$$\widehat{dc}(\lambda)\widehat{dk}(\lambda) = \frac{1}{\lambda}, \quad \lambda > 0, \tag{4.11}$$

hence by uniqueness of the Laplace transform

$$\int_0^t c(t-\tau)dk(\tau) = \int_0^t k(t-\tau)dc(\tau) = t, \quad t > 0, \tag{4.12}$$

i.e. $c(t)$ solves the Volterra equation of the first kind (4.12). We have proved the first part of

Proposition 4.4 *Let* $k \in BV_{loc}(\mathbb{R}_+)$ *be a creep function. Then there is a unique nondecreasing function* $c \in BV_{loc}(\mathbb{R}_+)$ *such that (4.12) holds. Moreover, we have*

$$k_\infty > 0 \Leftrightarrow c(\infty) < \infty, \quad k_\infty c(\infty) = 1, \tag{4.13}$$

and

$$c(0+)k_0 = 0, \quad c(0+)(k_1(0+) + k_\infty) = 1 \quad \text{if } k_0 = 0. \tag{4.14}$$

(4.13) follows from the identity

$$c(\infty) = \lim_{\lambda \to 0+} \widehat{dc}(\lambda) = \lim_{\lambda \to 0+} \frac{1}{k_\infty + \lambda k_0 + \lambda \hat{k}_1(\lambda)} = \frac{1}{k_\infty},$$

while the first part of (4.14) is implied by

$$k_0 c(0+) = k_0 \lim_{\lambda \to \infty} \widehat{dc}(\lambda) = \lim_{\lambda \to \infty} \frac{k_0}{\lambda k_0 + \lambda \hat{k}_1(\lambda) + k_\infty} = 0;$$

for the second part of (4.14) see below. Concerning the regularity of $c(t)$, we have to distinguish three cases.

Case 1: $k_0 > 0$.
Let c_2 denote the solution of the Volterra equation

$$k_0 c_2 + (k_1 + k_\infty) * c_2 = k_1 + k_\infty. \tag{4.15}$$

Then

$$\hat{c} = \frac{1}{\lambda^2 \widehat{dk}} = \frac{1}{\lambda^2 k_0}(1 - \hat{c}_2), \quad \text{i.e. } c(t) = k_0^{-1} t - k_0^{-1} t * c_2(t);$$

this implies that c belongs to $W_{loc}^{2,1}(\mathbb{R}_+)$, and

$$c_1(t) = \dot{c}(t) = k_0^{-1} - k_0^{-1} \int_0^t c_2(\tau) d\tau, \quad -\dot{c}_1(t) = k_0^{-1} c_2(t), \ t > 0,$$

Therefore we have the relations

$$c_1(0+) = k_0^{-1}, \quad \frac{-\dot{c}_1(0+)}{c_1(0+)^2} = k_1(0+) + k_\infty, \tag{4.16}$$

since

$$c_2(0+) = \lim_{t \to 0} c_2(t) = \lim_{t \to 0} \frac{(k_1(t) + k_\infty)}{k_0} = \frac{(k_1(0+) + k_\infty)}{k_0}. \quad \square$$

Case 2: $k_0 = 0, \quad k_1(0+) < \infty$.
Define $k_2(t) = k_1(0+) - k_1(t)$ for $t > 0$; then $k_2 \in BV^0(\mathbb{R}_+)$. We have

$$\widehat{dc}(\lambda) = \frac{1}{\lambda \widehat{dk}(\lambda)} = \frac{1}{\lambda \hat{k}_1(\lambda) + k_\infty} = \frac{1}{k_1(0+) + k_\infty - \widehat{dk}_2(\lambda)}$$

$$= \frac{1}{k_1(0+) + k_\infty} \cdot \sum_{n \geq 0} \left(\frac{\widehat{dk}_2(\lambda)}{k_1(0+) + k_\infty} \right)^n, \quad \lambda > 0,$$

i.e. $c(t)$ has a jump at $t = 0$ and

$$(k_1(0+) + k_\infty)c(0+) = 1. \tag{4.17}$$

If $k_1(t)$ is continuous on $(0, \infty)$, then $c(t)$ has no other jumps, and if $k_1(t)$ is even absolutely continuous on $(0, \infty)$ then $c(t)$ is as well.

Case 3: $k_0 = 0, \quad k_1(0+) = \infty$.

Then $c(t)$ cannot have jumps, since otherwise $dc(t) \geq \gamma\delta(t - t_0)$ for some $\gamma > 0$, $t_0 \geq 0$, hence

$$1 = (k_1 + k_\infty) * dc \geq \gamma(k_1 + k_\infty) * \delta(t - t_0) = \gamma(k_1(t - t_0) + k_\infty)$$

for $t > t_0$, which yields a contradiction to $k_1(0+) = \infty$. However, the singular part of $c(t)$ can still be nontrivial, as the example

$$k_1(t) = 1 + \sum_{n=1}^{\infty} e_0(2^{-n} - t), \quad k_\infty = k_0 = 0$$

shows. On the other hand, if $c(t)$ is absolutely continuous on $(0, \infty)$ then $\dot{c}(0+) = \infty$; in fact, otherwise we would have

$$1 = \lim_{\lambda\to\infty} \lambda\hat{c}(\lambda) \cdot (\hat{k}_1(\lambda) + \frac{k_\infty}{\lambda}) = \lim_{\lambda\to\infty} \lambda\hat{c}(\lambda) \cdot \lim_{\lambda\to\infty} (\hat{k}_1(\lambda) + \frac{k_\infty}{\lambda}) = \dot{c}(0+) \cdot 0 = 0.$$

If $k_1(t)$ is absolutely continuous on $(0, \infty)$, then $-t\dot{k}_1(t)$ is integrable near zero; via the representation

$$\widehat{tdc}(\lambda) = -\widehat{dc}'(\lambda) = -[\widehat{dc}(\lambda)]^2 [\widehat{t\dot{k}_1}(\lambda)], \quad \lambda > 0,$$

there follows that $c(t)$ is absolutely continuous on $(0, \infty)$. □

Definition 4.5 *Let $c \in BV_{loc}(\mathbb{R}_+)$. The measure dc is called* **completely positive** *if there is a creep function such that (4.12) holds. If c is absolutely continuous on each interval $[0, \tau]$ and dc is completely positive then we call $\dot{c}(t)$ a* **completely positive function**.

Observe that a completely positive measure dc is Laplace-transformable for $\lambda > 0$, $\widehat{dc}(\lambda)$ is positive, completely monotonic, and satisfies

$$\widehat{dc}(\lambda)\widehat{dk}(\lambda) = \frac{1}{\lambda}, \quad \lambda > 0. \tag{4.18}$$

We shall frequently make use of the function $s(t; \mu)$ introduced in Section 1.3.

$$s(t; \mu) + \mu \int_0^t s(t - \tau; \mu)dc(\tau) = 1, \quad t, \mu > 0. \tag{4.19}$$

The next proposition contains several different characterizations of completely positive measures.

Proposition 4.5 *Suppose $c \in BV_{loc}(\mathbb{R}_+)$ is Laplace transformable and such that $\widehat{dc}(\lambda) > 0$ for all $\lambda > 0$. Then the following assertions are equivalent:*
(i) dc is completely positive;
(ii) $\varphi(\lambda) = 1/\widehat{dc}(\lambda)$ is a Bernstein Function;

(iii) $\psi_\tau(\lambda) = \exp\left(-\tau/\widehat{dc}(\lambda)\right)$ *is completely monotonic, for every* $\tau > 0$;
(iv) $\varphi_\mu(\lambda) = \widehat{dc}(\lambda)/(1 + \mu\widehat{dc}(\lambda))$ *is completely monotonic, for every* $\mu > 0$;
(v) $s(t, \mu)$ *is positive and nonincreasing w.r.t.* $t > 0$, *for every* $\mu > 0$.

Proof: (i) $\Rightarrow$ (ii). Since dc is completely positive, there is a creep function $k(t)$, such that (4.18) holds. Then $\varphi(\lambda) = 1/\widehat{dc}(\lambda) = \lambda\widehat{dk}(\lambda)$ is a Bernstein function by Proposition 4.3.

(ii) $\Rightarrow$ (iii). If $\varphi(\lambda) = 1/\widehat{dc}(\lambda)$ is a Bernstein function, then

$$\psi_\tau(\lambda) = \exp(-\tau/\widehat{dc}(\lambda)) = (\exp(-\tau z) \circ \varphi)(\lambda)$$

is the composition of the completely monotonic function $\exp(-\tau z)$ and a Bernstein function; Proposition 2 yields $\psi_\tau \in \mathcal{CM}$ for every $\tau > 0$.

(iii) $\Rightarrow$ (iv). Let $\psi_\tau \in \mathcal{CM}$ for every $\tau > 0$. We have the identity

$$\varphi_\mu(\lambda) = \frac{\widehat{dc}(\lambda)}{1 + \mu\widehat{dc}(\lambda)} = \frac{1}{\mu + \varphi(\lambda)} = \int_0^\infty e^{-\tau\varphi(\lambda)}e^{-\mu\tau}d\tau = \int_0^\infty e^{-\mu\tau}\psi_\tau(\lambda)d\tau$$

in view of $\varphi(\lambda) > 0$ for $\lambda > 0$; so $\varphi_\mu \in \mathcal{CM}$ for each $\mu > 0$, since $\mathcal{CM}$ forms a closed convex cone.

(iv) $\Rightarrow$ (v). Suppose $\varphi_\mu \in \mathcal{CM}$ for all $\mu > 0$. Since $\mu\varphi_\mu \leq 1$ for all $\lambda, \mu > 0$,

$$\lambda\hat{s}(\lambda; \mu) = (1 + \mu\widehat{dc}(\lambda))^{-1} = 1 - \frac{\mu\widehat{dc}(\lambda)}{1 + \mu\widehat{dc}(\lambda)} = 1 - \mu\varphi_\mu(\lambda)$$

is a Bernstein function. By Proposition 4.3 there are creep functions $k^\mu(t)$ such that

$$\lambda\hat{s}(\lambda; \mu) = \lambda\widehat{dk^\mu}(\lambda) = \lambda k_0^\mu + k_\infty^\mu + \lambda\widehat{k_1^\mu}(\lambda), \lambda > 0.$$

By uniqueness of the Laplace transform this yields $k_0^\mu = 0$ and

$$s(t; \mu) = k_\infty^\mu + k_1^\mu(t),$$

i.e. $s(t; \mu)$ is positive and nonincreasing.

(v) $\Rightarrow$ (ii). Let $s(t; \mu)$ be positive and nonincreasing. Then

$$\mu\varphi(\lambda)/(\mu + \varphi(\lambda)) = \mu\lambda\hat{s}(\lambda; \mu)$$

is a Bernstein function, and so $\varphi(\lambda) = \lim_{\mu\to\infty} \mu\lambda\hat{s}(\lambda; \mu)$ is as well since $\mathcal{BF}$ is closed.

(ii) $\Rightarrow$ (i). Let $\varphi(\lambda) = 1/\widehat{dc}(\lambda)$ be a Bernstein function. Then by Proposition 4.3 there is a creep function $k(t)$ such that $\varphi(\lambda) = \lambda\widehat{dk}(\lambda)$, i.e. (4.18) holds. Hence dc is completely positive. $\square$

Corollary 4.2 *Let dc be a completely positive measure, and let $\phi_\mu(\lambda) = \lambda\hat{s}(\lambda; \mu) = \varphi(\lambda)/(\mu + \varphi(\lambda))$, $\lambda, \mu > 0$. Then $\phi_\mu(\lambda)$ is a Bernstein function; the corresponding completely positive measure is $dc_\mu = \delta_0 + \mu dc$, $s(0+; \mu) = (1 + \mu c(0+))^{-1}$, and*

$s(\infty; \mu) = (1 + \mu c(\infty))^{-1}$. Moreover, $1/\varphi_\mu(\lambda)$ is a Bernstein function and there is a completely positive measure $dr(\cdot; \mu)$ such that $\widehat{dr}(\lambda; \mu) = \varphi_\mu(\lambda)$.

Observe that we have the following relations between $s(t; \mu)$ and $r(t; \mu)$.

$$s(t; \mu) + \mu r(t; \mu) = 1, \quad r(t; \mu) = \int_0^t s(t - \tau; \mu) dc(\tau), \quad t, \mu > 0. \qquad (4.20)$$

The definition of completely positive measures is rather indirect, and therefore we now give sufficient conditions for measures to be completely positive. For this purpose we need

Lemma 4.1 Let $b \in L^1_{loc}(0, \infty)$ be positive, nonincreasing, and log-convex. Then there is $r \in L^1(\mathbb{R}^+) \cap C(0, \infty)$, $r(t) \geq 0$ for $t > 0$, such that $r + b * r = b$. Moreover, we have $0 \leq r(t) \leq b(t)$, for all $t > 0$ and

$$\int_0^\infty r(t) dt = \hat{r}(0) = \frac{\hat{b}(0)}{1 + \hat{b}(0)} \leq 1.$$

If $b \in \mathcal{CM}$ then $r \in \mathcal{CM}$.

For a proof of Lemma 4.1 we refer to the monograph of Gripenberg, Londen, Staffans [156] or to the original papers of Friedman [120], Miller [241] and Reuter [290].

Consider a creep function $c \in W^{1,1}_{loc}(\mathbb{R}_+)$ such that $\dot{c}(t)$ is in addition log-convex. With $b(t) = \mu\dot{c}/(1 + \mu c_0)$, (4.19) is equivalent to $s + b * s = \gamma$, where $\gamma = (1 + \mu c_0)^{-1}$. Since $b(t)$ satisfies the assumption of Lemma 4.1 we obtain $s(t; \mu) = \gamma(1 - \int_0^t r(\tau) d\tau)$, i.e. $s(t; \mu) \geq 0$ and $\dot{s}(t; \mu) = -\gamma r(t) \leq 0$ for all $t > 0$. Therefore $s(t; \mu)$ is nonnegative and nonincreasing, and so Proposition 4.5 yields dc completely positive.

If in addition $c_1 \in \mathcal{CM}$, then $r \in \mathcal{CM}$ as well, by Lemma 4.1, but also $s \in \mathcal{CM}$. Therefore $\mu \int_0^t s(\tau; \mu) dt$ is a Bernstein function, for each $\mu > 0$. From the identity

$$\frac{\partial}{\partial \mu} \hat{s}(\lambda; \mu) = \frac{1}{\lambda} \frac{\partial}{\partial \mu} \frac{1}{1 + \mu \widehat{dc}(\lambda)} = -\frac{1}{\lambda} \frac{\widehat{dc}(\lambda)}{(1 + \mu \widehat{dc}(\lambda))^2} = -\hat{s}(\lambda; \mu) \widehat{dr}(\lambda; \mu)$$

we obtain

$$\frac{\partial}{\partial \mu} s(t; \mu) = -s(t; \mu) * dr(t; \mu) \leq 0, \quad \text{for all } t, \mu > 0,$$

and so $s_\infty(t) = \lim_{\mu \to \infty} s(t; \mu)$ exists for each $t \geq 0$. Since $0 \leq s(t; \mu) \leq 1$ holds, the latter implies $\hat{s}(\lambda; \mu) \to \hat{s}_\infty(\lambda) = 0$ as $\mu \to \infty$, hence $s_\infty(t) = 0$ a.e. (4.20) implies therefore $\mu r(t; \mu) = dc * \mu s(t; \mu) \to 1$ pointwise as $\mu \to \infty$, hence $\mu \int_0^t s(\tau; \mu) d\tau = \mu r(\cdot; \mu) * dk(t) \to k(t)$ for a.a. $t > 0$, and so $k(t)$ is a Bernstein function, in particular $k_1(t)$ is completely monotonic. We have proved

Proposition 4.6 *Let $c \in BV_{loc}(\mathbb{R}_+)$ be a creep function such that $c_1(t)$ is log-convex. Then dc is completely positive. If in addition $c \in \mathcal{BF}$, then $k(t)$ defined by (4.12) is also a Bernstein function.*

4.3 The Subordination Principle

Let $b \in BV_{loc}(\mathbb{R}_+)$ be nondecreasing, such that $\int_0^\infty db(t)e^{-\lambda t} < \infty$ for each $\lambda > 0$, and let dc be a completely positive measure. Then $f(\lambda) = \widehat{db}(\lambda)$ is completely monotonic, and $\varphi(\lambda) = 1/\widehat{dc}(\lambda)$ is a Bernstein function. By Proposition 4.2, $f \circ \varphi$ is completely monotonic, hence there is $a \in BV_{loc}(\mathbb{R}_+)$ nondecreasing, such that

$$\widehat{da}(\lambda) = f(\varphi(\lambda)) = \widehat{db}(\frac{1}{\widehat{dc}(\lambda)}), \quad \lambda > 0. \tag{4.21}$$

This is the so-called *subordination principle* for completely positive measures. We want to use this principle to construct new resolvents $S_a(t)$ from a given resolvent $S_b(t)$. Before we can do this, the subordination principle has to be studied in some detail.

Proposition 4.7 *Let $b, c \in BV_{loc}(\mathbb{R}_+)$ be such that $\int_0^\infty |db(t)|e^{-\beta t} < \infty$, $\beta \geq 0$, and dc be completely positive. Define $\alpha \geq 0$ by*

$$\alpha = \widehat{dc}^{-1}(\frac{1}{\beta}) \quad if \ c(\infty) > \frac{1}{\beta}, \quad \alpha = 0 \ otherwise. \tag{4.22}$$

Then there is a function $a \in BV_{loc}(\mathbb{R}_+)$ with $\int_0^\infty |da(t)|e^{-\alpha t} < \infty$ such that (4.21) holds for all $\lambda > \alpha$. Moreover,
(a) b nondecreasing $\Rightarrow$ a nondecreasing;
(b) db completely positive $\Rightarrow$ da completely positive;
(c) b, c Bernstein functions $\Rightarrow$ a Bernstein function;
(d) $b, c \in W^{1,1}_{loc}(\mathbb{R}_+)$ $\Rightarrow$ $a \in W^{1,1}_{loc}(\mathbb{R}_+)$.

Proof: Let b and c satisfy the assumptions of Proposition 4.7; by means of the Jordan decomposition of b we may w.l.o.g. assume that $b(t)$ is nondecreasing. From the definition of α in (4.22) it is clear that $g(\lambda) = \widehat{db}(1/\widehat{dc}(\lambda))$ is well-defined for $\lambda \geq \alpha$ and $g(\lambda) \leq \widehat{db}(1/\widehat{dc}(\alpha)) = \widehat{db}(\beta) < \infty$ on $[\alpha, \infty)$. It follows from Proposition 4.2 that $g(\lambda)$ is completely monotonic on (α, ∞), hence Bernstein's theorem yields a nondecreasing function $a(t)$ such that in addition $\int_0^\infty da(t)e^{-\alpha t} = g(\alpha) < \infty$. This proves the statement as well as (a); (b) follows from Proposition 4.2 (ii) and from Proposition 4.5.

To prove (c) let $b, c \in \mathcal{BF}$. Then by Bernstein's theorem, there is a function $\beta \in BV_{loc}(\mathbb{R}_+)$ with $\int_1^\infty d\beta(s)/s < \infty$ such that

$$\widehat{db}(z) = b_0 + \int_0^\infty \frac{1}{z + \mu} d\beta(\mu), \quad z > 0;$$

hence we have

$$\hat{a}(\lambda) = \frac{1}{\lambda}\widehat{db}(\frac{1}{\widehat{dc}(\lambda)}) = \frac{b_0}{\lambda} + \int_0^\infty \frac{\widehat{c}(\lambda)}{1 + \mu\widehat{dc}(\lambda)} d\beta(\mu), \quad \lambda > 0.$$

This yields the representation

$$a(t) = b_0 + \int_0^\infty r(t;\mu)d\beta(\mu), \quad t > 0,$$

where $r(t;\mu)$ denotes as usual the solution of $r + \mu dc * r = c$. This integral converges absolutely for each $t > 0$, since $\mu r(t;\mu) \leq 1$ holds. By Lemma 4.1, $c \in \mathcal{BF}$ implies $r(\cdot;\mu) \in \mathcal{BF}$ for each $\mu > 0$, and so $a(t)$ is again a Bernstein function.

It remains to prove (d). For this purpose we let

$$BV_\beta(\mathbb{R}_+) = \{b \in BV_{loc}(\mathbb{R}_+) : |b|_\beta = \int_0^\infty |db(t)|e^{-\beta t} < \infty\};$$

$BV_\beta(\mathbb{R}_+)$ equipped with the norm $|b|_\beta$ is a Banach space. We have shown that the linear operator $C : BV_\beta(\mathbb{R}_+) \to BV_\alpha(\mathbb{R}_+)$ defined by $Cb = a$, with a given by (4.21) is well-defined. (4.21) also implies the closedness of C, hence C is bounded by the closed graph theorem. Since

$$AC_\beta(\mathbb{R}_+) = BV_\beta(\mathbb{R}_+) \cap W_{loc}^{1,1}(\mathbb{R}_+)$$

is a closed subspace of $BV_\beta(\mathbb{R}_+)$, it is therefore sufficient to show that C maps a subset $E \subset AC_\beta(\mathbb{R}_+)$ with $\overline{\text{span}E} = AC_\beta(\mathbb{R}_+)$ into $AC_\alpha(\mathbb{R}_+)$. So let $c \in AC_{loc}(\mathbb{R}_+)$ be such that dc is completely positive and set $E = \{e^{-\mu t} : \mu > -\beta\}$. Then $\overline{\text{span}E} = AC_\beta(\mathbb{R}_+)$ and for $b \in E$ we have

$$\widehat{da}(\lambda) = \frac{\widehat{c}(\lambda)}{1 + \mu\widehat{c}(\lambda)}, \quad \lambda \geq \alpha.$$

Since $1 + \mu\widehat{c}(\lambda) \neq 0$ for Re $\lambda \geq \alpha$, the Paley-Wiener lemma implies $a \in AC_\alpha(\mathbb{R}_+)$. This completes the proof. $\square$

Consider now the Volterra equation

$$u(t) = f(t) + \int_0^t b(t - \tau)Au(\tau)d\tau, \quad t > 0, \tag{4.23}$$

where A is closed linear and densely defined, the kernel $b \in L_{loc}^1(\mathbb{R}_+)$ is such that $\int_0^\infty |b(t)|e^{-\beta t}dt < \infty$, and let $c(t)$ be a completely positive function. Define $a \in L_{loc}^1(\mathbb{R}_+)$ by

$$\hat{a}(\lambda) = \hat{b}(\frac{1}{\hat{c}(\lambda)}), \quad \lambda \geq \alpha, \tag{4.24}$$

according to Proposition 4.7, where α is given by

$$\alpha = \hat{c}^{-1}(\frac{1}{\beta}) \text{ if } \int_0^\infty c(t)dt > \frac{1}{\beta}, \quad \alpha = 0 \text{ otherwise.} \tag{4.25}$$

The Volterra equation

$$v(t) = g(t) + \int_0^t a(t - \tau)Av(\tau)d\tau, \quad t > 0, \tag{4.26}$$

is then called *subordinate* to (4.23) via $c(t)$. We now prove that (4.26) admits a resolvent whenever (4.23) does; this is the general *subordination principle for resolvents*.

Theorem 4.1 *Let A be a closed linear and densely defined operator in X, and let $b, c \in L^1_{loc}(\mathbb{R}_+)$ such that $\int_0^\infty |b(t)| e^{-\beta t} dt < \infty$ for some $\beta \in \mathbb{R}$. Assume*
(i) $c(t)$ is a completely positive function;
(ii) there is a resolvent $S_b(t)$ of (4.23) of type (M, ω_b), $\omega_b \geq 0$. // Then there is a resolvent $S_a(t)$ for (4.26). Moreover, let ω_a be defined by

$$\omega_a = \hat{c}^{-1}(\frac{1}{\omega_b}) \text{ if } \int_0^\infty c(t) dt > \frac{1}{\omega_b}, \quad \omega_a = 0 \text{ otherwise;} \tag{4.27}$$

then there is a constant $M_a \geq 1$ such that

$$\begin{aligned}
|S_a(t)| &\leq M_a e^{\omega_a t}, & t > 0, & \text{ if } \omega_b = 0, & \text{ or } & \omega_b \hat{c}(0) \neq 1, \\
|S_a(t)| &\leq M_\varepsilon e^{\varepsilon t}, & t > 0, & \text{ if } \omega_b > 0 & \text{ and } & \omega_b \hat{c}(0) = 1,
\end{aligned} \tag{4.28}$$

where $\varepsilon > 0$ is arbitrary.

Proof: Let A, b, c satisfy the assumptions of Theorem 4.1 and let $a(t)$ be defined by (4.24), according to Proposition 4.7. Then

$$\hat{S}_b(z) = \int_0^\infty e^{-z\tau} S_b(\tau) d\tau = \frac{1}{z}(I - \hat{b}(z)A)^{-1}, \quad z > \omega_b,$$

and so we obtain

$$H(\lambda) = \frac{1}{\lambda}(I - \hat{a}(\lambda)A)^{-1} = \frac{1}{\lambda\hat{c}(\lambda)} \cdot \hat{c}(\lambda)(I - \hat{b}(\frac{1}{\hat{c}(\lambda)})A)^{-1}$$

$$= \frac{1}{\lambda\hat{c}(\lambda)} \int_0^\infty e^{-\tau/\hat{c}(\lambda)} S_b(\tau) d\tau$$

for λ sufficiently large. With

$$h(\lambda; \tau) = \frac{1}{\lambda\hat{c}(\lambda)} e^{-\tau/\hat{c}(\lambda)}, \quad \lambda > 0, \tau \geq 0,$$

this yields the representation

$$H(\lambda) = \int_0^\infty h(\lambda; \tau) S_b(\tau) d\tau, \quad \lambda > \omega_a.$$

Proposition 4.5 shows that $h(\lambda; \tau)$ is completely monotonic on $(0, \infty)$ for each $\tau > 0$, hence with $L^n_\lambda = (-1)^n n!^{-1} (d/d\lambda)^n$ we obtain

$$|L^n_\lambda H(\lambda)| \leq \int_0^\infty L^n_\lambda h(\lambda; \tau) |S_b(\tau)| d\tau \leq M \int_0^\infty L^n_\lambda h(\lambda; \tau) e^{\omega_b \tau} d\tau$$

$$= ML_\lambda^n \int_0^\infty h(\lambda; \tau) e^{\omega_b \tau} d\tau = ML_\lambda^n \left[\frac{1}{\lambda(1 - \omega_b \hat{c}(\lambda))} \right] = ML_\lambda^n [\hat{s}(\lambda; -\omega_b)], \quad \lambda > \omega_a.$$

Suppose for a moment that

$$s(t; -\omega_b) \le m e^{\omega t}, \quad t > 0, \tag{4.29}$$

is satisfied. Since $s(t; -\omega_b) \ge 0$ we obtain by (4.29)

$$0 \le L_\lambda^n [\hat{s}(\lambda; -\omega_b)] \le \frac{m}{(\lambda - \omega)^{n+1}}, \quad \lambda > \omega, \ n \in \mathbb{N}_0, \tag{4.30}$$

hence

$$|L_\lambda^n H(\lambda)| \le \frac{mM}{(\lambda - \omega)^{n+1}}, \quad \lambda > \omega, \ n \in \mathbb{N}_0.$$

The generation theorem, Theorem 1.3, then applies and therefore equation (4.26) admits a resolvent of type (mM, ω).

To verify (4.29) we have to consider several cases.

a) $\omega_b = 0$

This is the simplest case, since then $s(t; -\omega_b) \equiv 1$, i.e. (4.29) holds with $m = 1$ and $\omega = \omega_a = 0$.

b) $\omega_b \hat{c}(0) < 1$, $\omega_b > 0$.

In this case $c \in L^1(\mathbb{R}_+)$ and $1 - \omega_b \hat{c}(\lambda) \ne 0$ for all Re $\lambda \ge 0$ since $c(t)$ is nonnegative; the Paley-Wiener Lemma then yields $r(\cdot; -\omega_b) \in L^1(\mathbb{R}_+)$, where

$$\hat{r}(\lambda; -\omega_b) = \frac{\hat{c}(\lambda)}{1 - \omega_b \hat{c}(\lambda)}, \quad \text{Re } \lambda \ge 0.$$

From this we obtain

$$s(t; -\omega_b) = 1 + \omega_b \int_0^t r(\tau; -\omega_b) d\tau, \quad t > 0,$$

bounded, i.e. (4.29) holds with $\omega = \omega_a = 0$ and $m = (1 - \omega_b \hat{c}(0))^{-1}$; observe that $r(t; -\omega_b) \ge 0$, i.e. $s(t; -\omega_b)$ is nonnegative and nondecreasing.

c) $\omega_b \hat{c}(0) > 1, \omega_b > 0$.

In this case $\omega_a > 0$, and $\lambda = \omega_a$ is a simple pole of $\hat{s}(\lambda; -\omega_b)$, since this function is completely monotonic, and analytic in $\mathbb{C}_+$, except for poles, which are the zeros of $1 - \omega_b \hat{c}(\lambda)$. Since by positivity of $c(t)$ no other poles are on the line Re$\lambda = \omega_a$, and $s(t; -\omega_b)$ is nonnegative, nondecreasing, we may apply Ikehara's Theorem (cp. Widder [339], Thm V. 17]) to the result

$$\lim_{t \to \infty} s(t; -\omega_b) e^{-\omega_a t} = -\frac{\hat{c}(\omega_a)}{\omega_a \hat{c}'(\omega_a)}.$$

This shows that (4.29) holds with $\omega = \omega_a$ and some $m \ge 1$.

d) $\omega_b \hat{c}(0) = 1, \omega_b > 0$.

Then $c \in L^1(\mathbb{R}_+)$, hence $\hat{s}(\lambda; -\omega_b)$ is analytic in $\mathbb{C}_+$, continuous in $\overline{\mathbb{C}}_+ \setminus \{0\}$, and

$\lambda = 0$ is a singularity of higher order, therefore $s(t; -\omega_b)$ cannot be bounded. However, by the Paley-Wiener lemma, $r(t; -\omega_b)e^{-\omega t}$ is integrable for every $\omega > 0$. We conclude that (4.29) holds for every $\omega > 0$. $\square$

It should be noted that there is a result similar to Theorem 4.1 for the integral resolvents $R_b(t)$ and $R_a(t)$ defined in Section 1.6; see Section 4.7.

4.4 Equations with Completely Positive Kernels

Suppose A is the generator of a C_0-semigroup of type (M, ω) in X, and let $c(t)$ be a completely positive function. Then (4.23) with $b(t) \equiv 1$ admits a resolvent of type (M, ω), and we are in position to apply Theorem 4.1. For the kernel $a(t)$ defined by (4.24) we obtain with $\hat{b}(z) = 1/z$

$$\hat{a}(\lambda) = \hat{b}(1/\hat{c}(\lambda)) = \hat{c}(\lambda), \quad \lambda > 0,$$

i.e. (4.26) becomes

$$v(t) = g(t) + \int_0^t c(t - \tau)Av(\tau)d\tau, \quad t > 0. \tag{4.31}$$

We have proved

Theorem 4.2 *Let A be the generator a C_0-semigroup of type (M, ω) in X, and let $c \in L^1_{loc}(\mathbb{R}_+)$ be a completely positive function. Then (4.31) admits a resolvent $S(t)$ in X which is exponentially bounded. Moreover, if $\omega_c \geq 0$ is defined by*

$$\omega_c = \hat{c}^{-1}(\frac{1}{\omega}) \text{ if } \int_0^\infty c(t)dt > \frac{1}{\omega}, \quad \omega_c = 0 \text{ otherwise }, \tag{4.32}$$

then $S(t)$ is of type (M_c, ω_c) if $\omega = 0$ or $\omega\hat{c}(0) \neq 1$, and of type $(M_\varepsilon, \varepsilon)$ for every $\varepsilon > 0$ if $\omega > 0$, $\omega\hat{c}(0) = 1$.

Next suppose A generates a cosine family $T(t)$ in X such that $|T(t)| \leq Me^{\omega t}$ for $t > 0$ holds, and let $c(t)$ be again completely positive. Then (4.23) with $b(t) = t$ admits a resolvent of type (M, ω). For the kernel $a(t)$ defined by (4.24) we obtain in this case

$$\hat{a}(\lambda) = \hat{b}(\frac{1}{\hat{c}(\lambda)}) = \frac{1}{\hat{c}(\lambda)^{-2}} = \hat{c}(\lambda)^2, \quad \lambda > 0,$$

since $\hat{b}(z) = 1/z^2$; this implies $a(t) = (c * c)(t)$. By Theorem 4.1, (4.26) admits an exponential bounded resolvent. In practice, however, (4.26) is given, and so there remains the problem to determine which kernels $a(t)$ are squares (with convolution as the multiplication) of completely positive functions; this amounts to showing that $\varphi(\lambda) = 1/\sqrt{\hat{a}(\lambda)}$ is a Bernstein function. A necessary condition for this obviously is $\hat{a}(\lambda) > 0$ for $\lambda > 0$, but of course it is far from sufficient.

A first sufficient condition is prevailed by Proposition 4.2; if $a(t)$ itself is completely positive, then $1/\hat{a}(\lambda)$ is a Bernstein function by Proposition 4.5, hence

$1/\sqrt{\hat{a}(\lambda)}$ is again a Bernstein function by Proposition 4.2, since $\sqrt{\lambda}$ belongs to this class.

From the applications point of view, creep functions $a(t)$ are important; see Section 5. This is in general not enough for $\varphi(\lambda) = 1/\sqrt{\hat{a}(\lambda)}$ to belong to the class $\mathcal{BF}$, however, we have the following.

Lemma 4.2 *Suppose $a(t) \not\equiv 0$ is a creep function with $a_1(t)$ log-convex. Then $\varphi(\lambda) = (\hat{a}(\lambda))^{-1/2}$ is a Bernstein function, and there is a completely positive function $c \in L^1_{loc}(\mathbb{R}_+)$ such that the factorization $a = c * c$ holds. In addition, if $a \in \mathcal{BF}$ then $c \in \mathcal{CM}$.*

Proof: Let $a(t)$ be a creep function with $a_1(t)$ log-convex and let $s \geq 0$. By Proposition 4.6, $da_s = da + sdt$ is a completely positive measure, hence $\varphi_s(\lambda) = 1/\widehat{da_s}(\lambda)$ is a Bernstein function for each $s \geq 0$. The complex variable formula

$$\frac{1}{\sqrt{z}} = \frac{1}{\pi} \int_0^\infty \frac{1}{z+r} \cdot \frac{dr}{\sqrt{r}}, \quad z \notin \mathbb{R}_-$$

then yields the representation

$$\varphi(\lambda) = \frac{1}{\pi} \int_0^\infty \frac{1}{\hat{a}(\lambda) + r} \frac{dr}{\sqrt{r}}, \quad \lambda > 0,$$

for $\varphi(\lambda) = (\hat{a}(\lambda))^{-1/2}$. The change of variables $r = s/\lambda^2$ gives

$$\varphi(\lambda) = \frac{1}{\pi} \int_0^\infty \varphi_s(\lambda) \frac{ds}{\sqrt{s}}, \quad \lambda > 0; \tag{4.33}$$

observe that the integral (4.33) for $\varphi(\lambda)$ is absolutely convergent, for each $\lambda > 0$, as are those for its derivatives. Since $\varphi_s(\lambda)$ is a Bernstein function for every $s \geq 0$, and the class $\mathcal{BF}$ is a closed convex cone, $\varphi(\lambda)$ is also a Bernstein function. This proves the first part of the Lemma.

Thus by Proposition 4.5, there is a completely positive measure dc such that $a = dc * dc$. The measure dc, however, cannot have a singular part since $a(t)$ is absolutely continuous on $(0, \infty)$.

The last part of the lemma is a special case of Lemma 4.3 below.   □

Summarizing we have the following result.

Theorem 4.3 *Let A generate a cosine family $T(t)$ in X such that $|T(t)| \leq Me^{\omega t}$ for $t > 0$ holds, and assume any of the following.*
(i) $a \in L^1_{loc}(\mathbb{R}_+)$ is completely positive;
(ii) $a(t)$ is a creep function with $a_1(t)$ log-convex;
(iii) $a = c * c$ with some completely positive $c \in L^1_{loc}(\mathbb{R}_+)$.
Then (4.36) admits a resolvent $S(t)$ with growth bound ω_c determined by

$$\omega_c = \hat{a}^{-1/2}(\omega^{-1}) \text{ if } \hat{a}(0) > \omega^{-2}, \quad \omega_c = 0 \text{ otherwise.} \tag{4.34}$$

In passing we mention one particular case of Theorem 4.3 which is well-known.

Corollary 4.3 *Let A generate a cosine family of type (M, ω), $\omega \geq 0$. Then A generates a C_0-semigroup of type $(M_1, \sqrt{\omega})$, for some $M_1 \geq 1$.*

In fact, choose $a(t) \equiv 1$ in Theorem 4.3 (ii) to see this.

As another application of Theorem 4.1 consider $b \in L^1_{loc}(\mathbb{R}_+)$ which is 1-regular and such that (4.23) is parabolic. Then by Theorem 3.1, (4.23) admits a bounded resolvent. If $c(t)$ is a completely positive function and $a(t)$ is defined by (4.24), then (4.26) is again parabolic, and Theorem 4.1 yields a bounded resolvent $S(t)$.

Corollary 4.4 *Suppose $b \in L^1_{loc}(\mathbb{R}_+)$ is 1-regular and such that (4.23) is parabolic; let $c(t)$ be a completely positive function. Then (4.26), with $a(t)$ defined by (4.24), is parabolic and admits a resolvent $S(t)$ of type $(M, 0)$, for some $M \geq 1$.*

Observe that in the situation of Corollary 4.4, although (4.26) is parabolic, Theorem 3.1 does not apply since it is not clear whether $a(t)$ is 1-regular again.

The second part of Lemma 4.2 is a special case of the following composition rule involving three Bernstein functions. It will be very useful in applications.

Lemma 4.3 *Suppose a, b, c are Bernstein functions. Then there is a Bernstein function e such that*

$$\hat{e}(\lambda) = \hat{a}(\lambda)\hat{dc}\left(\frac{\hat{a}(\lambda)}{\hat{b}(\lambda)}\right), \quad \lambda > 0.$$

Moreover, $e_0 = a_0(c_0 + \hat{c}_1(a_0/b_0))$ for $a_0 > 0$, $e_0 = 0$ otherwise.

Proof: (a) We first prove the lemma for the case $c(t) = 1 - e^{-t}$, i.e. we show that $\hat{a}\hat{b}/(\hat{a} + \hat{b})$ is a complete Bernstein function. Let $k(t)$ and $l(t)$ denote the solutions of $k * da \equiv t \equiv l * db$; by Proposition 4.6, $k, l \in \mathcal{BF}$, hence also $k + l \in \mathcal{BF}$ and so there is $m \in \mathcal{BF}$ such that $m * (dk + dl) \equiv t$, by Proposition 4.6 again. The Laplace transform of m is given by

$$\hat{m}(\lambda) = \frac{1}{\lambda^2(\widehat{dk}(\lambda) + \widehat{dl}(\lambda))} = \left(\frac{1}{\hat{a}(\lambda)} + \frac{1}{\hat{b}(\lambda)}\right)^{-1} = \frac{\hat{a}(\lambda)\hat{b}(\lambda)}{\hat{a}(\lambda) + \hat{b}(\lambda)}, \quad \lambda > 0.$$

(b) Consider now the general case. Since $c(t)$ is a Bernstein function by assumption, there is a function $\beta \in BV_{loc}(\mathbb{R}_+)$ such that

$$c_1(t) = \int_0^\infty e^{-st} d\beta(s), \quad t > 0,$$

and $\int_1^\infty d\beta(s)/s < \infty$. Let $f(\lambda) = \hat{a}(\lambda)\widehat{dc}(\hat{a}(\lambda)/\hat{b}(\lambda))$, $\lambda > 0$; then we have the representation

$$f(\lambda) = c_0\hat{a}(\lambda) + c_\infty\hat{b}(\lambda) + \int_0^\infty \frac{\hat{a}(\lambda)\hat{b}(\lambda)}{\hat{a}(\lambda) + s\hat{b}(\lambda)}d\beta(s), \quad \lambda > 0.$$

Let $m_s \in \mathcal{BF}$ be defined by $\widehat{m_s} = s\hat{a}\hat{b}/(\hat{a} + s\hat{b})$, $s > 0$; then for each $t > 0$, $m_s(t)$ is nondecreasing w.r.t. $s > 0$ and $\lim_{s \to \infty} m_s(t) = a(t)$, as well as $\lim_{s \to 0+} m_s(t)/s = b(t)$. Therefore, $e(t)$ defined by

$$e(t) = c_0 a(t) + c_\infty b(t) + \int_0^\infty m_s(t) d\beta(s)/s, \quad t > 0$$

exists, the integral being absolutely convergent. Since $a, b, m_s \in \mathcal{BF}$ there follows $e \in \mathcal{BF}$, for $\mathcal{BF}$ is a closed convex cone. Obviously, $\hat{e}(\lambda) = f(\lambda)$ and so the assertion of the lemma follows. □

The choice $c(t) = (2/\sqrt{\pi})\sqrt{t}$ and $b(t) = t$ yields

$$\hat{a}(\lambda)\widehat{dc}(\frac{\hat{a}(\lambda)}{\hat{b}(\lambda)}) = \hat{a}(\lambda)[\lambda^2 \hat{a}(\lambda)]^{-\frac{1}{2}} = \sqrt{\hat{a}(\lambda)}/\lambda,$$

therefore by Lemma 4.3 the second part of Lemma 4.2 follows. As another application of Lemma 4.3, $\hat{a}(\lambda)^{1-\alpha}\hat{b}(\lambda)^\alpha$ is the Laplace transform of a Bernstein function, whenever $a, b \in \mathcal{BF}$ and $\alpha \in [0, 1]$.

4.5 Propagation Functions

Let dc be a completely positive measure, $\varphi(\lambda) = 1/\widehat{dc}(\lambda)$ its associated Bernstein function, and let $k \in BV_{loc}(\mathbb{R}_+)$ be the creep function such that $\varphi(\lambda) = \lambda\widehat{dk}(\lambda)$, $\lambda > 0$. $k(t)$ then has the unique representation (4.10) and for brevity we let $\kappa, \omega, \alpha \in \mathbb{R}_+$ in the sequel be defined by

$$\kappa = k_0, \quad \omega = k_\infty, \quad \alpha = k_1(0+) + k_\infty. \tag{4.35}$$

By Proposition 4.5, the functions $\psi_\tau(\lambda) = \exp(-\tau\varphi(\lambda))$ are completely monotonic with respect to $\lambda > 0$, for each fixed $\tau \geq 0$, and bounded by $e^{-\tau\varphi(0+)} = e^{-\tau\omega}$. From Bernstein's theorem it follows that there are unique nondecreasing functions $w(\cdot; \tau) \in BV(\mathbb{R}_+)$, normalized by $w(0; \tau) = 0$ and left-continuity, such that

$$\widehat{w}(\lambda; \tau) = \frac{\psi_\tau(\lambda)}{\lambda}, \quad \lambda > 0, \tau \in \mathbb{R}_+. \tag{4.36}$$

Obviously, $w(\cdot; \tau)$ enjoys the semigroup property

$$\int_0^t w(t - s; \tau_1) dw(s; \tau_2) = w(t; \tau_1 + \tau_2), \quad t, \tau_1, \tau_2 \geq 0,$$

$$w(t; 0) = e_0(t), \quad t \geq 0, \tag{4.37}$$

where $e_0(t)$ denotes the Heaviside function. In the sequel $w(t; \tau)$ will be called the *propagation function* associated with the completely positive measure dc (or with the Bernstein function φ). The properties of $w(t; \tau)$ will be studied in detail below.

The space $BV(\mathbb{R}_+)$ becomes a commutative Banach algebra with unit $e_0(t)$, when the multiplication $\bullet$ in $BV(\mathbb{R}_+)$ is defined by

$$(a \bullet b)(t) = (a * db)(t) = (b * da)(t), \quad t \geq 0, \tag{4.38}$$

and the norm in $BV(\mathbb{R}_+)$ is as usual

$$||a|| = \text{Var } a|_0^\infty = \int_0^\infty |da(t)|; \tag{4.39}$$

see e.g. Gelfand, Raikov and Shilov [128]. By $BV^+(\mathbb{R}_+)$ we denote the closed convex cone of nondecreasing functions in $BV(\mathbb{R}_+)$; this space then becomes a Banach lattice w.r.t. the ordering generated by $BV^+(\mathbb{R}_+)$. By virtue of the pairing

$$< p, a > = \int_0^\infty p(t) da(t), \quad p \in C_0(\mathbb{R}_+), \ a \in BV(\mathbb{R}_+), \tag{4.40}$$

$BV(\mathbb{R}_+)$ can be identified with the dual of $C_0(\mathbb{R}_+)$, and therefore carries the weak*-topology of $C_0(\mathbb{R}_+)$.

Consider the map $w : \mathbb{R}_+ \to BV(\mathbb{R}_+)$ defined by

$$w(\tau)(t) = w(t; \tau), \quad t, \tau \geq 0; \tag{4.41}$$

the identity (4.37) then shows that the family $\{w(\tau)\}_{\tau \geq 0}$ forms a semigroup in $BV(\mathbb{R}_+)$ which moreover is positive and satisfies $||w(\tau)|| \leq e^{-\tau\omega}$, $\tau \geq 0$, i.e. is of type $(1, -\omega)$. Besides, w is weak*-continuous; in fact, with $e_\lambda(t) = e^{-\lambda t}$, $t > 0$, we have

$$< e_\lambda, w(\tau) > = \psi_\tau(\lambda), \quad \lambda > 0, \ \tau \in \mathbb{R}_+, \tag{4.42}$$

hence the functions $< e_\lambda, w(\cdot) >$ are continuous on $\mathbb{R}_+$; since $||w(\tau)|| \leq 1$ and span $\{e_\lambda : \lambda > 0\}$ is dense in $C_0(\mathbb{R}_+)$ it follows that w is weak*-continuous. From the definition of $\psi_\tau(\lambda)$ we obtain

$$\widehat{dc}\frac{\partial}{\partial\tau}\hat{w}(\tau) + \hat{w}(\tau) = 0,$$
$$\hat{w}(0) = \hat{e}_0, \tag{4.43}$$

which formally is equivalent to the Volterra equation

$$w(t; \tau) + \int_0^t \frac{\partial}{\partial\tau}w(t - s; \tau) dc(s) = 0, \quad t, \tau \geq 0,$$
$$w(t; 0) = 1, \ t > 0; \ w(0; \tau) = 0, \quad \tau \geq 0. \tag{4.44}$$

Definition 4.6 *A function* $w : \mathbb{R}^+ \to BV(\mathbb{R}^+)$ *is called a* **solution** *of (4.44) if for each* $p \in C_c(\mathbb{R}_+)$, *the function* $< p, w(\cdot) >$ *is continuous on* $\mathbb{R}_+$, $< p, c \bullet w(\cdot) >$ *is continuously differentiable on* $\mathbb{R}_+$, *and*

$$\frac{d}{d\tau} < p, c \bullet w(\tau) > + < p, w(\tau) > = 0, \ w(0) = e_0, \quad \tau \geq 0, \tag{4.45}$$

is satisfied.

We now show that completely positive measures can be characterized in terms of (4.44).

Proposition 4.8 *Let $c \in BV_{loc}(\mathbb{R}_+)$ be of subexponential growth. Then the measure dc is completely positive iff (4.44) admits a solution $w : \mathbb{R}_+ \to BV^+(\mathbb{R}_+)$ such that $|w(\tau)| \leq 1$ for $\tau \geq 0$. If this is the case, the solution is unique.*

Proof: ($\Rightarrow$) Suppose dc is completely positive, and let $c \in BV(\mathbb{R}_+)$ first. Then (4.43) is equivalent to

$$< e_\lambda, c \bullet w(\tau) > + \int_0^\tau < e_\lambda, w(\sigma) > d\sigma = < e_\lambda, c > \qquad (4.46)$$

for all $\lambda > 0$. Since span$\{e_\lambda : \lambda > 0\}$ is dense in $C_0(\mathbb{R}_+)$, and $|w(\tau)| \leq 1$, (4.46) yields

$$< p, c \bullet w(\tau) > + \int_0^\tau < p, w(\sigma) > d\sigma = < p, c > \qquad (4.47)$$

for all $\tau \geq 0$ and $p \in C_0(\mathbb{R}_+)$. From this it follows that $w(\tau)$ is a solution of (4.44) in the sense of Definition 4.6.

If $c \notin BV(\mathbb{R}_+)$ we have to be a little more careful. Observe that (4.43) is equivalent to

$$< e_\lambda, [c \bullet w(\tau)]_\varepsilon > + \int_0^\tau < e_\lambda, [w(\sigma)]_\varepsilon > d\sigma = < e_\lambda, c_\varepsilon > \qquad (4.48)$$

for all $\lambda, \varepsilon > 0$, where we used the notation

$$a_\varepsilon(t) = \int_0^t e^{-\varepsilon s} da(s), \quad t \geq 0;$$

by assumption $c_\varepsilon \in BV(\mathbb{R}_+)$. Then (4.48) is equivalent to

$$< p, [c \bullet w(\tau)]_\varepsilon > + \int_0^\tau < p, [w(\sigma)]_\varepsilon > d\sigma = < p, c_\varepsilon > \qquad (4.49)$$

for all $p \in C_c(\mathbb{R}_+)$, and letting $\varepsilon \to 0$ we obtain

$$< p, c \bullet w(\tau) > + \int_0^\tau < p, w(\tau) > d\tau = < p, c > .$$

This implies that $w(\tau)$ is a solution of (4.44) in the sense of Definition 4.6.

($\Leftarrow$) Conversely, let $w : \mathbb{R}_+ \to BV^+(\mathbb{R}_+)$ be a solution of (4.44) with $|w(\tau)| \leq 1$, $\tau \geq 0$, and let $\int_0^\infty e^{-\varepsilon t}|dc(t)| < \infty$ for all $\varepsilon > 0$. Passing backwards from (4.45) to (4.43) we obtain

$$\widehat{dc} \cdot \widehat{w}(\tau) + \int_0^\tau \widehat{w}(\sigma)d\sigma = \widehat{c}, \quad \lambda, \tau > 0.$$

Let $\gamma = \widehat{dc}(\lambda)$ and $\varphi(\tau) = \widehat{w}(\tau)$, $\lambda > 0$ fixed. Then we obtain

$$\gamma\varphi'(\tau) + \varphi(\tau) = 0, \quad \tau \geq 0, \; \varphi(0) = 1;$$

therefore, $\gamma \neq 0$ since φ is of class C^1, and $\varphi(\tau) = e^{-\tau/\gamma}$, $\tau \geq 0$. Since $\|w(\tau)\| \leq 1$, $\tau \geq 0$, there follows $\gamma > 0$. Thus we have $\gamma = \widehat{dc}(\lambda) > 0$ for all $\lambda > 0$ and $\widehat{w}(\lambda;\tau) = \exp(-\tau/\widehat{dc}(\lambda))$ for all $\tau > 0$, $\lambda > 0$. Since $w(\tau) \in BV^+(\mathbb{R}_+)$, $\widehat{w}(\lambda;\tau)$ is completely monotonic w.r.t. $\lambda > 0$, for each $\tau \geq 0$. Proposition 4.5 now implies that dc is completely positive. Uniqueness of the solution follows from $\widehat{dc}(\lambda) > 0$ for all $\lambda > 0$. $\quad\square$

Some properties of the propagation function are collected in

Proposition 4.9 *Let dc be completely positive, and let w denote its associated propagation function. Then we have*
(i) $w(\cdot;\cdot)$ is Borel measurable on $\mathbb{R}_+ \times \mathbb{R}_+$;
(ii) $w(\cdot;\tau)$ is nondecreasing and left-continuous on $\mathbb{R}_+$, and

$$w(0;\tau) = 0, \; w(\infty;\tau) = e^{-\omega\tau}, \quad \text{for all } \tau > 0.$$

(iii) $w(t;\cdot)$ is nonincreasing and right-continuous on $\mathbb{R}_+$, and

$$w(t;0) = w(t;0+) = 1, \; w(t;\infty) = 0, \quad \text{for all } t > 0;$$

(iv) $w(t;\tau) = 0$ iff $t \leq \kappa\tau$, and $w(\kappa\tau;\tau-) = w(\kappa\tau+;\tau) = e^{-\alpha\tau}$, $\tau \geq 0$.
(v) $s(t;\mu)$ defined by (4.19) is represented as

$$s(t;\mu) = -\int_0^\infty e^{-\mu\tau} d_\tau w(t;\tau), \quad t, \mu > 0.$$

Proof: (i) follows from the Post-Widder inversion formula for the Laplace transform by virtue of

$$w(t;\tau) = \lim_{m\to\infty} w(t - \frac{1}{m};\tau) = \lim_{m\to\infty} \lim_{n\to\infty} L_\lambda^n \widehat{w}(\frac{n}{t - 1/m};\tau)(\frac{n}{t - 1/m})^{n+1}$$

which holds for every $t, \tau > 0$, $\tau \geq 0$; cp. Widder [339], Corollary VII. 6a.3.

(ii) is implied by Bernstein's theorem and by the normalization of functions of bounded variation; observe that $\lim_{t\to\infty} w(t;\tau) = \lim_{\lambda\to 0} \psi_\tau(\lambda) = e^{-\omega\tau}$, $\tau \geq 0$.

(iii) The semigroup property implies

$$w(t;\tau_1 + \tau_2) = \int_0^t w(t - s;\tau_1) dw(s;\tau_2) \leq w(\infty;\tau_1) w(t;\tau_2) \leq w(t;\tau_2)$$

for all $t, \tau_1, \tau_2 \geq 0$, since $w(\cdot;\tau)$ is nonnegative and nondecreasing; therefore $w(t;\cdot)$ is nonincreasing for fixed $t \geq 0$. Moreover, $w(t;0) = 1$ is obvious from $\psi_0(\lambda) \equiv 1$, and with $w(t;\infty) = \lim_{\tau\to\infty} w(t;\tau) \leq w(t;\tau)$ we obtain

$$0 \leq \widehat{w}(\lambda;\infty) \leq \widehat{w}(\lambda;\tau) = \frac{\psi_\tau(\lambda)}{\lambda} \to 0 \quad \text{as } \tau \to \infty, \; \lambda > 0,$$

hence $w(t; \infty) \equiv 0$ by uniqueness of the Laplace transform. Finally, observe that $w(\cdot; \tau + h) \xrightarrow{*} w(\cdot; \tau)$ as $h \to 0+$ implies $w(t; \tau + h) \to w(t; \tau)$ as $h \to 0$ for each $t \geq 0$, since $w(\cdot; \tau)$ is left-continuous; therefore $w(t; \cdot)$ is right-continuous.

(iv) Assume first $\kappa = 0$; we show that $w(t; \tau) > 0$ for all $t > 0$, $\tau \geq 0$. In fact, if $w(t; \tau) = 0$ for $0 < t \leq t_0$, and some fixed $\tau > 0$, then

$$\frac{1}{\lambda} e^{-\tau\varphi(\lambda)} = \int_0^\infty w(t; \tau) e^{-\lambda t} dt \leq \frac{1}{\lambda} e^{-\lambda t_0} w(\infty; \tau) = \frac{1}{\lambda} e^{-\lambda t_0 - \omega\tau},$$

hence $\tau\varphi(\lambda) \geq \lambda t_0 + \omega\tau$, which implies $\hat{k}_1(\lambda) \geq t_0/\tau > 0$ for all $\lambda > 0$; this yields a contradiction to $\hat{k}_1(\lambda) \to 0$ as $\lambda \to \infty$. Moreover, with $\lambda \hat{k}_1(\lambda) \to k_1(0+)$ as $\lambda \to \infty$ we obtain

$$\begin{aligned}
w(0+; \tau) &= \lim_{t \to 0+} w(t; \tau) = \lim_{\lambda \to \infty} \lambda \hat{w}(\lambda; \tau) = \exp(-\tau k_\infty - \tau \lim_{\lambda \to \infty} \lambda \hat{k}_1(\lambda)) \\
&= \exp(-(k_\infty + k_1(0+))\tau) = e^{-\alpha\tau}, \quad \tau \geq 0.
\end{aligned}$$

If $\kappa \neq 0$, then $\hat{w}(\lambda; \tau) = e^{-\kappa\lambda\tau} \hat{w}_1(\lambda; \tau)$, where $w_1(t; \tau)$ denotes the propagation function associated with the Bernstein function $\varphi_1(\lambda) = \varphi(\lambda) - \kappa\lambda$. Hence

$$w(t; \tau) = w_1(t - \kappa\tau; \tau) e_0(t - \kappa\tau), \quad t, \tau \geq 0,$$

and (iv) is proved.

(v) follows with an integration by parts from the identity

$$\left(\int_0^\infty e^{-\mu\tau} w(t, \tau) d\tau \right)^\wedge (\lambda) = \frac{1}{\lambda} \int_0^\infty e^{-\mu\tau} \psi_\tau(\lambda) d\tau = \frac{1}{\lambda} \frac{\widehat{dc}(\lambda)}{1 + \mu\widehat{dc}(\lambda)}. \quad \square$$

Compare the representation of $s(t; \mu)$ in Proposition 4.9 (v) with that in Section 3.6.

4.6 Structure of Subordinated Resolvents

Let $c \in L^1_{loc}(\mathbb{R}_+)$ be a completely positive function, $b \in L^1_{loc}(\mathbb{R}_+)$ Laplace transformable, and let A be a closed linear densely defined operator in X such that (4.23) admits a resolvent $S_b(t)$ of type (M, ω_b) in X. According to Theorem 4.1, there is an exponentially bounded resolvent $S_a(t)$ for (4.26), where $a \in L^1_{loc}(\mathbb{R}_+)$ is given by (4.24). Since

$$-\frac{\partial}{\partial\tau} \frac{e^{-\tau/\hat{c}(\lambda)}}{\lambda} = \frac{e^{-\tau/\hat{c}(\lambda)}}{\lambda\hat{c}(\lambda)}, \quad \tau, \lambda > 0,$$

and

$$\hat{S}_a(\lambda) = \frac{1}{\lambda\hat{c}(\lambda)} \int_0^\infty S_b(\tau) e^{-\tau/\hat{c}(\lambda)} d\tau = -\int_0^\infty S_b(\tau) \frac{\partial}{\partial\tau} \hat{w}(\lambda; \tau) d\tau,$$

with w as in Section 4.5, we obtain the following representation of $S_a(t)$ in terms of $S_b(t)$ and the propagation function $w(t; \tau)$ associated with c.

Corollary 4.5 *Let the assumptions of Theorem 4.1 be satisfied, and let $w(t;\tau)$ be the propagation function associated with $c \in L^1_{loc}(\mathbb{R}_+)$. Then the resolvents $S_a(t)$ and $S_b(t)$ of (4.26) and (4.23), respectively, are related by*

$$S_a(t) = -\int_0^\infty S_b(\tau)d_\tau w(t;\tau), \quad t > 0; \tag{4.50}$$

(4.50) holds in the strong sense.

Since $w(t;\cdot)$ is nonincreasing, and since $w(t;0) = 1$, $w(t;\infty) = 0$ holds, (4.51) shows that the resolvent $S_a(t)$ of (4.26) is the strong limit of convex combinations of $S_b(\tau)$. We note a simple consequence of this observation.

Corollary 4.6 *Under the assumptions of Corollary 4.5, let $D \subset X$ be a closed convex set, such that $S_b(t)D \subset D$ for all $t \geq 0$. Then $S_a(t)D \subset D$ for all $t \geq 0$. In particular, if $S_b(t)$ is positive with respect to some closed convex cone K, then $S_a(t)$ is positive as well.*

The structure of $S_a(t)$ will now be studied in more detail. For this purpose, we need the following regularity result for the propagation function.

Proposition 4.10 *Let dc be completely positive, let $w(t;\tau)$ denote the associated propagation function, and let κ, ω, α be as in (4.35). Then*

$$w(t;\tau) = e^{-\omega\tau}w_0(t - \kappa\tau;\tau) + e^{-\alpha\tau}e_0(t - \kappa\tau), \quad t,\tau > 0, \tag{4.51}$$

where $w_0(\cdot;\tau)$ is nondecreasing, left-continuous, $w_0(t;\cdot)$ is right-continuous, nonincreasing, and

$$w_0(t;\tau) > 0 \quad \text{iff } t > 0; \quad w_0(0+;\tau) = 0.$$

Moreover,
(i) $k_1 \in C(0,\infty)$ implies $w_0 \in C(\mathbb{R}_+ \times \mathbb{R}_+ \backslash \{(0,0)\})$;
(ii) $k_1 \in W^{1,1}_{loc}(0,\infty)$ implies $w_0 \in W^{1,1}_{loc}((0,\infty) \times (0,\infty))$.

Proof: Decomposition (4.51) and the properties of w_0 follows from Proposition 4.9; it remains to prove (i) and (ii). Observe first that we may assume $\kappa = \omega = 0$. In fact, $\kappa > 0$ only yields a shift of amount $\kappa\tau$ for w in t-direction, and $\omega > 0$ leads to multiplication with $e^{-\omega\tau}$; these operations do not change the regularity of w and w_0.

If $\alpha = \infty$, then $w = w_0$; if $\alpha < \infty$ then

$$\hat{w}_0(\lambda;\tau) = \frac{1}{\lambda}(e^{-\tau\varphi(\lambda)} - e^{-\alpha\tau}), \tag{4.52}$$

hence in both cases

$$\widehat{tw_0}(\lambda;\tau) = -\hat{w}'_0(\lambda;\tau) = \frac{1}{\lambda}\hat{w}_0(\lambda;\tau) + \tau\varphi'(\lambda)\hat{w}(\lambda;\tau), \quad \lambda,\tau > 0.$$

As in the proof of Proposition 4.3, we have the relation $\varphi'(\lambda) = (\lambda \hat{k}_1(\lambda))' = \widehat{dl}(\lambda)$, hence with $l(t) = -\int_0^t s \, dk_1(s)$

$$t w_0(t;\tau) = \int_0^t w_0(s;\tau) ds + \tau \int_0^t l(t-s) dw(s;\tau), \quad t, \tau \geq 0. \tag{4.53}$$

With $l(t) = \int_0^t k_1(s) ds - t k_1(t)$, we have $l(0) = l(0+) = 0$ and $l \in C(\mathbb{R}_+)$ in case $k_1 \in C(0,\infty)$. Then (4.53) implies $w_0(\cdot;\tau)$ continuous on $(0,\infty)$, uniformly for bounded τ; since $w(0+;\tau) = e^{-\alpha\tau}$ by Proposition 4.9 (iv), assertion (i) follows.

If $k_1 \in W^{1,1}_{loc}(0,\infty)$ then l is absolutely continuous, $\dot{l}(t) = -t \dot{k}_1(t)$ a.e., and so $w_0(\cdot;\tau)$ is absolutely continuous for each $\tau \geq 0$, $\dot{w}_0(\cdot;\tau) \in L^1(\mathbb{R}_+)$ for each $\tau \geq 0$, $|\dot{w}_0|_1 = w_0(\infty;\tau) \leq 1$. To prove that $\dot{w}_0(\cdot;\cdot)$ is product-measurable, let $\varphi_\varepsilon \in C_0^\infty(\mathbb{R})$ be mollifiers. Then $\varphi_\varepsilon * \dot{w}_0(\cdot;\tau) \to \dot{w}_0(\cdot;\tau)$ in $L^1(\mathbb{R})$ for each fixed $\tau > 0$, and from $\varphi_\varepsilon * \dot{w}_0(\cdot;\tau) = \dot{\varphi}_\varepsilon * w_0(\cdot;\tau)$ and continuity of w_0, the map $\tau \mapsto \varphi_\varepsilon * \dot{w}_0(\cdot;\tau)$ is continuous from $(0,\infty)$ to $L^1(\mathbb{R}_+)$. This implies that $\dot{w}_0 : (0,\infty) \to L^1(\mathbb{R}_+)$ is strongly measurable, hence $\dot{w}_0(\cdot;\cdot)$ is measurable on $\mathbb{R}_+ \times \mathbb{R}_+$. Finally,

$$-\frac{\partial}{\partial\tau}\hat{w}_0(\lambda;\tau) = \frac{\varphi(\lambda)}{\lambda}(e^{-\tau/\varphi(\lambda)} - e^{-\alpha\tau}) + \frac{\varphi(\lambda) - \alpha}{\lambda}e^{-\alpha\tau}, \quad \tau, \lambda > 0,$$

i.e. for each fixed $\tau > 0$ we have

$$-\frac{\partial}{\partial\tau}w_0(t;\tau) = k_1 * \dot{w}_0(t;\tau) + e^{-\alpha\tau}(k_1(t) - \alpha), \quad \text{for a.a. } \tau > 0.$$

This implies $w_0 \in W^{1,1}_{loc}((0,\infty) \times (0,\infty))$. □

Proposition 4.10 now allows for the following refinement of representation (4.51).

$$S_a(t) = S_b(t/\kappa)e^{-\alpha t/\kappa} + \int_0^{t/\kappa} S_b(\tau)v(t;\tau) d\tau, \quad t > 0, \tag{4.54}$$

where the function $v(t;\tau)$ is defined by

$$\begin{aligned} v(t;\tau) &= e^{-\omega\tau}(\kappa w_{0t}(t-\kappa\tau;\tau) - w_{0\tau}(t-\kappa\tau;\tau) + \omega w_0(t-\kappa\tau;\tau)) \\ &\quad + \alpha e^{-\alpha\tau}e_0(t-\kappa\tau), \quad t, \tau > 0; \end{aligned} \tag{4.55}$$

observe that $e^{-\alpha t/\kappa}$ resp. $\alpha e^{-\alpha\tau}$ have to be interpreted as zero in case $\kappa = 0$ or $\alpha = \infty$ resp. $\alpha = \infty$. The first term on the right hand side of (4.54) is just a damped dilatation of $S_b(t)$, and therefore has the same regularity properties as $S_b(t)$; however, the second term behaves better, we show that it is at least $\mathcal{B}(X)$-continuous.

Theorem 4.4 *Let A be a closed linear densely defined operator in X and $b, c \in L^1_{loc}(\mathbb{R}_+)$ Laplace transformable. Assume*
(i) $c(t)$ is a completely positive function, with creep function $k \in W^{2,1}_{loc}(0,\infty)$;

(ii) there is a resolvent $S_b(t)$ of (4.23) of type (M, ω_b).
Then the resolvent $S_a(t)$ for (4.26) is of the form

$$S_a(t) = S_b(t/\kappa)e^{-\alpha t/\kappa} + S_0(t), \quad t > 0, \tag{4.56}$$

where $S_0(t)$ is $\mathcal{B}(X)$-continuous on $(0, \infty)$, and even on $\mathbb{R}_+$ if $\kappa > 0$ and $\alpha < \infty$;
here $\kappa = k_0, \alpha = k_\infty + k_1(0+)$.

Proof: (a) We first prove the theorem for the special case $b(t) \equiv 1$, i.e. when A is the generator of a C_0-semigroup $T(t)$ in X. Define $\{R(t)\}_{t>0} \subset \mathcal{B}(X)$ by means of

$$R(t) = -\int_0^\infty \tau T(\tau)d_\tau w(t; \tau), \quad t > 0. \tag{4.57}$$

Since for $x \in D(A)$ we have

$$R(t)x = \int_0^\infty (\tau T(\tau)x)' w(t; \tau)d\tau = \int_0^\infty (T(\tau)x + \tau T(\tau)Ax)w(t; \tau)d\tau, \tag{4.58}$$

and since by Proposition 4.9

$$\begin{aligned}
|R(t)| &\leq -\int_0^\infty \tau |T(\tau)| d_\tau w(t; \tau) \leq -M \int_0^\infty \tau e^{\omega_b \tau} d_\tau w(t; \tau) \\
&= M\frac{\partial}{\partial \mu} \widehat{d_\tau w}(t; \mu)|_{\mu=-\omega_b} = -M\frac{\partial}{\partial \mu} s(t; \mu)|_{\mu=-\omega_b} \\
&= Ms(t; -\omega_b) * r(t; -\omega_b) \leq M_\varepsilon e^{(\omega_a + \varepsilon)t}, \quad t > 0,
\end{aligned}$$

holds for some constant $M_\varepsilon \geq 1$, it is clear from the uniform boundedness principle and Proposition 4.10 that $R(t)$ is strongly continuous on $\mathbb{R}_+$. By (4.55) and (4.57) we obtain

$$S_0(t) = \int_0^\infty T(\tau)v(t; \tau)d\tau, \quad t > 0,$$

hence by the measurability of $v(\cdot; \cdot)$

$$\begin{aligned}
\int_0^{t_0} |S_0(t+h) - S_0(t)|dt &\leq \int_0^\infty |T(\tau)| \int_0^{t_0} |v(t+h; \tau) - v(t; \tau)|dt d\tau \\
&\leq M \int_0^\infty e^{\omega_b \tau} \int_0^{t_0} |v(t+h; \tau) - v(t; \tau)|dt d\tau \to 0,
\end{aligned}$$

as $h \to 0$. This implies in particular $S_0 * S_a, S_a * S_0, S_0 * S_0 \in C(\mathbb{R}_+; \mathcal{B}(X))$, as direct estimates show. The relation

$$\begin{aligned}
\widehat{tS_a}(\lambda) &= -\int_0^\infty T(\tau)\frac{d}{d\lambda}(e^{-\tau/\hat{c}(\lambda)}/(\lambda\hat{c}(\lambda)))d\tau \\
&= \hat{S}(\lambda)/\lambda - (1/\hat{c})'(\lambda)\hat{c}(\lambda)\hat{S}(\lambda) + (1/\hat{c})'(\lambda)\hat{R}(\lambda),
\end{aligned}$$

as well as $(1/\hat{c})'(\lambda) = (\lambda\kappa + \omega + \lambda\hat{k}_1(\lambda))' = \kappa - t\hat{\dot{k}}_1(\lambda)$ then yield

$$tS_a(t) = [1 - \kappa c(t) + (t\dot{k}_1(t)) * c(t)] * S(t) - (t\dot{k}_1) * R(t) + \kappa R(t). \tag{4.59}$$

Similarly, the identity

$$
\begin{aligned}
\kappa\hat{R}(\lambda) &= \kappa(1/(\lambda\hat{c}(\lambda)))((z-A)^{-2}) \circ (1/\hat{c}(\lambda)) = \kappa(\hat{c}(\lambda)/\lambda)(I - \hat{c}(\lambda)A)^{-2} \\
&= \kappa\lambda\hat{c}(\lambda)\hat{S}_a(\lambda)^2 = \kappa(\kappa + \hat{k}_1(\lambda) + \omega/\lambda)^{-1}\hat{S}_a(\lambda)^2
\end{aligned}
$$

yields in case $\kappa > 0$

$$\kappa R(t) = (S_a * S_a)(t) + (q * S_a * S_a)(t), \quad t > 0, \tag{4.60}$$

where $q(t)$ is locally integrable on $\mathbb{R}_+$.

The semigroup property of $S_b(t) = T(t)$ together with (4.56) implies

$$S_a * S_a = S_a * S_0 + S_0 * S_a - S_0 * S_0 + tT(t/\kappa)e^{-\alpha t/\kappa},$$

hence combining (4.59), (4.60) with this equation, $\mathcal{B}(X)$-continuity of $tS_0(t)$ on $\mathbb{R}_+$ follows. In case $\kappa > 0$, $\alpha < \infty$, the continuity of $S_0(t)$ at $t = 0$ is a consequence of

$$\int_0^\infty v(t;\tau)d\tau = 1 - e^{-\alpha t/\kappa} \to 0 \quad \text{as } t \to 0.$$

(b) Consider $X = L^1(\mathbb{R}_+)$, $A = -d/d\tau$ with $\mathcal{D}(A) = W_0^{1,1}(\mathbb{R}_+)$; then $A + \omega_b I$ generates a C_0-semigroup $T(\tau)$ of type $(1,\omega_b)$ in X, which is given by

$$(T(\tau)f)(x) = e^{\omega_b\tau}f(x-\tau)e_0(x-\tau), \quad \tau, x > 0.$$

Therefore the resolvent $S(t)$ of (4.31), represented by (4.54) or (4.56) with $S_b(\tau) = T(\tau)$, results in

$$
\begin{aligned}
(S(t)f)(x) &= -\int_0^\infty (T(\tau)f)(x)d_\tau w(t;\tau) = -\int_0^x e^{\omega_b\tau}f(x-\tau)d_\tau w(t;\tau) \\
&= e^{(\omega_b-\alpha)t/\kappa}f(x-t/\kappa) + \int_0^x e^{\omega_b\tau}f(x-\tau)v(t;\tau)d\tau,
\end{aligned}
$$

i.e.

$$(S_0(t)f)(x) = \int_0^x e^{\omega_b\tau}f(x-\tau)v(t;\tau)d\tau, \quad t, x > 0.$$

$\mathcal{B}(X)$-continuity of $S_0(t)$ in this special case amounts to

$$\int_0^\infty e^{\omega_b\tau}|v(t;\tau) - v(\bar{t};\tau)|d\tau \to 0 \quad \text{as} \quad \bar{t} \to t > 0, \tag{4.61}$$

and also for $t = 0$ in case $\kappa > 0$.

(c) We consider now the general case. Then

$$S_0(t) = \int_0^\infty S_b(\tau)v(t;\tau)d\tau, \quad t > 0,$$

hence by step (b)

$$|S_0(t) - S_0(\bar{t})| \le M \int_0^\infty e^{\omega_b \tau} |v(t; \tau) - v(\bar{t}; \tau)| d\tau \to 0$$

as $\bar{t} \to t > 0$, and also for $t = 0$ in case $\kappa > 0$. The proof of Theorem 4.4. is complete. □

Uniform continuity of the resolvent $S_a(t)$ of (4.26) has a number of important consequences. For example, the variation of parameters formula (1.8) then yields mild solutions of (4.26) if only $f \in BV(J; X)$ rather than $f \in W^{1,1}(J; X)$; Proposition 1.2 holds with '$W^{1,1}$' replaced by 'BV' in this case. Another consequence is a simple characterization of compactness of subordinated resolvents.

Corollary 4.7 *Let the assumptions of Theorem 4.4 be satisfied. Then the subordinated resolvent $S_a(t)$ is compact for all $t > 0$ if and only if $(z_0 - A)^{-1}$ is compact for some $z_0 \in \rho(A)$ and either $S_b(t)$ is compact for all $t > 0$ or $1/\kappa + \alpha = \infty$.*

Proof: If $S_a(t)$ is compact for each $t > 0$ then the resolvent equation (1.6) for $t \to 0+$ with $x = (z_0 - A)^{-1}y$, $z_0 \in \rho(A)$ implies that $(z_0 - A)^{-1}$ is compact. This in turn implies $(a * S_a)(t)$ compact for all $t \ge 0$, and so $\hat{S}_a(\lambda) = 1/\hat{a}(\lambda)(a * S_a)^\wedge(\lambda)$ also has this property for all Re $\lambda > \omega_a$, since $a * S_a$ is $\mathcal{B}(X)$-continuous. If $1/\kappa + \alpha < \infty$ we have by (4.56)

$$\hat{S}_0(\lambda) = \hat{S}_a(\lambda) - \kappa \hat{S}_b(\lambda \kappa + \alpha) = \hat{S}_a(\lambda) - \frac{\kappa}{\lambda \kappa + \alpha}(I - \hat{b}(\lambda \kappa + \alpha)A)^{-1};$$

thus $\hat{S}_0(\lambda)$ is compact for all Re $\lambda > \omega_a$. The complex inversion formula for the Laplace transform

$$S_0(t) = \lim_{N \to \infty} \frac{e^{\gamma t}}{2\pi} \int_{-N}^N (1 - \frac{|\rho|}{N}) e^{i\rho t} \hat{S}_0(\gamma + i\rho) d\rho, \quad t > 0, \ \gamma > \omega_a,$$

then implies compactness of $S_0(t)$ for all $t > 0$, since this limit exists in the uniform operator topology, thanks to uniform continuity of $S_0(t)$. Therefore in case $1/\kappa + \alpha < \infty$, $S_b(t)$ must be compact for all $t > 0$.
Conversely, if $(z_0 - A)^{-1}$ is compact for some $z_0 \in \rho(A)$ then $\hat{S}_0(\lambda)$ and $\hat{S}_a(\lambda)$ are compact for all Re $\lambda > \omega_a$, hence as above $S_0(t)$ has this property for all $t > 0$. (4.56) then shows that $S_a(t)$ is compact for all $t > 0$ if either $1/\kappa + \alpha = \infty$ or $S_b(t)$ is compact for all $t > 0$. □

4.7 Comments
a) For a proof of Bernstein's Theorem see e.g. Widder [339], which contains also Proposition 4.1. The notion of *Bernstein functions* has been introduced by Berg and Forst [21], although Proposition 4.2 and a representation similar to Proposition 4.3 is already in Feller [115]. Proposition 4.3 is taken from Clément and Prüss [54].

b) The notion of *completely positive functions* has been introduced by Clément and Nohel [50], [51] where also the equivalence (i) $\Leftrightarrow$ (v) in Proposition 4.5 appears for the first time. The importance of this class of kernels has also been recognized in other fields; Berg and Forst [21] study *potential kernels*, Kingman [198] *p-standard functions*, and it turns out that up to certain normalizations these classes coincide. The equivalences of (i) $\sim$ (iv) appear also in Berg and Forst [21]. Proposition 4.4 is originally due to Gripenberg [152]; for the present approach see Clément and Prüss [54]. Lemma 4.1 is independently due to Friedman [120] and to Miller [241], while its last part was obtained by Reuter [290] before; see also Hirsch [182] and Gripenberg [153].

c) The *subordination principle* for completely positive measures was first observed by Bochner [26], its application to vector-valued Volterra equations appears in Prüss [276] where also Theorem 4.1 in a weaker form was obtained. Theorem 4.2 is due to Clément and Nohel [51] while Lemma 4.2 and Theorem 4.3 (ii) are proved in Prüss [274]. The present approach to Theorems 4.2, 4.3 and Corollary 4.4 based on the subordination principle is taken from Prüss [276].

d) In Friedman [121] it was observed that in case X is a Hilbert space, A negative definite, and $a = c$ is nonnegative, nonincreasing, and log-convex, the resolvent $S(t)$ of (1.1) has the property that $(S(t)x, x)$ is nonnegative, nonincreasing, for each $x \in X$. This is implied by the representation (1.24) of $S(t)$ via the functional calculus of selfadjoint operators in Theorem 1.1. Obviously this assertion remains valid if A is merely negative semidefinite and $a(t)$ a completely positive function, by Proposition 4.5.

e) The propagation function $w(t; \tau)$ associated with a completely positive measure dc is the main subject in potential theory, where the semigroups in $C_0(\mathbb{R})$ and $L^p(\mathbb{R})$ defined by $w(t; \tau)$ by means of

$$(T(\tau)f)(t) = \int_0^\infty f(t - s) d_s w(s; \tau), \quad t \in \mathbb{R} \qquad (4.62)$$

are studied. It has been shown that a positive translation invariant C_0-semigroup in $L^p(\mathbb{R})$ of contractions leaving invariant the subspace $L^p(\mathbb{R}_+)$ of $L^p(\mathbb{R})$ is given by (4.62), where $w(t; \tau)$ is the propagation function associated with a completely positive measure; see Berg and Forst [21] and Clément and Prüss [54]. The connection of the propagation function with Volterra equations was first observed and exploited in Prüss [274] which also contains Proposition 4.9 and Theorem 4.4 in weaker form. Proposition 4.8 is taken from Clément and Prüss [54]. A physical interpretation of $w(t; \tau)$ will be given in Section 5.4.

f) The following example shows that for each $k \in \mathbb{N}$ there are k-monotone functions $c(t)$ which are not completely positive.

Example 4.1: Define $c(t) = (1 - t)^l$ for $t \leq 1, c(t) = 0$ for $t > 1$; obviously $c(t)$ is $(l + 1)$-monotone. Suppose that $c(t)$ is also completely positive. Then there are

$k_0, k_\infty \geq 0, k_1(t)$ nonnegative and nonincreasing, such that $k_0 c + (k_1 + k_\infty) * c = 1$. This implies

$$k_0 \dot{c}(t) + (k_1 + k_\infty) * \dot{c}(t) = -(k_1(t) + k_\infty)c(0), \quad t \geq 0,$$

and

$$k_0 \ddot{c}(t) + (k_1 + k_\infty) * \ddot{c}(t) = -\dot{k}_1(t)c(0) - (k_1(t) + k_\infty)\dot{c}(0), \quad t \geq 0.$$

With $c(0) = 1, \dot{c}(0) = -l, \ddot{c}(0) = l(l-1)$ these equations yield

$$-k_0 l = -(k_1(0+) + k_\infty) \quad \text{and} \quad k_0 l(l-1) = -\dot{k}_1(0+) + l(k_1(0+) + k_\infty),$$

hence

$$0 \leq -\dot{k}_1(0+) = k_0 l(l-1) - l(k_1(0+) + k_\infty) = k_0 l(l-1-l) = -k_0 l,$$

i.e $k_0 l = 0$ and so $k_1(0+) + k_\infty = 0$, which contradicts $k_0 c + (k_1 + k_\infty) * c = 1$. $\square$

It is not known whether any completely positive function is 1-regular.

g) It follows from Proposition 4.5 that $\varepsilon\delta_0 + dc$ is completely positive whenever $\varepsilon > 0$ and dc is completely positive. However, the class of completely positive kernels is not closed under addition; in fact, $\varepsilon dt + dc$ is completely positive for each $\varepsilon > 0$ only if dc is completely positive and $c(t)$ is a creep function with $c_1(t)$ convex. On the other hand, the class of creep functions $c(t)$ with $c_1(t)$ log-convex - according to Proposition 4.6 dc is then completely positive - is closed under addition. It would be interesting to know whether the class of creep functions $c(t)$ such that $dc(t)$ is completely positive is closed under addition. This question arises naturally in viscoelasticity; see Section 5. In the same connection extensions of Lemma 4.2 and 4.3 in this direction would be of importance.

h) The subordination principle for integral resolvents mentioned at the end of Section 4.3 reads as follows.

Theorem 4.5 *Let A be a closed linear densely defined operator in X and $b, c \in L^1_{loc}(\mathbb{R}_+)$ such that $\int_0^\infty |b(t)| e^{-\beta t} dt < \infty$ for some $\beta \in \mathbb{R}$. Assume*
(i) $c(t)$ is a completely positive function;
(ii) there is an integral resolvent $R_b(t)$ of (4.23) such that $|R_b(t)| \leq \varphi_b(t) e^{\omega_b t}$ for a.a. $t \geq 0$, where $\omega_b \geq 0$ and $\varphi_b \in L^1(\mathbb{R}_+)$.
*   Then there is an integral resolvent $R_a(t)$ for (4.26). If moreover ω_a is defined according to (4.27) then*

$$|R_a(t)| \leq \varphi_a(t) e^{\omega_a t} \quad \text{for a.a. } t \geq 0, \text{ for some } \varphi_a \in L^1(\mathbb{R}_+). \tag{4.63}$$

For the proof we only verify $(K2)$ of Theorem 1.6; $(K1)$ follows from the definition of ω_a. With $K_a(\lambda) = (1/\hat{a}(\lambda) - A)^{-1}$ we obtain

$$\sum_{n \geq 0} (\lambda - \omega_a)^n |L_\lambda^n K_a(\lambda)| \leq \sum_{n \geq 0} (\lambda - \omega_a)^n \int_0^\infty |R_b(\tau)| L_\lambda^n e^{-\tau/\hat{c}(\lambda)} d\tau$$

$$\leq \int_0^\infty \varphi_b(\tau) e^{\omega_b \tau} \sum_{n \geq 0} (\lambda - \omega_a)^n L_\lambda^n e^{-\tau/\hat{c}(\lambda)} d\tau$$

$$= \int_0^\infty \varphi_b(\tau) e^{\omega_b \tau} e^{-\tau/\hat{c}(\omega_a)} d\tau = |\varphi_b|_1 < \infty, \quad \lambda > \omega_a.$$

Hence Theorem 1.6 implies existence of the integral resolvent $R_a(t)$ for (4.26) with property (4.63).

By means of the propagation function a representation of the integral resolvent can be given as well.

$$R_a(t) = \frac{d}{dt} \int_0^\infty R_b(\tau) w(t; \tau) d\tau, \quad t \geq 0;$$

observe that an invariance result similar to Corollary 4.6 is valid for the integral resolvent, too, as soon as $0 \in D$, or more generally, for $R_a(t)/c(t)$.

i) A compactness result similar to Corollary 4.7 is derived in Hannsgen and Wheeler [168] in case X is a Hilbert space, A is negative definite with A^{-1} compact, and $a(t)$ 2-monotone. Then $S(t)$ is compact for each $t > 0$ provided $-\dot{a}(0+) = \infty$.

j) If the kernel $c(t)$ appearing in the subordination principle is completely monotonic rather than completely positive, much more can be said about the regularity behaviour of the resolvent $S(t)$. A typical result in this direction, using the framework of resolvents, is the following result which is taken from Prüss [277]. It deals with the second order case, i.e. $b(t) \equiv t$, which is of interest in viscoelasticity; see Section 5.

Theorem 4.6 *Suppose A generates a cosine family $T(t)$ in X, and let $a(t)$ be a Bernstein function; define*

$$\chi(\rho) = \frac{-\mathrm{Im}\widehat{da}(i\rho)}{\rho \mathrm{Re}\widehat{da}(i\rho)}, \quad \rho \in \mathbb{R}, \tag{4.64}$$

and let $S(t)$ denote the resolvent for (4.26). Then
(i) if $\overline{\lim}_{\rho \to \infty} \rho \chi(\rho) < \infty$, then $S(t)$ is an analytic resolvent for (4.26);
(ii) if $\lim_{\rho \to \infty}[\chi(\rho) \log \rho] = 0$, then $S \in C^\infty((0, \infty); \mathcal{B}(X, X_A))$;
(iii) if $t_0 = 2\overline{\lim}_{\rho \to \infty} \chi(\rho) \log \rho > 0$, then $S \in C^n((t_n, \infty); \mathcal{B}(X))$ and $AS \in C^n((t_{n+2}, \infty); \mathcal{B}(X))$, where $t_n = (n+1)t_0$.

Observe that by Proposition 3.10,

$$c\chi(\rho) \leq \frac{a_\infty/2\rho^2 + \int_0^{1/\rho} ta_1(t)dt}{a_0 - \int_0^{1/\rho} t\dot{a}_1(t)dt} \leq c^{-1}\chi(\rho), \quad \rho \in \mathbb{R}, \tag{4.65}$$

for some constant $c > 0$. Therefore examples for (i) of Theorem 4.6 are $a_0 > 0$ or $a_1(t) \sim t^{\alpha-1}$ as $t \to 0$, for some $\alpha \in (0,1)$, while (ii) resp. (iii) are implied by $a_0 = 0$, $a_1(0+) < \infty$, and $-\dot{a}_1(t) \sim t^{\alpha-1}$ resp. $\dot{a}_1(t) \sim \log t$ as $t \to 0$.

For an equation occuring in one-dimensional viscoelasticity, a result similar to Theorem 4.6 was first proved by Hrusa and Renardy [185]. Desch and Grimmer [84] consider hyperbolic systems of first order partial integrodifferential equations in one space dimension, and present conditions for gradual smoothing of the solutions in time, like (ii) and (iii) of Theorem 4.6, in an L^2-setting. While Theorem 4.6 covers linear isotropic viscoelasticity, by means of semigroup methods, Desch and Grimmer [83] obtained a corresponding result also for nonisotropic materials which, however, must be completely monotonic; cp. Section 5.

Appendix: Some Common Bernstein Functions
For the convenience of the reader we list below some common Bernstein functions φ as well as the quantities c, k, r, s, w related to them, according to Sections 4.2 and 4.5.

If $\varphi(\lambda)$ is a Bernstein function, recall the following relations:

$$\hat{c}(\lambda) = \frac{1}{\lambda\varphi(\lambda)}, \quad \hat{k}(\lambda) = \frac{\varphi(\lambda)}{\lambda^2}$$

$$\hat{s}(\lambda; \mu) = \frac{\varphi(\lambda)}{\lambda} \cdot \frac{1}{\varphi(\lambda) + \mu}, \quad \hat{w}(\lambda; \tau) = \frac{e^{-\tau\varphi(\lambda)}}{\lambda};$$

the function $r(t; \mu)$ is obtained from $s(t; \mu)$ via $r(t; \mu) = (1 - s(t; \mu))/\mu$; see Section 4.2. We use the standard notation for the *complementary error function*

$$\mathrm{erfc}(x) = \frac{2}{\sqrt{\pi}} \int_x^\infty e^{-r^2} dr, \quad x \in \mathbb{R},$$

the *gamma function*

$$\Gamma(x) = \int_0^\infty e^{-t} t^{x-1} dt, \quad x > 0,$$

the *exponential integral*

$$E_1(x) = \int_x^\infty e^{-t} dt/t, \quad x > 0,$$

and the *modified Bessel function of order* 0

$$I_0(x) = \frac{1}{\pi} \int_0^\pi \cosh(x\cos\theta) d\theta, \quad x \in \mathbb{R}.$$

In Table 4.1 $c(t)$ and $s(t; \mu)$ are expressed in terms of elementary functions and integrals for diverse Bernstein functions $\varphi(\lambda)$; here $\alpha \in (0, 1)$.

$\varphi(\lambda)$	$c(t)$	$s(t; \mu)$
1	1	$(1 + \mu)^{-1}$
λ	t	$e^{-\mu t}$
$\sqrt{\lambda}$	$2\sqrt{t/\pi}$	$e^{\mu^2 t}\operatorname{erfc}(\mu\sqrt{t})$
$1 + \lambda$	$1 - e^{-t}$	$(1 + \mu)^{-1}[1 + \mu e^{-(1+\mu)t}]$
$\frac{\lambda}{1+\lambda}$	$1 + t$	$(1 + \mu)^{-1}[\mu + e^{-t\mu/(1+\mu)}]$
$\lambda + \sqrt{\lambda}$	$2\sqrt{t/\pi} + e^t\operatorname{erfc}\sqrt{t} - 1$	$\mu\pi^{-1}\int_0^\infty e^{-rt}(r + (r - \mu)^2)^{-1}dr/\sqrt{r}$
$\frac{\lambda}{\lambda+\sqrt{\lambda}}$	$1 + 2\sqrt{t/\pi}$	$(1 + \mu)^{-1}e^{\mu^2 t/(1+\mu)^2}\operatorname{erfc}\left(\frac{\mu\sqrt{t}}{1+\mu}\right)$
$1 - e^{-\lambda}$	$\sum_{n=0}^\infty e_0(t - n)$	$1 - \mu\sum_{n=0}^\infty (\mu + 1)^{-(n+1)}e_0(t - n)$
$\log(1 + \lambda)$	$\int_0^\infty [\int_0^t e^{-\tau}\tau^{\rho-1}d\tau]d\rho/\Gamma(\rho)$	$1 - \int_0^\infty \mu e^{-\mu\rho}[\int_0^t e^{-\tau}\tau^{\rho-1}d\tau]d\rho/\Gamma(\rho)$
λ^α	$t^\alpha/\Gamma(\alpha + 1)$	$\frac{\sin\alpha\pi}{\pi}\int_0^\infty e^{-rt}\frac{\mu r^{\alpha-1}}{\mu^2 + r^{2\alpha} + 2\mu r^\alpha\cos\alpha\pi}dr$

Table 4.1

Table 4.2 contains expressions for the creep functions $k(t)$ and the propagation functions $w(t; \tau)$ associated with several Bernstein functions $\varphi(\lambda)$.

$\varphi(\lambda)$	$k(t)$	$w(t; \tau)$
1	t	$e^{-\tau}$
λ	1	$e_0(t - \tau)$
$\sqrt{\lambda}$	$2\sqrt{t/\pi}$	$\operatorname{erfc}(\tau/2\sqrt{t})$
$1 + \lambda$	$1 + t$	$e^{-\tau}e_0(t - \tau)$
$\frac{\lambda}{1+\lambda}$	$1 - e^{-t}$	$e^{-\tau}[I_0(2\sqrt{t\tau})e^{-t} + \int_0^t e^{-s}I_0(2\sqrt{s\tau})ds]$
$\lambda + \sqrt{\lambda}$	$1 + 2\sqrt{t/\pi}$	$\operatorname{erfc}(\tau/2\sqrt{t - \tau})e_0(t - \tau)$
$\frac{\lambda}{\lambda+\sqrt{\lambda}}$	$2\sqrt{t/\pi} + e^t\operatorname{erfc}\sqrt{t} - 1$	$1 - \int_0^\infty e^{-rt-r\tau/(1+r)}\sin(\tau\frac{\sqrt{r}}{1+r})dr/\pi r$
$1 - e^{-\lambda}$	$\min\{t, 1\}$	$e^{-\tau}\sum_{n=0}^\infty \tau^n e_0(t - n)/n!$
$\log(1 + \lambda)$	$1 - e^{-t} + tE_1(t)$	$\int_0^t e^{-s}\frac{s^{\tau-1}}{\Gamma(\tau)}ds$
λ^α	$t^{1-\alpha}/\Gamma(2 - \alpha)$	$1 - \pi^{-1}\int_0^\infty e^{-rt-\tau r^\alpha c_\alpha}\sin(\tau r^\alpha s_\alpha)dr/\pi r$

Table 4.2

Here we employed the abbreviations $c_\alpha = \cos\alpha\pi$ and $s_\alpha = \sin\alpha\pi$.

In Figure 4.1 the propagation function $w(t; \tau)$ is depicted for $\varphi = \sqrt{\lambda}$, $\varphi = \lambda + \sqrt{\lambda}$, and $\varphi(\lambda) = \lambda + \lambda/(1 + \lambda)$, respectively.

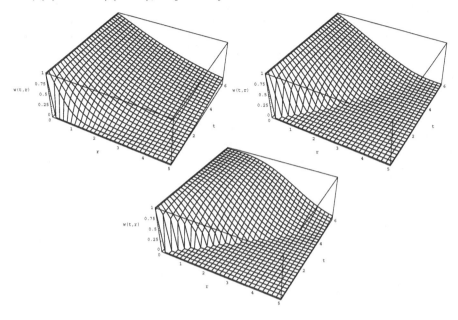

Figure 4.1 Typical Propagation Functions

These Bernstein functions serve as typical representatives for the three different types of qualitative behaviour of $w(t; \tau)$: infinite propagation speed, finite propagation speed but no wavefront, and finite propagation speed with wave front, respectively.

5   Linear Viscoelasticity

A rich source for vector-valued Volterra equations is the continuum mechanics for materials with memory, i.e. the theory of viscoelastic materials. In this section the basic concepts of this theory are introduced and the resulting boundary value problems are formulated. The discussion of the involved material functions shows that the notion of creep functions introduced in Section 4 appears here naturally. The well-posedness of some special problems which lead to Volterra equations of scalar type will be discussed, but also the limits of this class of equations become apparent.

5.1 Balance of Momentum and Constitutive Laws

Consider a 3-dimensional body which is represented by an open set $\Omega \subset \mathbb{R}^3$ with boundary $\partial\Omega$ of class C^1. Points in Ω (i.e. material points) will be denoted by $x, y, \dots$. Associated with this body there is a strictly positive function $\rho_0 \in C(\overline{\Omega})$ called the *density of mass*. Acting forces will deform the body, and the material point x will be displaced to its new position $x + u(t,x)$ at time t; the vector field $u(t,x)$ is called the displacement field, or briefly *displacement*. The *velocity* of the material point $x \in \Omega$ at time t is then given by $v(t,x) = \dot{u}(t,x)$, where the dot indicates partial derivative with respect to t. The linearized *strain* in the body due to a deformation is defined by

$$\mathcal{E}(t,x) = \frac{1}{2}(\nabla u(t,x) + (\nabla u(t,x))^T), \quad t \in \mathbb{R}, \ x \in \Omega, \tag{5.1}$$

i.e. $\mathcal{E}(t,x)$ is the symmetric part of the *displacement gradient* ∇u.

A given strain-history of the body causes *stress* in a way to be specified, expressing the properties of the material the body is made of. The stress tensor will be denoted by $\mathcal{S}(t,x)$; both, $\mathcal{E}(t,x)$ and $\mathcal{S}(t,x)$ are symmetric. Let $g(t,x)$ be an external body force field like gravity. Then balance of momentum in the body becomes

$$\rho_0(x)\ddot{u}(t,x) = \operatorname{div}\mathcal{S}(t,x) + \rho_0(x)g(t,x), \quad t \in \mathbb{R}, \ x \in \Omega; \tag{5.2}$$

in components (5.2) reads

$$\rho_0(x)\ddot{u}_i(t,x) = \sum_{j=1}^n \frac{\partial}{\partial x_j}\mathcal{S}_{ij}(t,x) + \rho_0(x)g_i(t,x), \quad i = 1,2,3.$$

(5.2) has to be supplemented by boundary conditions; these are basically either 'prescribed displacement' or 'prescribed normal stress (traction)' at the surface $\partial\Omega$ of the body. Let $\partial\Omega = \Gamma_d \cup \Gamma_s$, where Γ_d, Γ_s are closed, $\overline{\overset{\circ}{\Gamma}_d} = \Gamma_d, \overline{\overset{\circ}{\Gamma}_s} = \Gamma_s$, and such that $\overset{\circ}{\Gamma}_d \cap \overset{\circ}{\Gamma}_s = \emptyset$; let $n(x)$ denote the outer normal of Ω at $x \in \partial\Omega$. The boundary conditions then can be stated as follows.

$$
\begin{aligned}
u(t,x) &= u_d(t,x), \quad t \in \mathbb{R}, \ x \in \overset{\circ}{\Gamma}_d, \\
\mathcal{S}(t,x)n(x) &= g_s(t,x), \quad t \in \mathbb{R}, \ x \in \overset{\circ}{\Gamma}_s.
\end{aligned}
\tag{5.3}
$$

Observe that the functions $g(t,x), u_d(t,x), g_s(t,x)$ here are considered as known. In many practical problems, however, the force field $g(t,x)$ and the traction $g_s(t,x)$ are actually depending on u or on its derivatives $\dot{u}$ and ∇u. For example if the body is subject to an external force field h which depends on the spatial position only, then $f(t,x) = h(x + u(t,x))$; other examples are boundary feedback laws like $g_s(t,x) = -\gamma v(t,x)$.

Suppose the body is at rest up to time $t = 0$, but is then exposed to forces $g(t,x)$ and $g_s(t,x)$ and to a sudden velocity change $\dot{u}(0,x) = u_1(x)$. Taking the inner product of (5.2) with $\dot{u}$ and integrating over Ω, an integration by parts yields the *energy equality*, formally at least.

$$\int_\Omega |\dot{u}(t,x)|^2 \rho_0(x)dx + \int_0^t \int_\Omega \mathcal{S}(\tau,x) : \dot{\mathcal{E}}(\tau,x)dxd\tau = \int_\Omega |u_1(x)|^2 \rho_0(x)dx$$

$$+ \int_0^t \int_\Omega g(\tau,x) \cdot \dot{u}(\tau,x)\rho_0(x)dxd\tau + \int_0^t \int_\Omega g_s(\tau,y) \cdot \dot{u}(\tau,y)dyd\tau. \qquad (5.4)$$

Since the total kinetic energy in the body at time t cannot exceed its initial value plus the work done by the acting body and surface forces, the inequality

$$\int_0^t \int_\Omega \mathcal{S}(\tau,x) : \dot{\mathcal{E}}(\tau,x)dxd\tau \geq 0 \qquad (5.5)$$

must hold for all values of $t > 0$, and for any choice of initials values and forces.

A material is called *incompressible*, if there are no changes of volume in the body Ω during a deformation, i.e. if

$$\det(I + \nabla u(t,x)) = 1, \quad t \in \mathbb{R}, \ x \in \Omega, \qquad (5.6)$$

is satisfied; otherwise the material is called *compressible*. (5.6) is a nonlinear constraint for the system (5.2), (5.3); for the linear theory (5.6) can be simplified to

$$\operatorname{div} u(t,x) = 0, \quad t \in \mathbb{R}, \ x \in \Omega. \qquad (5.7)$$

In fact, if the material is linear, then $u(t,x)$ in (5.6) can be replaced by $\mu u, \mu > 0$; taking the derivative in (5.6) w.r.t. μ at $\mu = 0$, (5.7) follows. As a consequence of the constraints (5.6) or (5.7), for incompressible materials the stress tensor $\mathcal{S}(t,x)$ cannot be completely determined by u and its derivatives; it is then convenient to decompose $\mathcal{S}(t,x)$ as

$$\mathcal{S}(t,x) = -\pi(t,x)I + \mathcal{S}_e(t,x), \quad t \in \mathbb{R}, \ x \in \Omega, \qquad (5.8)$$

where the scalar function π is called *pressure*, and $\mathcal{S}_e(t,x)$ *extra stress*, which satisfies $\operatorname{tr} \mathcal{S}_e(t,x) = 0$. Thus we have the relation

$$-\pi(t,x) = \frac{1}{3}\operatorname{tr} \mathcal{S}(t,x), \quad t \in \mathbb{R}, \ x \in \Omega. \qquad (5.9)$$

While for compressible materials (5.9) is just a definition, for the incompressible case $\pi(t,x)$ is an unknown function, which has to be determined by (5.2), (5.3) and (5.7), too. For a mathematically oriented introduction to the basic concepts of continuum mechanics we refer e.g. to the monograph Gurtin [159].

To make the system complete we have to add an equation which relates the stress $\mathcal{S}(t, x)$ (or the extra stress for the incompressible case) to u and its derivatives; such relations will be referred to as *constitutive relations* or *constitutive laws* or *stress-strain relations*. Here we concentrate on linear materials. Since the stress should only depend on the history of the strain, the general constitutive law in the compressible case is given by

$$\mathcal{S}(t, x) = \int_0^\infty d\mathcal{A}(\tau, x)\dot{\mathcal{E}}(t - \tau, x), \quad t \in \mathbb{R}, \ x \in \Omega, \tag{5.10}$$

where $\mathcal{A} : \mathbb{R}_+ \times \Omega \to \mathcal{B}(\text{Sym}\{3\})$ is locally of bounded variation w.r.t. $t \geq 0$; $\text{Sym}\{N\}$ denotes the space of N-dimensional real symmetric matrices. In components the latter means

$$\mathcal{A}_{ijkl}(t, x) = \mathcal{A}_{jikl}(t, x) = \mathcal{A}_{ijlk}(t, x), \quad t \in \mathbb{R}_+, \ x \in \Omega, \tag{5.11}$$

for all $i, j, k, l \in \{1, 2, 3\}$. The component functions $\mathcal{A}_{ijkl}(t, x)$ are called the *stress relaxation moduli* of the material.

Alternatively, (5.10) can be written in the form

$$\mathcal{E}(t, x) = \int_0^\infty d\mathcal{K}(\tau, x)\mathcal{S}(t - \tau, x), \quad t \in \mathbb{R}, \ x \in \Omega, \tag{5.12}$$

where $\mathcal{K} : \mathbb{R}_+ \times \Omega \to \mathcal{B}(\text{Sym}\{3\})$ is locally of bounded variation in $t \in \mathbb{R}_+$. The components of $\mathcal{K}$ are called the *creep moduli* of the material and also enjoy the symmetries (5.11). Relations (5.10), (5.12) yield the identity

$$(\mathcal{A} * d\mathcal{K})(t, x) = (\mathcal{K} * d\mathcal{A})(t, x) = t\mathcal{I}, \tag{5.13}$$

where $\mathcal{I}$ denotes the identity tensor.

Let us consider some special cases which have been studied extensively in classical continuum mechanics.

(i) *Ideally elastic solids*
Classical elasticity postulates the stress-strain relations

$$\mathcal{S}(t, x) = \mathcal{A}^\infty(x)\mathcal{E}(t, x), \quad t \in \mathbb{R}, \ x \in \Omega,$$

with $\mathcal{A}^\infty \in L^\infty(\Omega; \mathcal{B}(\text{Sym}\{3\}))$ positive definite, uniformly in Ω. This corresponds to the stress relaxation tensor

$$\mathcal{A}(t, x) = t\mathcal{A}^\infty(x), \quad t > 0, \ x \in \Omega,$$

and to a creep tensor of the form

$$\mathcal{K}(t, x) = \mathcal{K}^0(x), \quad t > 0, \ x \in \Omega,$$

where
$$\mathcal{A}^{\infty}(x)\mathcal{K}^0(x) = \mathcal{K}^0(x)\mathcal{A}^{\infty}(x) = \mathcal{I}, \quad x \in \Omega.$$

(ii) *Ideally viscous fluids*
For such fluids the stress strain relations are

$$\mathcal{S}(t,x) = \mathcal{A}^0(x)\dot{\mathcal{E}}(t,x), \quad t \in \mathbb{R}, \; x \in \Omega,$$

with $\mathcal{A}^0 \in L^{\infty}(\Omega; \mathcal{B}(\mathrm{Sym}\{3\}))$ positive definite, uniformly in Ω. Here we obtain

$$A(t,x) = \mathcal{A}^0(x), \quad t > 0, \; x \in \Omega,$$

and

$$K(t,x) = t\mathcal{K}^{\infty}(x), \quad t > 0, \; x \in \Omega,$$

where

$$\mathcal{A}^0(x)\mathcal{K}^{\infty}(x) = \mathcal{K}^{\infty}(x)\mathcal{A}^0(x) = \mathcal{I}, \quad x \in \Omega.$$

(iii) *Kelvin-Voigt solids*
Such materials are governed by constitutive laws of the form

$$\mathcal{S}(t,x) = \mathcal{A}^0(x)\dot{\mathcal{E}}(t,x) + \mathcal{A}^{\infty}(x)\mathcal{E}(t,x), \quad t \in \mathbb{R}, \; x \in \Omega,$$

with $\mathcal{A}^0, \mathcal{A}^{\infty} \in L^{\infty}(\Omega; \mathcal{B}(\mathrm{Sym}\{3\}))$ positive definite, uniformly in Ω. This gives

$$A(t,x) = \mathcal{A}^0(x) + t\mathcal{A}^{\infty}(x), \quad t \in 0, \; x \in \Omega.$$

(iv) *Maxwell fluids*
This class of materials is characterized by the stress-strain relation

$$\mathcal{S}(t,x) = \int_0^{\infty} \mu e^{-\mu s} \mathcal{A}^1(x)\dot{\mathcal{E}}(t-s,x)ds, \quad t \in \mathbb{R}, \; x \in \Omega,$$

with $\mathcal{A}^1 \in L^{\infty}(\Omega; \mathcal{B}(\mathrm{Sym}\{3\}))$ positive definite, uniformly in Ω, and $\mu > 0$. In this case we have
$$A(t,x) = (1 - e^{-\mu t})\mathcal{A}^1(x), \quad t > 0, \; x \in \Omega. \quad \square$$

A material is called *homogeneous* if $\rho_0, \mathcal{A}, \mathcal{K}$ do not depend on the material points $x \in \Omega$. It is called *isotropic* if the constitutive law is invariant under the group of rotations. It can be shown that the general isotropic stress relaxation tensor is given by

$$A_{ijkl}(t,x) = \frac{1}{3}(3b(t,x) - 2a(t,x))\delta_{ij}\delta_{kl} + a(t,x)(\delta_{ik}\delta_{jl} + \delta_{il}\delta_{jk});$$

where δ_{ij} denotes Kronecker's symbol; similarly

$$K_{ijlk}(t,x) = \frac{1}{3}(\frac{1}{3}l(t,x) - \frac{1}{2}k(t,x))\delta_{ij}\delta_{kl} + \frac{1}{4}k(t,x)(\delta_{ik}\delta_{jl} + \delta_{il}\delta_{jk}).$$

This results in the constitutive laws

$$\begin{aligned}
\mathcal{S}(t,x) &= 2\int_0^\infty da(\tau,x)\dot{\mathcal{E}}(t-\tau,x) \\
&\quad + \frac{1}{3}\mathcal{I}\int_0^\infty (3db(\tau,x)-2da(\tau,x))\operatorname{tr}\dot{\mathcal{E}}(t-\tau,x), \quad (5.14) \\
\mathcal{E}(t,x) &= \frac{1}{2}\int_0^\infty dk(\tau,x)\mathcal{S}(t-\tau,x) \\
&\quad + \frac{1}{3}\mathcal{I}\int_0^\infty (\frac{1}{3}dl(\tau,x)-\frac{1}{2}dk(\tau,x))\operatorname{tr}\mathcal{S}(t-\tau,x).
\end{aligned}$$

Taking traces in (5.14) we obtain

$$\begin{aligned}
\operatorname{tr}\mathcal{S}(t,x) &= 3\int_0^\infty db(\tau,x)\operatorname{tr}\dot{\mathcal{E}}(t-\tau,x), \\
\operatorname{tr}\mathcal{E}(t,x) &= \frac{1}{3}\int_0^\infty dl(\tau,x)\operatorname{tr}\mathcal{S}(t-\tau,x). \quad (5.15)
\end{aligned}$$

Introducing the *deviatoric stress* and the *deviatoric strain*

$$\mathcal{S}_d(t,x) = \mathcal{S}(t,x) - \frac{1}{3}\mathcal{I}\operatorname{tr}\mathcal{S}(t,x), \quad \mathcal{E}_d(t,x) = \mathcal{E}(t,x) - \frac{1}{3}\mathcal{I}\operatorname{tr}\mathcal{E}(t,x), \quad (5.16)$$

(5.14) and (5.15) yield

$$\begin{aligned}
\mathcal{S}_d(t,x) &= 2\int_0^\infty da(\tau,x)\dot{\mathcal{E}}_d(t-\tau,x), \\
\mathcal{E}_d(t,x) &= \frac{1}{2}\int_0^\infty dk(\tau,x)\mathcal{S}_d(t-\tau,x). \quad (5.17)
\end{aligned}$$

Thus the kernels b and l describe the behaviour of the material under compression, while a and k determine its response to shear. Therefore, db is called *compression modulus* and da *shear modulus*; from (5.15) resp. (5.17) there follow the identities

$$(da * k)(t) = t = (db * l)(t), \quad t > 0. \quad (5.18)$$

In general, a and b are independent functions, however, if $b(t,x) = \beta a(t,x)$ for some constant $\beta > 0$ then the material is called *synchronous*.

For isotropic incompressible materials the corresponding stress-strain relation becomes

$$\mathcal{S}(t,x) = -\pi(t,x)\mathcal{I} + 2\int_0^\infty da(\tau,x)\dot{\mathcal{E}}(t-\tau,x), \quad t \in \mathbb{R}, \ x \in \Omega, \quad (5.19)$$

hence in virtue of the incompressibility conditon (5.7)

$$\pi(t,x) = -\frac{1}{3}\operatorname{tr}\mathcal{S}(t,x), \quad t \in \mathbb{R}, \ x \in \Omega. \quad (5.20)$$

The inverse relation then reads

$$\mathcal{E}(t,x) = \frac{1}{2} \int_0^\infty dk(\tau,x) \mathcal{S}_d(t-\tau,x), \quad t \in \mathbb{R}, \ x \in \Omega; \tag{5.21}$$

in other words $l(t,x) \equiv 0$ for the incompressible case. Observe that incompressible materials can be considered as the limiting case of synchronous materials when $\beta \to \infty$.

Inequality (5.5) leads to the following restriction on the stress relaxation tensor which will be called *dissipation inequality* in the sequel.

$$\int_0^T \int_0^t [d\mathcal{A}(\tau,x) : F(t-\tau)] : F(\tau)d\tau dt \geq 0, \quad T > 0, \ F \in C(\mathbb{R}_+; \mathrm{Sym}\{3\}). \tag{5.22}$$

In other words, the tensor-valued measure $d\mathcal{A}$ is of *positive type*. A special case of this are *completely monotonic* relaxation tensors, i.e. $d\mathcal{A}$ is of the form

$$d\mathcal{A}(t,x) = \mathcal{A}^0(x)\delta_0 + \mathcal{A}^\infty(x)dt + \left(\int_0^\infty e^{-t\xi}\mathcal{A}^1(\xi,x)d\alpha(\xi)\right)dt, \quad t > 0, \ x \in \Omega, \tag{5.23}$$

with $\mathcal{A}^0, \mathcal{A}^\infty \in L^\infty(\Omega; \mathcal{B}(\mathrm{Sym}\{3\}))$ positive semidefinite three-dimensional tensors, $\mathcal{A}^1 : \mathbb{R}_+ \to L^\infty(\Omega; \mathcal{B}(\mathrm{Sym}\{3\}))$ Borel-measurable, bounded, and positive semidefinite for each $t \geq 0$, and $\alpha : \mathbb{R}_+ \to \mathbb{R}_+$ nondecreasing, $\alpha(0) = \alpha(0+) = 0$, $\int_0^\infty (1+\xi)^{-1}d\alpha(\xi) < \infty$.

In the homogeneous and isotropic case, $d\mathcal{A}$ will be of positive type iff da and db are of positive type, and $d\mathcal{A}$ will be completely monotonic iff a, b are Bernstein functions. Observe also that the stress relaxation tensors from Examples (i)∼(iv) above are completely monotonic.

Summarizing, we obtain the equations for linear isothermal viscoelasticity.

$$\rho_0(x)\ddot{u}(t,x) = \mathrm{div}\left(\int_0^\infty d\mathcal{A}(\tau,x)\nabla\dot{u}(t-\tau,x)\right) + \rho_0(x)g(t,x), \tag{5.24}$$

for all $t \in \mathbb{R}$ and $x \in \Omega$. For homogenenous and isotropic materials (5.24) can be simplified to

$$\begin{aligned}
\ddot{u}(t,x) &= \int_0^\infty da(\tau)\Delta\dot{u}(t-\tau,x) \\
&+ \int_0^\infty \left(db(\tau) + \frac{1}{3}da(\tau)\right)\nabla\nabla\cdot\dot{u}(t-\tau,x) + g(t,x),
\end{aligned} \tag{5.25}$$

for all $t \in \mathbb{R}$ and $x \in \Omega$, where we have put $\rho_0(x) \equiv \rho_0 = 1$ for simplicity. If in addition the material is incompressible, then (5.25) becomes

$$\ddot{u}(t,x) = \int_0^\infty da(\tau)\Delta\dot{u}(t-\tau,x) - \nabla\pi(t,x) + g(t,x), \tag{5.26}$$

$$\nabla \cdot u(t,x) = 0, \quad t \in \mathbb{R}, \ x \in \Omega.$$

These equations have to be supplemented by the boundary conditions (5.3).

5.2 Material Functions

Homogeneous isotropic linear materials are described by means of two material functions, the shear modulus da and the compression modulus db. If the material is in addition synchronous or incompressible, the only function needed is the shear modulus. In this subsection we briefly want to exhibit the physical meaning of these material functions and to motivate some of their properties which are used below.

Let da be the shear modulus of a homogeneous isotropic linear material and dk the corresponding creep modulus, i.e.

$$(da * k)(t) = (a * dk)(t) = t, \quad t > 0. \tag{5.27}$$

We will now consider the behaviour of the material in simple shearing motion. Imagine two parallel plates (of zero mass), let the space between them be filled with the material under consideration. Keep one of these plates fixed, while a parallel force $f(t)$ is applied to the second plate.

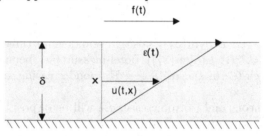

Figure 5.1: Creep and relaxation experiments

If the distance δ of these plates is small enough then inertial effects can be neglected, and displacement is only a linear function of x, the variable orthogonal to the plates. Stress and strain then only depend on time and have only one nontrivial entry (up to symmetry) $\sigma(t)$ resp. $\varepsilon(t)$, and the following relations hold.

$$\sigma(t) = (da * \dot{\varepsilon})(t), \quad \varepsilon(t) = (dk * \sigma)(t). \tag{5.28}$$

Observe also that $\sigma(t) = -f(t)$, and $\varepsilon(t)\delta$ is the horizontal displacement of the second plate. We describe now two experiments from which the material functions can be visualized.

Creep Experiment: Apply a sudden but then constant force at $t = 0$, and observe the displacement of the plate. This means $\sigma(t) = e_0(t)$, the Heaviside function, and $\varepsilon(t) = k(t)$; thus the creep modulus $k(t)$ describes, how the material 'creeps' to its final position, in this experiment. It is natural to expect that $k(t)$ is nonnegative, nondecreasing and concave, in other words $k(t)$ is a *creep function* in the terminology of Section 4.

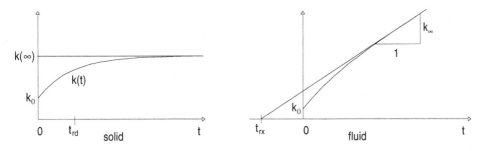

Figure 5.2: Creep response

Stress Relaxation Experiment: Displace the material at $t = 0$ and keep it there; observe the force $f(t) = -\sigma(t)$. This means $\sigma(t) = da(t)$. The responding force may have an initial Dirac measure - like in the case of a Newtonian fluid - but otherwise it should be a nonnegative, nonincreasing function. Therefore $a(t)$ itself is again a *creep function*.

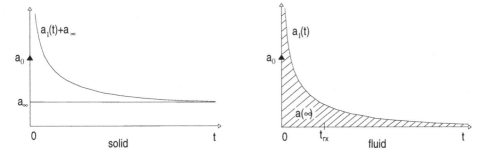

Figure 5.3: Stress relaxation

There are two ways to classify materials, according to the behaviour of $a(t)$ and $k(t)$ at $t = 0$ or at $t = \infty$.

Behaviour at $t = 0$: A material is called *rigid* if $k_0 > 0$; k_0 is the *rigidity*. A material is called *viscous* if $a_0 > 0$; a_0 is termed *dynamic viscosity*. Since $a_0 \cdot k_0 = 0$ by Proposition 4.4, these notions are exclusive but not exhaustive; there is a third class corresponding to the case $a_0 = k_0 = 0$. As shown in Section 4.2, a material is rigid with rigidity $k_0 > 0$ iff $a_0 = 0$ and $a_1(0+) < \infty$; we then have

$$a_1(0+) + a_\infty = \frac{1}{k_0}.$$

Similarly, a material is viscous with dynamic viscosity $a_0 > 0$ iff $k_0 = 0$ and $k_1(0+) < \infty$; we then have

$$k_1(0+) + k_\infty = \frac{1}{a_0}.$$

Nonrigid, nonviscous materials satisfy $a_0 = k_0 = 0$ which is equivalent to $a_1(0+) = k_1(0+) = \infty$.

Behaviour at $t = \infty$: A material is called a *solid* if $a_\infty > 0$; a_∞ is the *elasticity modulus*. It is called a *fluid* if $k_\infty > 0$; $1/k_\infty$ is termed *static viscosity*. Since $a_\infty \cdot k_\infty = 0$ these notions are also exclusive, but not exhaustive; there is a third class according to $a_\infty = k_\infty = 0$. Solids are characterized by $k(t)$ bounded, i.e. by

$$k_\infty = 0, \quad k_1 \in L^1(\mathbb{R}_+), \quad \text{and} \quad k(\infty) = k_0 + \int_0^\infty k_1(\tau)d\tau = \frac{1}{a_\infty}.$$

Similarly, fluids are equally described by $a(t)$ bounded, i.e. by

$$a_\infty = 0, \quad a_1 \in L^1(\mathbb{R}_+), \quad \text{and} \quad a(\infty) = a_0 + \int_0^\infty a_1(\tau)d\tau = \frac{1}{k_\infty}.$$

Materials which are neither solids nor fluids are characterized by $a_\infty = k_\infty = 0$ which is equivalent to $a_1, k_1 \notin L^1(\mathbb{R}_+)$.

For a fluid the *mean relaxation time* t_{rx} is defined by

$$t_{rx} = \int_0^\infty t\,da(t) / \int_0^\infty da(t) = k_\infty \int_0^\infty ta_1(t)dt.$$

A *Newtonian fluid* is characterized by $t_{rx} = 0$; in general t_{rx} can be taken as an estimate for how fast stress in a fluid relaxes.

Similarly, for a solid the *mean retardation time* t_{rd} is defined by

$$
\begin{aligned}
t_{rd} &= \int_0^\infty t(k(\infty) - k(t))dt / \int_0^\infty (k(\infty) - k(t))dt \\
&= \frac{1}{2} \int_0^\infty t^2 k_1(t)dt / \int_0^\infty tk_1(t)dt \\
&= (a_0 + \int_0^\infty a_1(t)dt)/a_\infty + \int_0^\infty ta_1(t)dt/(a_0 + \int_0^\infty a_1(t)dt).
\end{aligned}
$$

A *Hookean solid* is characterized by $t_{rd} = 0$; the mean retardation time gives an estimate for the time the material needs to creep to its final position.

Let us briefly mention some well known standard models.

(i) Hookean Solid

$$a(t) = \mu t, \quad k(t) = \frac{1}{\mu}, \quad t > 0.$$

(ii) Newtonian Fluid

$$a(t) = \nu, \quad k(t) = \frac{t}{\nu}, \quad t > 0.$$

(iii) Kelvin-Voigt Solid

$$a(t) = \nu + \mu t, \quad k(t) = \frac{1}{\mu}(1 - e^{-\mu t/\nu}), \quad t > 0.$$

(iv) Maxwell Fluid

$$a(t) = \nu(1 - e^{-\mu t/\nu}), \quad k(t) = \frac{1}{\mu} + \frac{t}{\nu}, \quad t > 0.$$

(v) Poynting-Thompson Solid

$$a(t) = \mu_0 t + \nu(1 - e^{-\mu t/\nu}), \quad k(t) = \mu_0^{-1}[1 - \mu(\mu_0 + \mu)^{-1} e^{-\mu \mu_0 t/\nu(\mu + \mu_0)}].$$

(vi) Power Type Materials

$$a(t) = \frac{t^\alpha}{\Gamma(\alpha + 1)}, \quad k(t) = \frac{t^{1-\alpha}}{\Gamma(2 - \alpha)}, \quad t > 0,$$

where $\alpha \in (0, 1)$.

Observe that in any of these examples $a(t)$ and $k(t)$ are Bernstein functions. Let the two basic materials - Hookean solid and the Newtonian fluid - be represented by the symbols *spring* and *dashpot*. A Kelvin-Voigt solid can then be viewed as a spring and a dashpot in parallel - in this case the forces (stresses) acting on either element are additive while the Maxwell fluid corresponds to a spring and a dashpot in series - here the displacements (strains) are additive. Mechanical engineers like to think of a linear viscoelastic material as a finite network consisting of the basic elements spring and dashpot.

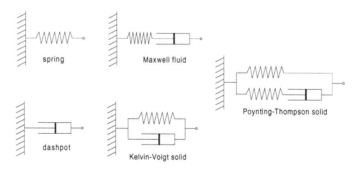

Figure 5.4: Mechanical models

The resulting material functions $a(t)$ and $k(t)$ will then be Bernstein functions of the form

$$a(t) = a_0 + a_\infty t + \sum_{j=1}^{N} a_j(1 - e^{-t/t_j}), \quad t > 0, \tag{5.29}$$

where $a_j, t_j > 0$ for all j. More rigorously, physical chemists have derived (5.29) from various molecular models. In their approach, the number N of independent *relaxation times* t_j corresponds to the degrees of freedom of a molecule. Therefore, for complicated materials like high polymers, N will be very large and it becomes impractical to work with expressions like (5.29); nevertheless $a(t)$ and $k(t)$ can still be expected to be Bernstein functions.

Here the so-called *fractional derivative models* should be mentioned, which are frequently used to describe the behaviour of polymeric materials. The five parameter fractional derivative model is given by the stress-strain law

$$\sigma(t) + \beta(\frac{d}{dt})^\mu \sigma(t) = \gamma(\varepsilon(t) + \alpha(\frac{d}{dt})^\nu \varepsilon(t)), \quad t \in \mathbb{R},$$

where $\alpha, \beta \geq 0$, $\gamma > 0$, $0 \leq \mu \leq \nu \leq 1$. This leads to the moduli da, dk via their Laplace transforms

$$\widehat{da}(\lambda) = \frac{\gamma}{\lambda} \cdot \frac{1 + \alpha\lambda^\nu}{1 + \beta\lambda^\mu}, \quad \widehat{dk}(\lambda) = \frac{1}{\gamma} \cdot \frac{1 + \beta\lambda^\mu}{1 + \alpha\lambda^\nu};$$

hence $a(t)$ and $k(t)$ are Bernstein functions. Such materials are solid, with $a_\infty = \gamma$, and they are rigid iff $\mu = \nu$, with rigidity $k_0 = \beta/\alpha\gamma$; for $\mu < \nu$ they are neither rigid nor viscous.

Finally, let us mention that the compression modulus $db(t)$ enjoys the same properties as the shear modulus $da(t)$, hence $b(t)$ and $l(t)$ can be expected to be creep functions or even Bernstein functions, as well.

5.3 Energy Balance and Thermoviscoelasticity

Consider first a rigid body $\Omega \subset \mathbb{R}^3$ which is subject to temperature changes. Let $\varepsilon(t, x)$ denote the density of internal energy at time $t \in \mathbb{R}$, $q(t, x)$ the heat flux vector field, $\theta(t, x)$ the temperature, and $r(t, x)$ the external heat supply. Balance of energy then reads as

$$\dot{\varepsilon}(t, x) + \mathrm{div}\, q(t, x) = r(t, x), \quad t \in \mathbb{R},\ x \in \Omega, \tag{5.30}$$

with boundary conditions basically either prescribed temperature or prescribed heat flux through the boundary, i.e.

$$
\begin{aligned}
\theta(t, x) &= \theta_b(t, x), & t \in \mathbb{R},\ x \in \overset{\circ}{\Gamma}_b, \\
-q(t, x) \cdot n(x) &= q_f(t, x), & t \in \mathbb{R},\ x \in \overset{\circ}{\Gamma}_f,
\end{aligned}
\tag{5.31}
$$

where $\Gamma_b, \Gamma_f \subset \partial\Omega$ are closed, $\overline{\overset{\circ}{\Gamma}_b} = \Gamma_b, \overline{\overset{\circ}{\Gamma}_f} = \Gamma_f, \overset{\circ}{\Gamma}_b \cap \overset{\circ}{\Gamma}_f = \emptyset$ and $\Gamma_b \cup \Gamma_f = \partial\Omega$.

For the constitutive laws we shall use the linear relations

$$
\begin{aligned}
\varepsilon(t, x) &= \int_0^\infty dm(\tau, x)\theta(t - \tau, x) + \varepsilon_\infty(x), & t \in \mathbb{R},\ x \in \Omega, \\
q(t, x) &= -\int_0^\infty d\mathcal{C}(\tau, x)\nabla\theta(t - \tau, x), & t \in \mathbb{R},\ x \in \Omega,
\end{aligned}
\tag{5.32}
$$

where $m \in BV_{loc}(\mathbb{R}_+; L^\infty(\Omega))$, and $\mathcal{C} \in BV_{loc}(\mathbb{R}_+; L^\infty(\Omega; \mathrm{Sym}\{3\}))$. In the isotropic and homogeneous case these relations simplify to

$$\varepsilon(t,x) = \int_0^\infty dm(\tau)\theta(t-\tau,x) + \varepsilon_\infty, \quad t \in \mathbb{R}, \ x \in \Omega,$$

$$q(t,x) = -\int_0^\infty dc(\tau)\nabla\theta(t-\tau,x), \quad t \in \mathbb{R}, \ x \in \Omega, \tag{5.33}$$

where $m, c \in BV_{loc}(\mathbb{R}_+)$ are scalar functions. Thus (5.30), (5.31) becomes in this case

$$dm * \dot\theta(t,x) = dc * \Delta\theta(t,x) + r(t,x), \quad t \in \mathbb{R}, \ x \in \Omega,$$

$$\theta(t,x) = \theta_b(t,x), \quad t \in \mathbb{R}, \ x \in \overset{\circ}{\Gamma}_b, \tag{5.34}$$

$$dc * \frac{\partial\theta}{\partial n}(t,x) = q_f(t,x), \quad t \in \mathbb{R}, \ x \in \overset{\circ}{\Gamma}_f .$$

From the literature one can infer that $m(t)$ is a creep function which is also bounded; thus $m_0 \geq 0$, $m_\infty = 0$ and $m_1 \in L^1(\mathbb{R}_+)$. $m_0 = m(0+)$ is called the *instantaneous heat capacity*, while $m(\infty) = m_0 + \int_0^\infty m_1(\tau)d\tau$ is termed *equilibrium heat capacity*. One should have at least $m(\infty) > 0$ (i.e. $m(t) \not\equiv 0$), but most authors also inquire $m_0 > 0$. $m_1(t)$ is called *energy-temperature relaxation function*.

Concerning the function $c(t)$, however, the literature is somewhat controversial. From Meixner [240], Gurtin and Pipkin [157], and Nunziato [261] one can expect that $c(t)$ is a bounded creep function as well, in particular $c_\infty = 0$ and $c_1 \in L^1(\mathbb{R}_+)$. $c_0 = c(0+) \geq 0$ is called *instantaneous conductivity*, $c(\infty) = c_0 + \int_0^\infty c_1(\tau)d\tau$ *equilibrium conductivity* and c_1 is termed *heat conduction relaxation function*. At least on should have $c(\infty) > 0$, and there result two different theories according to $c_0 > 0$ or $c_0 = 0$. Meixner [240] calls a material of *relaxation type*, if $m(t)$ and $c(t)$ are Bernstein functions, i.e. if m_1, c_1 are completely monotonic.

On the other hand, Clément and Nohel [51], Clément and Prüss [54], and Lunardi [226], write

$$c(t) = c_0 - \int_0^t \gamma(s)ds, \quad t > 0, \tag{5.35}$$

with $c_0 > \int_0^\infty \gamma(s)ds$, $\gamma(t) \geq 0$ nonincreasing; in this case $c(t)$ is 2-monotone. Thus in this theory the equilibrium conductivity $c(\infty) = c_0 - \int_0^\infty \gamma(s)ds > 0$ is smaller than the instantaneous conductivity, in contrast to Nunziato [261].

Next let us consider *linear thermoviscoelasticity*, i.e. the coupling of balances of momentum and energy. For this we have to replace the isothermal stress-strain law (5.10) by

$$\mathcal{S}(t,x) = \int_0^\infty d\mathcal{A}(\tau,x)\dot{\mathcal{E}}(t-\tau,x) - \int_0^\infty d\mathcal{B}(\tau,x)\dot\theta(t-\tau,x), \tag{5.36}$$

and the first equation in (5.32) by

$$\varepsilon(t,x) = \int_0^\infty dm(\tau,x)\theta(t-\tau,x) + \int_0^\infty d\mathcal{B}(\tau,x):\dot{\mathcal{E}}(t-\tau,x) + \varepsilon_\infty(x), \tag{5.37}$$

for all $t \in \mathbb{R}$, $x \in \Omega$. Here the second term in (5.36) corresponds to the stress generated by a nonuniform temperature-history, while the second term in (5.37) represents the conversion of strain energy into internal energy. In case $\mathcal{A}(t, x) = t\mathcal{A}^\infty(x)$, $\mathcal{B}(t, x) = t\mathcal{B}^\infty(x)$, $m(t, x) = m^0(x)$, $t > 0$, then (5.36) and (5.37) yield the corresponding relations for thermoelasticity.

The equations of linear anisotropic thermoviscoelasticity read now

$$
\rho_0(x)\ddot{u}(t, x) = \mathrm{div}\Big[\int_0^\infty d\mathcal{A}(\tau, x)\nabla\dot{u}(t - \tau, x)
$$

$$
- \int_0^\infty d\mathcal{B}(\tau, x)\dot{\theta}(t - \tau, x)\Big] + \rho_0(x)g(t, x),
$$

$$
\int_0^\infty dm(\tau, x)\dot{\theta}(t - \tau, x) = \mathrm{div}\Big[\int_0^\infty d\mathcal{C}(\tau, x)\nabla\theta(t - \tau, x)\Big] \qquad (5.38)
$$

$$
- \int_0^\infty d\mathcal{B}(\tau, x) : \nabla\ddot{u}(t - \tau, x) + r(t, x),
$$

for $t \in \mathbb{R}$ and $x \in \Omega$. Observe that $\mathcal{B} \in BV_{loc}(\mathbb{R}_+; L^\infty(\Omega; \mathrm{Sym}\{3\}))$. (5.38) has to be supplemented with the boundary conditions (5.3) and (5.31).

In the case of isotropic materials, the deviatoric stress $\mathcal{S}_d(t, x)$ is only temperature-dependent through the shear modulus da, however, this is a second order effect which has to be neglected in the linear theory. The normal stresses are on the other hand influenced by the change in volume due to change in temperature. This leads to the constitutive law

$$
\mathrm{tr}\,\mathcal{S}(t, x) = 3\int_0^\infty db(\tau, x)\{\mathrm{tr}\,\dot{\mathcal{E}}(t - \tau, x) - \int_0^\infty d\alpha(s, x)\dot{\theta}(t - \tau - s, x)\}, \quad (5.39)
$$

i.e.

$$
\mathcal{B}(t, x) = \mathcal{I}\int_0^\infty db(\tau, x)\alpha(t - \tau, x), \quad t \in \mathbb{R}_+, \ x \in \Omega.
$$

Here $\alpha(t, x)$ represents the expansion of a volume element induced by a sudden temperature change; α is expected to be a creep function, which in addition is bounded.

We summarize the equations of linear thermoviscoelasticity for homogeneous isotropic materials; w.l.o.g. let $\rho_0(x) = \rho_0 = 1$.

$$
\ddot{u}(t, x) = \int_0^\infty da(\tau)\Delta\dot{u}(t - \tau, x) + \int_0^\infty (db(\tau) + \frac{1}{3}da(\tau))\nabla\nabla \circ \dot{u}(t - \tau, x)
$$

$$
- \int_0^\infty db(\tau)[\int_0^\infty d\alpha(s)\nabla\dot{\theta}(t - \tau - s, x)] + g(t, x) \qquad (5.40)
$$

$$
\int_0^\infty dm(\tau)\dot{\theta}(t - \tau, x) = \int_0^\infty dc(\tau)\Delta\theta(t - \tau, x)
$$

$$
- \int_0^\infty db(\tau)[\int_0^\infty d\alpha(\rho)\nabla \cdot \ddot{u}(t - \tau - \rho, x)] + r(t, x),
$$

for all $t \in \mathbb{R}$ and $x \in \Omega$. Observe that the equations of classical linear thermoelasticity are obtained from (5.40) with

$$a(t) \equiv a_\infty t, \quad b(t) \equiv b_\infty t, \quad m(t) \equiv m_0,$$
$$c(t) \equiv c_0, \quad \alpha(t) \equiv \alpha_0 \quad \text{for all} \quad t > 0; \tag{5.41}$$

the equations then read

$$\ddot{u} = a_\infty \Delta u + (b_\infty + a_\infty/3)\nabla\nabla \cdot u - b_\infty \alpha_0 \nabla\theta + f$$
$$m_0 \dot{\theta} = c_0 \Delta\theta - b_\infty \alpha_0 \nabla \cdot \dot{u} + r. \tag{5.42}$$

Here the constants a_∞, b_∞, m_0, c_0, α_0 are all positive and are the shear modulus, compression modulus, heat capacity, heat conductivity, coefficient of thermal expansion, respectively.

5.4 Some One-dimensional Problems
This subsection is devoted to some specific problems in isothermal viscoelasticity which lead to one-dimensional boundary values problems.

(i) **Simple Shear. The Rayleigh Problem**
Consider a homogeneous isotropic viscoelastic fluid filling a halfspace, and which is at rest up to time $t = 0$. But then the bounding plane is suddenly moved tangentially to itself with constant speed 1. The induced velocity field of the fluid will be parallel to that of the boundary and will depend only on t and the distance $x > 0$ to the bounding plane. Assuming non-slip boundary conditions and the stress-strain relation of homogeneous isotropic materials (5.14) or (5.19), we obtain the following boundary value problem for the velocity $w(t; x)$ of the fluid:

$$w_t(t; x) = \int_0^t da(\tau)w_{xx}(t - \tau; x), \quad t, x > 0,$$
$$w(t; 0) = 1, \quad t > 0, \tag{5.43}$$
$$w(0; x) = 0, \quad x > 0.$$

This is Stokes' first problem or the *Rayleigh problem* of viscoelasticity. Essentially the same problem arises if the material is a solid, when the boundary is suddenly displaced; $w(t; x)$ then means the displacement in the body.

Let $h(\lambda; x)$ denote the Laplace transform of $w(t; x)$ w.r.t. $t > 0$; a simple computation yields

$$h(\lambda; x) = \frac{1}{\lambda}\exp(-\frac{x}{\sqrt{\hat{a}(\lambda)}}), \quad \lambda, x > 0. \tag{5.44}$$

Assume that $a(t)$ is a creep function with $a_1(t)$ log-convex; then by Lemma 4.2, there is a completely positive function $c(t)$ such that $\sqrt{\hat{a}(\lambda)} = \hat{c}(\lambda), \lambda > 0$. Thus $w(t; x)$ must be the propagation function associated with $c(t)$; c.p. Section 4.5.

We may now apply Propositions 4.9 and 4.10 for the description of the solution of the Rayleigh problem. For this purpose we have to determine the characteristic numbers κ, ω, α of $c(t)$, defined in (4.35), in terms of those of $a(t)$. With $\varphi(\lambda) = 1/\sqrt{\hat{a}(\lambda)}$ and by the relations $\lim_{\lambda \to 0+} \lambda \hat{a}_1(\lambda) = a_1(\infty) = 0$, $\lim_{\lambda \to \infty} \lambda \hat{a}_1(\lambda) = a_1(0+)$, and $\lim_{\lambda \to \infty} \lambda \hat{\dot{a}}_1(\lambda) = \dot{a}_1(0+)$, we obtain

$$\omega = \lim_{\lambda \to 0+} \varphi(\lambda) = 0,$$

$$\kappa = \lim_{\lambda \to \infty} \varphi(\lambda)/\lambda = \begin{cases} 0 & \text{if } a_0 > 0 \\ (a_\infty + a_1(0+))^{-1/2} & \text{if } a_0 = 0, \end{cases}$$

$$\alpha = \lim_{\lambda \to \infty} (\varphi(\lambda) - \kappa\lambda) = \begin{cases} \infty & \text{if } \kappa = 0 \\ \dfrac{-\dot{a}_1(0+)}{2(a_\infty + a_1(0+))^{3/2}} & \text{if } \kappa > 0. \end{cases}$$

By Proposition 4.10 the velocity field $w(t; x)$ admits the decomposition

$$w(t; x) = w_0(t - \kappa x; x) + e^{-\alpha x} e_0(t - \kappa x), \quad t, x > 0,$$

where w_0 is continuous and even of class $W^{1,1}_{loc}$, provided $\kappa = 0$ or $\dot{a}_1(t)$ is locally absolutely continuous on $(0, \infty)$. Since $w(t; x) = 0$ iff $t \leq \kappa x$, the perturbation propagates with speed $1/\kappa$ into the material; this speed is finite iff $a_0 = 0$ and $a_1(0+) < \infty$, i.e. iff the material is rigid, according to the classification in Section 5.2, and $\kappa = \sqrt{k_0}$, where k_0 denotes the rigidity of the material. If $\alpha < \infty$ then the wave $w(t; x)$ exhibits a *wave front*, i.e. a discontinuity at $t = \kappa x$, which is damped exponentially with exponent α w.r.t. $x > 0$. There is no wave front iff $\alpha = \infty$, i.e. if $\kappa = 0$ or $-\dot{a}_1(0+) = \infty$; the wave front is undamped iff $\alpha = 0$, i.e. iff $-\dot{a}_1(0+) = 0$, $\kappa > 0$ which corresponds to the purely elastic case $a(t) \equiv a_\infty t$. Observe the coexistence of finite wave speed and absence of the wave front in case $a_0 = 0$, $a_1(0+) < \infty$ but $-\dot{a}_1(0+) = \infty$.

We now consider briefly the general problem

$$\begin{aligned} u_t(t, x) &= \int_0^t da(\tau) u_{xx}(t - \tau, x) + f(t, x), \quad t, x > 0 \\ u(t, 0) &= g(t), \quad t > 0, \\ u(0, x) &= u_0(x), \quad x > 0. \end{aligned} \tag{5.45}$$

Let X denote any of the spaces $C_0(\mathbb{R}_+)$ or $L^q(\mathbb{R}_+)$, $1 \leq q < \infty$, and define A by means of $(Au)(x) = u''(x), x > 0$, where $\mathcal{D}(A) = \{u \in X : u', u'' \in X, u(0) = 0\}$. It is well known that A generates a bounded cosine family $\mathrm{Co}(\tau)$ in X which is given by

$$(\mathrm{Co}(\tau)u)(x) = \frac{1}{2}(u(x + \tau) + u(|x - \tau|)\mathrm{sgn}(x - \tau)), \quad \tau, x > 0.$$

Therefore, by Theorem 4.3 we may conclude that (5.45) admits a bounded resolvent $S(t)$ in X, if $a(t)$ is a creep function such that $a_1(t)$ is log-convex, in particular if $a(t)$ is a Bernstein function. Via the representation (4.50) of $S(t)$ and the variation of parameters formula, the solution of (5.45) can be represented explicitly in terms of the propagation function $w(t; \tau)$, i.e. the solution of the Rayleigh Problem.

(ii) **Torsion of a Rod**

Consider a rod of length l and of uniform circular cross section with radius $r > 0$ made of a homogeneous isotropic viscoelastic solid. Let the left end $(x = 0)$ of the rod be fixed, and let at the right end $(x = l)$ of the rod a tip mass be attached, with moment of inertia $\beta \geq 0$ around the axis of the rod, which is subject to a torque. After normalization this leads to the following initial boundary value problem for the angular displacement $\Theta(t, x)$ of the rod.

$$
\begin{aligned}
\Theta_t(t, x) &= \int_0^t da(\tau)\Theta_{xx}(t - \tau, x) + h(t, x), \\
\Theta(t, 0) &= 0, \ \Theta(0, x) = \Theta_0(x), \quad t > 0, \ x \in (0, 1), \qquad (5.46) \\
\beta\Theta_t(t, 1) &= -\int_0^t da(\tau)\Theta_x(t - \tau, 1) + g(t).
\end{aligned}
$$

Here $h(t, x)$ contains the given strain history as well as a probably also present distributed torque, and similarly $g(t)$ corresponds to the given boundary strain history and the torque applied to the tip mass. Observe that such torsional problems only involve the shear modulus da, the compression modulus does not enter the analysis, at least not to first order. We concentrate now on the case $\beta > 0$; for $\beta = 0$ see (iii) below.

Let $X = L^2(0, 1) \times \mathbb{C}$ and consider the operator A in X defined by

$$
\begin{aligned}
Au &= A(\Theta, \vartheta) = (\Theta'', -\beta^{-1}\Theta'(1)), \quad u \in \mathcal{D}(A), \\
\mathcal{D}(A) &= \{(\Theta, \vartheta) \in X : \Theta \in W^{2,2}(0, 1), \Theta(0) = 0, \Theta(1) = \vartheta\}.
\end{aligned}
$$

With respect to the inner product

$$
(u_1, u_2)_X = \int_0^1 \Theta_1(x)\overline{\Theta_2(x)}dx + \beta\vartheta_1\overline{\vartheta_2},
$$

this operator is selfadjoint and negative definite, as can be shown in the usual way, hence generates a bounded cosine family. After an integration, with $u(t) = (\Theta(t, \cdot), \Theta(t, 1))$, (5.46) can be written in the abstract form (1.1) where $f(t) = (\Theta_0(\cdot) + \int_0^t h(s, \cdot)ds, \Theta_0(1) - \beta^{-1}\int_0^t g(s)ds)$. Thus we may apply Theorem 4.3 to obtain a bounded resolvent, if the kernel $a(t)$ is a creep function with $a_1(t)$ log-convex. Alternatively, we may also apply Corollary 1.2 since we are in a Hilbert space setting, to the result that (5.46) admits a bounded resolvent if the kernel $a(t)$ is merely a creep function with $da(t)$ of positive type.

(iii) **Simple Tension. The Tensile Modulus**

As in (ii) we consider a viscoelastic rod which is fixed at the left end $(x = 0)$, but now is subject to a force in axial direction at the right end ($x = l$). Then the stress tensor to first order has only one nontrivial component $\sigma(t, x) = S_{11}(t, x)$; this situation is referred to as a problem of *simple tension*. Let da, db denote the shear and compression moduli of the material and let dk and dl be defined according to (5.18). The stress-strain relations (5.14) then imply

$$\mathcal{E} = \frac{1}{2} dk * \mathcal{S} + \frac{1}{3} \mathcal{I} \left(\frac{1}{3} dl - \frac{1}{2} dk \right) * \sigma,$$

hence $\mathcal{E}_{ij} \equiv 0$ for $i \neq j$, $\mathcal{E}_{11} = \frac{1}{3} dk * \sigma + \frac{1}{9} dl * \sigma$, and $\mathcal{E}_{22} = \mathcal{E}_{33} = \frac{1}{3} (\frac{1}{3} dl - \frac{1}{2} dk) * \sigma$, where $\mathcal{E} = (\mathcal{E}_{ij})$. With $\varepsilon = \mathcal{E}_{11}$ this yields the stress-strain relations

$$\varepsilon = \frac{1}{3} \left(dk + \frac{1}{3} dl \right) * \sigma, \quad \sigma = de * \dot{\varepsilon}, \tag{5.47}$$

where the *tensile modulus* $de(t)$ is defined by

$$\widehat{de}(\lambda) = \frac{3}{\lambda} \frac{1}{\widehat{dk}(\lambda) + \widehat{dl}(\lambda)/3} = 9 \frac{\widehat{db}(\lambda) \widehat{da}(\lambda)}{3 \widehat{db}(\lambda) + \widehat{da}(\lambda)}, \quad \lambda > 0. \tag{5.48}$$

Observe that de is at least a completely positive measure if k, l are creep functions or, equivalently da, db are completely positive. However, it seems not to be clear that $e(t)$ is also a creep function without further assumptions on a, b, although physically it should be. On the other hand, by Lemma 4.3 we see that $e(t)$ is even a Bernstein function if $a, b \in \mathcal{BF}$ or $k, l \in \mathcal{BF}$. If the material is *synchronous*, then $b(t) = \beta a(t)$ for some $\beta > 0$, hence $e(t) = [9\beta/(3\beta + 1)]a(t)$.

The equations of motion for the axial extension of a viscoelastic rod after normalization become

$$u_t(t, x) = \int_0^t de(\tau) u_{xx}(t - \tau, x) + h(t, x),$$

$$u(t, 0) = 0, \quad u(0, x) = u_0(x), \tag{5.49}$$

$$\int_0^t de(\tau) u_x(t - \tau, 1) = g(t), \quad t > 0, \ x \in (0, 1).$$

The linear operator defined by

$$Au(x) = u''(x), \quad x \in [0, 1], \ u \in \mathcal{D}(A),$$

$$\mathcal{D}(A) = \{ u \in X : u', u'' \in X, u(0) = 0, u'(1) = 0 \}$$

is well known to generate a bounded cosine family in any of the spaces $L^q(0, 1), 1 \leq q < \infty$ and in the subspace $X_0 = \{ u \in C[0, 1] : u(0) = 0 \}$ of $C[0, 1]$. Thus if $e(t)$ is a Bernstein function (or merely a creep function with $e_1(t)$ log-convex) then Theorem 4.3 shows that (5.49) admits a resolvent $S(t)$, uniformly bounded by 1, in any of these spaces. For $q = 2$ we may apply Corollary 1.2 instead, to the result that a resolvent exists if $de(t)$ merely is of positive type, and $|S(t)| \leq 1$ then holds again.

5.5 Heat Conduction in Materials with Memory

Consider the problem of heat conduction in materials with memory described in Section 5.3, where the rigid body Ω is assumed to be homogeneous and isotropic. Here we want to discuss the well-posedness of the resulting boundary value problem (5.34) in the usual function spaces.

Let $X_q = L^q(\Omega)$, $1 \leq q < \infty$, $X_0 = C_{\Gamma_b}(\overline{\Omega}) = \{u \in C(\overline{\Omega}) : u \mid_{\Gamma_b} \equiv 0\}$, and define

$$(A_q v)(x) = \Delta v(x), \quad x \in \Omega, \ v \in \mathcal{D}(A_q),$$

where Δ denotes the Laplace operator, and for $1 < q < \infty$

$$\mathcal{D}(A_q) = \{v \in W^{2,q}(\Omega) : v \mid_{\Gamma_b} \equiv 0, \ \frac{\partial v}{\partial n} \mid_{\Gamma_f} \equiv 0\},$$

$$\mathcal{D}(A_0) = \{v \in \cap_{q>1} \mathcal{D}(A_q) : A_0 v \in C_{\Gamma_b}(\overline{\Omega})\},$$

and in $X_1 = L^1(\Omega)$, A_1 is defined e.g. by the closure of A_2. If $\Omega \subset \mathbb{R}^N$ is bounded, $\partial\Omega$ is smooth, and $\Gamma_b \cap \Gamma_f = \emptyset$, then it is well known that A_q is closed linear densely defined, and $\sigma(A_q) \subset (-\infty, 0]$ for each $q \in \{0\} \cup [1, \infty)$. Moreover, for each q and $\phi \in [0, \pi/2)$ there is a constant $M(\phi, q)$ such that

$$|(z - A_q)^{-1}| \leq \frac{M(\phi, q)}{|z|}, \quad z \in \Sigma(0, \phi + \pi/2),$$

holds, i.e. A_q generates a C_0-semigroup which is analytic in $\mathbb{C}_+$ and bounded on each sector $\Sigma(0, \phi)$, $\phi < \pi/2$. In the Hilbert space case $q = 2$, the operator A_2 is even selfadjoint and negative semidefinite, hence A_2 generates a bounded cosine family in $X_2 = L^2(\Omega)$.

As motivated by thermodynamics, we assume that $m(t)$ is a bounded creep function with $m(0+) = m_0 > 0$, and denote by $b(t)$ the corresponding completely positive function, i.e. the solution of $dm * b = 1$; cp. Proposition 4.4. Without specifying more properties of $c \in BV_{loc}(\mathbb{R}_+)$ the initial value problem in X_q for (5.34) with homogeneous boundary conditions is then equivalent to

$$u(t) = (a * A_q u)(t) + f(t), \quad t > 0, \tag{5.50}$$

where $a = b * dc$ and f contains $b * r$ as well as the temperature history.

Following the discussion of the physical properties of $c(t)$ in Section 5.3, we divide into two cases.

Case 1: $c_0 = c(0+) > 0$; $c \in BV_{loc}(\mathbb{R}_+)$ arbitrary otherwise.
This means that the instantaneous heat conductivity is nonvanishing; w.l.o.g. we may assume $c_0 = 1$. In this case $a = b + dk * b$, where $k \in BV_{loc}(\mathbb{R}_+)$ is continuous at $t = 0$; this shows that the perturbed equation (3.18) arises naturally in this context. Since $b(t)$ is completely positive, it is of positive type. Hence the equation

$$u(t) = (b * A_q u)(t) + f(t), \quad t > 0, \tag{5.51}$$

is *parabolic* by Corollary 3.1. To show that $b(t)$ is also 1-regular, write

$$-\lambda \frac{\hat{b}'(\lambda)}{\hat{b}(\lambda)} = \frac{m_0 + (\lambda \hat{m}_1(\lambda))'}{m_0 + \hat{m}_1(\lambda)} = \frac{m_0 - \widehat{tdm_1}(\lambda)}{m_0 + \hat{m}_1(\lambda)}, \quad \text{Re } \lambda > 0.$$

Since $m_1(t)$ is 1-monotone and belongs to $L^1(\mathbb{R}_+)$ by assumption, we have

$$\mid \widehat{tdm_1}(\lambda) \mid \leq \int_0^\infty -tdm_1(t) = \int_0^\infty m_1(t)dt < \infty, \quad \text{Re } \lambda > 0.$$

On the other hand, $m_0 + \hat{m}_1(\lambda) \neq 0$ for Re $\lambda \geq 0$, and converges to m_0 as $|\lambda| \to \infty$, hence is bounded from below. This proves that $b(t)$ is 1-regular. Theorem 3.1 then shows that (5.51) admits a bounded resolvent $S(t)$, and Theorem 3.2 yields a resolvent for (5.50). Thus in case $c_0 = c(0+) > 0$, (5.34) is well-posed in any space X_q, $q \in \{0\} \cup [1, \infty)$. Moreover, if $b(t)$ is in addition 2-regular, e.g. if $m_1(t)$ is also convex, then Theorem 3.3 shows that (5.50) has the maximal regularity property of type C^α in any of the spaces X_q.

Observe that classical heat conduction belongs to Case 1, since then $m(t) \equiv m_0$ and $c(t) \equiv c_0$ for $t > 0$.

Case 2: $m, c \in \mathcal{BF}$ bounded; materials of relaxation type.
Having the case $c_0 = 0$ in mind, one cannot expect well-posedness in each space X_q since $m(t) \equiv m_0$, $c(t) = (1 - e^{\alpha t})/\alpha$, $\alpha > 0$ corresponds to the damped wave equation

$$u_{tt} + \alpha u_t = \Delta u + g,$$

which for $q \neq 2$ is not well-posed in X_q, at least for dimensions $N > 1$, A_q then does not generate a cosine family. Therefore, we restrict to $q = 2$ in this case.

Well-posedness in $L^2(\Omega)$ of (5.50) is an easy consequence of Corollary 1.2. In fact, since $m_0 > 0$, and $m_1 \in L^1(\mathbb{R}_+)$ we obtain $b \in W^{1,1}_{loc}(\mathbb{R}_+)$, and even $\dot{b} \in L^1(\mathbb{R}_+)$; therefore $a \in BV(\mathbb{R}_+)$ and

$$\begin{aligned}
\text{Re } \widehat{da}(i\rho) &= \text{Re } (i\rho \hat{b}(i\rho)\hat{dc}(i\rho)) = \text{Re } \frac{c_0 + \hat{c}_1(i\rho)}{m_0 + \hat{m}_1(i\rho)} \\
&= \frac{[c_0 + \text{Re } \hat{c}_1(i\rho)][m_0 + \text{Re } \hat{m}_1(i\rho)] + \text{Im } \hat{c}_1(i\rho) \text{ Im } \hat{m}_1(i\rho)}{\mid m_0 + \hat{m}_1(i\rho) \mid^2} \geq 0,
\end{aligned}$$

and so Corollary 1.2 yields a resolvent $S(t)$ for (5.50) in $L^2(\Omega)$, bounded by 1, if only $m, c \in \mathcal{CF}$ are bounded, $m_0 > 0$, and m_1, c_1 are of positive type.

However, if $m, c \in \mathcal{BF}$ we may also apply the results of Section 4, in particular Theorems 4.3 and 4.4. In fact, by Lemma 4.3

$$\frac{1}{\lambda}\sqrt{\hat{b}(\lambda)\hat{dc}(\lambda)} = \hat{c}(\lambda)/\sqrt{\frac{\hat{c}(\lambda)}{\hat{b}(\lambda)/\lambda}}$$

is the Laplace transform of a Bernstein function $e(t)$; observe that $1 * b \in \mathcal{BF}$ by Proposition 4.6. Therefore, Theorem 4.3 (ii) as well as Theorem 4.4 apply, hence the resolvent $S(t)$ exists and is of the form

$$S(t) = \mathrm{Co}(t/\kappa)e^{-\alpha t/\kappa} + S_0(t), \quad t > 0,$$

where $\mathrm{Co}(t)$ denotes the cosine family generated by A_2. For the characteristic numbers κ, α, ω (cp. Section 4.5) we obtain

$$\kappa = \lim_{\lambda \to \infty} 1/\lambda \sqrt{\hat{b}(\lambda)\widehat{dc}(\lambda)} = \begin{cases} 0 & \text{if } c_0 > 0 \\ \sqrt{m_0/c_1(0+)} & \text{if } c_0 = 0 \end{cases}$$

$$\omega = \lim_{\lambda \to 0+} (\hat{b}(\lambda)\widehat{dc}(\lambda))^{-1/2} = \lim_{\lambda \to 0+} (\lambda \frac{m_0 + \hat{m}_1(\lambda)}{c_0 + \hat{c}_1(\lambda)})^{1/2} = 0,$$

$$\alpha = \lim_{\lambda \to \infty} (\hat{b}(\lambda)\widehat{dc}(\lambda))^{-1/2} - \kappa\lambda = \lim_{\lambda \to \infty} [(\lambda \frac{m_0 + \hat{m}_1(\lambda)}{c_0 + \hat{c}_1(\lambda)})^{1/2} - (\frac{m_0}{c_1(0+)})^{1/2}\lambda]$$

$$= \frac{1}{2}[\frac{m_1(0+)}{(m_0 c_1(0+))^{1/2}} - \frac{m_0^{1/2}\dot{c}_1(0+)}{c_1(0+)^{3/2}}] \quad \text{in case } \kappa > 0,$$

$$\alpha = \infty \quad \text{otherwise.}$$

Thus (5.34) has the finite propagation property iff $c_0 = 0, m_0 > 0, c_1(0+) < \infty$, and no wavefront iff $\kappa = 0$ or $m_1(0+) - \dot{c}_1(0+) = \infty$.

One of the motivations of the paper of Gurtin and Pipkin [157] was to remove the defect of the infinite propagation speed of classical heat flow. They already observed that this corresponds to $m_0 > 0, c_0 = 0, c_1(0+) < \infty$. On the other hand, the solutions of (5.34) should have the property of *positivity*, i.e. if $r(t, x) \geq 0, \theta_b(t, x) \geq 0, q_f(t, x) \geq 0$ for all t, x, then the solution $\theta(t, x)$ of (5.34) should satisfy $\theta(t, x) \geq 0$ for all t, x.

Assume that (5.34) enjoys the property of positivity for all domains $\Omega \subset \mathbb{R}^3$ which are bounded with $\partial\Omega$ smooth. Since the eigenvector corresponding to the first eigenvalue $\lambda_1 = \lambda_1(\Omega) > 0$ of $-\Delta$ with say Dirichlet boundary conditions is positive, and as Ω varies, the range of $\lambda_1(\Omega)$ is all of $(0, \infty)$ this implies that $(\lambda \widehat{dm}(\lambda) + \mu \widehat{dc}(\lambda))^{-1}$ is completely monotonic w.r.t. $\lambda > 0$, for each $\mu > 0$. As a result we then obtain with $\mu \to 0+$ that $(\lambda \widehat{dm}(\lambda))^{-1}$ is completely monotonic on $(0, \infty)$ - which is consistent with $m \in \mathcal{CF}$ - but also that $1/\widehat{dc}(\lambda)$ belongs to $\mathcal{CM}$, as $\mu \to \infty$. This obviously implies $c_0 > 0$, i.e. we are in Case 1; if one even assumes $c \in \mathcal{CF}$ like is the case for materials of relaxation type, then $c(t) \equiv c_0 > 0$ is constant. By Theorem 4.2 and Corollary 4.6 we then obtain the positivity alone from $m \in \mathcal{CF}$, but on the other hand the wave speed is in this case necessarily infinite, the problem is parabolic. Thus there is no coexistence of positivity and finite wave speed for the linear heat conduction problem.

5.6 Synchronous and Incompressible Materials

In this subsection we want to discuss the well-posedness of some problems of three dimensional viscoelasticity and thermoviscoelasticity.

(i) **Isotropic Incompressible Fluids**

Consider a homogeneous isotropic incompressible viscoelastic fluid which occupies a region $\Omega \subset \mathbb{R}^3$. According to Section 5.1, the velocity field $v(t, x)$ of the fluid is governed by (5.26), hence the corresponding initial value problem becomes

$$
\begin{aligned}
v_t(t, x) &= \int_0^t da(\tau)\Delta v(t - \tau, x) - \nabla \pi(t, x) + g(t, x), \quad t > 0, \ x \in \Omega, \\
\nabla \cdot v(t, x) &= 0, \quad t > 0, \ x \in \Omega, \\
v(t, x) &= 0, \quad t > 0, \ x \in \partial\Omega, \\
v(0, x) &= v_0(x), \quad x \in \Omega,
\end{aligned} \tag{5.52}
$$

where we employed the nonslip boundary conditions; here $\pi(t, x)$ denotes the hydrostatic pressure in the fluid. The kernel $da(t)$ is the shear modulus introduced in Section 5.1; according to Section 5.2, $a(t)$ should be at least a creep function with $da(t)$ completely positive. The case $a(t) \equiv a_0$ for $t > 0$ corresponds to a Newtonian fluid with viscosity $a_0 > 0$, and (5.52) then becomes the well known linear Navier-Stokes system.

We want to discuss the well-posedness of (5.52) in $L^q(\Omega)^3$ where $1 < q < \infty$, in particular for $q = 2$. Define

$$
\begin{aligned}
E^q(\Omega) &= \{\nabla\pi : \pi \in W^{1,q}_{loc}(\Omega), \nabla\pi \in L^q(\Omega)^3\}, \\
C^\infty_{0,\sigma}(\Omega) &= \{u \in C^\infty_0(\Omega)^3 : \nabla \circ u = 0\}, \quad \text{and} \quad L^q_\sigma(\Omega) = \overline{C^\infty_{0,\sigma}(\Omega)},
\end{aligned}
$$

where the closure is taken in $L^q(\Omega)^3$. It is known that

$$
L^q(\Omega)^3 = E^q(\Omega) \oplus L^q_\sigma(\Omega) \tag{5.53}
$$

holds topologically if Ω is open and connected, and $\partial\Omega$ is compact and of class C^1. This is the so-called *Helmholtz decomposition* of $L^q(\Omega)^3$. Let P_q denote the projection of $L^q(\Omega)^3$ onto $L^q_\sigma(\Omega)$ along $E^q(\Omega)$ associated with the Helmholtz decomposition; P_q is bounded and even orthogonal for $q = 2$. Let the *Stokes operator* A_q be defined in $X_q = L^q_\sigma(\Omega)$ according to

$$
(A_q u)(x) = (P_q \Delta u)(x), \quad x \in \Omega, \ u \in \mathcal{D}(A_q), \tag{5.54}
$$

with domain

$$
\mathcal{D}(A_q) = \{u \in W^{2,q}(\Omega)^3 \cap X_q : u \mid_{\partial\Omega} = 0\}.
$$

It is known that A_q is a closed linear densely defined operator in X_q with $\sigma(A_q) \subset \mathbb{R}_-$, and for each $0 \le \phi < \pi/2$ there is a constant $M(\phi, q) \ge 1$ such that

$$
|(z - A_q)^{-1}|_{X_q} \le \frac{M(\phi, q)}{|z|}, \quad z \in \Sigma(0, \phi + \pi/2). \tag{5.55}
$$

Thus A_q generates a C_0-semigroup which is analytic on $\mathbb{C}_+$ and uniformly bounded on each sector $\Sigma(0, \phi)$, $\phi < \pi/2$. In the Hilbert space case $q = 2$, A_2 is even selfadjoint and negative semidefinite. Since $P_q \nabla \pi(t; \cdot) = 0$ by the definition of P_q, application of P_q to (5.52) shows that this equation can be written in abstract form as

$$v(t) = \int_0^t a(t - \tau) A_q v(\tau) d\tau + f(t), \quad t > 0, \tag{5.56}$$

where

$$f(t) = v_0 + \int_0^t P_q g(\tau) d\tau, \quad t > 0;$$

here we assumed $v_0 \in X_q$.

We are now in position to apply the existence results of Sections 1~4. Since completely positive measures are of positive type by Corollary 4.1, Corollary 1.2 yields a resolvent $S_2(t)$ in X_2 which is bounded by 1. If in addition $\log a_1(t)$ is convex, we may apply Theorem 4.3 instead, as well as Theorem 4.4 to obtain the decomposition

$$S_2(t) = \mathrm{Co}(t/\kappa) e^{-\alpha t/\kappa} + S_0(t), \tag{5.57}$$

where $\mathrm{Co}(t)$ denotes the cosine family generated by A_2, and α, κ are the characteristic numbers of $a(t)$ defined in Section 5.4(i).

For $q \neq 2$, (5.56) is not well-posed in general, restrictions on the kernel $a(t)$ are needed. Suppose $a_1(t)$ is 3-monotone,

$$\varlimsup_{t \to 0} \frac{t^{-1} \int_0^t \tau a_1(\tau) d\tau}{a_0 - \int_0^t \tau \dot{a}_1(\tau) d\tau} < \infty, \tag{5.58}$$

and consider the case of a fluid $a_\infty = 0$, $a_1 \in L^1(\mathbb{R}_+)$. Then we are in the situation of Corollary 3.3, to the result that the resolvent $S_q(t)$ of (5.52) is bounded on $\mathbb{R}_+$, as is $t \dot{S}_q(t)$, and (5.56) has the maximal regularity property of type C^α, by Theorem 3.3. The remark following the proof of Corollary 3.3 yields the same result also in case $a_\infty \neq 0$ or $a_1 \notin L^1(\mathbb{R}_+)$ if the domain Ω is bounded, since then A_q is invertible. If $a(t)$ is even a Bernstein function and (5.58) holds then it is not difficult to show that $|\arg \hat{a}(\lambda)| \leq \pi/2 + \phi < \pi$ for $\lambda \in \Sigma(0, \pi/2 + \varepsilon), \varepsilon > 0$ sufficiently small, and so Theorem 2.1 shows that the resolvent $S_q(t)$ is even analytic. This discussion covers the case when (5.52) is parabolic.

(ii) Isotropic Synchronous Materials

Equations (5.25) for general homogenous isotropic viscoelastic materials in three dimensions involve two kernels, the shear modulus da and the compression modulus db. A priori there are no physical reasons why these kernels should be related, and so (5.25) cannot be transformed into an equation of the form (1.1) in some function space, except for very special configurations. For example, if $\Omega = \mathbb{R}^3$ then decomposing $u(t, x)$ as

$$u(t, x) = \nabla \varphi(t, x) + \nabla \times v(t, x)$$

where $\nabla \cdot v(t, x) = 0$, it is possible to reduce (5.25) to two independent equations for φ and v which separately only involve the kernels $(4/3)da + db$ and da, respectively.

This shows the limitations of the theory of equations of scalar type. However, if the material is *synchronous*, i.e. if there is $\beta > 0$ such that $b(t) = \beta a(t)$, $t > 0$, then (5.25) can be rewritten as an equation of the form (1.1).

In fact, let Ω be a bounded domain with boundary $\partial\Omega$ of class C^1 and define in $X = L^2(\Omega)^3$ an operator A by means of

$$(Au)(x) = \Delta u(x) + (\beta + 1/3)\nabla\nabla \cdot u(x), \quad x \in \Omega \tag{5.59}$$

with domain

$$D(A) = \{u \in W^{2,2}(\Omega)^3 : u|_{\Gamma_d} = 0, \, (\frac{\partial u}{\partial n} + (\nabla u)^T \cdot n + n(\beta - \frac{2}{3})\nabla \cdot u)|_{\Gamma_s} = 0\}.$$

It is well known that the elasticity operator A is selfadjoint and negative semidefinite and $0 \in \rho(A)$ iff $\Gamma_d \neq \emptyset$, $\mathcal{N}(A) \oplus \mathcal{R}(A) = X$ if $\Gamma_d = \emptyset$. Thus we can write the initial value problem for (5.25) as

$$v(t) = \int_0^t a(t - \tau)Av(\tau) + f(t), \quad t > 0,$$

in the space $\mathcal{R}(A) \subset X$ and apply Corollary 1.2 or Theorems 4.3 and 4.4 to obtain well-posedness of linear isotropic viscoelasticity in $L^2(\Omega)^3$ for the case of synchronous materials, where $a \in \mathcal{CF}$ is such that a_1 is log-convex, say. For $X = L^q(\Omega)^3, q \neq 2$, remarks similar to the case of incompressible materials apply.

(iii) Synchronous Isotropic Thermoviscoelasticity

In the simplest case of thermoviscoelasticity, the equations for homogeneous isotropic materials (5.40) already involve five kernels which in general should be considered as independent. Therefore a reduction of (5.40) to a Volterra equation of scalar type is in general not accessible. However, such a reformulation is possible in the case of *synchronous* materials which are defined by the relations

$$\widehat{db} = \beta\widehat{da}, \quad \hat{c}^2 = \nu^2\widehat{dm}^2 \hat{a}, \quad \widehat{da}^2 = \gamma^2\frac{\widehat{dm}}{\lambda^2\hat{a}}, \tag{5.60}$$

with some positive constants β, ν, γ. Recall that a, b, c, m, α should be creep functions with c, m, α in addition bounded. Considering the shear modulus da and the heat capacity dm as the principal quantities, (5.60) are then definitions for the remaining material functions. Observe that classical isotropic thermoelasticity is covered by (5.60); cp. with (5.41).

If $a(t)$ and $m(t)$ are Bernstein functions and $b(t), \alpha(t), c(t)$ are defined by (5.60), then $b, \alpha \in \mathcal{BF}$. In fact, this is trivial for $b(t)$; for $\alpha(t)$ we obtain

$$\hat{\alpha}(\lambda) = \gamma\sqrt{\hat{m}(\lambda)\hat{k}(\lambda)}, \quad \hat{k}(\lambda) = 1/(\lambda^3\hat{a}(\lambda)), \quad \lambda > 0,$$

hence $k, \alpha \in \mathcal{BF}$ by Proposition 4.6 and Lemma 4.3, respectively. However, $\alpha(t)$ will be bounded iff $\widehat{da}(0) = \alpha(\infty) < \infty$ which is equivalent to $m(\infty) < \infty$ and

$a_\infty > 0$; the last relation in (5.60) is consistent with physics only for solids. However, $c(t)$ defined by (5.60) will not always be a Bernstein function again, for this to happen, $a(t)$ and $m(t)$ cannot be completely independent. For example,

$$\widehat{dc}(\lambda) = \widehat{dl}(1/\sqrt{\hat{a}(\lambda)}), \quad \widehat{dm}(\lambda) = \nu \widehat{dl}(1/\sqrt{\hat{a}(\lambda)})/(\lambda\sqrt{\hat{a}(\lambda)}), \quad \lambda > 0,$$

with some bounded $l \in \mathcal{BF}$ is consistent with (5.60), and $m, c \in \mathcal{BF}$ are bounded if $a_\infty > 0$, by Proposition 4.7 and Lemma 4.3, since $\sqrt{\hat{a}(\lambda)}/\lambda$ is a complete Bernstein function, as Lemma 4.2 shows. Observe that $\widehat{dc}(\lambda) \equiv 1$, i.e. $l(t) = e_0(t)$, is a possible choice and compare the discussion at the end of Section 5.5.

Now suppose the material is synchronous; choose $X = W^{1,2}_{\Gamma_d}(\Omega)^3 \times L^2(\Omega)^3 \times L^2(\Omega)$ with inner product

$$(z_1, z_2)_X = \int_\Omega (\nabla w_1 : \nabla w_2 + (\beta + \frac{1}{3})(\nabla \cdot w_1)(\nabla \cdot w_2) + v_1 \cdot v_2 + \vartheta_1 \vartheta_2) dx,$$

where $z_i = (w_i, v_i, \vartheta_i)^T$. Here we assume $\mathrm{mes}(\Gamma_d) \neq \emptyset$ for simplicity and use the notation

$$W^{1,2}_{\Gamma_d}(\Omega)^3 = \{w \in W^{1,2}(\Omega)^3 : w|_{\Gamma_d} = 0\}.$$

Define A in X by means of

$$Az = \begin{pmatrix} v \\ \Delta w + (\beta + \frac{1}{3})\nabla\nabla \cdot w - \beta\gamma\nabla\vartheta \\ \nu\Delta\vartheta - \beta\gamma\nabla \cdot v \end{pmatrix}, \tag{5.61}$$

for

$$z \in \mathcal{D}(A) = \{(w, v, \vartheta) \in W^{2,2}(\Omega)^3 \times W^{1,2}(\Omega)^3 \times W^{2,2}(\Omega) :$$

$$v|_{\Gamma_d} = w|_{\Gamma_d} = 0, \vartheta|_{\Gamma_b} = 0, \frac{\partial\vartheta}{\partial n}|_{\Gamma_f} = 0, \tag{5.62}$$

$$(\frac{\partial w}{\partial n} + (\nabla w)^T \cdot n + (\beta - \frac{2}{3})\nabla \circ w - \beta\gamma\frac{\partial\vartheta}{\partial n})|_{\Gamma_s} = 0\}.$$

Then A is precisely the classical thermoelasticity operator in X. The variables w, v, ϑ in our setting differ from the classical ones. We define

$$w = e * v, \quad v = \dot{u}, \quad \vartheta = dl * \Theta, \tag{5.63}$$

where the kernels $e(t)$ and $l(t)$ are related to $a(t)$ resp. $m(t)$ by

$$\hat{e}(\lambda) = \sqrt{\hat{a}(\lambda)}, \quad \widehat{dl}(\lambda) = \sqrt{\widehat{dm}(\lambda)}, \quad \lambda > 0; \tag{5.64}$$

observe that $e(t)$ is completely monotonic by Lemma 4.2 and l is a Bernstein function by Proposition 4.7, in case $a, m \in \mathcal{BF}$. A simple computation then shows that the initial boundary value problem for (5.40) becomes

$$z(t) = \int_0^t e(t - \tau)Az(\tau)d\tau + g(t), \quad t > 0. \tag{5.65}$$

Since A generates a C_0-semigroup of contractions, and $e(t)$ is completely monotonic, Theorem 4.2 applies and yields a resolvent $S(t)$ which is bounded by 1. Concerning wave propagation, observe that the characteristic numbers ω, κ, α of $e(t)$ are important; these are precisely those which we computed in Section 5.4.

5.7 A Simple Control Problem

In manufacturing polymeric materials there arises frequently the problem of glueing. For this purpose the surfaces of the pieces first have to be heated in such a way that a surface layer of prescribed depth exceeds a certain given temperature, but stays below another critical temperature, beyond which the material would be destroyed. An effective means which meets these requirements is heating by means of radiation, e.g. infrared or microwave radiation. In a very simplified model this leads to a control problem for a one-dimensional heat equation for materials with memory. It is in particular assumed that the absorption of the radiation in the material follows an exponential distribution $\alpha e^{-\alpha x}$, where x denotes distance from the surface and $\alpha > 0$ is a constant. If $\theta(t,x)$ denotes the temperature, $\theta_\infty = 0$ the (constant) temperature of the environment, and if the constitutive laws for isotropic homogeneous materials with memory (5.33) with $m \in \mathcal{BF}$, $m_0 > 0$, $m(\infty) < \infty$, $c(t) \equiv c_0 > 0$ are employed, this leads to the problem

$$
\begin{aligned}
dm * \dot{\theta} &= c_0 \theta_{xx} + i(t)\alpha e^{-\alpha x}, \quad t, x > 0, \\
\theta(0, x) &= \theta(t, \infty) = 0, \quad t, x > 0, \\
c_0 \theta_x(t, 0) &= \beta \theta(t, 0), \quad t > 0.
\end{aligned}
\tag{5.66}
$$

Here $i(t)$ denotes the intensity of the radiation which serves as the control variable; it is subject to the constraints

$$
0 \le i(t) \le i_0, \quad t > 0, \tag{5.67}
$$

where i_0 denotes the maximal available intensity. The constant $\beta > 0$ accounts for heat radiation of the surface of the body into its enviroment. Given a temperature $\theta_0 > 0$, the surface temperature to be reached, the problem consists of two parts.

Part 1: Let $i(t) \equiv i_0$; find $\theta(t, x)$, and the first time $0 < t_0 < \infty$, such that $\theta(t_0, 0) = \theta_0$, i.e. the desired surface temperature θ_0 is reached at time $t = t_0$.

Part 2: Find a control $i(t)$ for $t \ge t_0$, subject to the constraints (5.67) such that $\theta(t, 0) \equiv \theta_0$ for $t \ge t_0$, i.e. the surface temperature θ_0 is maintained, and determine $\theta(t, x)$.

Obviously, for $t_0 < \infty$ to exist, the maximal intensity must be large enough. But, concerning the second part, it is not at all clear whether such a control exists. However, we show that this problem is in fact uniquely solvable.

To obtain an abstract reformulation, we let $a \in \mathcal{CM}$ be defined by $\hat{a}(\lambda) = 1/\lambda \widehat{dm}(\lambda)$, $\lambda > 0$, and $f(x) = \alpha e^{-\alpha x}$, $x > 0$. A good choice for the state space is $X = C_0(\mathbb{R}_+)$, and with $A\theta = c_0\theta''$, $\mathcal{D}(A) = \{u \in C_0^2(\mathbb{R}_+) : c_0\theta'(0) = \beta\theta(0)\}$, (5.66) can be rewritten as

$$\theta(t) = a * A\theta(t) + (a * i)(t)f, \quad t \geq 0. \tag{5.68}$$

With $f^* \in C_0(\mathbb{R}_+)^*$ defined by $< \theta, f^* > = \theta(0)$, the observation $y(t)$ is given by

$$y(t) = < \theta(t), f^* >, \quad t \geq 0. \tag{5.69}$$

Since A generates a bounded analytic C_0-semigroup in X and $a(t)$ is completely monotonic, (5.68) is a well-posed parabolic problem, its solution is given by the variation of parameters formula involving the integral resolvent $R(t)$ for (5.68)

$$\theta(t) = \int_0^t R(t - \tau)fi(\tau)d\tau, \quad t \geq 0. \tag{5.70}$$

Hence the observation becomes

$$y(t) = \int_0^t c(t - \tau)i(\tau)d\tau, \quad t \geq 0, \tag{5.71}$$

where the scalar function $c(t)$ is given by

$$c(t) = < R(t)f, f^* >, \quad t \geq 0. \tag{5.72}$$

Observe that by maximal regularity of type C^η, θ will be a strong solution on $\mathbb{R}_+$ whenever $i \in C^\eta(\mathbb{R}_+)$. It is not difficult to compute $\hat{c}(\lambda)$; in fact, with

$$\begin{aligned} \varphi(z) &= < (z - A)^{-1}f, f^* > \\ &= \frac{\alpha}{(\sqrt{z} + \alpha\sqrt{c_0})(\sqrt{z} + \beta/\sqrt{c_0})}, \quad z \in \Sigma(0, \pi), \end{aligned} \tag{5.73}$$

one obtains

$$\begin{aligned} \hat{c}(\lambda) &= < \hat{R}(\lambda)f, f^* > = < (1/\hat{a}(\lambda) - A)^{-1}f, f^* > \\ &= \varphi(1/\hat{a}(\lambda)), \quad \lambda > 0. \end{aligned} \tag{5.74}$$

Next

$$\frac{1}{\varphi(z)} = \frac{z}{\alpha} + (\sqrt{c_0} + \beta/(\alpha\sqrt{c_0}))\sqrt{z} + \beta, \quad z \in \Sigma(0, \pi),$$

shows that $1/\varphi(z)$ is a complete Bernstein function, and so $\varphi(z)$ itself is the Laplace transform of a completely monotonic function, cp. Proposition 4.6. The subordination principle Proposition 4.7 then implies that $c(t)$ is completely monotonic as well. It is this property of the involved kernel which leads to the well-posedness of the control problem in question.

For Part 1 of the problem we see that $y(t) = i_0 \int_0^t c(\tau)d\tau$, $t > 0$, is a Bernstein function; hence $t_0 < \infty$ exists if and only if

$$\beta\theta_0 < i_0, \tag{5.75}$$

since

$$y(\infty) = i_0 \lim_{t \to \infty} \int_0^t c(\tau)d\tau = i_0\hat{c}(0) = i_0\varphi(1/\hat{a}(0)) = i_0\varphi(0) = i_0/\beta.$$

For Part 2 we have to solve the Volterra equation of the first kind for the function $j(t) = i(t + t_0)$ resulting from (5.71) with $y(t + t_0) \equiv \theta_0$, i.e.

$$\int_0^t c(t - \tau)j(\tau)d\tau = i_0 \int_0^{t_0} (c(\tau) - c(t + \tau))d\tau, \quad t > 0. \tag{5.76}$$

Defining the Bernstein function $k(t)$ by $\widehat{dk}(\lambda) = 1/\lambda\hat{c}(\lambda)$, i.e.

$$\widehat{dk}(\lambda) = \alpha^{-1}\widehat{dm}(\lambda) + (\sqrt{c_0} + \beta/(\alpha\sqrt{c_0}))\sqrt{\hat{m}(\lambda)} + \beta/\lambda, \quad \lambda > 0,$$

the unique solution of (5.76) is given by

$$\begin{aligned}
j(t) &= i_0 \int_0^t dk(\tau)\frac{d}{dt}\int_0^{t_0}(c(s) - c(t - \tau + s))ds \\
&= i_0 \int_0^t dk(\tau)\int_0^{t_0} -\dot{c}(t - \tau + s))ds \\
&= i_0 \int_0^t dk(\tau)(c(t - \tau) - c(t + t_0 - \tau)) \\
&= i_0(1 - \int_0^t dk(\tau)c(t + t_0 - \tau)).
\end{aligned}$$

Since $dk(t) \geq 0$ and $c(t) \geq 0$ is nonincreasing there follows easily $0 \leq j(t) \leq 1$, i.e. $j(t)$ is subject to the constraints (5.67).

5.8 Comments
a) For the early history of linear viscoelasticity one should consult the papers of Boltzmann [28], Maxwell [236], and Volterra [332], [333]. These papers date back to the end of the last turn of the century and contain the isothermal field theory formulation of viscoelasticity.

b) For today's theory of viscoelasticity we refer to the monographs of Bland [24], Christensen [42]; Flügge [117], Mainardi [233], Pipkin [269], and Renardy, Hrusa, and Nohel [289], for thermoviscoelasticity see Christensen [42] and Pipkin [269], and for heat conduction in materials with memory see Gurtin and Pipkin [157], Meixner [240], and Nunziato [261].

c) The network approach to linear isothermal viscoelasticity mentioned at the end of Section 5.2 is worked out in detail in Bland [24]. For an approach based on molecular theory see e.g. Bird, Armstrong and Hassager [23].

d) Usually a viscoelastic material is called a *solid* if it is not a *fluid*. Our definition of a *solid* is slightly different from this; however, for symmetry reasons it seems to be more natural.

e) The discussion of the Rayleigh problem is standard in viscoelasticity. Pipkin [269] mentions that $w(\cdot; x)$ is increasing if a is a Bernstein function. Renardy [288] first observed the possibility of the finite propagation speed together with absence of the wave front; see also Hrusa and Renardy [185], where the regularity of $w(t; \tau)$ away from $\{t = \kappa\tau\}$ is studied. The continuity across this line in case $-\dot{a}_1(0+) = \infty, a_1(t)$ log-convex as well as the monotonicity of w was first proved in Prüss [274].

Torsion of a rod is also one of the standard problems in viscoelasticity. It has been discussed in detail in the papers of Hannsgen, Renardy and Wheeler [166], and Hannsgen and Wheeler [167], [170], [171], in particular stabilization and destabilization by means of boundary feedbacks; see also Desch and Miller [89].

f) Composed relaxation moduli like the tensile modulus de defined in (5.48) occur in many other problems involving linear viscoelastic rods, beams and plates; see Bland [24], Hannsgen [165], and Noren [259]. Here Lemma 4.3 proves its usefulness if $a(t)$ and $b(t)$ are Bernstein functions, at least. It would be of interest to know whether this lemma remains true if $a, b \in \mathcal{CF}$ are such that da and db are only completely positive.

g) There has been a considerable interest in heat conduction in materials with memory in the 1960s and 1970s as the many publications from this period show. The main emphasis at that time was on the derivation of the equations from basic principles and on uniqueness results as well as on wave propagation. We refer to the papers of Coleman [57], Coleman and Gurtin [58], Gurtin [158], Gurtin and Pipkin [157], Nunziato [261], [262], Meixner [240], Finn and Wheeler [116], Nachlinger and Wheeler [254], and Davis [78], [79]. The first proofs for well-posedness in the linear case for very special configurations appear in Grabmüller [134]; Miller [244] contains the first general treatment based on perturbation techniques. Since then the heat equation with memory has been used as motivation and application in many research articles.

h) The properties of the Laplacian in $L^q(\Omega), 1 < q < \infty$, used in Section 5.5 are wellknown; cp. e.g. Fattorini [113] or Tanabe [319]. The analyticity of the C_0-semigroups in $C_{\Gamma_b}(\overline{\Omega})$ and in $L^1(\Omega)$ generated by the Laplacian is more recent, it is due to Stewart [317], [318], and to Amann [8], respectively.

The Helmholtz decomposition of $L^q(\Omega)^3$ for bounded domains has been obtained by Fujiwara and Morimoto [126]; for $N = 3$ and exterior domains see Solonnikov [311], Miyakawa [253], and von Wahl [334]. The most general result which covers domains in $\mathbb{R}^N$ with compact boundary of class C^1 seems to be the paper of Simader and Sohr [304].

Boundedness and analyticity of the semigroup generated by the Stokes operator was proved by Giga [129] and Solonnikov [311] for bounded domains, and by Borchers and Sohr [29], and Giga and Sohr [131] for exterior domains.

The selfadjointness of the elasticity operator in Section 5.6(ii) is also wellknown; see e.g. Marsden and Hughes [234] or Leis [208]. These references contain also proofs for the m-dissipativity of the thermoelasticity operator defined by (5.61) and (5.62) in Section 5.6(iii).

i) Concerning the positivity of the solutions of (5.34) in Case 1 of Section 5.5 we refer to Clément and Nohel [51] and to Lunardi [226] where sufficient conditions in terms of $m(t)$ and $c(t)$ are given. A complete characterization of positivity seems to be unknown.

Chapter II

Nonscalar Equations

6 Hyperbolic Equations of Nonscalar Type

The basic properties of Volterra equations of nonscalar type are discussed in this section. Resolvents for such problems are introduced and their relations to well-posedness and variation of parameters formulae are studied. The latter are then used for perturbation results which yield several well-known existence theorems. The generation theorem for the nonscalar case is proved and then applied to the convergence of resolvents and to existence theorems for equations in Hilbert spaces involving operator-valued kernels of positive type.

6.1 Resolvents of Nonscalar Equations

Throughout this section X and Y denote Banach spaces such that $Y \overset{d}{\hookrightarrow} X$, and $A \in L^1_{loc}(\mathbb{R}_+; \mathcal{B}(Y, X))$. We consider the linear Volterra equation

$$u(t) = f(t) + \int_0^t A(t - \tau)u(\tau)d\tau, \quad t \in J, \tag{6.1}$$

where $f \in C(J; X)$, $J = [0, T]$. Observe that the scalar case which has been considered so far corresponds to $A(t)$ of the form $A(t) = a(t)A$, where $a \in L^1_{loc}(\mathbb{R}_+)$, A a closed linear densely defined operator in X, and $Y = X_A$, the domain $D(A)$ of A equipped with the graph norm of A.

Definition 6.1 *A function $u \in C(J; X)$ is called*
(a) a **strong solution** *of (6.1) if $u \in L^\infty(J; Y)$ and (6.1) holds in X on J;*
(b) a **mild solution** *of (6.1) if there are $(f_n) \subset C(J; X)$ and strong solutions $(u_n) \subset C(J; X)$ of (6.1) with f replaced by f_n, such that $f_n \to f$ and $u_n \to u$ in $C(J; X)$.*

Comparing these solution concepts with the corresponding notions for equations of scalar type (see Definition 1.1) it is apparent that 'strong solution' in the sense of Definition 6.1 is the natural extension of Definition 1.1 (a). However, there is no direct analog of Definition 1.1 (b) for (6.1), and so a new concept for 'mild solution' had to be introduced. It is clear that strong solutions are also mild ones.

As before, the central object associated with (6.1) will be the resolvent $S(t)$. However, its definition differs somewhat from the corresponding concept introduced in Section 1.1 since (S2) no longer has a meaning. This leads to a number of complications.

Definition 6.2 *A family $\{S(t)\}_{t \geq 0} \subset \mathcal{B}(X)$ is called* **pseudo-resolvent** *for (6.1) if the following conditions are satisfied.*
(S1) $S(t)$ is strongly continuous in X on $\mathbb{R}_+$, and $S(0) = I$;
(S2) $U(t) = \int_0^t S(\tau)d\tau$ is leaving Y invariant, and $\{U(t)\}_{t \geq 0} \subset \mathcal{B}(Y)$ is locally Lipschitz on $\mathbb{R}_+$;
(S3) the resolvent equations

$$S(t)y = y + \int_0^t A(t - \tau)dU(\tau)y, \quad t \geq 0, \, y \in Y, \tag{6.2}$$

$$S(t)y = y + \int_0^t S(t-\tau)A(\tau)y d\tau, \quad t \geq 0, \ y \in Y, \tag{6.3}$$

hold. (6.2) resp. (6.3) are called the **first** *resp.* **second resolvent equation** *for (6.1). A pseudo-resolvent $S(t)$ is called* **resolvent** *for (6.1) if in addition* **(S4)** *for $y \in Y$, $S(\cdot)y \in Y$ a.e. and $S(\cdot)y$ is Bochner-measurable in Y on $\mathbb{R}_+$ is satisfied. A resolvent or pseudo-resolvent $S(t)$ is called a-**regular**, where $a \in L^1_{loc}(\mathbb{R}_+)$, if $a * Sx \in C(\mathbb{R}_+; Y)$ for each $x \in X$.*

Observe that for each $y \in Y$, the function $A(\cdot)y$ belongs to $L^1_{loc}(\mathbb{R}_+; X)$; therefore the convolution $S * Ay$ appearing in (6.3) is well-defined by the strong continuity of $S(t)$ in X, even pointwise as a Bochner integral. On the other hand, the definition of the convolution $A * dUy$ in (6.2) needs some care, since $A(t)$ is not assumed to be continuous. However, we may proceed as follows. For $y \in Y$ the function $g(t) = U(t)y$ is locally Lipschitz in Y by (S2), hence belongs to $BV_{loc}(\mathbb{R}_+; Y)$. Since $U(0) = 0$ this shows that the convolution $F * dg$ for each $F \in C(\mathbb{R}_+; \mathcal{B}(Y, X))$ is well-defined pointwise, e.g. as a Riemann-Stieltjes integral, and $F * dg$ is continuous on $\mathbb{R}_+$, $(F * dg)(0) = 0$. Moreover, the estimate

$$| (F * dg)(t) | \leq L(t) \int_0^t | F(\tau) | \, d\tau, \quad t > 0, \tag{6.4}$$

holds on $\mathbb{R}_+$, where $L(t)$ denotes the smallest Lipschitz constant of g in Y on $[0, t]$. By means of $(G_0 F)(t) = (F * dg)(t), t \geq 0$, the function g defines a linear operator $G_0 : C(\mathbb{R}_+; \mathcal{B}(Y, X)) \to C(\mathbb{R}_+; X)$; (6.4) shows that G_0 admits a unique bounded extension G to $L^1_{loc}(\mathbb{R}_+; \mathcal{B}(Y, X))$. We then define

$$(A * dg)(t) = (GA)(t), \quad t \geq 0, \ A \in L^1_{loc}(\mathbb{R}_+; \mathcal{B}(Y, X));$$

observe that $A * dg$ is continuous on $\mathbb{R}_+$, $(A * dg)(0) = 0$, and (6.4) holds with F replaced by A. These arguments also show that

$$(A * g)(t) = (1 * (A * dg))(t), \quad t \geq 0,$$

holds. The resolvent equations can therefore be rewritten as

$$U(t)y = ty + \int_0^t A(t-\tau)U(\tau)y d\tau, \quad t \geq 0, \ y \in Y; \tag{6.5}$$

$$U(t)y = ty + \int_0^t U(t-\tau)A(\tau)y d\tau, \quad t \geq 0, \ y \in Y. \tag{6.6}$$

Sometimes it is more convenient to work with (6.5), (6.6) instead of (6.2), (6.3). Let us show that pseudo-resolvents are always unique.

Proposition 6.1 *Suppose $S_1(t), S_2(t)$ are pseudo-resolvents for (6.1). Then $S_1(t) = S_2(t)$ for all $t \geq 0$.*

Proof: Let $y \in Y$ be fixed and let $U_i(t) = \int_0^t S_i(\tau)d\tau, i = 1, 2$. Since $U_i(t)$ satisfies (6.5) and (6.6) we obtain

$$
\begin{aligned}
t * U_2 y &= U_2 * (U_1 - A * U_1)y \\
&= U_2 * U_1 y - (U_2 * A) * U_1 y \\
&= U_2 * U_1 y - (U_2 - t) * U_1 y = t * U_1 y,
\end{aligned}
$$

hence $t * U_1 y = t * U_2 y$ on $\mathbb{R}_+$. Differentiating twice this yields $S_1(t)y = S_2(t)y$ for all $t \geq 0$, $y \in Y$, and since Y by assumption is dense in X we obtain $S_1(t) = S_2(t)$ for all $t \geq 0$. $\square$

For the case of scalar equations $A(t) = a(t)A$, a moment of reflection shows that a pseudo-resolvent is already a resolvent, and also a-regular by Proposition 1.1. Therefore Definition 6.2 is an extension of the previous case. We are mainly interested in resolvents rather than pseudo-resolvents, so let us show that under some additional restrictions on $A(t)$ or on the space Y a pseudo-resolvent is already a resolvent.

Proposition 6.2 *Suppose $S(t)$ is a pseudo-resolvent for (6.1) and assume one of the following conditions.*
(i) Y has the Radon-Nikodym property.
(ii) There is a dense subset $Z \subset Y$ such that $A(t)z \in Y$ for a.a. $t > 0$ and $A(\cdot)z \in L^1_{loc}(\mathbb{R}_+; Y)$ for each $z \in Z$.
(iii) $S(t)$ is a-regular, $A(t) = (a * dB)(t)$ for $t > 0$, where $a \in L^1_{loc}(\mathbb{R}_+)$ and $B \in BV_{loc}(\mathbb{R}_+; \mathcal{B}(Y, X))$ is such that $B(\cdot)y$ has a locally bounded Radon-Nikodym derivative w.r.t. $b(t) = Var B|_0^t$, for each $y \in Y$.
Then $S(t)$ is a resolvent for (6.1). In cases (ii) and (iii), $S(t)$ is even strongly continuous in Y.

Proof: Let $S(t)$ be a pseudo-resolvent for (6.1) and $U(t) = \int_0^t S(\tau)d\tau$. Consider first case (i). Since for each $y \in Y$ the function $g(t) = U(t)y$ is locally Lipschitz in Y, and Y has the Radon-Nikodym property (c.p. Diestel [98], Chapter 6), $g(t)$ is a.e. differentiable in Y and $\dot{g} \in L^\infty_{loc}(\mathbb{R}_+; Y)$. Since $Y \hookrightarrow X$ we obtain $\dot{g}(t) = S(t)y \in Y$ for all $t \in \mathbb{R}_+ \setminus N(y)$, $y \in Y$, where $N(y)$ has Lebesgue-measure zero, and so (S4) follows.

Next assume (ii), and let $z \in Z$. The second resolvent equation then shows $S(t)z \in Y$ for all $t \geq 0$ and $S(\cdot)z \in C(\mathbb{R}_+; Y)$; in fact, by the remarks following Definition 6.2 the function $S * Az = dU * Az$ belongs to Y and is continuous there. Thus by (S2), $U_h(t) = h^{-1}(U(t + h) - U(t))$ is locally bounded in Y and $U_h(t)z \to S(t)z$ in $C(\mathbb{R}_+; Y)$, hence by the Banach-Steinhaus theorem, $U_h(t) \to S(t)$ strongly in Y, uniformly on bounded intervals. This proves (S4).

For the third case, fix $y \in Y$ and write

$$
(S * A)(t)y = ((a * S) * dB)(t)y = \int_0^t (a * S)(t - \tau)h(\tau)db(\tau),
$$

where $h(t) = dB(t)y/db(t)$ denotes the Radon-Nikodym derivative of the function $B(t)y$ in X. Since $a * Sx$ is strongly continuous in Y, we see that $(S * Ay)(t)$ belongs to Y and is continuous there. The second resolvent equation then implies (S4). □

Note that reflexive spaces have the Radon-Nikodym property, as well as separable dual spaces. If Y is reflexive and $S(t)$ is a pseudo-resolvent then we even have $S(t)Y \subset Y$ for each $t \geq 0$ and $S(\cdot)y$ is weakly continuous in Y; $S(t)$ inherits some regularity in Y from the embedding $Y \hookrightarrow X$, strong continuity of $S(t)$ in X and reflexivity of Y.

Concerning (iii), observe that kernels $B(t)$ of the form

$$B(t) = \sum_{i=1}^{\infty} B_i \chi_{(t_i, \infty)}(t) + \int_0^t B_0(s)ds$$

with $B_i \in \mathcal{B}(Y, X)$ and $B_0(\cdot)y \in L^1_{loc}(\mathbb{R}_+; X), |B_0(t)y|_X \leq b_0(t)|y|_Y$ such that $b(t) = b_0(t) + \sum_{t_i < t} |B_i|_{\mathcal{B}(Y,X)} < \infty$ for each $t > 0$ are covered.

6.2 Well-posedness and Variation of Parameters Formulae

Suppose $S(t)$ is a pseudo-resolvent for (6.1), let $f \in C(J; X)$ and $u \in L^\infty(J; Y)$ be a strong solution of (6.1). Then by (6.1) and (6.3)

$$S * u = 1 * u + (S * A) * u = 1 * u + S * (u - f) = 1 * u + S * u - S * f,$$

hence

$$1 * u = S * f.$$

If u is merely a mild solution, then there are $f_n \in C(J; X), f_n \to f$ in $C(J; X)$, and strong solutions $u_n \in L^\infty(J; Y)$ of (6.1) with f replaced by f_n, such that $u_n \to u$ in $C(J; X)$. The identities

$$1 * u_n = S * f_n$$

then imply $1 * u = S * f$ again. Thus, if $u \in C(J; X)$ is a mild solution of (6.1) then $S * f \in C^1(J; X)$ and

$$u(t) = \frac{d}{dt} \int_0^t S(t - \tau)f(\tau)d\tau, \quad t \in J, \tag{6.7}$$

i.e. the variation of parameters formula (1.8) also holds for the general equation (6.1), in particular mild solutions are again unique.

Conversely, (6.7) defines a function $u \in C(J; X)$ whenever $f \in W^{1,1}(J; X)$ and we then have

$$\begin{aligned} u(t) &= S(t)f(0) + \int_0^t S(t - \tau)f'(\tau)dt \tag{6.8} \\ &= S(t)f(0) + \int_0^t dU(\tau)f'(t - \tau), \quad t \in J. \end{aligned}$$

Thus if even $f \in W^{1,1}(J;Y)$ and $f(0) = 0$ then $u \in C(J;Y)$ is a strong solution of (6.1). Approximating $f \in W^{1,1}(J;X)$ with $f(0) = 0$ in this space by functions $f_n \in W^{1,1}(J;Y)$ with $f_n(0) = 0$, it becomes apparent that (6.8) yields a mild solution of (6.1). Thus Proposition 1.2 (i), (ii), (iv) remain valid for (6.1). However, in this general framework there is no analog of (iii) of this proposition, unless the resolvent $S(t)$ has more special properties, e.g. is a-regular in the sense of Definition 6.2. Let us summarize.

Proposition 6.3 *Suppose $S(t)$ is a pseudo-resolvent for (6.1), and let $f \in C(J;X)$. Then*
(i) If $u \in C(J;X)$ is a mild solution of (6.1), then $S * f$ is continuously differentiable on J and

$$u(t) = \frac{d}{dt} \int_0^t S(t - \tau) f(\tau) d\tau, \quad t \in J; \tag{6.9}$$

in particular, mild solutions of (6.1) are unique.
(ii) If $f \in W^{1,1}(J;X)$, $f(0) = 0$, then

$$u(t) = \int_0^t S(t - \tau) f'(\tau) d\tau, \quad t \in J, \tag{6.10}$$

is a mild solution of (6.1).
(iii) If $f \in W^{1,1}(J;Y)$, $f(0) = 0$, then

$$u(t) = \int_0^t dU(\tau) f'(t - \tau), \quad t \in J, \tag{6.11}$$

is a strong solution of (6.1).
(iv) If $S(t)$ is a-regular and $f = a * g$, $g \in W^{1,1}(J;X)$, then $u(t)$ given by (6.10) is a strong solution of (6.1).

Observe that in case $S(t)$ is a resolvent for (6.1), $S(t)x$ is a mild solution of (6.1) with $f(t) = x$, and a strong solution if $x \in Y$. In this situation, the assumption $f(0) = 0$ in (ii) and (iii) of this proposition can be dropped.

The following example shows that a resolvent $S(t)$ for (6.1) is in general only 0-regular in which case (iv) of Proposition 6.3 is empty, and that time-regularity of $f(t)$ alone cannot be sufficient for existence of strong solutions, in general.

Example 6.1 Let $a \in BV_{loc}(\mathbb{R}_+)$ be such that da is of positive type and let $s(t) = s(t; 1)$ denote the solution of (1.18) with $\mu = 1$; observe that $|s(t)| \leq 1$. Consider $X = l^2(\mathbb{N})$, $Y = X_A$, where $(Ax)_n = nx_n$, $n \geq 1$, $D(A) = \{(x_n) \in X : (nx_n) \in l^2(\mathbb{N})\}$. Then both spaces X and Y are separable Hilbert spaces and $Y \overset{d}{\hookrightarrow} X$. Define $A \in L^1_{loc}(\mathbb{R}_+; \mathcal{B}(Y, X))$ by means of $(A(t)x)_n = -a(t)x_n$, $t \geq 0$, $n \in \mathbb{N}$; (6.1) then admits a resolvent $S(t)$ which is given by $(S(t)x)_n = s(t)x_n$, $t \geq 0$, $n \in \mathbb{N}$. Suppose $u(t)$ is a strong solution of (6.1) on an interval $J = [0, T]$;

then

$$|f(t)|_Y \leq |u(t)|_Y + |(A * u)(t)|_Y$$

$$\leq |u(t)|_Y + |(a * u)(t)|_Y \leq \left(1 + \int_0^T |a(t)| dt\right)|u|_{Y,\infty} < \infty$$

for all $t \in J$, i.e. $f \in L^\infty(J; Y)$. Thus if $f \in C(J; X)$ but $f \notin L^\infty(J; Y)$ there cannot be a strong solution of (6.1). □

For a large class of equations (6.1) existence of a pseudo-resolvent $S(t)$ already implies a-regularity of $S(t)$. Since this class is important for applications we state this result as

Proposition 6.4 *Suppose* $A \in L^1_{loc}(\mathbb{R}_+; \mathcal{B}(Y, X))$ *is of the form*

$$A(t) = a(t)A + \int_0^t a(t - \tau) dB(\tau), \quad t > 0, \tag{6.12}$$

where A is a closed linear densely defined operator in X such that $Y = X_A$ and $\rho(A) \neq \emptyset$, and $B \in BV_{loc}(\mathbb{R}_+; \mathcal{B}(Y, X))$ is left-continuous and $B(0) = B(0+) = 0$. Then each pseudo-resolvent $S(t)$ for (6.1) is a-regular, and it is even a resolvent if B satisfies in addition (iii) of Proposition 6.2.

Proof: Let $\mu \in \rho(A)$ be fixed and let $K(t) = -B(t)(\mu - A)^{-1}$, $t \geq 0$. Then $K \in BV_{loc}(\mathbb{R}_+; \mathcal{B}(X))$, hence by Theorem 0.5 there is $L \in BV_{loc}(\mathbb{R}_+; \mathcal{B}(X))$ such that

$$L = K - dK * L = K - L * dK$$

holds; observe that $K(0) = K(0+) = 0$. Rewrite the first resolvent equation as

$$Sy - y - \mu a * Sy = (\delta + dK) * (A - \mu)(a * dU)y, \quad y \in Y,$$

where $\delta_0(t)$ denotes Dirac's distribution, and convolve with $\delta_0 - dL$ to the result

$$A(a * dUy) = Sy - y - dL * (Sy - y - \mu a * Sy).$$

Since A is closed, $\rho(A) \neq \emptyset$ and $a * dUy = a * Sy$ in X, we obtain $(a * Sx)(t) \in Y$ for each $t \geq 0$, $x \in X$ and $A(a * Sx)(t)$ is continuous in X, i.e. $a * S$ is strongly continuous in Y. □

In the sequel, (6.1) will be said to have a *main part* if $A(t)$ is of the form (6.12).

In analogy to the case of scalar equations, well-posedness of (6.1) is defined as follows.

Definition 6.3 *(i) Equation (6.1) is called* **well-posed**, *if for each $y \in Y$ there is a unique strong solution $u(t; y)$ of*

$$u(t) = y + \int_0^t A(t - \tau) u(\tau) d\tau, \quad t \geq 0, \tag{6.13}$$

and $u(t; y_n) \to 0$ in X, uniformly on compact intervals, whenever $(y_n) \subset Y$ and $y_n \to 0$ in X.

*(ii) Let $a \in L^1_{loc}(\mathbb{R}_+)$. (6.1) is called a-**regularly well-posed**, if it is well-posed and if the equation*

$$u(t) = (\int_0^t a(\tau)d\tau)x + \int_0^t A(t - \tau)u(\tau)d\tau, \quad t \geq 0, \qquad (6.14)$$

admits a strong solution for every $x \in X$.

Suppose $S(t)$ is a resolvent for (6.1). From its definition and Proposition 6.2 it is then evident that (6.1) is well-posed, and even a-regularly well-posed if $S(t)$ is also a-regular. Conversely, suppose (6.1) is well-posed. Then we define

$$S(t)y = u(t; y), \quad t \geq 0, \ y \in Y;$$

as in Section 1.1 it then follows that $S(t)$ extends to all of X, is strongly continuous in X, strongly measurable and locally bounded in Y, and satisfies the first resolvent equation. Thus $S(t)$ satisfies (S1), (S4), the first part of (S3), hence also (S2); it remains to show that $S(t)$ satisfies the second resolvent equation, or equivalently that mild solutions of (6.1) are unique. However, it does not seem possible to derive the latter from the well-posedness of (6.1) alone; further properties are needed.

Proposition 6.5 *(a) If (6.1) admits a resolvent $S(t)$ then (6.1) is well-posed.*
(b) If (6.1) is well-posed and (ii) of Proposition 6.2 holds, then (6.1) admits a resolvent $S(t)$.
(c) Let $a \in L^1_{loc}(\mathbb{R}_+), a \not\equiv 0$. Then (6.1) is a-regularly well-posed if and only if there is an a-regular resolvent $S(t)$ for (6.1).

Proof: (a) is already clear. To prove (b) we are left with the verification of the second resolvent equation. Suppose (ii) of Proposition 6.2 holds, let $z \in Z$ and consider (6.1) with $f(t) = \int_0^t A(\tau)zd\tau$. Then $f \in C(J; X) \cap W^{1,1}(J; Y)$, hence by Proposition 6.3 the function $u = S * f' = S * Az$ is a strong solution of (6.1). On the other hand, we have for $v = Sz - z$

$$v = A * Sz = 1 * Az + A * v,$$

i.e. v is also a strong solution of (6.1) and so by uniqueness of strong solutions

$$Sz - z = v = u = S * Az.$$

This shows that the second resolvent equation holds for each $z \in Z$; since by assumption $Z \subset Y$ is dense, it follows also for each $y \in Y$.

(c) It is clear that existence of an a-regular resolvent $S(t)$ implies that (6.1) is a-regularly well-posed. Conversely, to prove the second resolvent equation let $f(t) = a * 1 * Ay, y \in Y$. Then by Proposition 6.3, $u = S * f' = a * S * Ay$ is a strong solution of (6.1); on the other hand $v = a * Sy - 1 * ay$ is also a strong solution of (6.1), hence by uniqueness of strong solutions we obtain $u = v$, i.e.

$$a * (Sy - y - S * Ay) \equiv 0, \quad y \in Y.$$

Titchmarsh's theorem implies then the second resolvent equation since $a \neq 0$.

To show that $S(t)$ is also a-regular, observe that by Definition 6.3 and a-regularity of (6.1), $a * U(t)$ is locally Lipschitz in $\mathcal{B}(X, Y)$ and $(d/dt) a * Uy(t) = a * dU(t)y$ is continuous in Y for each $y \in Y$. The Banach-Steinhaus theorem then implies that $a * dUx = a * Sx$ is continuous in Y for each $x \in X$, i.e. $S(t)$ is a-regular. $\square$

6.3 Hyperbolic Perturbation Results
Consider the perturbed equation

$$u(t) = f(t) + (A * u)(t) + (B * u)(t), \quad t \in J, \tag{6.15}$$

where $X, Y, A(t), f(t)$ are as before and $B \in L^1_{loc}(\mathbb{R}_+; \mathcal{B}(Y, X))$. Suppose (6.1) admits a resolvent $S(t)$ and (6.15) a resolvent $R(t)$; then by the variation of parameters formula (6.9) we have the relations

$$R(t)y = S(t)y + \frac{d}{dt}(S * B * R)(t)y, \quad t \in \mathbb{R}, \ y \in Y, \tag{6.16}$$

$$R(t)y = S(t)y + \frac{d}{dt}(R * B * S)(t)y, \quad t \in \mathbb{R}, \ y \in Y. \tag{6.17}$$

Under appropriate assumptions on $B(t)$, these identities can be used as in Section 1.4 to show that (6.16) admits an a-(pseudo-)resolvent iff (6.1) does.

Theorem 6.1 *Let $a \in L^1_{loc}(\mathbb{R}_+)$, and suppose $B \in L^1_{loc}(\mathbb{R}_+; \mathcal{B}(Y, X))$ is of the form*

$$B(t)y = B_0(t)y + (a * B_1)(t)y, \quad t \in \mathbb{R}, \ y \in Y, \tag{6.18}$$

where $\{B_0(t)\}_{t \geq 0} \subset \mathcal{B}(X) \cap \mathcal{B}(Y)$ and $\{B_1(t)\}_{t \geq 0} \subset \mathcal{B}(Y, X)$ satisfy
(i) $B_0(\cdot)y \in BV_{loc}(\mathbb{R}_+; Y)$ for each $y \in Y$; $B_0(\cdot)x \in BV_{loc}(\mathbb{R}_+; X)$ for each $x \in X$;
(ii) $B_1(\cdot)y \in BV_{loc}(\mathbb{R}_+; X)$ for each $y \in Y$.
Then (6.15) admits an a-regular (pseudo-)resolvent $R(t)$ if and only if there is an a-regular (pseudo-)resolvent $S(t)$ for (6.1)

For the proof we need the following simple

Lemma 6.1 Let $X_i, i = 1, 2$, be Banach spaces, $J = [0, a]$. Then
(i) Suppose $\{T(t)\}_{t \geq 0} \subset \mathcal{B}(X_1, X_2)$ is strongly continuous and $f \in BV(J; X_1)$. Then $u = T * f$ is Lipschitz in X_2 on J.
(ii) Suppose $\{U(t)\}_{t \geq 0} \subset \mathcal{B}(X_1, X_2)$ is locally Lipschitz, $U(0) = 0$, and $f \in BV(J; X_1)$. Then $v = U * df$ is Lipschitz in X_2 on J.

Proof: (i) Define $f(t) = 0$ for $t < 0$, and let $\varphi(t) = \mathrm{Var}\, f|_{-\infty}^t$; then $\varphi(t)$ is nondecreasing and

$$|f(t) - f(s)| \leq \varphi(t) - \varphi(s), \quad s \leq t \leq a.$$

Therefore

$$
\begin{aligned}
|u(t) - u(s)| &\leq \int_0^\infty |T(\tau)||f(t - \tau) - f(s - \tau)|d\tau \\
&\leq M \int_0^\infty (\varphi(t - \tau) - \varphi(s - \tau))d\tau \\
&= M \int_s^t \varphi(\tau)d\tau \leq M\varphi(a)(t - s), \quad s \leq t \leq a,
\end{aligned}
$$

where $M = \sup_J |T(t)|$.

(ii) Define $U(t) = 0$ for $t < 0$, the convolution $v = U * df$ exists even as a Riemann-Stieltjes integral and we obtain

$$
\begin{aligned}
|v(t) - v(s)| &= |\int_0^\infty (U(t - \tau) - U(s - \tau))df(\tau)| \\
&\leq L(t - s) \int_0^a |df(\tau)| = L \cdot \mathrm{Var}\, f|_0^a (t - s), \quad s \leq t \leq a,
\end{aligned}
$$

where $L = \sup\{|\frac{U(t) - U(s)}{t - s}| : 0 \leq s \leq t \leq a\}$. $\square$

Proof of Theorem 6.1: Suppose (6.1) admits an a-regular (pseudo-)resolvent $S(t)$, and let $U = 1 * S$. For each $x \in X$, $S * B_0 x$ is locally Lipschitz in X, by Lemma 6.1, hence $K_0 = S * B_0$ is so in $\mathcal{B}(X)$, as the uniform boundedness principle shows. Similarly, $K_0 = U * dB_0$ is also Lipschitz on compact intervals in $\mathcal{B}(Y)$, by Lemma 6.1 (ii). The same argument yields $K_1 = S * B_1$ locally Lipschitz in $\mathcal{B}(Y, X)$ and by a-regularity of $S(t)$, $a * K_1 = (a * S) * B_1$ is locally Lipschitz in $\mathcal{B}(Y)$. Note that neither K_0 nor K_1 have a jump at $t = 0$.

Consider the equation

$$Wx = (a * S)x + dK_0 * Wx + (d(a * K_1)) * Wx, \quad x \in X; \tag{6.19}$$

since $(a * S)x$ is strongly continuous in Y for each $x \in X$, by Corollary 0.3, there is a unique family of operators $\{W(t)\}_{t \geq 0} \subset \mathcal{B}(X, Y)$ which is strongly continuous and satisfies (6.19). (W will be $a * R$ lateron.) Next we consider the equation

$$Rx = Sx + dK_0 * Rx + dK_1 * Wx, \quad x \in X; \tag{6.20}$$

since $S(t)$ as well as $dK_1 * W$ are strongly continuous in X, by Corollary 0.3 there is a unique strongly continuous solution $\{R(t)\}_{t \geq 0} \subset \mathcal{B}(X)$ of (6.20). Convolving (6.20) with a and comparing the result with (6.19) in the space X, we obtain $W = a * R$ by uniqueness. In the next step we consider

$$Vy = Uy + dK_0 * Vy + (d(a * K_1)) * Vy, \quad y \in Y; \tag{6.21}$$

since $U(t)$ is locally Lipschitz in Y, by Corollary 0.3 there is a unique family of operators $\{V(t)\}_{t\geq 0} \subset \mathcal{B}(Y)$, strongly continuous in Y but also locally Lipschitz, satisfying (6.21). Integrating (6.20) for $x = y \in Y$ and from $W = a * R$, by uniqueness of the solution of (6.21), we then obtain $Vy = 1 * Ry$ for each $y \in Y$.

If $S(t)$ is even a resolvent, we may consider the equation

$$u = Sy + dK_0 * u + d(a * K_1) * u, \tag{6.22}$$

for a fixed $y \in Y$. Since $S(\cdot)y \in L^{\infty}_{loc}(\mathbb{R}_+; Y)$, there is a unique solution $u \in L^{\infty}_{loc}(\mathbb{R}_+; Y)$ of (6.22), and again by uniqueness we obtain $u(t) = R(t)y$ a.e.

So far we have shown that our candidate $R(t)$ for the resolvent of (6.15) satisfies (S1), (S2), and also (S4) if $S(t)$ is a resolvent, and is a-regular. It remains to verify the resolvent equations; we use their integrated forms (6.5), (6.6). Here we restrict attention to the case $a \not\equiv 0$, the other case is even simpler. (6.21) implies for $y \in Y$ by the definition of K_0 and K_1

$$a * 1 * [(A + B) * Vy - Vy + ty]$$
$$= a * 1 * [A * Uy - Uy + ty] + [A * a * S + a * 1 - a * S] * B * Vy = 0,$$

since $S(t)$ is a pseudo-resolvent for (6.1); hence by Titchmarsh's theorem

$$Vy = ty + (A + B) * Vy, \quad t \geq 0, \ y \in Y,$$

which is the first resolvent equation for (6.15). To prove the second resolvent equation observe that for $y \in Y$ we have $W * (A + B)y = (a * S) * (A + B)y + d(S * B) * (W * (A + B)y)$ by (6.19), but also

$$(W - 1 * a)y = (a * S - 1 * a)y + d(S * B)y * 1 * a + d(S * B) * (W - 1 * a)y,$$

hence by uniqueness of the solution and in virtue of

$$(a * S - 1 * a)y + d(S * B)y * 1 * a = a * (S * A + S * B),$$

we may conclude

$$W * (A + B)y = Wy - 1 * ay, \quad y \in Y, \ t \in \mathbb{R}_+;$$

since $W = a * R$, again Titchmarsh's theorem implies

$$R * (A + B)y = Ry - y, \quad y \in Y, \ t \in \mathbb{R}_+,$$

i.e. $R(t)$ satisfies the second resolvent equation for (6.15).

Interchanging the roles of (6.1) and (6.15), the 'only if' part follows as well, and the proof is complete. □

Observe that Theorem 1.2, first part, is contained in Theorem 6.1. Let us consider some special cases which combined with the results of Chapter 1 lead to first existence results for resolvents of nonscalar equations.

Corollary 6.1 *Let A be a closed linear densely defined operator in X, set $Y = X_A$, let $a \in L^1_{loc}(\mathbb{R}_+)$, B_0, B_1 as in Theorem 6.1, and let $A(t)$ be of the form*

$$A(t) = a(t)A + (a * B_1)(t) + B_0(t), \quad t \in \mathbb{R}_+. \tag{6.23}$$

Suppose (1.1) admits a resolvent. Then (6.1) admits an a-regular resolvent.

For $a(t) \equiv 1$, $t > 0$, in the situation of Corollary 6.1, (6.15) is formally equivalent to

$$\dot{u} = g + Au + dB_0 * u + B_1 * u, \quad u(0) = x. \tag{6.24}$$

As a consequence of Corollary 6.1 we then obtain

Corollary 6.2 *Let A, Y, B_0, B_1 be as in Corollary 6.1.*
Then (6.24) is well-posed if and only if A generates a C_0-semigroup in X. If this is the case, the resolvent $S(t)$ for (6.24) is a-regular with $a(t) \equiv 1$, and $S(\cdot)y$ is continuously differentiable in X for each $y \in Y$.

For $a(t) \equiv t$, in the situation of Corollary 6.1, (6.15) is formally equivalent to

$$\ddot{u} = h + Au + dB_0 * \dot{u} + B_1 * u, \quad u(0) = x, \ \dot{u}(0) = y. \tag{6.25}$$

For this case Corollary 6.1 yields

Corollary 6.3 *Let A, Y, B_0, B_1 be as in Corollary 6.1.*
Then (6.25) is well-posed if and only if A generates a cosine family in X. If this is the case, the resolvent $S(t)$ for (6.25) is t-regular, and $S(\cdot)y$ is continuously differentiable in X for each $y \in Y$, even twice a.e. if $B_0 \in W^{1,1}_{loc}(\mathbb{R}_+; \mathcal{B}(X))$.

Observe that for equations with main part one of the conditions in (i) of Theorem 6.1 on $B_0(t)$ can be replaced by $B_0 = a * dC$, for some $C \in BV_{loc}(\mathbb{R}_+; \mathcal{B}(X))$; this follows from the resolvent equations. In particular, for the first order case considered in Corollary 6.2, one of the conditions in (i) is sufficient. This remark will be useful in Section 9.

6.4 The Generation Theorem
In this subsection the extension of the generation theorem, Theorem 1.3, to the case of nonscalar equations is considered. This discussion is restricted to pseudo-resolvents and kernels which are of exponential growth, w.o.l.g. of subexponential growth.

So we let $X, Y, A(t)$ as before and assume in addition

$$\int_0^\infty e^{-\varepsilon t} |A(t)|_{\mathcal{B}(Y,X)} dt < \infty \quad \text{for each } \varepsilon > 0, \tag{6.26}$$

i.e. $A(t)$ is growing subexponentially; $a \in L_{loc}^1(\mathbb{R}_+)$ is also assumed to be of subexponential growth. Suppose $S(t)$ is a pseudo-resolvent for (6.1) and for each $\varepsilon > 0$ there is a constant M_ε such that

$$|S(t)|_{\mathcal{B}(X)} + \sup_{0<s<t} (t-s)^{-1}|U(t) - U(s)|_{\mathcal{B}(Y)} \le M_\varepsilon e^{\varepsilon t} \quad \text{for each } t > 0; \quad (6.27)$$

then for each $y \in Y$ the resolvent equations in integrated form yield the relations

$$(I - \hat{A}(\lambda))\hat{U}(\lambda)y = \hat{U}(\lambda)(I - \hat{A}(\lambda))y = \frac{y}{\lambda^2}, \quad \mathrm{Re}\, \lambda > 0.$$

With $H(\lambda) = \hat{S}(\lambda)$ and $\hat{U}(\lambda) = \hat{S}(\lambda)/\lambda$ this equation can be rewritten as

$$\lambda(I - \hat{A}(\lambda))H(\lambda)y = H(\lambda)\lambda(I - \hat{A}(\lambda))y = y, \quad \mathrm{Re}\, \lambda > 0. \quad (6.28)$$

Observe that $\{H(\lambda)\}_{\mathrm{Re}\,\lambda>0} \subset \mathcal{B}(X) \cap \mathcal{B}(Y)$ is holomorphic in both spaces, $\mathcal{B}(X)$ and $\mathcal{B}(Y)$, and $\{\hat{A}(\lambda)\}_{\mathrm{Re}\,\lambda>0} \subset \mathcal{B}(Y,X)$ is holomorphic in $\mathcal{B}(Y,X)$. (6.28) implies that $(I - \hat{A}(\lambda))$ is invertible in Y (more precisely, the part of $I - \hat{A}(\lambda)$ in Y is invertible), and

$$H(\lambda)y = \frac{1}{\lambda}(I - \hat{A}(\lambda))^{-1}y, \quad \mathrm{Re}\, \lambda > 0, \ y \in Y.$$

Since Y is dense in X, and $H(\lambda)$ is also bounded in X, this implies that $(I-\hat{A}(\lambda))^{-1}$ is closable in X and

$$H(\lambda) = \frac{1}{\lambda}\overline{(I - \hat{A}(\lambda))^{-1}}, \quad \mathrm{Re}\, \lambda > 0,$$

in X. Note that it is unclear in general, whether the operators $\hat{A}(\lambda)$ considered as unbounded operators in X with domain $D(\hat{A}(\lambda)) = Y$ are closed or even closable!

However, if $S(t)$ is in addition a-regular for some nontrivial function a, and for each $\varepsilon > 0$ there is a constant N_ε such that

$$|a * S(t)|_{\mathcal{B}(X,Y)} \le N_\varepsilon e^{\varepsilon t} \quad \text{for each } t > 0, \quad (6.29)$$

holds, then $\hat{a}(\lambda)H(\lambda) = (a*S)\hat{}(\lambda)$ forms a holomorphic operator family in $\mathcal{B}(X,Y)$. Since the zeros of $\hat{a}(\lambda)$ in $\mathrm{Re}\,\lambda > 0$ are isolated, the latter implies that $H(\lambda) \in \mathcal{B}(X,Y)$ for each $\mathrm{Re}\,\lambda > 0$. (6.28) then implies that the operators $\hat{A}(\lambda)$ are closed in X with constant domain Y.

After this preliminary discussion the extension of Theorem 1.3 to the nonscalar case can be stated as follows.

Theorem 6.2 *Suppose $A \in L_{loc}^1(\mathbb{R}_+; \mathcal{B}(Y,X))$ satisfies (6.26) and $a \in L_{loc}^1(\mathbb{R}_+)$ is of subexponential growth, $a \not\equiv 0$. Then (6.1) admits a pseudo-resolvent $S(t)$ such that (6.27) holds if and only if the following conditions are satisfied.*
(N1) *For each $\lambda > 0$ the operators $I - \hat{A}(\lambda)$ are injective, the part of $I - \hat{A}(\lambda)$ in Y is surjective, and $(I - \hat{A}(\lambda))^{-1}$ is bounded in X and in Y;*

we let $H(\lambda) = \overline{(I - \hat{A}(\lambda))^{-1}}/\lambda$, $\lambda > 0$.
(N2) *For each $\varepsilon > 0$ there is a constant $M_\varepsilon \geq 1$ such that*

$$|(\frac{d}{d\lambda})^n H(\lambda)|_{\mathcal{B}(X)} + |(\frac{d}{d\lambda})^n H(\lambda)|_{\mathcal{B}(Y)} \leq M_\varepsilon n! (\lambda - \varepsilon)^{-(n+1)}, \qquad (6.30)$$

for all $n \in \mathbb{N}_0$ and $\lambda > \varepsilon$. If this is the case, then $S(t)$ is a-regular and (6.29) holds if and only if
(N3) *for each $\lambda > 0$ the operators $\hat{A}(\lambda)$ in X with domain $D(\hat{A}(\lambda)) = Y$ are closed, and for each $\varepsilon > 0$ there is a constant $N_\varepsilon \geq 1$ such that*

$$|(\frac{d}{d\lambda})^n [\hat{a}(\lambda) H(\lambda)]|_{\mathcal{B}(X,Y)} \leq N_\varepsilon n! (\lambda - \varepsilon)^{-(n+1)}, \qquad n \in \mathbb{N}_0, \ \lambda > \varepsilon. \qquad (6.31)$$

is satisfied.

Proof: (Necessity.) For the necessity it only remains to prove the estimates (6.30) and (6.31); these follow directly from Theorem 0.3 by (6.27) and (6.29).
(Sufficiency.) If (N1) and (N2) hold, by Theorem 0.3 there are locally Lipschitz functions $U \in C(\mathbb{R}_+; \mathcal{B}(X))$, $V \in C(\mathbb{R}_+; \mathcal{B}(Y))$, $U(0) = V(0) = 0$ with

$$\hat{U}(\lambda) = H(\lambda)/\lambda \ \text{in} \ X, \quad \hat{V}(\lambda) = H(\lambda)/\lambda \ \text{in} \ Y, \quad \lambda > 0,$$

and for each $\varepsilon > 0$ there is M_ε such that

$$\sup_{0<s<t} (t-s)^{-1} |U(t) - U(s)|_{\mathcal{B}(X)}$$
$$+ \sup_{0<s<t} (t-s)^{-1} |V(t) - V(s)|_{\mathcal{B}(Y)} \leq M_\varepsilon e^{\varepsilon t}, \quad t > 0. \qquad (6.32)$$

Since Y is dense in X, we have $U(t)y = V(t)y$ for all $y \in Y$, $t \geq 0$, hence $U(t)Y \subset Y$ and $U(t)$ is locally Lipschitz in Y, i.e. (S2) holds. The identities (6.28) further imply

$$U(t)y = ty + (A * U)(t)y = ty + (U * A)(t)y, \quad t > 0, \ y \in Y, \qquad (6.33)$$

i.e. the resolvent equations hold in integrated form. Since $U(t)$ is locally Lipschitz in X, the family $U_h(t) = (U(t+h) - U(t))/h$ is uniformly bounded for $0 < h < 1$, and bounded $t > 0$. For $y \in Y$ we have

$$(U * A)(t)y = 1 * (dU * A)(t)y, \quad t > 0,$$

and $(dU * A)(t)y$ is continuous in X; see the remarks following Definition 6.2. Therefore, (6.33) shows that the functions $U(\cdot)y$ are continuously differentiable in X, for each $y \in Y$. Defining $S(t)y = U'(t)y$, $t \geq 0$, $y \in Y$, $S(t)$ is bounded in X, uniformly for bounded $t \geq 0$, hence extends to a strongly continuous operator family in X, i.e. (S1) is satisfied. Differentiating (6.33) yields (S3), the resolvent equations. Finally (6.32) yields (6.27).

Suppose in addition that (N3) holds. Then, by Theorem 0.3 there is a locally Lipschitz family $\{W(t)\}_{t \geq 0} \subset \mathcal{B}(X, Y))$ with the property that for each $\varepsilon > 0$ there is a constant N_ε such that

$$\sup_{0 < s < t} (t - s)^{-1} |W(t) - W(s)|_{\mathcal{B}(X,Y)} \leq N_\varepsilon e^{\varepsilon t} \quad \text{for each } t > 0, \tag{6.34}$$

and $\hat{W}(\lambda) = \hat{a}(\lambda) H(\lambda)/\lambda$, $\lambda > 0$. By uniqueness of the Laplace transform, we have $W(t) = (a * U)(t), t > 0$. Then $W'(t)y = (a * dU)(t)y$ exists for each $y \in Y$ and is continuous in Y. By the uniform boundedness principle $W'(t)x$ exists in Y for each $x \in X$ and is strongly continuous on $\mathbb{R}_+$. Finally, we have $W'(t)x = (a * dU)(t)x = (a * S)(t)x$ for each $x \in X$, $t \geq 0$, i.e. $S(t)$ is a-regular; (6.34) implies (6.29). The proof is complete. $\square$

The following example shows that not every resolvent can be obtained via the Generation Theorem; even if $A \in BV(\mathbb{R}_+; \mathcal{B}(Y, X))$ the resolvent $S(t)$ if it exists need not be exponentially bounded.

Example 6.2 Let $X = l^2(\mathbb{N})$, A the skew-adjoint operator defined by $(Ax)_n = inx_n$, $n \in \mathbb{N}$, with $D(A) = \{x \in l^2(\mathbb{N}) : (nx_n) \in l^2(\mathbb{N})\}$, and $Y = X_A$. Define $B(t) \in \mathcal{B}(Y, X))$ by means of

$$(B(t)x)_n = ne^{int}\chi_{[t_n, \infty)}(t)x_n, \quad n \in N, \ t \geq 0,$$

where $(t_n) \subset \mathbb{R}_+$ is an increasing sequence with $t_n \to \infty$ which will be chosen later. We consider (6.1) in differentiated form, i.e. the equation

$$\dot{u} = g + Au + B * u, \quad u(0) = x.$$

Observe that $(B(t)x)_n \neq 0$ only for finitely many n, independent of X, hence we even have $B \in BV_{loc}(\mathbb{R}_+; \mathcal{B}(X, Y))$; Theorem 6.1 therefore implies that a unique resolvent $S(t)$ exists, which is even 1-regular, since A generates a bounded C_0-group in X. The important point is that $|B(t)|_{\mathcal{B}(Y,X)} \leq 1$ holds, but $|B(t)|_{\mathcal{B}(X)} \sim e^{t^2}$, as we shall see.

The resolvent $S(t)$ can be explicitly computed, it is given by $(S(t)x)_n = e^{int}r_n(t)x_n$, $n \in \mathbb{N}$, $t > 0$ where

$$r_n(t) = \sum_{j=0}^{\infty} n^j (t - jt_n)_+^{2j}/(2j)!, \quad t \geq 0, \ n \in \mathbb{N};$$

observe that for a given t, we have $r_n(t) = 1$ for $n \geq n(t)$, where $n(t)$ is defined by $n(t) = \max\{n : t_n \leq t\}$. The Laplace-transforms of r_n are easily computed to the result

$$\hat{r}_n(\lambda) = \lambda(\lambda^2 - ne^{-\lambda t_n})^{-1}, \quad \lambda > \sqrt{n}, \ n \in \mathbb{N};$$

If $S(t)$ is exponentially bounded, say by ω, then $\hat{r}_n(\lambda)$ cannot have poles in a halfplane $\text{Re}\lambda > \omega$. Now choose $t_n > 0$ such that $t_n^2 e^{t_n} = n$, $n \in \mathbb{N}$; then $\lambda = t_n$

is a pole of $\hat{r}_n(\lambda)$. Since the function xe^x is strictly increasing to infinity, we have (t_n) strictly increasing and $t_n \to \infty$. Thus $S(t)$ cannot be exponentially bounded.

It is easy to see that $\log n > t_n^2 > \log n - 2\log\log n$, i.e. $t_n \sim \sqrt{\log n}$, hence $n(t) \sim e^{t^2}$. This implies for the norm of $B(t)$ in X and also in Y the asymptotic relations $|B(t)|_{\mathcal{B}(X)} \sim |B(t)|_{\mathcal{B}(Y)} \sim e^{t^2}$. $\quad\square$

It is very unpleasant that in the sufficiency part of Theorem 6.2 one has to verify the estimates on the derivatives of $H(\lambda)$ not only in X but also in Y, and maybe even for $\hat{a}(\lambda)H(\lambda)$ in the a-regular case. However, if $A(t)$ has a main part as in (6.12) then the estimates for $H(\lambda)$ are sufficient, as we show now.

Corollary 6.4 *Suppose $A(t)$ is of the form*

$$A(t) = a(t)A + \int_0^t a(t-\tau)dB(\tau), \quad t > 0, \tag{6.35}$$

where A is a closed linear densely defined operator in X with $\rho(A) \neq \emptyset$, $a \in L^1_{loc}(\mathbb{R}_+)$, $a(t) \not\equiv 0$, and $B \in BV_{loc}(\mathbb{R}_+; \mathcal{B}(Y,X))$ is left-continuous and $B(0) = B(0+) = 0$, where $Y = X_A$. Assume that a and B are of subexponential growth.

Then (6.1) admits a pseudo-resolvent $S(t)$ such that for some $M \geq 1$, $\omega > 0$,

$$|S(t)|_{\mathcal{B}(X)} \leq Me^{\omega t}, \quad t \geq 0,$$

if and only if the following two conditions are satisfied.
(N1') *For each $\lambda > \omega$ the operators $\hat{A}(\lambda)$ with domain $Y = D(A)$ are closed, $I - \hat{A}(\lambda)$ is bijective.*
(N2') *$H(\lambda)$ defined by $H(\lambda) = (I - \hat{A}(\lambda))^{-1}/\lambda$, $\lambda > \omega$, satisfies*

$$|(\frac{d}{d\lambda})^n H(\lambda)|_{\mathcal{B}(X)} \leq n!M(\lambda - \omega)^{-(n+1)}, \quad n \in \mathbb{N}_0, \lambda > \omega. \tag{6.36}$$

If this is the case then $S(t)$ is already a-regular.

Proof: If $A(t)$ has the form (6.35) then by Proposition 6.4 any pseudo-resolvent of (6.1) is already a-regular. Thus the necessity part as well as the last statement follow from Theorem 6.2. For the sufficiency part, it remains to show that (N2') implies (N2), probably with λ replaced by $\lambda + \alpha$, for some $\alpha \geq \omega$.

For this purpose, we fix any $\mu \in \rho(A)$ and let $K(t) = -B(t)(\mu - A)^{-1}$, and denote by $L(t)$ the solution of the Volterra equation $L = K - dK * L$. Then we have

$$(\hat{a}(\lambda)A + \hat{a}(\lambda)\widehat{dB}(\lambda))H(\lambda) = H(\lambda)(\hat{a}(\lambda)A + \hat{a}(\lambda)\widehat{dB}(\lambda)),$$

hence

$$(I + \widehat{dK}(\lambda))(\mu - A)H(\lambda) = H(\lambda)(I + \widehat{dK}(\lambda))(\mu - A)$$

for λ sufficiently large. From these relations we obtain

$$(\mu - A)H(\lambda)(\mu - A)^{-1} = (I - \widehat{dL}(\lambda))H(\lambda)(I + \widehat{dK}(\lambda)).$$

It is not difficult to show that the estimates (N2') for $H(\lambda)$ imply (N2') for the similarity transform $(\mu - A)H(\lambda)(\mu - A)^{-1}$ of $H(\lambda)$ with ω replaced by some α, since $K(t)$ and $L(t)$ are exponentially bounded. Since $\mu - A$ induces an isomorphism between Y and X, (N2) for $H(\lambda + \alpha)$ follows. $\square$

6.5 Convergence of Resolvents

Consider the sequence of problems

$$u_n(t) = f_n(t) + \int_0^t A_n(t - \tau)u_n(\tau)d\tau, \quad t \in J, \, n \in \mathbb{N}_0, \tag{6.37}$$

where $f_n \to f_0$ and $A_n \to A_0$ in an appropriate sense as $n \to \infty$. Assuming existence of the pseudo-resolvents $S_n(t)$ for (6.37) as well as the *stability condition*

$$|S_n(t)|_{\mathcal{B}(X)} + \sup_{0 < s < t} \{(t-s)^{-1}|U_n(t) - U_n(s)|_{\mathcal{B}(Y)}\} \leq Me^{\omega t}, \quad t > 0, \, n \in \mathbb{N}_0, \tag{6.38}$$

as in the case of the Trotter-Kato theorem on convergence of C_0-semigroups, the strong convergence $S_n(t) \to S_0(t)$ in X can be shown.

Theorem 6.3 *Let $\{A_n\}_{n \geq 0} \subset L^1_{loc}(\mathbb{R}_+; \mathcal{B}(Y, X))$ satisfy*
(i) there is $\varphi \in L^1_{loc}(\mathbb{R}_+)$, $\int_0^\infty e^{-\omega t}\varphi(t)dt < \infty$ such that

$$|A_n(t)y|_X \leq \varphi(t)|y|_Y, \quad \text{for } y \in Y, \, n \in \mathbb{N}, \, a.a. \, t > 0;$$

and
(ii) there is a dense subset $Z \subset Y$ such that

$$A_n(t)z \to A_0(t)z \text{ in } X \text{ as } n \to \infty, \quad \text{for } z \in Z, \, a.a. \, t > 0.$$

Suppose $S_n(t)$ are pseudo-resolvents for (6.37), $n \in \mathbb{N}_0$, such that the stability condition (6.38) holds. Then $S_n(t)x \to S_0(t)x$ as $n \to \infty$, for each $x \in X$, $t \geq 0$, and the convergence is uniform on compact subsets of $\mathbb{R}_+ \times X$.

Proof: Without loss of generality, we may assume $\omega = 0$. Next observe that (i) and (ii) imply $A_n(t)y \to A_0(t)y$ in X for a.a. $t > 0$, $y \in Y$ as well as (i) also for $n = 0$. We therefore may set $Z = Y$. By virtue of (i), the Laplace transforms $\hat{A}_n(\lambda)$ of $A_n(t)$, $n \in \mathbb{N}_0$, exist for $\lambda \in \mathbb{C}_+$, are holomorphic on $\mathbb{C}_+$ with values in $\mathcal{B}(Y, X)$, and $\hat{A}_n(\lambda)y \to \hat{A}_0(\lambda)y$ in X as $n \to \infty$, for each $y \in Y$ and uniformly on right halfplanes Re $\lambda \geq \eta > 0$.

By the generation theorem, Theorem 6.2, $I - \hat{A}_n(\lambda)$ are invertible in X but also in Y and with $H_n(\lambda) = \hat{S}_n(\lambda)$ we have

$$\lambda H_n(\lambda) = \overline{(I - \hat{A}_n(\lambda))^{-1}} \quad \text{in } X$$

as well as

$$|\lambda H_n(\lambda)|_{\mathcal{B}(X)} + |\lambda H_n(\lambda)|_{\mathcal{B}(Y)} \leq M, \quad \lambda > 0, \, n \in \mathbb{N}_0, \tag{6.39}$$

note that the choice $\varepsilon = 0$ in Theorem 6.2 is possible, due to the stability condition (6.38).

We show next

$$H_n(\lambda)x \to H_0(\lambda)x \quad \text{as } n \to \infty \text{ in } X, \quad \text{for all } \lambda > 0, \ x \in X. \tag{6.40}$$

In fact, for each $y \in Y$ we have

$$(I - \hat{A}_n(\lambda))^{-1}y - (I - \hat{A}_0(\lambda))^{-1}y = (I - \hat{A}_n(\lambda))^{-1}(\hat{A}_n(\lambda) - \hat{A}_0(\lambda))(I - \hat{A}_0(\lambda))^{-1}y,$$

hence

$$H_n(\lambda)y - H_0(\lambda)y = \lambda H_n(\lambda)(\hat{A}_n(\lambda) - \hat{A}_0(\lambda))H_0(\lambda)y, \quad \lambda > 0, \ y \in Y.$$

Since $H_n(\lambda)Y \subset Y$ for each $n \in \mathbb{N}_0$, $\lambda > 0$, by the strong convergence $\hat{A}_n(\lambda) \to \hat{A}_0(\lambda)$ and by (6.39) this identity implies (6.40) for each $y \in Y$, hence also for each $x \in X$, since Y is dense in X, and by (6.39) again. Observe that the convergence (6.40) is even uniform on compact subsets of $(0, \infty)$, hence also uniform on compact subsets of $(0, \infty) \times X$.

Define operators $T_n \in \mathcal{B}(L^1(\mathbb{R}_+; X), X)$ by means of

$$T_n f = \int_0^\infty S_n(\tau) f(\tau) d\tau, \quad f \in L^1(\mathbb{R}_+; X), \ n \in \mathbb{N}_0. \tag{6.41}$$

If f is an exponential polynomial of the form

$$f(t) = \sum_{i=1}^N e^{-\lambda_i t} x_i, \quad t > 0, \tag{6.42}$$

where $\{\lambda_i\}_1^N \subset (0, \infty)$ and $\{x_i\}_1^N \subset X$, then by (6.40)

$$T_n f = \sum_{i=1}^N H_n(\lambda_i) x_i \to \sum_{i=1}^N H_0(\lambda_i) x_i = T_0 f, \quad \text{as } n \to \infty.$$

The estimate

$$|T_n f|_X \le M |f|_1, \quad f \in L^1(\mathbb{R}_+; X), \ n \in \mathbb{N}_0,$$

implies uniform boundedness of the family $\{T_n\}_{n\ge 0} \subset \mathcal{B}(L^1(\mathbb{R}_+; X), X)$. In Lemma 6.2 below we show that exponential polynomials of the form (6.42) are dense in $L^1(\mathbb{R}_+; X)$, hence by the Banach-Steinhaus theorem we obtain $T_n f \to T_0 f$ as $n \to \infty$ for each $f \in L^1(\mathbb{R}_+; X)$.

To prove the convergence of $S_n(t)$, choose

$$f_{t,n}(s) = A_n(t - s) y \cdot \chi_{(0,t]}, \quad s > 0,$$

where $y \in Y$. By assumptions (i) and (ii) it is not difficult to show that the set $\{f_{t,n} : 0 \le t \le \tau, n \in \mathbb{N}_0\} \subset L^1(\mathbb{R}_+; X)$ is relatively compact as long as $y \in Y$ is fixed. Thus we obtain

$$(S_n * A_n y)(t) = T_n f_{t,n} \to T_0 f_{t,0} = (S_0 * A_0 y)(t) \quad \text{as } n \to \infty,$$

uniformly for $t \in [0, T]$. The second resolvent equations for (6.37) finally implies $S_n(t)y \to S_0(t)y$ as $n \to \infty$ for each fixed $y \in Y$, uniformly for $t \in [0, \tau]$, and the stability condition (6.38) then yields the strong convergence $S_n(t)x \to S_0(t)x$ for each $x \in X$, uniformly for $t \in [0, \tau]$, by the Banach-Steinhaus theorem, since Y is dense in X. The proof is complete. $\square$

In the proof of Theorem 6.3 we used the following

Lemma 6.2 *Consider $EP \subset L^1(\mathbb{R}_+; X)$ defined by*

$$EP = \{f : f(t) = \sum_{i=1}^{N} e^{-\lambda_i t} x_i, \text{ for some } N \in \mathbb{N}, \lambda_i > 0, x_i \in X\}.$$

Then EP is dense in $L^1(\mathbb{R}_+; X)$.

Proof: (a) We first derive a representation for functionals $f^* \in (L^1(\mathbb{R}_+; X))^*$. Define $x^*(t) \in X^*$ by means of

$$< x, x^*(t) >_X = < \chi_{(0,t]} x, f^* >, \quad x \in X, \ t \geq 0;$$

obviously, we have $x^*(0) = 0$, and for $t > s$

$$|x^*(t) - x^*(s)| = \sup_{|x| \leq 1} | < \chi_{(s,t]} x, f^* > | \leq |t - s| |f^*|.$$

This shows $x^* \in Lip(\mathbb{R}_+; X^*)$. Moreover, for a step function $f \in L^1(\mathbb{R}_+; X)$ of the form

$$f(t) = \sum_{i=1}^{N} \chi_{(t_{i-1}, t_i]}(t) x_i, \quad t > 0,$$

where $0 = t_0 < t_1 < \cdots < t_N < \infty$, by linearity we obtain

$$
\begin{aligned}
< f, f^* > &= \sum_{i=1}^{N} < \chi_{(t_{i-1}, t_i]} x_i, f^* > = \sum_{i=1}^{N} < x_i, x^*(t_i) - x^*(t_{i-1}) >_X \\
&= \int_0^{\infty} < f(t), dx^*(t) > .
\end{aligned}
$$

Since such step functions are dense in $L^1(\mathbb{R}_+; X)$, by continuity we obtain the representation

$$< f, f^* > = \int_0^{\infty} < f(t), dx^*(t) > \quad \text{for all } f \in L^1(\mathbb{R}_+; X). \tag{6.43}$$

It is clear that the function $x^*(t)$ is uniquely determined by f^*, and by the convention $x^*(0) = 0$.

(b) Suppose $EP \subset L^1(\mathbb{R}_+; X)$ is not dense. By the Hahn-Banach theorem, there is $f^* \in (L^1(\mathbb{R}_+; X))^*$, $f^* \neq 0$, such that $f^* \perp EP$. Let $x^* \in Lip(\mathbb{R}_+; X^*)$

denote the representation of f^* from (a). Then for all functions of the form
$f(t) = e^{-\lambda t}x$, $\lambda > 0$, $x \in X$ we obtain with (6.43)

$$
\begin{aligned}
0 = <f, f^*> \;\; &= \;\; \int_0^\infty <f(t), dx^*(t)> \\
&= \;\; \int_0^\infty <e^{-\lambda t}x, dx^*(t)> \;=\; <x, \int_0^\infty e^{-\lambda t}dx^*(t)>,
\end{aligned}
$$

hence $\widehat{dx^*}(\lambda) = 0$ for all $\lambda > 0$, which implies $x^*(t) \equiv x^*(0) = 0$; (6.43) then yields
$f^* = 0$, a contradiction. Thus EP is dense in $L^1(\mathbb{R}_+; X)$. $\square$

Observe that in Step (a) of this proof we have shown

$$
(L^1(\mathbb{R}_+; X))^* \cong Lip_0(\mathbb{R}_+; X^*)
$$

via the duality given by (6.43), where the subscript 0 refers to $u(0) = 0$. If X^* has
the Radon-Nikodym property, then every Lipschitz-function $x^* \in Lip_0(\mathbb{R}_+; X^*)$
has an a.e. derivative $\dot{x}^* \in L^\infty(\mathbb{R}_+; X^*)$; hence in this case $Lip_0(\mathbb{R}_+; X^*) \cong$
$L^\infty(\mathbb{R}_+; X^*)$, and the usual duality $(L^1)^* \cong L^\infty$ applies again.

Let us briefly specialize to the case of scalar equations with a fixed operator
A, considered in Chapter 1, i.e.

$$
u_n(t) = f_n(t) + (a_n * Au_n)(t), \quad t \in J, \; n \in \mathbb{N}_0. \tag{6.44}
$$

In this situation, Theorem 6.3 can be strengthened and simplified as follows.

Corollary 6.5 *Let* $\{a_n\}_{n\geq 0} \in L^1_{loc}(\mathbb{R}_+)$, A *closed linear and densely defined, such
that* $\int_0^\infty |a_n(t) - a(t)|e^{-\omega t}dt \to 0$ *as* $n \to \infty$. *Assume (6.44) admits a resolvent
$S_n(t)$ in X for each $n \in \mathbb{N}$ and that the stability condition*

$$
|S_n(t)| \leq Me^{\omega t}, \quad t \in \mathbb{R}_+, \; n \in \mathbb{N}, \tag{6.45}
$$

*holds. Then there is a resolvent $S_0(t)$ of type (M, ω) for (6.44) with $n = 0$ and
$S_n(t)x \to S_0(t)x$ as $n \to \infty$, uniformly on compact subsets of $\mathbb{R}_+ \times X$.*

Observe that in contrast to the general case, existence of $S_0(t)$ need not to be
assumed, but can be proved. In fact, the convergence of a_n implies

$$
(I - \widehat{a_n}(\lambda)A)^{-1} \to (I - \widehat{a_0}(\lambda)A)^{-1} \quad \text{in } \mathcal{B}(X)
$$

for all Re $\lambda > \omega$ with $\hat{a}_0(\lambda) \neq 0$. Since points with $\hat{a}_0(\lambda) = 0$ are isolated in Re
$\lambda > \omega$, one then shows as in Section 1.5 that there are none. The stability condition
yields (H1) and (H2) of the generation theorem for the scalar case, Theorem 1.3,
and existence of $S_0(t)$ follows. The asserted convergence finally is obtained as in
the proof of Theorem 6.3.

6.6 Kernels of Positive Type in Hilbert spaces
In the following, the spaces X and Y will always be Hilbert spaces.

Definition 6.4 Let $A \in BV_{loc}(\mathbb{R}_+; \mathcal{B}(Y, X))$ be of subexponential growth, i.e.

$$\int_0^\infty e^{-\varepsilon t} |dA(t)|_{\mathcal{B}(Y,X)} < \infty \quad \text{for each } \varepsilon > 0.$$

The operator-valued kernel $dA(t)$ is called of **positive type**, if

$$\text{Re} \int_0^T \int_0^t (dA(\tau)u(t-\tau), u(t))_X \, dt \geq 0 \tag{6.46}$$

for each $u \in C(\mathbb{R}_+; Y)$ and $T > 0$.

For instance, if A is a positive semidefinite selfadjoint operator in X, $Y = X_A$, and $a \in BV_{loc}(\mathbb{R}_+)$ is such that $da(t)$ is of positive type in the scalar sense (see Section 1.3), then $dA(t)$ is of positive type in the sense of Definition 6.4, as we shall see below. As another example consider $A(t) \equiv A$ where A is linear, densely defined and accretive; then $dA(t) = \delta(t)A$ is of positive type. Examples of nonscalar type will be given in Section 9, where kernels of positive type appear naturally.

Kernels of positive type give rise to quite strong a priori estimates, the *energy inequality*.

Proposition 6.6 Let $A \in BV_{loc}(\mathbb{R}_+; \mathcal{B}(Y, X))$, $f \in W^{1,1}(J; X)$, and suppose $-dA(t)$ is of positive type. Then any mild solution of (6.1) verifies the energy inequality

$$|u(t)|_X \leq |f(0)|_X + \int_0^t |\dot{f}(\tau)|_X \, d\tau, \quad t \in J. \tag{6.47}$$

In particular, mild solutions of (6.1) are unique.

Proof: Since $u \in C(J; X)$ is by assumption a mild solution of (6.1), there are $(f_n) \subset C(J; X)$ and strong solutions $(u_n) \subset C(J; X)$ of (6.1) with f replaced by f_n, such that $f_n \to f$ and $u_n \to u$ in $C(J; X)$ as $n \to \infty$. Choose mollifiers $\rho_\varepsilon \in C_0^\infty(0; \varepsilon)$ with $\rho_\varepsilon \geq 0$ and $\int_{-\infty}^\infty \rho_\varepsilon(\tau) d\tau = 1$, and define $f_{n\varepsilon} = f_n * \rho_\varepsilon$, $u_{n\varepsilon} = u_n * \rho_\varepsilon$, $f_\varepsilon = f * \rho_\varepsilon$, $u_\varepsilon = u * \rho_\varepsilon$. Then $f_{n\varepsilon} \in C^1(J; X)$, $u_{n\varepsilon} \in C^1(J; Y)$ and we have

$$\dot{u}_{n\varepsilon} = dA * u_{n\varepsilon} + \dot{f}_{n\varepsilon}, \quad u_{n\varepsilon}(0) = f_{n\varepsilon}(0) = 0.$$

From this we obtain

$$\frac{d}{dt} |u_{n\varepsilon}(t)|^2 = 2\text{Re}(\dot{u}_{n\varepsilon}(t), u_{n\varepsilon}(t)) = 2\text{Re}(dA * u_{n\varepsilon}(t), u_{n\varepsilon}(t)) + 2\text{Re}(\dot{f}_{n\varepsilon}(t), u_{n\varepsilon}(t)),$$

hence after integration and by (6.46)

$$|u_{n\varepsilon}(t)|^2 \leq 2\text{Re} \int_0^t (dA * u_{n\varepsilon}(s), u_{n\varepsilon}(s)) ds + 2 \int_0^t |\dot{f}_{n\varepsilon}(s)| |u_{n\varepsilon}(s)| ds$$

$$\leq 2 \int_0^t |\dot{f}_{n\varepsilon}(\tau)| |u_{n\varepsilon}(\tau)| d\tau, \quad t \in J.$$

Denote the right hand side of this inequality by $\psi(t)$; then

$$\frac{d}{dt}\sqrt{\psi(t)} = \frac{\dot{\psi}(t)}{2\sqrt{\psi(t)}} = \frac{2|\dot{f}_{n\varepsilon}(t)||u_{n\varepsilon}(t)|}{2\sqrt{\psi(t)}} \leq |\dot{f}_{n\varepsilon}(t)|$$

for each $t \in J$, such that $\psi(t) > 0$. Therefore we conclude

$$|u_{n\varepsilon}(t)| \leq \sqrt{\psi(t)} \leq \int_0^t |\dot{f}_{n\varepsilon}(\tau)| d\tau, \quad t \in J.$$

Letting first $n \to \infty$ we obtain

$$|u_\varepsilon(t)| \leq \int_0^t |\dot{f}_\varepsilon(\tau)| d\tau \leq (\int_0^t \rho_\varepsilon(\tau) d\tau)|f(0)| + \int_0^t (\rho_\varepsilon * |\dot{f}|)(\tau) d\tau$$

and then with $\varepsilon \to 0+$ Inequality (6.47) follows. $\quad\square$

As in the scalar case, operator-valued kernels can be characterized in terms of their Laplace transforms. This characterization is useful in theory but even more in applications.

Proposition 6.7 *Let* $A \in BV_{loc}(\mathbb{R}_+; \mathcal{B}(Y, X))$ *be of subexponential growth. Then* $dA(t)$ *is of positive type if and only if*

$$\text{Re}\,(\widehat{dA}(\lambda)y, y)_X \geq 0 \quad \text{for each} \quad y \in Y, \ \text{Re}\,\lambda > 0. \tag{6.48}$$

Proof: ($\Rightarrow$) Suppose $dA(t)$ is of positive type, and let $y \in Y$, $\lambda \in \mathbb{C}_+$ be given. Choosing $u(t) = e^{-\bar{\lambda}t}y$, $t > 0$, (6.46) yields

$$\text{Re}\int_0^T \int_0^t (dA(\tau)e^{-\bar{\lambda}(t-\tau)}y, e^{-\bar{\lambda}t}y) dt \geq 0, \quad \text{for each } T > 0;$$

since $A(t)$ is of subexponential growth, the limit as $T \to \infty$ of the left hand side exists. From this we obtain

$$\begin{aligned}
0 &\leq \text{Re}\int_0^\infty \int_0^t (dA(\tau)e^{-\bar{\lambda}(t-\tau)}y, e^{-\bar{\lambda}t}y) dt \\
&= \text{Re}\int_0^\infty (\int_\tau^\infty e^{-(\lambda+\bar{\lambda})t} dt) e^{\bar{\lambda}\tau}(dA(\tau)y, y) \\
&= \frac{1}{2\text{Re}\lambda}\text{Re}\int_0^\infty e^{-(\lambda+\bar{\lambda})\tau}e^{\bar{\lambda}\tau}(dA(\tau)y, y) = \frac{1}{2\text{Re}\lambda}\text{Re}(\widehat{dA}(\lambda)y, y),
\end{aligned}$$

and so (6.48) follows.

($\Leftarrow$) Conversely, assume (6.48) holds and let $u \in C(J; Y)$ be given. Extending $u(t)$ by zero to all of $\mathbb{R}$ and using Parseval's identity in the Hilbert space $L^2(\mathbb{R}; X)$ we obtain for any $\mu > 0$

$$2\pi \mathrm{Re} \int_0^T \int_0^t e^{-\mu\tau}(dA(\tau)u(t-\tau), u(t))dt$$

$$= \mathrm{Re} \int_{-\infty}^{\infty} ([\int_0^t dA(\tau)e^{-\mu\tau}u(t-\tau)]\widetilde{\ }(\rho), \tilde{u}(\rho))d\rho$$

$$= \mathrm{Re} \int_{-\infty}^{\infty} (\widehat{dA}(\mu+i\rho)\tilde{u}(\rho), \tilde{u}(\rho))d\rho \geq 0,$$

and with $\mu \to 0+$, (6.46) follows. $\square$

As a consequence of Proposition 6.7, if $-dA(t)$ is of positive type then the operators $\widehat{dA}(\lambda)$, $\mathrm{Re}\,\lambda > 0$, with domain Y are dissipative in X; in particular they are closable, and the inequality

$$\mathrm{Re}\,\lambda|y|_X \leq |(\lambda - \widehat{dA}(\lambda))y|_X, \quad y \in Y, \quad \mathrm{Re}\,\lambda > 0$$

is satisfied. If in addition the range of $\lambda - \widehat{dA}(\lambda)$ is dense in X, then $H(\lambda) = \overline{(\lambda - \widehat{dA}(\lambda))^{-1}} = (\lambda - \overline{\widehat{dA}(\lambda)})^{-1}$ is well-defined and we have the estimate $|H(\lambda)| \leq 1/\mathrm{Re}\,\lambda$. If in addition $\widehat{dA}(\lambda)$ with domain Y is closed in X for each $\lambda > 0$, or equivalently $\lambda - \widehat{dA}(\lambda)$ is surjective for each $\lambda > 0$ then the following result holds.

Theorem 6.4 *Let $A \in BV_{loc}(\mathbb{R}_+; \mathcal{B}(Y, X))$ be of subexponential growth, suppose that $-dA(t)$ is of positive type and that $\lambda - \widehat{dA}(\lambda)$ is surjective for each $\lambda > 0$. Then*
(i) $H(\lambda) = (\lambda - \widehat{dA}(\lambda))^{-1}$ satisfies

$$|(\frac{d}{d\lambda})^n H(\lambda)|_{\mathcal{B}(X)} \leq n!\lambda^{-(n+1)}, \quad \text{for all } \lambda > 0, \ n \in \mathbb{N}_0. \tag{6.49}$$

*(ii) There is a unique **weak resolvent** $S(t)$ for (6.1), i.e. a family $\{S(t)\}_{t\geq 0} \subset \mathcal{B}(X)$ satisfying (S1) and the second resolvent equation (6.3).*
(iii) $|S(t)|_{\mathcal{B}(X)} \leq 1$ for each $t \geq 0$, and $S(\cdot)y \in W^{1,\infty}_{loc}(\mathbb{R}_+; X)$ for each $y \in Y$.

Proof: (i) Define operators A_n^0 in X with domain $D(A_n^0) = Y$ by means of $A_n^0 = \widehat{dA}(n)$, $n \in \mathbb{N}$. Then A_n^0 is m-dissipative, hence generates a C_0-semigroup of contractions in X; note that $Y = X_{A_n^0}$ for each $n \in \mathbb{N}$. Next we define $B_n \in L^1_{loc}(\mathbb{R}_+; \mathcal{B}(Y, X))$ by means of

$$B_n(t) = \sum_{k=1}^{\infty} \frac{n^{2k}}{k!} \widehat{dA}^{(k)}(n)(-1)^k \frac{t^{k-1}}{(k-1)!} e^{-nt}, \quad t > 0, \ n \in \mathbb{N}; \tag{6.50}$$

a simple estimate yields

$$\int_0^\infty e^{-\varepsilon t}|B_n(t)|_{\mathcal{B}(Y,X)}dt \le \sum_{k=1}^\infty \frac{n^{2k}}{k!}|\widehat{dA}^{(k)}(n)|_{\mathcal{B}(Y,X)} \cdot (n+\varepsilon)^{-k}$$

$$\le \sum_{k=0}^\infty \frac{n^{2k}}{k!}(\int_0^\infty t^k e^{-nt}|dA(t)|_{\mathcal{B}(Y,X)})(n+\varepsilon)^{-k}$$

$$= \int_0^\infty (\sum_{k=0}^\infty \frac{1}{k!}(\frac{n^2 t}{n+\varepsilon})^k)e^{-nt}|dA(t)|_{\mathcal{B}(Y,X)} = \int_0^\infty e^{-\frac{n\varepsilon t}{n+\varepsilon}}|dA(t)|_{\mathcal{B}(Y,X)} < \infty,$$

for each $\varepsilon > 0$, $n \in \mathbb{N}$, showing that B_n is of subexponential growth. A straight-forward computation gives

$$\hat{B}_n(\lambda) = \widehat{dA}(\frac{\lambda n}{\lambda+n}) - \widehat{dA}(n), \quad \lambda > 0, \; n \in \mathbb{N}. \tag{6.51}$$

Differentiating (6.50) we see that $B_n \in W_{loc}^{1,1}(\mathbb{R}_+; \mathcal{B}(Y,X))$ and

$$\int_0^\infty e^{-\varepsilon t}|\dot{B}_n(t)|_{\mathcal{B}(Y,X)}dt \le (2n+\varepsilon)\int_0^\infty e^{-\frac{n\varepsilon t}{n+\varepsilon}}|dA(t)|_{\mathcal{B}(Y,X)} < \infty$$

for each $\varepsilon > 0$, $n \in \mathbb{N}$.

By Corollary 6.2, there is a resolvent $S_n(t)$ for the problem

$$\dot{u} = A_n^0 u + B_n * u + f, \quad u(0) = x,$$

since A_n^0 generates a C_0-semigroup in X and $B_n \in W_{loc}^{1,1}(\mathbb{R}_+; \mathcal{B}(Y,X))$. Let $A_n(t)$ be defined by $A_n(t) = A_n^0 + \int_0^t B_n(s)ds$; then $\hat{A}_n(\lambda) = (A_n^0 + \hat{B}_n(\lambda))/\lambda = \widehat{dA}(\frac{\lambda n}{\lambda+n})/\lambda$, by (6.51), and so

$$\mathrm{Re}\,(\widehat{dA}_n(\lambda)y, y) = \mathrm{Re}\,(\widehat{dA}(\frac{\lambda n}{\lambda+n})y, y) \le 0 \quad \text{for each } y \in Y, \; \mathrm{Re}\,\lambda > 0;$$

observe that $\mathrm{Re}\,\frac{\lambda n}{\lambda+n} > 0$ for each $\mathrm{Re}\,\lambda > 0$, $n \in \mathbb{N}$. By Proposition 6.7, $-dA_n(t) = -A_n^0\delta(t) - B_n(t)dt$ is of positive type, and so from Proposition 6.6 we obtain the uniform bound $|S_n(t)|_{\mathcal{B}(X)} \le 1$, $t \ge 0$, $n \in \mathbb{N}$. Taking Laplace transforms this gives

$$|(\frac{d}{d\lambda})^m \hat{S}_n(\lambda)|_{\mathcal{B}(X)} \le m!(\mathrm{Re}\,\lambda)^{-(m+1)} \quad \text{for all} \quad \mathrm{Re}\,\lambda > 0, \; m, n \in \mathbb{N}. \tag{6.52}$$

Theorem 6.2 yields $\hat{S}_n(\lambda) = (I - \hat{A}_n(\lambda))^{-1}/\lambda = (I - \hat{A}(\frac{\lambda n}{\lambda+n}))^{-1}/\lambda$, $\lambda > 0$, $n \in \mathbb{N}$, and therefore $\hat{S}_n(\lambda) \to H(\lambda)$ as $n \to \infty$, for each $\lambda > 0$. From the holomorphy of $\hat{S}_n(\lambda)$ on $\mathrm{Re}\,\lambda > 0$ and (6.52) we now conclude that (6.49) is valid.

(ii) Theorem 0.2 and (6.49) imply the existence of a Lipschitz family $\{U(t)\}_{t\ge0} \subset \mathcal{B}(X)$ such that $\hat{U}(\lambda) = H(\lambda)/\lambda$, $\lambda > 0$ holds. The definition of $U(t)$ yields the second resolvent equation in integrated form (6.6). Since $A \in BV_{loc}(\mathbb{R}_+; \mathcal{B}(Y,X))$ by assumption, $U(t)x$ is C^1 for each $x \in Y$, hence even for each $x \in X$, by the Banach-Steinhaus Theorem. Thus $S(t) = \dot{U}(t)$ satisfies (S1) and (6.6), hence also (6.3) by differentiation, i.e. $S(t)$ is a weak resolvent.

(iii) The first assertion is obvious, and the second follows from the second resolvent equation. □

It is clear that for weak resolvents $S(t)$ the variation of parameters formula (6.7) is still valid, since in its derivation at the beginning of Section 6.2 only (S1) and the second resolvent equation were used. In particular, mild solutions are then unique. On the other hand, (6.10) makes sense for such $S(t)$ if $f \in W^{1,1}(J;X)$, however, it is not clear that the resulting function $u \in C(J;X)$ is a mild solution, even if f is much better, say $f \in W^{1,1}(J;Y)$. The reason for this is of course that $S(t)$ or even $U(t)$ cannot be expected to leave the space Y invariant, unless some further conditions are satisfied. We mention two such conditions in the following corollaries.

Corollary 6.6 *In the situation of Theorem 6.4, assume in addition that $A(t)$ has a main part, i.e. is of the form (6.35) of Corollary 6.4. Then the weak resolvent $S(t)$ of Theorem 6.4 is an a-regular resolvent for (6.1).*

Proof: Combine Theorem 6.4 with Corollary 6.4 and Proposition 6.2; observe that X has the Radon-Nikodym property since it is even a Hilbert space. □

Corollary 6.7 *In the situation of Theorem 6.4, assume in addition that there is a third Hilbert space $Z \overset{d}{\hookrightarrow} Y$, such that the restriction $A_z(t)$ of $A(t)$ to Z fulfills the assumptions of Theorem 6.4 with X and Y replaced by Y and Z, respectively. Then the weak resolvent $S(t)$ of Theorem 6.4 is a resolvent in X.*

Proof: By Theorem 6.4, $H(\lambda)$ verifies (6.30), hence there is a pseudo-resolvent $S(t)$ in X for (6.1) by Theorem 6.2, which is already a resolvent by Proposition 6.2. By uniqueness, $S(t)$ coincides with the weak resolvent from Theorem 6.4 in X and also in Y. □

6.7 Hyperbolic Problems of Variational Type

Let V and H denote Hilbert spaces such that $V \overset{d}{\hookrightarrow} H$, and let $((\cdot, \cdot))$ resp. $(\cdot, \cdot)$ denote the inner products in V resp. H, and $|| \cdot ||$ resp. $| \cdot |$ the corresponding norms. Identifying the antidual $\overline{H}^*$ of H with H, by duality we also get $H \overset{d}{\hookrightarrow} \overline{V}^*$, where $\overline{V}^*$ means the antidual of V. The relation between the antiduality $< \cdot, \cdot >$ between V and $\overline{V}^*$ and the inner product in H is given by

$$(v,h) = \, < v, h >, \quad v \in V, \ h \in H.$$

The norm in $\overline{V}^*$ will be denoted by $|| \cdot ||_*$.

Let $\alpha : \mathbb{R}_+ \times V \times V \to \mathbb{C}$ be such that $\alpha(t, \cdot, \cdot)$ is a bounded sesquilinear form on V for each $t \geq 0$, and

$$|\alpha(t, u, v) - \alpha(s, u, v)| \leq (\alpha_0(t) - \alpha_0(s))\|u\|\,\|v\|, \quad \text{for all } t \geq s \geq 0, \ u, v \in V, \tag{6.53}$$

where α_0 is nondecreasing. Suppose $f \in C(J; \overline{V}^*)$ is given and consider the Volterra equation of variational type

$$(w, u(t)) + \int_0^t \alpha(t - s, w, u(s))ds \,=\, < w, f(t) >, \quad t \in J, \ w \in V. \tag{6.54}$$

Representing the forms $\alpha(t, \cdot, \cdot)$ by bounded linear operators $A(t) \in \mathcal{B}(V, \overline{V}^*)$ as

$$\alpha(t, w, u) = - < w, A(t)u >, \quad t \geq 0, \ w, u \in V, \tag{6.55}$$

it is clear that (6.54) can be written as (6.1) in $X = \overline{V}^*$ with $Y = V$; observe $A \in BV_{loc}(\mathbb{R}_+; \mathcal{B}(V, \overline{V}^*))$ by (6.53).

Exploiting the variational structure of (6.54), by the methods introduced in Section 6.6 it is possible to prove existence of the resolvent $S(t)$ of (6.54) in $\overline{V}^*$ and to derive a number of further properties of $S(t)$, provided the form α is *coercive* in the sense of the following definition.

Definition 6.5 *A form $\alpha : \mathbb{R}_+ \times V \times V \to \mathbb{C}$ as above is called **coercive** if there is a constant $\gamma > 0$ such that*

$$2\,\mathrm{Re} \int_0^T \left(\int_0^t d\alpha(s, u(t), u(t - s)) \right) dt \geq \gamma \| \int_0^T u(t)dt \|^2, \tag{6.56}$$

for all $u \in C(\mathbb{R}_+; V)$ and $T > 0$.

Suppose $u \in C(J; V)$ is a solution of (6.54) and let $f \in W^{1,1}(J; \overline{V}^*)$, $f(0) \in H$. Similarly to Proposition 6.6, the coercive estimate (6.56) then yields the inequality

$$|u(t)|^2 + \gamma \| \int_0^t u(\tau)d\tau \|^2 \leq |f(0)|^2 + 2\,\mathrm{Re} \int_0^t < u(\tau), \dot{f}(\tau) > d\tau, \tag{6.57}$$

for all $t \geq 0$. This is the basic *energy inequality* for (6.54) in the case of coercive forms. If α is only positive, i.e. $\gamma = 0$, (6.57) is still valid; however, we then do not obtain bounds on any quantity related to the solution $u(t)$ in V. It turns out that (6.57) implies estimates for mild solutions as well.

Proposition 6.8 *Suppose $u \in C(\mathbb{R}_+; \overline{V}^*)$ is a mild solution of (6.54). Then*
(i) $f \in W^{1,1}_{loc}(\mathbb{R}_+; H)$ implies $u \in C(\mathbb{R}_+; H)$, $1 * u \in C(\mathbb{R}_+; V)$, and

$$|u(t)|^2 + \gamma \| \int_0^t u(\tau)d\tau \|^2 \leq (|f(0)| + \int_0^t |\dot{f}(\tau)|d\tau)^2, \quad t \geq 0; \tag{6.58}$$

(ii) $f \in W_{loc}^{2,1}(\mathbb{R}_+; \overline{V}^*)$, $f(0) = 0$, imply $u \in C(\mathbb{R}_+; H)$, $1 * u \in C(\mathbb{R}_+; V)$, and for each $\delta \in (0, \gamma)$

$$|u(t)|^2 + (\gamma - \delta)||\int_0^t u(\tau)d\tau||^2 \leq c(\gamma, \delta)^2 (||\dot{f}(0)||_* + \int_0^t |\ddot{f}(t)||_* d\tau)^2, \qquad (6.59)$$

for all $t \geq 0$, where $c(\gamma, \delta) = \gamma^{1/2} + (\gamma - \delta)^{-1/2}$.

Proof: Since $u \in C(\mathbb{R}_+; \overline{V}^*)$ is by assumption a mild solution of (6.54), there are $(f_n) \subset C(\mathbb{R}_+; \overline{V}^*)$ and strong solutions $(u_n) \subset C(\mathbb{R}_+; V)$ of (6.54) with f replaced by f_n, such that $f_n \to f$ and $u_n \to u$ in $C(\mathbb{R}_+; \overline{V}^*)$ as $n \to \infty$. Choose mollifiers $\rho_\varepsilon \in C_0^\infty(0, \varepsilon)$ with $\rho_\varepsilon \geq 0$ and $\int_{-\infty}^\infty \rho_\varepsilon(\tau)d\tau = 1$, and define $f_{n\varepsilon} = f_n * \rho_\varepsilon$, $u_{n\varepsilon} = u_n * \rho_\varepsilon$, $f_\varepsilon = f * \rho_\varepsilon$, $u_\varepsilon = u * \rho_\varepsilon$. Then $f_{n\varepsilon} \in C^1(\mathbb{R}_+; \overline{V}^*)$, $u_{n\varepsilon} \in C^1(\mathbb{R}_+; V)$ and we have from (6.57)

$$|u_{n\varepsilon}(t)|^2 + \gamma||\int_0^t u_{n\varepsilon}(\tau)d\tau||^2 \leq 2Re \int_0^t < u_{n\varepsilon}(\tau), \dot{f}_{n\varepsilon}(\tau) > d\tau, \qquad (6.60)$$

for all $t \geq 0$, since $f_{n\varepsilon}(0) = 0$.

(i) Let $\psi(t) = 2\int_0^t |\dot{f}_{n\varepsilon}(\tau)||u_{n\varepsilon}(\tau)|d\tau$ for $t > 0$; then

$$\frac{d}{dt}\sqrt{\psi(t)} = \frac{\dot{\psi}(t)}{2\sqrt{\psi(t)}} = \frac{2|\dot{f}_{n\varepsilon}(t)||u_{n\varepsilon}(t)|}{2\sqrt{\psi(t)}} \leq |\dot{f}_{n\varepsilon}(t)|,$$

for each $t \in \mathbb{R}_+$, such that $\psi(t) > 0$. Therefore we conclude

$$(|u_{n\varepsilon}(t)|^2 + \gamma|\int_0^t u_{n\varepsilon}(\tau)d\tau|^2)^{1/2} \leq \sqrt{\psi(t)} \leq \int_0^t |\dot{f}_{n\varepsilon}(\tau)|d\tau, \quad t \in \mathbb{R}_+.$$

Letting first $n \to \infty$ we obtain

$$|u_\varepsilon(t)|^2 + \gamma||\int_0^t u_\varepsilon(\tau)d\tau||^2 \leq (\int_0^t |\dot{f}_\varepsilon(\tau)|d\tau)^2 \leq (\int_0^t (\rho_\varepsilon * |df|)(\tau)d\tau)^2,$$

and then with $\varepsilon \to 0+$, inequality (6.58) follows.

(ii) For the proof of (ii), let $w(t) = 1 * u(t)$, $w_n(t) = 1 * u_n(t)$, and $\delta \in (0, \gamma)$; we integrate by parts in (6.60), rearrange, and estimate.

$$|u_{n\varepsilon}(t)|^2 + \gamma||w_{n\varepsilon}(t)||^2 \leq 2Re \int_0^t < u_{n\varepsilon}(\tau), \dot{f}_{n\varepsilon}(\tau) > d\tau$$

$$= 2Re < w_{n\varepsilon}(t), \dot{f}_{n\varepsilon}(t) > -2Re \int_0^t < w_{n\varepsilon}(\tau), \ddot{f}_{n\varepsilon}(\tau) > d\tau$$

$$\leq \delta||w_{n\varepsilon}(t)||^2 + \delta^{-1}||\dot{f}_{n\varepsilon}(t)||_*^2 + 2\int_0^t ||w_{n\varepsilon}(\tau)|| \cdot ||\ddot{f}_{n\varepsilon}(\tau)||_* d\tau.$$

Setting

$$\psi(t) = \delta^{-1}||\dot{f}_{n\varepsilon}(t)||_*^2 + 2\int_0^t ||w_{n\varepsilon}(\tau)|| \cdot ||\ddot{f}_{n\varepsilon}(\tau)||_* d\tau,$$

we obtain

$$\begin{aligned}
\dot{\psi}(t) &\leq 2\delta^{-1}||\dot{f}_{n\varepsilon}(t)||_* \cdot ||\ddot{f}_{n\varepsilon}(t)||_* + 2||w_{n\varepsilon}(t)|| \cdot ||\ddot{f}_{n\varepsilon}(t)||_* \\
&\leq 2||\ddot{f}_{n\varepsilon}(t)||_* \cdot (\delta^{-1/2} + (\gamma - \delta)^{-1/2})\sqrt{\psi(t)},
\end{aligned}$$

hence we conclude as in (i)

$$(|u_{n\varepsilon}(t)|^2 + (\gamma - \delta)||\int_0^t u_{n\varepsilon}(\tau)d\tau||^2)^{1/2} \leq (\delta^{-1/2} + (\gamma - \delta)^{-1/2})\int_0^t ||\ddot{f}_{n\varepsilon}(\tau)||_* d\tau,$$

for all $t \geq 0$. Letting first $n \to \infty$ and then $\varepsilon \to 0$, assertion (ii) follows. $\square$

Since α_0 is assumed to be of subexponential growth, coerciveness can be characterized in terms of Laplace transforms; similar to kernels of positive type. For the sake of completeness, we include an indication of proof, although it is similar to that of Proposition 6.7.

Proposition 6.9 *Let $\alpha(t; \cdot, \cdot)$ be a sesquilinear form on V such that (6.53) is satisfied, where α_0 is nondecreasing and of subexponential growth. Then α satisfies (6.56) iff*

$$Re \, \widehat{d\alpha}(\lambda; u, u) \geq \gamma Re(1/\lambda)||u||^2, \quad \text{for each } u \in V, \text{ and } Re \, \lambda > 0. \qquad (6.61)$$

Proof: Replacing $\alpha(t; \cdot, \cdot)$ by $\alpha(t; \cdot, \cdot) - (\gamma t/2)((\cdot, \cdot))$, it is sufficient to verify the equivalence of (6.56) and (6.61) for $\gamma = 0$.

($\Rightarrow$) Suppose α is of positive type, and let $u \in V$, $\lambda \in \mathbb{C}_+$ be given. Choosing $u(t) = e^{-\lambda t}u$, $t > 0$, (6.56) yields

$$2 \, Re \int_0^T \int_0^t d\alpha(\tau; e^{-\lambda t}u, e^{-\lambda(t-\tau)}u)dt \geq 0, \quad \text{for each } T > 0;$$

since $\alpha_0(t)$ is of subexponential growth, the limit as $T \to \infty$ of the left hand side exists. From this we obtain for $T \to \infty$

$$\begin{aligned}
0 &\leq 2Re \int_0^\infty \int_0^t d\alpha(\tau; e^{-\lambda t}u, e^{-\lambda(t-\tau)}u)dt \\
&= 2Re \int_0^\infty (\int_\tau^\infty e^{-(\lambda+\bar{\lambda})t}dt)e^{\bar{\lambda}\tau}d\alpha(\tau; u, u) \\
&= \frac{1}{Re \, \lambda}Re \int_0^\infty e^{-(\lambda+\bar{\lambda})\tau}e^{\bar{\lambda}\tau}d\alpha(\tau; u, u) = \frac{1}{Re \, \lambda}Re \, \widehat{d\alpha}(\lambda; u, u),
\end{aligned}$$

and so (6.61) with $\gamma = 0$ follows.

($\Leftarrow$) Conversely, assume (6.61) holds and let $u \in C(\mathbb{R}_+; V)$ be given. Redefining $u(t) = 0$ for $t \notin [0, T]$ and using the Fourier inversion formula and Fubini's theorem, we obtain for any $\mu > 0$

$$2\pi \mathrm{Re} \int_0^T \int_0^t e^{-\mu\tau} d\alpha(\tau; u(t), u(t-\tau)) dt$$

$$= \mathrm{Re} \int_{-\infty}^\infty < \tilde{u}(-t), \int_0^t e^{-\mu t} dA(\tau) u(t-\tau) > dt$$

$$= \mathrm{Re} \int_{-\infty}^\infty < \tilde{u}(\rho), \widehat{dA}(\mu + i\rho) \tilde{u}(\rho) > d\rho$$

$$= \mathrm{Re} \int_{-\infty}^\infty \widehat{d\alpha}(\mu - i\rho; \tilde{u}(\rho), \tilde{u}(\rho)) d\rho \geq 0,$$

and with $\mu \to 0+$, (6.56) with $\gamma = 0$ follows. $\square$

Examples for coercive forms in linear thermoviscoelaticity will be given in Section 9.4.

The result announced above reads as follows.

Theorem 6.5 *Suppose $\alpha : \mathbb{R}_+ \times V \times V \to \mathbb{C}$ satisfies*
(V1) *$\alpha(t, \cdot, \cdot)$ is a bounded sesquilinear form on V, for $t \geq 0$, and $\alpha(0, \cdot, \cdot) = 0$;*
(V2) *$\alpha(\cdot, u, v) \in W_{loc}^{1,\infty}(\mathbb{R}_+)$ for each $u, v \in V$, and*

$$|\dot{\alpha}(t, u, v) - \dot{\alpha}(s, u, v)| \leq (\alpha_1(t) - \alpha_1(s))\|u\| \|v\|, \quad u, v \in V, \ t \geq s \geq 0,$$

where $\alpha_1(t)$ is nondecreasing and of subexponential growth, $\alpha_1(0) = 0$;
(V3) *α is coercive with coercivity constant $\gamma > 0$.*
Define $A \in W_{loc}^{1,\infty}(\mathbb{R}_+; \mathcal{B}(V, \overline{V}^*))$ by means of (6.55). Then for $X = \overline{V}^*$ and $Y = V$, (6.1) admits a t-regular resolvent $S(t)$. Moreover, with $R = 1 * S$ and $T = t * S$ we have in addition
(a) $\{S(t)\}_{t \geq 0} \subset \mathcal{B}(V) \cap \mathcal{B}(\overline{V}^*) \cap \mathcal{B}(H)$ is strongly continuous in $\overline{V}^*, H$, and V, and

$$|S(t)|_{\mathcal{B}(V)}, \ |S(t)|_{\mathcal{B}(\overline{V}^*)} \leq 1 + 2\gamma^{-1}\alpha_1(t), \ |S(t)|_{\mathcal{B}(H)} \leq 1, \ \text{for all } t \geq 0;$$

(b) $\{R(t)\}_{t \geq 0} \subset \mathcal{B}(\overline{V}^*, H) \cap \mathcal{B}(H, V)$, $\{T(t)\}_{t \geq 0} \subset \mathcal{B}(\overline{V}^*, V)$ are strongly continuous, and

$$|R(t)|_{\mathcal{B}(\overline{V}^*, H)}, \ |R(t)|_{\mathcal{B}(H, V)} \leq \gamma^{-1/2}, \ |T(t)|_{\mathcal{B}(\overline{V}^*, V)} \leq 2\gamma^{-1}, \ \text{for all } t \geq 0;$$

(c) $\{S(t)\}_{t \geq 0} \subset \mathcal{B}(V, H) \cap \mathcal{B}(H, \overline{V}^*)$ is strongly continuously differentiable, even twice a.e. in $\mathcal{B}(V, \overline{V}^*)$, and for a.a. $t \geq 0$ we have

$$|\dot{S}(t)|_{\mathcal{B}(V, H)}, \ |\dot{S}(t)|_{\mathcal{B}(H, \overline{V}^*)} \leq \gamma^{-1/2}\alpha_1(t),$$

and

$$|\ddot{S}(t)|_{\mathcal{B}(V; \overline{V}^*)} \leq \alpha_1(t)(1 + 2\gamma^{-1}\alpha_1(t)).$$

Before we turn to the proof of Theorem 6.5 several remarks are in order. Assumption (V2) means $\dot{A} \in BV_{loc}(\mathbb{R}_+; \mathcal{B}(V, \overline{V}^*))$, even in $BV(\mathbb{R}_+; \mathcal{B}(V, \overline{V}^*))$ if α_1 is bounded. Therefore (6.54) is equivalent to the equation of second order

$$\ddot{v}(t) = \int_0^t d\dot{A}(\tau)v(t - \tau) + g(t), \quad t \geq 0, \tag{6.62}$$

$$v(0) = v_0, \quad \dot{v}(0) = v_1.$$

For the solution of (6.62) we have the following variation of parameters formula.

$$v(t) = S(t)v_0 + R(t)v_1 + \int_0^t R(t - \tau)g(\tau)d\tau, \quad t \geq 0. \tag{6.63}$$

Thus the resolvent $S(t)$ corresponds to the cosine family, $R(t)$ to the sine family of second order differential equations. The following corollary describes the solvability behaviour of (6.62) implied by Theorem 6.5.

Corollary 6.8 *Let the assumptions of Theorem 6.5 be satisfied, $v_0, v_1 \in \overline{V}^*$, $g \in L^1_{loc}(\mathbb{R}_+; \overline{V}^*)$, and let $v(t)$ be given by (6.63). Then*
(i) $v_0 \in V$, $v_1 \in H$, $g \in C(\mathbb{R}_+; H)$ imply $v \in C(\mathbb{R}_+; V)$, $\dot{v} \in C(\mathbb{R}_+; H)$, $\ddot{v} - \dot{A}(\cdot)v_0 \in C(\mathbb{R}_+; \overline{V}^*)$, and $v(t)$ is a strong solution of (6.62);
(ii) $v_0 \in V$, $v_1 \in H$, $g \in W^{1,1}(\mathbb{R}_+; \overline{V}^*)$ imply $v \in C(\mathbb{R}_+; V)$, $\dot{v} \in C(\mathbb{R}_+; H)$, $\ddot{v} - \dot{A}(\cdot)v_0 \in C(\mathbb{R}_+; \overline{V}^*)$, and $v(t)$ is a strong solution of (6.62);
(iii) $v_0 \in H$, $v_1 \in \overline{V}^*$, $g \in C(\mathbb{R}_+; \overline{V}^*)$ imply $v \in C(\mathbb{R}_+; H)$, $\dot{v} \in C(\mathbb{R}_+; \overline{V}^*)$, and $v(t)$ is a mild solution of (6.62).

Proof: The construction is similar to the proof of Theorem 6.4, therefore the exposition will be kept brief, here. Introduce sesquilinear forms α_n by means of

$$\dot{\alpha}_n(t, u, v) = \widehat{\alpha}(n, u, v) + \sum_{k=1}^{\infty} \frac{n^{2k}}{k!} \widehat{\alpha}^{(k)}(n, u, v)(-1)^k \frac{t^{k-1}}{(k-1)!} e^{-nt},$$

for $t \geq 0$, $u, v \in V$, $n \in \mathbb{N}$; these forms α_n are obviously bounded and sesquilinear; they are exactly those forms corresponding to the operator $A_n(t)$ in the proof of Theorem 6.4, and so the approximating equations with $A(t)$ replaced by $A_n(t)$ admit resolvents $S_n(t)$ in $\overline{V}^*$. Since

$$\widehat{d\alpha_n}(\lambda, u, v) = \widehat{d\alpha}(\frac{\lambda n}{\lambda + n}, u, v), \quad \text{Re } \lambda > 0, \; n \in \mathbb{N}, \; u, v \in V,$$

α_n is coercive with the same constant γ of coercivity. A priori estimates (6.58) and (6.59) yield for $f(t) \equiv h \in H$ resp. $f(t) = tv^* \in \overline{V}^*$

$$|S_n(t)|_{\mathcal{B}(H)} \leq 1, \; |R_n(t)|_{\mathcal{B}(H,V)}, \; |R_n(t)|_{\mathcal{B}(\overline{V}^*,H)} \leq \gamma^{-1/2}, \; |T_n(t)|_{\mathcal{B}(\overline{V}^*,V)} \leq 2\gamma^{-1},$$

for all $t > 0$, where $R_n = 1 * S_n$, $T_n = t * S_n$. These inequalities imply

$$|(\frac{d}{d\lambda})^k \hat{S}_n(\lambda)|_{\mathcal{B}(H)} \leq k!\lambda^{-(k+1)},$$

$$|(\frac{d}{d\lambda})^k [\hat{S}_n(\lambda)/\lambda]|_{\mathcal{B}(\overline{V}^*,H)} \leq \gamma^{-1/2} k!\lambda^{-(k+1)},$$

$$|(\frac{d}{d\lambda})^k [\hat{S}_n(\lambda)/\lambda]|_{\mathcal{B}(H,\overline{V}^*)} \leq \gamma^{-1/2} k!\lambda^{-(k+1)},$$

$$|(\frac{d}{d\lambda})^k [\hat{S}_n(\lambda)/\lambda^2]|_{\mathcal{B}(\overline{V}^*,V)} \leq 2\gamma^{-1} k!\lambda^{-(k+1)},$$

for all $n \geq 1$, $k \geq 0$, $\lambda > 0$. Hence passing to the limit $n \to \infty$ the same estimates are valid for $H(\lambda) = (\lambda - \widehat{\dot{A}}(\lambda))^{-1}$; note that $\{H(\lambda)\}_{\text{Re } \lambda > 0} \subset \mathcal{B}(\overline{V}^*, V)$ exists and is holomorphic, as standard coerciveness arguments show, and that $\hat{S}_n^{(k)}(\lambda) \to H^{(k)}(\lambda)$ in $\mathcal{B}(\overline{V}^*, V)$ uniformly on compact subsets of $\mathbb{C}_+$, and for all k. The identities

$$H(\lambda) = I/\lambda + \widehat{d\dot{A}}(\lambda)H(\lambda)/\lambda^2 = I/\lambda + [H(\lambda)/\lambda^2]\widehat{d\dot{A}}(\lambda), \quad \lambda > 0,$$

by (V2) imply for each $\varepsilon > 0$

$$|(\frac{d}{d\lambda})^k H(\lambda)|_{\mathcal{B}(\overline{V}^*)}, |(\frac{d}{d\lambda})^k H(\lambda)|_{\mathcal{B}(V)} \leq (1 + 2\gamma^{-1}\widehat{d\alpha_1}(\varepsilon))k!(\lambda - \varepsilon)^{-k-1}, \quad \lambda > \varepsilon;$$

this follows from the estimate for $H(\lambda)/\lambda^2$ and from

$$\sum_{k=0}^{\infty} (\lambda - \varepsilon)^k |\widehat{d\dot{A}}^{(k)}(\lambda)|/k! \leq \int_0^{\infty} e^{-\varepsilon t} |d\dot{A}(t)| \leq \widehat{d\alpha_1}(\varepsilon), \quad \varepsilon > 0.$$

Therefore, the assumptions of the generation theorem, Theorem 6.2, are satisfied in $X = \overline{V}^*$ with $Y = V$, hence with $a(t) = t$, there is a t-regular resolvent $S(t)$ for (6.1). The remaining assertions follow easily from the a priori estimates (6.58) and (6.59) and from the resolvent equations in $\overline{V}^*$. $\square$

To compare Theorem 6.5 with Corollary 1.2, let $\alpha(t, u, v) = a(t)(u, Av)$ for some positive definite selfadjoint operator A in H, $V = D(A^{1/2})$ equipped with the inner product $((u, v)) = (A^{1/2}u, A^{1/2}v)$, and $a(t) = a_\infty t + \int_0^t a_1(\tau)d\tau$, where $a_\infty \geq 0$, and a_1 of positive type. Then α is coercive if $\gamma = a_\infty > 0$, and Theorem 6.5 applies if in additon $a_1 \in BV_{loc}(\mathbb{R}_+)$ and $\alpha_1(t) = \text{Var } a_1|_0^t$ is of subexponential growth.

Another coercivity concept different from Definition 6.5 is based on an inequality of the form

$$\text{Re } \widehat{d\alpha}(\lambda, v, v) \geq \gamma \text{Re } \frac{1}{\lambda + \eta}||v||^2, \quad v \in V, \text{ Re } \lambda > 0, \tag{6.64}$$

where γ and η are positive constants; compare with (6.61). Forms satisfying (6.64) will be called η-coercive in the sequel. By the same methods as in the proof of Theorem 6.5 the following result for η-coercive forms is obtained.

Corollary 6.9 *Suppose* $\alpha : \mathbb{R}_+ \times V \times V \to \mathbb{C}$ *is η-coercive for some $\eta > 0$, and satisfies (V1), (V2) of Theorem 6.5. Let $A(t)$ be defined by (6.55) again. Then for $X = \overline{V}^*$, $Y = V$, (6.1) admits a t-regular resolvent $S(t)$. Moreover, with $R_\eta(t) = (e^{-\eta t} * S)(t)$, $T_\eta(t) = (e^{-\eta t} * R_\eta)(t)$ we have the estimates*

$$|S(t)|_{\mathcal{B}(H)} \leq 1; \quad |R_\eta(t)|_{\mathcal{B}(\overline{V}^*,H)}, \; |R_\eta(t)|_{\mathcal{B}(H,V)} \leq \gamma^{-1/2}; \quad |T_\eta(t)|_{\mathcal{B}(\overline{V}^*,V)} \leq 2\gamma^{-1},$$

and

$$||R_\eta(\cdot)x||_2 \leq (2\eta\gamma)^{-1/2}|x|; \quad ||T_\eta(\cdot)x||_2 \leq 2(2\eta\gamma^2)^{-1/2}||x||_*.$$

In the situation of Corollary 6.9 one can also obtain bounds for the remaining quantities, i.e. $S(t)$ in $\mathcal{B}(V)$ and $\mathcal{B}(\overline{V}^*)$, $\dot{S}(t)$ in $\mathcal{B}(V,H)$ and $\mathcal{B}(H,\overline{V}^*)$, as well as $\ddot{S}(t)$ in $\mathcal{B}(V,\overline{V}^*)$, to the result that similar estimates as in Theorem 6.5 are valid; in particular, all of these operator families are bounded on $\mathbb{R}_+$ if α_1 is bounded.

Concerning the proof of Corollary 6.9 we only mention the basic *energy inequality* corresponding to (6.58) which is implied by η-coercivity.

$$|v(t)|^2 + \gamma||w(t)||^2 + 2\eta\gamma \int_0^t ||w(s)||^2 ds \leq |f(0)|^2 + 2\,\text{Re} \int_0^t < v(s), \dot{f}(s) > ds,$$
(6.65)

for all $t \geq 0$, where $w(t) = e^{-\eta t} * v(t)$.

6.8 Comments

a) The setting for (6.1), i.e. $Y \overset{d}{\hookrightarrow} X, A \in L^1_{loc}(\mathbb{R}_+; \mathcal{B}(Y; X))$, is fairly general and seems to cover all abstract treatments of Volterra equations which have been considered so far. Of course, this is not the most general approach one can think of, e.g. the domains $D(A(t))$ of the operators $A(t)$ could be imagined to vary with t in a much stronger way. However, then it becomes very difficult to develop a reasonable theory. From the applications point of view, our setting covers most of the equations arising in linear viscoelasticity, at least for the variational approach.

b) By far the most papers in the literature deal with first order equations of the form

$$\dot{u} = Au + B * u + f, \; u(0) = x,$$
(6.66)

where A generates a C_0-semigroup in X or even an analytic semigroup, and $B(t)$ is 'dominated' by A in the sense that at least $B(\cdot)x$ is measurable in X for each $x \in D(A)$, and $|B(t)x| \leq \varphi(t)(|x| + |Ax|)$ a.e. for some $\varphi \in L^1_{loc}(\mathbb{R}_+)$; see e.g. Chen and Grimmer [40], [41], Grimmer [139], Grimmer and Kappel [140], Grimmer and Pritchard [144], Grimmer and Schappacher [146], Miller [243], Miller and Wheeler [245], Da Prato and Iannelli [65], Tsuruta [329], and many others. Further references are given below. Grimmer and Prüss [145] contains the first general treatment of (6.66), i.e. the connection between well-posedness and resolvents, the generation theorem, the 1-regularity of the resolvent and several counterexamples, some of which have been mentioned before. Observe that (6.66) is of the form (6.12) with $a(t) \equiv 1$, hence Propositions 6.2, 6.4 and 6.5(c) as well as Corollaries 6.2 and 6.4 are applicable.

c) Although Corollary 6.2 with $B_0 = 0$ has many predecessors, its present form is originally due to Desch and Schappacher [93], who used the semigroup approach via history spaces for the proof. Grimmer and Prüss [145] gave an independent argument based on operator-valued bounded Volterra equations. This approach was later improved and extended in Prüss [276], and by Pugliese [286]. Theorem 6.1 and its corollaries unify most results of this nature, although in special situations better results are known.

d) The scalar version of the generation theorem, i.e. Theorem 1.3 for the case $a \in BV_{loc}(\mathbb{R}_+)$, is due to Da Prato and Iannelli [65]. Grimmer and Prüss [145] generalized their result to first order equations of the form (6.66), while Pugliese [286] contains a version of Theorem 6.2 for the case $A \in BV_{loc}(\mathbb{R}_+; \mathcal{B}(Y, X))$; see also Sforza [300]. The simple but nevertheless interesting Example 6.2 is a slightly modified version of an example due to Desch and Schappacher [93].

e) A second class of problems of the form (6.1) which has been considered by many authors is the class of second order equations

$$\ddot{u} = Au + B * u + f, \ u(0) = x, \ \dot{u}(0) = y, \tag{6.67}$$

where A is the generator of a strongly continuous cosine family in X, and the operator family $B(t)$ belongs to $BV_{loc}(\mathbb{R}_+; \mathcal{B}(X_A, X))$; see e.g. Adali [2], Dafermos [76], [75], Desch and Grimmer [82], and Tsuruta [330]. The results in the literature as well as Corollary 6.3 for (6.54) are not optimal for the complete second order differential equation

$$\ddot{u} = Au + B\dot{u} + f, \ u(0) = x, \ \dot{u}(0) = y;$$

see Clément and Prüss [53] Fattorini [114], Obrecht [263], [264] and Watanabe [335] for stronger results.

f) The convergence theorem, Theorem 6.3, seems to be new. The proof follows an idea of F. Neubrander, who proved a general convergence theorem for vector-valued Laplace transforms. It would be interesting to have a version of this result which does not assume the existence of the limiting pseudo-resolvent, and only involves the convergence of the Laplace-transforms of the approximating resolvents, not of the kernels itself, similar to the classical Trotter-Kato theorem; see e.g. Goldstein [133] for the latter. Other versions of Corollary 6.5, in which the kernel $a(t)$ is fixed but A depends on $n \in \mathbb{N}_0$, can be found in Lizama [216].

g) It is of course very well known that the duality (6.43) yields $L^1(\mathbb{R}_+; X)^* \cong Lip_0(\mathbb{R}_+; X)$; see Diestel and Uhl [99] and Hashimoto and Oharu [173].

h) The results in Section 6.6 on operator-valued kernels of positive type are different from but similar to those obtained by Da Prato and Iannelli, [65]; see also Barbu, Da Prato and Iannelli [18] for results on asymptotic behaviour in the setting of the paper just cited. An interesting special case of such kernels is $A(t) = a(t)A + b(t)B$, where a, b are scalar kernels such that da, db are of positive type and A, B are selfadjoint and positive semidefinite. Even in this situation the regularity of the weak resolvent $S(t)$ is unknown except if A, B commute.

i) A theory for *nonautonomous* linear problems has been developed for first order equations of the form

$$\dot{u}(t) = A(t)u(t) + \int_s^t B(t,\tau)u(\tau)d\tau, \quad u(s) = x, \ t \geq s, \tag{6.68}$$

where the operator family $A(t)$ generates an evolution operator in X, and $B(t, s)$ is subordinate to $A(t)$ in a suitable sense; see Chen and Grimmer [40], [41], Grimmer [139], and Prüss [270] for the hyperbolic case, and Acquistapache [1], Di Blasio [95], Friedman and Shinbrot [123], Lunardi [225], Prüss [270], and Tanabe [320], [321] in parabolic situations. In all of these papers the idea is to consider the integral term as a perturbation of the evolution equation $\dot{u}(t) = A(t)u(t) + f(t)$ and to use the variation of parameters formula. This is similar to the approach taken in Sections 6.3 for the hyperbolic case, and to that of Sections 7.4, 7.5, 7.6 as well as of Sections 8.5 and 8.6 below in the parabolic case.

7 Nonscalar Parabolic Equations

The results on existence of analytic resolvents in Section 2 and on resolvents for parabolic problems in Section 3 as well as the corresponding results on maximal regularity of type C^α are here extended to nonscalar equations. Particularly easy to verify are the conditions in the variational approach in Section 7.3 which continues the discussion begun in Section 6.7. The remaining subsections are devoted to a far reaching improvement of the perturbation theorem from Section 6.3 in the parabolic case.

7.1 Analytic Resolvents

Consider again (6.1) in a Banach space X, i.e.

$$u(t) = f(t) + \int_0^t A(t - \tau)u(\tau)d\tau, \quad t \in J, \tag{7.1}$$

where $Y \overset{d}{\hookrightarrow} X$, $A \in L^1_{loc}(\mathbb{R}_+; \mathcal{B}(Y, X))$, $f \in C(J; X)$, and $J = [0, T]$. We begin the study of (7.1) in the parabolic case with an extension of Theorem 2.1 on analytic resolvents for scalar equations to (7.1). In analogy to Definition 2.1 we introduce

Definition 7.1 *(i) A weak resolvent $S(t)$ for (7.1) is called* **analytic** *if the function $S(\cdot) : \mathbb{R}_+ \to \mathcal{B}(X)$ admits analytic extension to a sector $\Sigma(0, \theta_0)$, $\theta_0 \leq \pi/2$. Then $S(t)$ will be said to be of type (ω_0, θ_0), if for each $\omega > \omega_0$, $\theta < \theta_0$, there is a constant $M \geq 1$ such that*

$$|S(z)|_{\mathcal{B}(X)} \leq Me^{\omega Re z}, \quad z \in \Sigma(0, \theta). \tag{7.2}$$

(ii) A resolvent $S(t)$ for (7.1) is called **analytic** *if the function $S(\cdot) : \mathbb{R}_+ \to \mathcal{B}(X) \cap \mathcal{B}(Y)$ admits analytic extension to a sector $\Sigma(0, \theta_0)$, $\theta_0 \leq \pi/2$. Then $S(t)$ will be said to be* **of type** (ω_0, θ_0), *if for each $\omega > \omega_0$, $\theta < \theta_0$ there is a constant $M \geq 0$ such that*

$$|S(z)|_{\mathcal{B}(X)} + |S(z)|_{\mathcal{B}(Y)} \leq Me^{\omega Re z}, \quad z \in \Sigma(0, \theta). \tag{7.3}$$

Obviously, Corollary 2.1 remains valid for weak analytic resolvents of type (ω_0, θ_0) since its proof only involves Cauchy's integral formula and Estimate (7.2). If $S(t)$ is an analytic resolvent we have even

$$|S^{(n)}(t)|_{\mathcal{B}(X)} + |S^{(n)}(t)|_{\mathcal{B}(Y)} \leq Mn!e^{\omega t(1+\alpha)}(\alpha t)^{-n}, \quad t > 0, \ n \in \mathbb{N} \tag{7.4}$$

where $M = M(\omega, \theta)$, $\alpha = \sin\theta$, and $\omega > \omega_0$, $\theta < \theta_0$.

Theorem 2.1 can be extended directly to the case of nonscalar equations of the form (7.1).

Theorem 7.1 *Suppose $A \in L^1_{loc}(\mathbb{R}_+; \mathcal{B}(Y, X))$ and $\int_0^\infty |A(t)|_{\mathcal{B}(Y,X)}e^{-\omega_A t}dt < \infty$, for some $\omega_A \geq 0$. Then Equation (7.1) admits an analytic resolvent $S(t)$ of type (ω_0, θ_0) if and only if the following conditions are satisfied.*

(AN1) *There is $\lambda_0 > \omega_A$ such that $I - \hat{A}(\lambda)$ is injective, the part of $I - \hat{A}(\lambda)$ in Y is surjective, and $(I - \hat{A}(\lambda))^{-1}$ is bounded in X and in Y, for each $\lambda > \lambda_0$; we let $H(\lambda) = (I - \hat{A}(\lambda))^{-1}/\lambda,\ \lambda > \lambda_0$.*

(AN2) *The family $\{H(\lambda)\}_{\lambda > \lambda_0} \subset \mathcal{B}(X) \cap \mathcal{B}(Y)$ admits analytic extension to the sector $\Sigma(\omega_0, \pi/2 + \theta_0)$.*

(AN3) *For each $\omega > \omega_0,\ \theta > \theta_0$ there is a constant $C = C(\omega, \theta)$ such that*

$$|H(\lambda)|_{\mathcal{B}(X)} + |H(\lambda)|_{\mathcal{B}(Y)} \le C/|\lambda - \omega|, \quad \lambda \in \Sigma(\omega, \pi/2 + \theta). \tag{7.5}$$

Proof: Necessity. Let $S(t)$ be an analytic resolvent of type (ω_0, θ) of (7.1). Theorem 6.2 implies then (AN1) as well as $H(\lambda) = \hat{S}(\lambda)$ for $\lambda > \lambda_0$. As in Section 2.1, Cauchy's formula and the holomorphy of $S(t)$ in X and in Y yield

$$\hat{S}(\lambda) = \int_0^\infty S(te^{i\theta}) e^{-\lambda t e^{i\theta}} e^{i\theta} dt, \quad \lambda > \omega_0,\ |\theta| < \theta_0.$$

This gives the analytic continuation of $H(\lambda)$ to $\Sigma(\omega_0, \pi/2 + \theta_0)$ as well as the estimate (7.5), i.e (AN2) and (AN3) hold.

Sufficiency. Conversely, suppose (AN1) $\sim$ (AN3) hold. Then (N1) of Theorem 6.2 holds for $\lambda > \lambda_0$, and Cauchy's formula yields the representation

$$H^{(n)}(\lambda) = \frac{(-1)^n n!}{2\pi i} \int_\Gamma H(z)(z - \lambda)^{-(n+1)} dz, \quad \lambda > \omega,\ n \in \mathbb{N}, \tag{7.6}$$

where $\omega > \omega_0,\ \theta < \theta_0$, and Γ denotes the contour

$$\{\omega + (-\infty, R]e^{-i(\pi/2+\theta)}\} \cup \{\omega + Re^{i[-\pi/2-\theta, \pi/2+\theta]}\} \cup \{\omega + [R, \infty)e^{i(\pi/2+\theta)}\}.$$

Choosing e.g. $R = (\lambda - \omega)/(n+1)$ a direct estimate then yields (N2), hence there is a pseudo-resolvent $S(t)$ for (7.1), by Theorem 6.2. However, by uniqueness of the Laplace transform, as in the proof of Theorem 2.1, $S(t)$ is given by the contour integral

$$S(z) = \frac{1}{2\pi i} \int_\Gamma e^{\lambda z} H(\lambda) d\lambda, \quad |\arg z| < \theta_0, \tag{7.7}$$

and this representation formula yields analyticity of $S(z)$ in $\Sigma(0, \theta_0)$, as well as the type estimates, in X and also in Y. $\square$

Similarly, one can also characterize analyticity of weak resolvents of (7.1). We state this as

Corollary 7.1 *Suppose $A \in L^1_{loc}(\mathbb{R}_+; \mathcal{B}(Y, X))$ and $\int_0^\infty |A(t)|_{\mathcal{B}(Y,X)}\, e^{-\omega_A t} dt < \infty$. Then (7.1) admits a weak analytic resolvent of type (ω_0, θ_0) if and only if the following conditions are satisfied.*

(WA1) *There is $\lambda_0 > \omega_A$ such that $I - \hat{A}(\lambda)$ is injective and admits a left-inverse $\lambda H(\lambda) \in \mathcal{B}(X)$ for each $\lambda > \lambda_0$.*

(WA2) *The operator family* $\{H(\lambda)\}_{\lambda > \lambda_0} \subset \mathcal{B}(X)$ *admits analytic extension to* $\Sigma(\omega_0, \pi/2 + \theta_0)$.
(WA3) *For each* $\omega > \omega_0$, $\theta > \theta_0$ *there is a constant* $C = C(\omega, \theta)$ *such that*

$$|H(\lambda)|_{\mathcal{B}(X)} \leq C/|\lambda - \omega|, \quad \lambda \in \Sigma(\omega, \pi/2 + \theta). \tag{7.8}$$

The proof is similar to that of Theorem 7.1 based on (7.6) and (7.7) in X and is therefore omitted.

In practice, the analytic extensions of $H(\lambda)$ required in (AN2) and (WA2) are obtained by extending $\hat{A}(\lambda)$ analytically to the required sector, and then to study invertibility of $I - \hat{A}(\lambda)$ in X and in Y. As seen in Corollary 2.3, this usually will only work if the family $\{A(t)\}_{t>0} \subset \mathcal{B}(Y, X)$ itself has some very strong smoothness properties, like analyticity in $\Sigma(0, \theta_0)$.

Next we consider a-regularity of weak analytic resolvents.

Corollary 7.2 *Under the assumptions of Corollary 7.1, let in addition* $a \in L^1_{loc}(\mathbb{R}_+)$ *be of exponential growth, such that* $\hat{a}(\lambda)$ *admits analytic extension to* $\Sigma(\omega_0, \pi/2 + \theta_0)$, *and that in addition the following condition holds.*
(AN4) *For each* $\omega > \omega_0$, $\theta > \theta_0$ *there is a constant* $C = C(\omega, \theta)$ *such that*

$$|\hat{a}(\lambda)H(\lambda)|_{\mathcal{B}(X,Y)} \leq C/|\lambda - \omega|, \quad \lambda \in \Sigma(\omega, \pi/2 + \theta). \tag{7.9}$$

Then the weak analytic resolvent $S(t)$ *for (7.1) is a-regular.* $T(t) = (a * S)(t)$ *admits analytic extension to* $\Sigma(0, \theta_0)$, *and for each* $\omega > \omega_0$, $\theta > \theta_0$ *there is a constant* $M > 0$ *such that*

$$|T(z)|_{\mathcal{B}(X,Y)} \leq Me^{\omega Rez}, \quad z \in \Sigma(0, \theta). \tag{7.10}$$

The proof employs the same arguments as that of Theorem 7.1 and is therefore omitted.

If (7.1) admits an analytic resolvent, the Cauchy estimates (7.4) show that the arguments in the proof of Theorem 2.4 still can be applied. This leads to the following result on maximal regularity of type C^α for (7.1).

Corollary 7.3 *Let* $S(t)$ *be a weak analytic resolvent for (7.1),* $f : J \to X$ *continuous,* $J = [0, T]$, *and let* $u(t)$ *formally be defined by the variation of parameters formula*

$$u(t) = f(t) + \int_0^t \dot{S}(t - s)f(s)ds, \quad t \in J. \tag{7.11}$$

Then (i) If $f \in C_0^\alpha(J; X)$ *then* $u \in C_0^\alpha(J; X)$; *if in addition* $S(t)$ *is a pseudo-resolvent then* u *is a mild solution of (7.1).*
(ii) If $f \in C_0^\alpha(J; Y)$ *and if* $S(t)$ *is an analytic resolvent for (7.1) then* $u \in C_0^\alpha(J; Y)$ *and* u *is a strong solution of (7.1).*

(iii) If (AN4) holds and if $f = a * g$ with $g \in C_0^\alpha(J; X)$ then $u \in C_0^\alpha(J; X)$ is a strong solution of (7.1).

For the proof one only has to follow the proof of Theorem 2.4 to obtain $u \in C^\alpha$ in X resp. Y, and then apply the usual approximation arguments.

7.2 Parabolic Equations

For parabolic equations of the form (7.1) the methods leading to Theorem 3.1 can also be employed, we briefly discuss this now. Definition 3.1 is extended to nonscalar equations by the following

Definition 7.2 *Let $A \in L_{loc}^1(\mathbb{R}_+; \mathcal{B}(Y, X))$ be of subexponential growth. (7.1) is called* **weakly parabolic** *if the following two conditions are satisfied*
(NP1) *For each Re $\lambda > 0$, the operator $\hat{A}(\lambda)$ with domain Y is closed in X, and $I - \hat{A}(\lambda)$ is invertible in X.*
(NP2) *There is a constant $M \geq 1$ such that*

$$|(I - \hat{A}(\lambda))^{-1}|_{\mathcal{B}(X)} \leq M, \quad \text{for all Re } \lambda > 0.$$

(7.1) is called **parabolic** *if it is weakly parabolic and in addition*
(NP3) *There is a constant $M \geq 1$ such that*

$$|(I - \hat{A}(\lambda))^{-1}|_{\mathcal{B}(Y)} \leq M, \quad \text{for all Re } \lambda > 0,$$

is valid.

The second notion needed for the announced extension of Theorem 3.1 is introduced in

Definition 7.3 *Let $A \in L_{loc}^1(\mathbb{R}_+; \mathcal{B}(Y, X))$ be of subexponential growth and suppose $\hat{A}(\lambda)$ is closed with domain Y, for each Re $\lambda > 0$; let $k \in \mathbb{N}$. Then $A(t)$ is called* **weakly k-regular**, *if there is a constant $C > 0$ such that*

$$|\lambda^n \hat{A}^{(n)}(\lambda)y|_X \leq C(|y|_X + |\hat{A}(\lambda)y|_X), \quad \text{Re } \lambda > 0, \ y \in Y, \ n \leq k. \quad (7.12)$$

holds. $A(t)$ is called **k-regular** *if it is weakly k-regular and*

$$|\lambda^n \hat{A}^{(n)}(\lambda)|_Y y|_Y \leq C(|y|_Y + |\hat{A}(\lambda)|_Y y|_Y), \quad \text{Re } \lambda > 0, \ y \in D(\hat{A}(\lambda)|_Y), \ n \leq k.$$
$$(7.13)$$

is satisfied for some constant $C \geq 0$.

Here $\hat{A}^{(n)}(\lambda)|_Y$ denotes the part of $\hat{A}^{(n)}(\lambda)$ in Y, as before. Obviously, in the scalar case $A(t) = a(t)A$, $Y = X_A$, $A(t)$ is k-regular if $a(t)$ is k-regular in the sense of Definition 3.3.

Theorem 7.2 *Let $A \in L_{loc}^1(\mathbb{R}_+; \mathcal{B}(Y, X))$ be of subexponential growth, and assume that (7.1) is parabolic and k-regular for some $k \geq 1$. Then there is a resolvent*

$S \in C^{k-1}((0,\infty); \mathcal{B}(X) \cap \mathcal{B}(Y))$ *for (7.1), and there is a constant* $N \geq 1$ *such that*

$$|t^n S^{(n)}(t)|_{\mathcal{B}(X) \cap \mathcal{B}(Y)} \leq N, \quad t \geq 0, \ n \leq k-1, \tag{7.14}$$

and

$$|t^k S^{(k-1)}(t) - s^k S^{(k-1)}(s)|_{\mathcal{B}(X) \cap \mathcal{B}(Y)} \leq N|t-s|[1+\log \frac{t}{t-s}], \quad 0 \leq s < t. \tag{7.15}$$

If (7.1) is merely weakly parabolic and $A(t)$ weakly k-regular for some $k \geq 1$, then there is a weak resolvent $S \in C^{k-1}((0,\infty); \mathcal{B}(X))$ for (7.1), and (7.14), (7.15) hold in $\mathcal{B}(X)$, instead of $\mathcal{B}(X) \cap \mathcal{B}(Y)$.

Proof: The proof of Theorem 7.2 is based on Theorem 6.2 and results on the inversion of the Laplace transform in Section 0. As before we let

$$H(\lambda) = (I - \hat{A}(\lambda))^{-1}/\lambda, \quad \mathrm{Re}\,\lambda > 0,$$

which exists by weak parabolicity and satisfies the parabolic estimate

$$|H(\lambda)|_{\mathcal{B}(X)} \leq \frac{M}{|\lambda|}, \quad \mathrm{Re}\,\lambda > 0.$$

For $H'(\lambda)$ we have

$$H'(\lambda) = -H(\lambda)/\lambda + H(\lambda)\hat{A}'(\lambda)(I - \hat{A}(\lambda))^{-1}, \quad \mathrm{Re}\,\lambda > 0,$$

hence by weak 1-regularity and weak parabolicity

$$|H'(\lambda)|_{\mathcal{B}(X)} \leq \frac{M}{|\lambda|^2}, \quad \mathrm{Re}\,\lambda > 0.$$

By induction we obtain

$$|H^{(n)}(\lambda)|_{\mathcal{B}(X)} \leq \frac{M_n}{|\lambda|^{n+1}}, \quad \mathrm{Re}\,\lambda > 0, \ n \leq k,$$

for some constants M_n. Theorem 0.4 now implies

$$|H^{(n)}(\lambda)|_{\mathcal{B}(X)} \leq \frac{M_0 n!}{\lambda^{n+1}}, \quad \text{for all } \lambda > 0, \ n \in \mathbb{N}_0,$$

and as in the proof of Theorem 6.4 this yields a weak resolvent $S(t)$ for (7.1), which by Theorem 0.4 satisfies estimates (7.14), (7.15) in $\mathcal{B}(X)$. If (7.1) is even parabolic and k-regular, we may repeat the arguments just given in Y, and conclude the existence of a pseudo-resolvent $S(t)$ by Theorem 6.2 which is subject to (7.14), (7.15), and a resolvent by Theorem 0.4. $\square$

By similar arguments also a-regularity of (weak) resolvents for parabolic problems can be obtained. Without proof we state this as

Corollary 7.4 *Assume that (7.1) is weakly parabolic, $A(t)$ weakly k-regular for some $k \geq 1$, and let $a \in L^1_{loc}(\mathbb{R}_+)$ be k-regular and such that*
(NP4) $|\hat{a}(\lambda)(I - \hat{A}(\lambda))^{-1}|_{\mathcal{B}(X,Y)} \leq M$, $Re\ \lambda > 0$, *for some constant M.*
Then the weak resolvent $S(t)$ for (7.1) is a-regular and $T(t) = (a * S)(t)$ satisfies in addition

$$|t^n T^{(n)}(t)|_{\mathcal{B}(X,Y)} \leq N, \quad t \geq 0,\ n \leq k-1, \tag{7.16}$$

and

$$|t^k T^{(k-1)}(t) - s^k T^{(k-1)}(s)|_{\mathcal{B}(X,Y)} \leq N|t - s|[1 + \log \frac{t}{t-s}], \quad 0 \leq s < t, \tag{7.17}$$

with some constant $N \geq 1$.

The (weak) resolvent constructed in Theorem 7.2 can be used as in the proof of Theorem 3.3 to obtain maximal regularity of type C^α.

Corollary 7.5 *Let the assumptions of Theorem 7.2 be satisfied with $k = 2$, $f : J \to X$ be continuous, and let $u(t)$ formally be defined by (7.11). Then*
(i) If $f \in C_0^\alpha(J; X)$ then $u \in C^\alpha(J; X)$ is a mild solution of (7.1).
(ii) If $f \in C_0^\alpha(J; Y)$ then $u \in C^\alpha(J; Y)$ is a strong solution of (7.1).
(iii) If (7.1) is weakly parabolic and $A(t)$ weakly 2-regular, then $f \in C_0^\alpha(J; X)$ implies $u \in C^\alpha(J; X)$.
(iv) If the assumptions of Corollary 7.4 hold, then $f = a * g$, $g \in C_0^\alpha(J; X)$ implies that $u \in C^\alpha(J; Y)$ is a strong solution of (7.1).

7.3 Parabolic Problems of Variational Type

As in Section 6.7, let V and H denote Hilbert spaces such that $V \overset{d}{\hookrightarrow} H$, and let $((\cdot, \cdot))$ resp. $(\cdot, \cdot)$ denote the inner products in V resp. H, and $|| \cdot ||$ resp. $| \cdot |$ the corresponding norms. Again we identify the antidual $\overline{H}^*$ of H with H and by duality we also get $H \overset{d}{\hookrightarrow} \overline{V}^*$; the relation between the duality $< \cdot, \cdot >$ of $V, \overline{V}^*$ and the inner product in H is

$$(v, h) = < v, h >, \quad \text{for all } v \in V,\ h \in H.$$

As before, the norm in $\overline{V}^*$ will be denoted by $|| \cdot ||_*$.

Let $\alpha : \mathbb{R}_+ \times V \times V \to \mathbb{C}$ be such that $\alpha(t, \cdot, \cdot)$ is a bounded sesquilinear form on V for a.a. $t \geq 0$. α is said to be *locally integrable*, if there is $\alpha_0 \in L^1_{loc}(\mathbb{R}_+)$ such that

$$|\alpha(t, u, v)| \leq \alpha_0(t)||u||\,||v|| \quad \text{for all } u, v \in V,\ t \notin E_0, \tag{7.18}$$

where E_0 has Lebesgue-measure zero. α is called *uniformly measurable* if for each $T > 0$, $\varepsilon > 0$ there are $N + 1$ measurable disjoint subsets E_j of $J = [0, T]$ with $\cup_{j=0}^N E_j = J$, mes$(E_0) < \varepsilon$, and bounded sesquilinear forms α_j on V such that

$$|\alpha(t, u, v) - \sum_{j=1}^N \chi_j(t)\alpha_j(u, v)| \leq \varepsilon||u||\,||v|| \quad \text{for all } u, v \in V,\ t \in J \setminus E_0. \tag{7.19}$$

By the Riesz-representation theorem there are operators $A(t) \in \mathcal{B}(V, \overline{V}^*)$ such that

$$\alpha(t, u, v) = - < u, A(t)v > \qquad \text{for all } u, v \in V, \text{ a.a.} t \geq 0; \qquad (7.20)$$

It is readily verified that uniform measurability and local integrability of α imply $A \in L^1_{loc}(\mathbb{R}_+; \mathcal{B}(V; \overline{V}^*))$.

Consider the Volterra equation of variational type (6.54), i.e.

$$(w, u(t)) + \int_0^t \alpha(t - s, w, u(s))ds = < w, f(t) >, \quad t \in J, \ w \in V, \qquad (7.21)$$

where $f \in C(J; \overline{V}^*)$. As in Section 6.7, with $X = \overline{V}^*$, $Y = V$, and $A(t)$ defined by (7.20), Equation (7.21) can be rewritten in the form (7.1). Assuming a parabolic coercive estimate as well as some regularity of α, the results of Sections 7.1 and 7.2 can be directly applied to (7.21). The main assumptions will be stated in the frequency domain; they read

(PV1) *There is a function* $\varphi : \Sigma(0, \theta) \to (0, \infty)$ *such that*

$$|\hat{\alpha}(\lambda, u, v)| \leq \varphi(\lambda)||u|| \, ||v||, \qquad \text{for all } u, v \in V, \ \lambda \in \Sigma(0, \theta).$$

(PV2) *There is a constant* $\gamma > 0$ *such that*

$$|\hat{\alpha}(\lambda, u, u)| \geq \gamma \varphi(\lambda)||u||^2, \qquad \text{for all } u \in V, \ \lambda \in \Sigma(0, \theta).$$

(PV3) *There is* $\phi \in (0, \pi)$ *such that*

$$|\arg \hat{\alpha}(\lambda, u, u)| \leq \phi, \qquad \text{for all } u \in V, \ \lambda \in \Sigma(0, \theta).$$

Here we tacitly assumed that $\alpha_0 \in L^1_{loc}(\mathbb{R}_+)$ appearing in (7.18) is of supexponential growth, so that $\hat{\alpha}(\lambda, u, v)$ is well-defined for Re $\lambda > 0$, at least. We want to stress that the function $\varphi(\lambda)$ appearing in (PV1) and (PV2) is meant to be the same; this will be crucial in the sequel.

Assume now that (PV1), (PV2), (PV3) are satisfied. We are going to show that (7.1) resulting from the variational equation (7.21) becomes parabolic in the sense of Definition 7.2. Since for each $t \geq 0$ and $z \in \Sigma(0, \pi)$ we have with $\psi = \arg z$

$$|t + z|^2 = t^2 + |z|^2 + 2t|z|\cos\psi \geq c(\psi)^2(t^2 + |z|^2),$$

where

$$c(\psi) = \begin{cases} 1 & \text{for } |\psi| \leq \pi/2 \\ \sqrt{2}\cos(\psi/2) & \text{for } |\psi| > \pi/2, \end{cases}$$

(PV2) and (PV3) yield

$$\begin{aligned} ||u|| \cdot ||tu - \hat{A}(\lambda)u||_* &\geq |< u, tu - \hat{A}(\lambda)u >| = |t|u|^2 + \hat{\alpha}(\bar{\lambda}, u, u)| \\ &\geq c(\phi)(t^2|u|^2 + |\hat{\alpha}(\lambda, u, u)|^2)^{1/2} \\ &\geq c(\phi)\gamma\varphi(\bar{\lambda})||u||^2, \qquad \text{for all } u \in V, \ \lambda \in \Sigma(0, \theta). \end{aligned}$$

Therefore, for $t \geq 0$ the operators $t - \hat{A}(\lambda)$ are bijective and boundedly invertible, we obtain the estimates

$$|(t - \hat{A}(\lambda))^{-1}|_{\mathcal{B}(\overline{V}^*, V)} \leq \frac{1}{\gamma \varphi(\bar{\lambda}) c(\phi)}, \quad \lambda \in \Sigma(0, \theta), \ t \geq 0. \tag{7.22}$$

On the other hand, (PV1) implies

$$|\hat{A}(\lambda)|_{\mathcal{B}(V, \overline{V}^*)} = \sup_{||u||, ||v|| \leq 1} | < u, \hat{A}(\lambda) v > | = \sup_{||u||, ||v|| \leq 1} |\hat{a}(\bar{\lambda}, u, v)| \leq \varphi(\bar{\lambda}), \tag{7.23}$$

hence by (7.22),

$$|(t - \hat{A}(\lambda))^{-1}|_{\mathcal{B}(\overline{V}^*)} \leq \frac{1 + \gamma c(\phi)}{\gamma c(\phi) t} \quad \lambda \in \Sigma(0, \theta), \ t \geq 0. \tag{7.24}$$

Combining (7.24) with (7.23), and (7.22) for $t = 0$ yields

$$|(t - \hat{A}(\lambda))^{-1}|_{\mathcal{B}(V)} = |(\hat{A}(\lambda))^{-1}(t - \hat{A}(\lambda))^{-1}\hat{A}(\lambda)|_{\mathcal{B}(V)}$$

$$\leq |\hat{A}(\lambda)^{-1}|_{\mathcal{B}(\overline{V}^*, V)} \cdot |(t - \hat{A}(\lambda))^{-1}|_{\mathcal{B}(\overline{V}^*)} \cdot |\hat{A}(\lambda)|_{\mathcal{B}(V, \overline{V}^*)}$$

$$\leq (\gamma \varphi(\bar{\lambda}) c(\phi))^{-1} (1 + \gamma c(\phi))(\gamma c(\phi) t)^{-1} \varphi(\bar{\lambda}) = \frac{1 + \gamma c(\phi)}{t(\gamma c(\phi))^2}, \quad t > 0,$$

i.e.

$$|(t - \hat{A}(\lambda))^{-1}|_{\mathcal{B}(V)} \leq c_1(\phi)/t, \quad \lambda \in \Sigma(0, \theta), \ t > 0. \tag{7.25}$$

In particular, (7.24) and (7.25) with $t = 1$ imply the parabolic estimates, provided $\theta \geq \pi/2$; if $\theta > \pi/2$ we even see that the assumptions of Theorem 7.1, in particular (7.5), with $\omega_0 = 0$, $\theta_0 = \theta - \pi/2$ are satisfied, hence in this situation we obtain an analytic resolvent of type $(0, \theta_0)$ for (7.21), without any further assumption.

An interesting special case arises if $\varphi(\lambda)$ can be chosen as

$$\varphi(\bar{\lambda}) = |\hat{a}(\lambda)|, \quad \lambda \in \Sigma(0, \theta), \tag{7.26}$$

for some $a \in L_{loc}^1(\mathbb{R}_+)$ of subexponential growth. Then (7.22) implies

$$|\hat{a}(\lambda) H(\lambda)|_{\mathcal{B}(\overline{V}^*, V)} \leq c_2(\phi)/|\lambda|, \quad \lambda \in \Sigma(0, \theta), \tag{7.27}$$

which is the basic estimate for a-regularity of the resolvent for (7.21).

In case $\theta = \pi/2$ only, Theorem 7.2 can be applied if in addition $A(t)$ is k-regular for some $k \geq 1$. For the variational setting this assumption seems to be too stringent; however, the condition

(PV4) *there is $C > 0$ such that*

$$|\lambda^n (\frac{d}{d\lambda})^n \hat{a}(\lambda, u, v)| \leq C \varphi(\lambda) ||u|| \, ||v||, \quad \text{for all } u, v \in V, \ \text{Re } \lambda > 0, \ 1 \leq n \leq k,$$

is easily seen to imply weak k-regularity of $A(t)$, by (PV2). It turns out that (PV4) is in fact enough to obtain a resolvent, as a small modification of the proof of Theorem 7.2 shows; namely to estimate $|H(\lambda)|_{\mathcal{B}(Y)}$ we write

$$H'(\lambda) = -H(\lambda)/\lambda + H(\lambda)\hat{A}'(\lambda)(I - \hat{A}(\lambda))^{-1}$$

and obtain with the aid of (7.25), (PV4), and (7.22)

$$
\begin{aligned}
|H'(\lambda)|_{\mathcal{B}(Y)} &\leq |H(\lambda)|_{\mathcal{B}(Y)}/|\lambda| + |H(\lambda)|_{\mathcal{B}(\overline{V}^*,V)}|\hat{A}'(\lambda)|_{\mathcal{B}(V,\overline{V}^*)}|\lambda H(\lambda)|_{\mathcal{B}(V)} \\
&\leq |c_1(\phi)|/|\lambda|^2 + (\gamma\varphi(\bar{\lambda})c(\phi))^{-1}(C\varphi(\bar{\lambda})/|\lambda|)c_1(\phi)/|\lambda| \\
&= c_3(\phi)/|\lambda|^2.
\end{aligned}
$$

Similarly, the estimates for $H^{(n)}(\lambda)$, $n \leq k$, are obtained, and then the remaining part of the proof of Theorem 7.2 applies again.

Summarizing we have the following result on resolvents and maximal regularity of type C^α for parabolic problems of variational type.

Theorem 7.3 *Let $\alpha : \mathbb{R}_+ \times V \times V \to \mathbb{C}$ be such that $\alpha(t,\cdot,\cdot)$ is a bounded sesquilinear form on V for a.a. $t \geq 0$, uniformly measurable and locally integrable with α_0 from (7.18) of subexponential growth.*

(i) Suppose $\hat{\alpha}(\lambda, u, v)$ admits analytic extension to the sector $\Sigma(0,\theta)$, $\theta > \pi/2$, and assume (PV1), (PV2) and (PV3) on $\Sigma(0,\theta)$. Then there is an analytic resolvent $S(t)$ for (7.21) in $\overline{V}^*$ of type $(0,\theta_0)$, where $\theta_0 = \theta - \pi/2$.

(ii) Assume (PV1), (PV2) and (PV3) with $\theta = \pi/2$, as well as (PV4) for some $k \geq 1$. Then there is a resolvent $S(t)$ for (7.21) in $\overline{V}^*$ of class $C^{(k-1)}$ on $(0,\infty)$, and Estimates (7.14) and (7.15) hold in $\overline{V}^*$ and V.

(iii) If $\varphi(\lambda) = |\hat{a}(\lambda)|$, $\lambda \in \Sigma(0,\theta)$, for some k-regular $a \in L^1_{loc}(\mathbb{R}_+)$ then the resolvent $S(t)$ for (7.21) in $\overline{V}^*$ is a-regular, and $(a * S)(t)$ is of class C^{k-1} on $(0,\infty)$ in $\mathcal{B}(\overline{V}^*,V)$.

(iv) In case (i) and in case (ii) for $k \geq 2$, (7.21) has the maximal regularity property of type C^α, $\alpha \in (0,1)$, as stated e.g. in Corollary 7.5.

7.4 Maximal Regularity of Perturbed Parabolic Problems
In the remainder of Section 7 we consider the perturbed problem

$$u(t) = \int_0^t A(t-\tau)u(\tau)d\tau + \int_0^t a(t-s)\left(\int_0^s dB(\tau)u(s-\tau)\right)ds + f(t), \quad t \in J, \quad (7.28)$$

where $a \in L^1_{loc}(\mathbb{R}_+)$, $A \in L^1_{loc}(\mathbb{R}_+; \mathcal{B}(Y,X))$, $B \in BV(\mathbb{R}_+; \mathcal{B}(Y,X))$ is such that $B(0) = B(0+) = 0$, $f \in C(J;X)$, and $J = [0,t_0]$. The unperturbed problem (7.1) is assumed to admit an a-regular resolvent $S_0 \in C^1((0,\infty); \mathcal{B}(X) \cap \mathcal{B}(Y))$ such that

$$|S_0(t)|_{\mathcal{B}(X)\cap\mathcal{B}(Y)} + |t\dot{S}_0(t)|_{\mathcal{B}(X)\cap\mathcal{B}(Y)} \leq N, \quad t \in J, \quad (7.29)$$

$$|\dot{S}_0(t) - \dot{S}_0(s)|_{\mathcal{B}(X)\cap\mathcal{B}(Y))} \leq N\frac{|t-s|}{ts}\left[1 + \log\frac{t}{t-s}\right], \quad 0 \leq s < t \leq t_0,$$

and $T_0 = a * S_0 \in C^1((0, \infty); \mathcal{B}(X, Y))$ is subject to

$$|T_0(t)|_{\mathcal{B}(X,Y)} + |t\dot{T}_0(t)|_{\mathcal{B}(X,Y)} \leq N, \quad t \in J, \tag{7.30}$$

$$|\dot{T}_0(t) - \dot{T}_0(s)|_{\mathcal{B}(X,Y)} \leq N \frac{|t-s|}{ts} [1 + \log \frac{t}{t-s}], \quad 0 \leq s < t \leq t_0.$$

The resolvents obtained in Theorems 2.1 and 7.1, as well as those from Theorems 3.1 and 7.2 for $k \geq 2$ have these properties and are good candidates for the sequel. An important special case is $A(t) \equiv A$ the generator of an analytic semigroup, and $a(t) \equiv 1$; then $S_0(t) = e^{At}$, $Y = X_A$, and (7.28) is equivalent to

$$\dot{u}(t) = Au(t) + \int_0^t dB(\tau)u(t-\tau) + g(t), \ u(0) = u_0, \quad t \in J,$$

where $g(t) = \dot{f}(t)$ and $u_0 = f(0)$.

We begin with an extension of the maximal regularity results of type C^α to the perturbed problem (7.28).

Theorem 7.4 *Under the assumptions stated above, (7.28) has the maximal regularity property of type C^α. More precisely, with $\alpha \in (0,1)$,*
(i) if $f \in C_0^\alpha(J; X)$ there is a unique mild solution $u \in C_0^\alpha(J; X)$ of (7.28);
(ii) if $f \in C_0^\alpha(J; Y)$ there is a unique strong solution $u \in C_0^\alpha(J; Y)$ of (7.28);
(iii) if $f = a * g$, $g \in C_0^\alpha(J; X)$ there is a unique strong solution $u \in C_0^\alpha(J; Y)$ of (7.28).

Proof: By the variation of parameters formula, (7.28) is equivalent to

$$u(t) = \frac{d}{dt}(S_0 * f)(t) + \frac{d}{dt}(S_0 * a * dB * u)(t), \quad t \in J. \tag{7.31}$$

The latter will be considered as a linear equation

$$u = u_0 + Ku \tag{7.32}$$

in the Banach space $U = C_0^\alpha(J; Y)$ with norm

$$||u||_{Y,\alpha,\omega} = \sup_{0 \leq s < t \leq t_0} e^{-\omega t} |u(t) - u(s)|_Y (t-s)^{-\alpha},$$

where $u_0 = (S_0 * f)'$ in case $f \in C_0^\alpha(J; Y)$, and $u_0 = \dot{T}_0 * g$ if $g \in C_0^\alpha(J; X)$. The operator K is defined by

$$Ku = \dot{T}_0 * dB * u, \quad u \in U. \tag{7.33}$$

Given $u \in U$, we obtain for $0 \leq s < t \leq t_0$ with $u(s) = 0$ for $s \leq 0$

$$|(dB * u)(t) - (dB * u)(s)|_X = |\int_0^t dB(\tau)(u(t-\tau) - u(s-\tau))|_X$$

$$\leq \int_0^t |dB(\tau)|_{\mathcal{B}(Y,X)} \cdot e^{\omega(t-\tau)} (t-s)^\alpha ||u||_{Y,\alpha,\omega}$$

$$= e^{\omega t}(t-s)^\alpha ||u||_{Y,\alpha,\omega} \cdot (\int_0^{t_0} e^{-\omega t} |dB(\tau)|_{\mathcal{B}(Y,X)}),$$

i.e. $dB * u \in C_0^\alpha(J; X)$ and

$$\|dB * u\|_{X,\alpha,\omega} \leq \eta(\omega)\|u\|_{Y,\alpha,\omega}, \quad u \in U. \tag{7.34}$$

Here $\eta(\omega) = \int_0^{t_0} e^{-\omega t}|dB(t)| \to 0$ as $\omega \to \infty$ since $B(t)$ has no jump at zero, by assumption.

On the other hand, following the proof of Theorem 2.4, by (7.29) we obtain $(S_0 * f)^{\cdot} \in U$ for each $f \in U$, and

$$\|\frac{d}{dt}(S_0 * f)\|_{Y,\alpha,\omega} \leq c(\alpha)\|f\|_{Y,\alpha,\omega}, \quad f \in U, \tag{7.35}$$

where

$$c(\alpha) \leq N(1 + 2/\alpha + 3\int_0^\infty (1 + \log(1 + r))r^{\alpha-1}dr/(1 + r)) < \infty$$

is independent of ω. Similarly, we also get

$$\|\dot{T}_0 * u\|_{Y,\alpha,\omega} \leq c(\alpha)\|u\|_{X,\alpha,\omega}, \quad u \in C_0^\alpha(J; X). \tag{7.36}$$

Therefore u_0 defined above belongs to U and $K \in \mathcal{B}(U)$ satisfies

$$\|K\|_{\mathcal{B}(U)} \leq c(\alpha)\eta(\omega) < 1, \tag{7.37}$$

provided ω is chosen large enough. Hence $I - K$ is invertible, and so (7.32) admits a unique solution $u \in U = C_0^\alpha(J; Y)$; this proves (ii) and (iii).

Concerning assertion (i), observe that by (iii) there is a unique strong solution v of (7.28) with f replaced by $a * f$. But then there is a unique mild solution u of

$$u = A * u + dB * v + f,$$

since $dB * v \in C_0^\alpha(J; X)$; u is given by

$$u = \frac{d}{dt}(S_0 * f) + \frac{d}{dt}(S_0 * dB * v).$$

Convolving this identity with a we obtain $v = a * u$ by uniqueness, hence u is a mild solution of (7.28). $\square$

There are many other function spaces for which the solution map $f \mapsto u$ of (7.28) has the maximal regularity property, namely the classes $B_p^{\alpha,q}(J; X)$ defined by

$$B_p^{\alpha,q}(J; X) = \{f \in L^p(J; X) : [f]_{\alpha,q;p}^X < \infty\}, \tag{7.38}$$

where, with $f(t) \equiv 0$ for $t < 0$,

$$[f]_{\alpha,q;p}^X = |f|_p + (\int_0^\infty (h^{-\alpha}|f(\cdot - h) - f(\cdot)|_p)^q dh/h)^{1/q}, \tag{7.39}$$

and

$$[f]_{\alpha,\infty;p}^X = |f|_p + \sup_{h>0} h^{-\alpha} |f(\cdot - h) - f(\cdot)|_p; \qquad (7.40)$$

$1 \le p < \infty$, $1 \le q \le \infty$, $\alpha \in (0,1)$. It is well known that

$$B_p^{\alpha,q}(J;X) = (L^p(J;X), W_0^{1,p}(J;X))_{\alpha,q},$$

are the *real interpolation spaces* between $L^p(J;X)$ and $W_0^{1,p}(J;X)$; cp. Butzer and Berens [36]. Similarly, we let

$$B_\infty^{\alpha,q}(J;X) = (C_0(J;X), C_0^1(J;X))_{\alpha,q}, \qquad \alpha \in (0,1), \ 1 \le q \le \infty, \qquad (7.41)$$

and with this notation we have

$$B_\infty^{\alpha,\infty}(J;X) = C_0^\alpha(J;X), \qquad \alpha \in (0,1).$$

Observe that the index 0 always refers to vanishing at $t = 0$.

The following maximal regularity result is a considerable extension of Theorem 7.4.

Theorem 7.5 *Under the assumptions stated above, let* $S : f \mapsto u$ *denote the solution map of (7.28), and let* $\alpha \in (0,1)$, $p, q \in [1, \infty]$. *Then* S *enjoys the following properties.*
(i) $S \in \mathcal{B}(B_p^{\alpha,q}(J;X)) \cap \mathcal{B}(B_p^{\alpha,q}(J;Y))$;
(ii) $a * S = S(a * \cdot) \in \mathcal{B}(B_p^{\alpha,q}(J;X), B_p^{\alpha,q}(J;Y))$.

Proof: The proof parallels that of Theorem 7.4, the estimates (7.34) on $dB * u$ and (7.35), (7.36) being the main ingredients. Therefore we only have to establish the analogues of these estimates in $B_p^{\alpha,q}(J;X)$.

Choosing the norm on the base spaces $L^p(J;X)$ resp. $C_0(J;X)$ as

$$|f|_{p,\omega} = \left(\int_0^{t_0} |f(t)|^p e^{-\omega p t} dt \right)^{1/p}, \qquad 1 \le p < \infty,$$

resp.

$$|f|_{\infty,\omega} = \sup_{0 < t < t_0} |f(t)| e^{-\omega t},$$

it follows by a direct estimate (or by interpolation)

$$[dB * u]_{\alpha,q;p}^X \le \eta(\omega)[u]_{\alpha,q;p}^Y, \qquad u \in B_p^{\alpha,q}(J;Y), \qquad (7.42)$$

with $\eta(\omega)$ as in the proof of Theorem 7.4. Thus it remains to prove the estimates for $(d/dt)(S_0 * f)$ and $\dot{T}_0 * f$. We concentrate here on $[(d/dt)S_0 * f]_{\alpha,q;p}^X$, since the others are analogues, by assumptions (7.29) and (7.30). Writing as in the proof of Theorem 2.4

$$u(t) = S_0(t)f(t) + \int_0^t \dot{S}_0(\tau)(f(t - \tau) - f(t))d\tau,$$

we obtain for $q > 1$ with $1/q + 1/q' = 1$

$$\begin{aligned}
|u|_{p,\omega} &\leq N|f|_{p,\omega} + N\int_0^{t_0} |f(\cdot - \tau) - f(\cdot)|_{p,\omega} d\tau/\tau \\
&\leq N|f|_{p,\omega} + N\left(\int_0^{t_0} \tau^{\alpha q'-1} d\tau\right)^{1/q'} [f]_{\alpha,q;p}^X \\
&\leq c(\alpha, t_0, q)[f]_{\alpha,q;p}^X,
\end{aligned}$$

and similarly for $q = 1$

$$\begin{aligned}
|u|_{p,\omega} &\leq N|f|_{p,\omega} + N\int_0^{t_0} \tau^\alpha \cdot \tau^{-\alpha}|f(\cdot - \tau) - f(\cdot)|_{p,\omega} d\tau/\tau \\
&\leq N|f|_{p,\omega} + Nt_0^\alpha [f]_{\alpha,1;p}^X \leq c(\alpha, t_0, 1)[f]_{\alpha,1;p}^X.
\end{aligned}$$

This shows $u \in L^p(J; X)$ for $1 \leq p \leq \infty$. Next we have with the decomposition

$$\begin{aligned}
u(t) - u(t-h) &= S_0(h)(f(t) - f(t-h)) + (S_0(t_0) - S_0(t_0 - h))f(t-h) \\
&+ \int_0^h \dot{S}_0(\tau)(f(t-\tau) - f(t))d\tau \\
&+ \int_h^{t_0} (\dot{S}_0(\tau) - \dot{S}_0(\tau - h))(f(t-\tau) - f(t-h))d\tau
\end{aligned}$$

and by (7.29)

$$\begin{aligned}
|u(\cdot) - u(\cdot - h)|_{p,\omega} &\leq N|f(\cdot) - f(\cdot - h)|_{p,\omega} + \varphi(h)|f|_{p,\omega} \\
+N\int_0^h |f(\cdot) - f(\cdot - \tau)|_{p,w} d\tau/\tau &+ \int_0^\infty \psi(\tau/h)|f(\cdot - \tau) - f(\cdot)|_{p,\omega} d\tau/\tau,
\end{aligned}$$

where $\varphi(h) = 2N \min(1, 2h/t_0)$ and $\psi(s) = N(1 + s)^{-1}(1 + \log(1 + s))$ for $s > 1$. Since

$$\int_0^\infty (h^{-\alpha}\varphi(h))^q dh/h \leq 2Nt_0^{-1} \int_0^{t_0} h^{(1-\alpha)q-1} dh + 2N \int_{t_0}^\infty h^{-\alpha q-1} dh < \infty,$$

and

$$\int_0^\infty s^\alpha \psi(s) ds/s = N \int_0^\infty (1 + s)^{-1}(1 + \log(1 + s))s^{\alpha-1} ds < \infty,$$

there follows the estimate

$$[u]_{\alpha,q;p}^X \leq c(\alpha, t_0, q)[f]_{\alpha,q;p}^X, \quad f \in B_p^{\alpha,q}(J; X), \tag{7.43}$$

with some constant $c(\alpha, t_0, q) < \infty$, thereby proving the theorem. □

7.5 Resolvents for Perturbed Parabolic Problems
Consider again (7.28) under the assumptions stated at the beginning of Section

7.4. We now show that there is an a-regular resolvent for (7.28). The resolvent will be constructed as follows. If (7.28) admits a resolvent $S(t)$ then it must be the solution of

$$S(t) = S_0(t) + \frac{d}{dt}(a * S_0 * dB * S)(t), \quad t \in \mathbb{R}_+. \tag{7.44}$$

From (7.44) we obtain with $T = a * S$ the following three integral equations.

$$T(t) = T_0(t) + (\dot{T}_0 * dB * T)(t), \quad t \in \mathbb{R}_+. \tag{7.45}$$

$$S(t) = S_0(t) + \frac{d}{dt}(S_0 * dB * T)(t), \quad t \in \mathbb{R}_+. \tag{7.46}$$

$$S(t) = S_0(t) + (\dot{T}_0 * dB * S)(t), \quad t \in \mathbb{R}_+. \tag{7.47}$$

Since $T_0 \in C^1((0,\infty); \mathcal{B}(X,Y))$, (7.45) will turn out to admit a solution with values in $\mathcal{B}(X,Y)$. (7.46) is merely the definition of $S(t)$ in $\mathcal{B}(X)$ after $T(t)$ is known to exist, and (7.47) shows that $S(t)$ is leaving Y invariant.

To solve (7.45) in $\mathcal{B}(X,Y)$ we employ the maximal regularity result of type $B_p^{\alpha,q}$, Theorem 7.4. This approach works, since by (7.30)

$$|T_0(t) - T_0(t-h)|_{\mathcal{B}(X,Y)} \le N \min\{2, \frac{h}{t-h}\}, \quad t > h,$$

hence

$$
\begin{aligned}
|T_0(\cdot) - T_0(\cdot - h)|_p &\le N[\int_0^{\frac{3}{2}h} 2^p dt + h^p \int_{\frac{3}{2}h}^\infty \frac{dt}{(t-h)^p}]^{1/p} \\
&= N[2^{p-1} \cdot 3h + h^p \cdot (h/2)^{1-p}(p-1)^{-1}]^{1/p} = C_p h^{1/p},
\end{aligned}
$$

i.e. $T_0(\cdot) \in B_p^{1/p,\infty}(J; \mathcal{B}(X,Y))$ for each $p \in (1,\infty)$. Similarly, one can show $T_0 \in B_p^{\alpha,q}(J; \mathcal{B}(X,Y))$ whenever $\alpha p < 1$, $p, q \in [1,\infty]$, $\alpha \in (0,1)$. Clearly this excludes $p = \infty$, and so by means of this method one cannot obtain uniform L^∞-bounds on $T(t)$, for this a different approach must be used. The L^∞-bounds are quite delicate to obtain, Section 7.6 is devoted to this aim. Here we construct the resolvent and show its continuity properties assuming L^∞-bounds. The result reads as follows.

Theorem 7.6 *Let* $A \in L_{loc}^1(\mathbb{R}_+; \mathcal{B}(Y,X))$, $B \in BV_{loc}(\mathbb{R}_+; \mathcal{B}(Y,X))$ *satisfy* $B(0) = B(0+) = 0$, $b(t) = \mathrm{Var}B|_0^t$, *and let* $a \in L_{loc}^1(\mathbb{R}_+)$. *Assume that (7.1) admits an a-regular resolvent* $S_0 \in C^1((0,\infty); \mathcal{B}(X) \cap \mathcal{B}(Y))$ *with* $T_0 = a * S_0 \in C^1((0,\infty);$ $\mathcal{B}(X,Y))$ *such that (7.29) and (7.30) hold on each interval* $J = [0,t_0]$, *with constant* $N = N(t_0)$. *Then there is an a-regular resolvent* $S(t)$ *for (7.28). $S(t)$ and* $T(t) = (a * S)(t)$ *have the following additional properties.*
(i) $S \in L_{loc}^\infty(\mathbb{R}_+; \mathcal{B}(X) \cap \mathcal{B}(Y))$, $T \in L_{loc}^\infty(\mathbb{R}_+; \mathcal{B}(X,Y))$;
(ii) $S(t)$ *resp.* $T(t)$ *is left-continuous in* $\mathcal{B}(X) \cap \mathcal{B}(Y)$ *resp. in* $\mathcal{B}(X,Y)$ *on* $(0,\infty)$;
(iii) $S(t)$ *resp.* $T(t)$ *is right-continuous in* $\mathcal{B}(X) \cap \mathcal{B}(Y)$ *resp. in* $\mathcal{B}(X,Y)$ *for* $t \notin N_r$, *where* N_r *denotes the set of jump points of the resolvent kernel r of b;*

(iv) S and T have the following additional regularity:

$$S \in \bigcap_{p\in(1,\infty)} B_{p,loc}^{1/p,\infty}(\mathbb{R}_+; \mathcal{B}(X) \cap \mathcal{B}(Y)),$$

and

$$T \in \bigcap_{p\in(1,\infty)} B_{p,loc}^{1/p,\infty}(\mathbb{R}_+; \mathcal{B}(X,Y)).$$

Note that the resolvent kernel $r(t)$ of $b(t)$ is defined as the solution of $r = b + db * r$; therefore $N_r = \{t > 0 : r(t) \neq r(t+)\}$ is the additive semimodule generated by the jump points of b, i.e. $N_b = \{t > 0 : b(t) \neq b(t+)\}$. Recall our convention that BV-functions b are always normalized by left-continuity and by $b(0) = 0$. Example 1.1 shows that Theorem 7.6 is optimal; the discontinuity of $B(t) = Ae_0(t-1)$ is reproduced in the resolvent $S(t)$ of this example at $t_n = n$, i.e. at each $t \in N_r = \mathbb{N}$.

Proof of Theorem 7.6: (a) By the estimates derived in the proof of Theorem 7.5, i.e. (7.42) and (7.43) and its analogs in Y and for $\dot{T}_0 * f$, it is readily seen that there is a unique solution $T \in \bigcap_{p\in(1,\infty)} B_{p,loc}^{1/p,\infty}(\mathbb{R}_+; \mathcal{B}(X,Y))$ of (7.45) as well as a unique solution $S_1 \in \bigcap_{p\in(1,\infty)} B_{p,loc}^{1/p,\infty}(\mathbb{R}_+; \mathcal{B}(Y))$ of (7.47). Then S defined by (7.46) belongs to $\bigcap_{p\in(1,\infty)} B_{p,loc}^{1/p,\infty}(\mathbb{R}_+; \mathcal{B}(X))$. Integrating (7.45), (7.46) and (7.47) and convolving the latter two with $a(t)$ we obtain

$$1 * T = 1 * T_0 + T_0 * dB * T \quad \text{in } \mathcal{B}(X,Y),$$

$$1 * a * S = 1 * a * S_0 + a * S_0 * dB * T \quad \text{in } \mathcal{B}(X),$$

$$1 * a * S_1 = 1 * a * S_0 + a * T_0 * dB * S_1 \quad \text{in } \mathcal{B}(Y),$$

hence by uniqueness and by $Y \overset{d}{\hookrightarrow} X$ we obtain $1 * Ty = 1 * a * Sy = 1 * a * S_1 y$ for each $y \in Y$, i.e. $S_1 = S|_Y$, and $T = a * S$; this proves (iv). It is also clear from the construction that S satisfies the resolvent equations for (7.28), e.g. in integrated form.

(b) Let $\rho_\varepsilon \in C_0^\infty(\mathbb{R})$ be nonnegative, supp $\rho_\varepsilon \subset (0, \varepsilon)$, $\int_{-\infty}^\infty \rho_\varepsilon(t)dt = 1$, and let

$$B_\varepsilon(t) = \int_0^\infty dB(\tau)\rho_\varepsilon(t-\tau), \quad t \in \mathbb{R}; \tag{7.48}$$

then $B_\varepsilon \in C^\infty(\mathbb{R}; \mathcal{B}(Y,X))$, supp $B_\varepsilon \subset (0, \infty)$, and $\int_0^t |B_\varepsilon(\tau)|dt \leq b(t)$, as well as $\int_0^{t_0} e^{-\omega t}|B_\varepsilon(t)|dt \leq \int_0^{t_0} db(\tau)e^{-\omega\tau}$.

By Theorem 6.1, there is an a-regular resolvent $S_\varepsilon(t)$ of (7.28) with dB replaced by B_ε, and as solution of (7.45)~(7.47), with dB replaced by B_ε, S_ε and $T_\varepsilon = a * S_\varepsilon$ also have property (iv). In Section 7.6 we shall prove the following

L^∞-estimates: *For each $t_0 > 0$ there is a constant $C = C(t_0)$ such that*

$$|S_\varepsilon(t)|_{\mathcal{B}(X)} + |S_\varepsilon(t)|_{\mathcal{B}(Y)} + |T_\varepsilon(t)|_{\mathcal{B}(X,Y)} \leq C, \quad t \in J, \ \varepsilon \in (0,1).$$

Assume that these L^∞-estimates are true. Then we have

$$T - T_\varepsilon = \dot{T}_0 * (dB * T_\varepsilon - B_\varepsilon * T_\varepsilon) + \dot{T}_0 * dB * (T - T_\varepsilon) \quad \text{in } \mathcal{B}(X,Y),$$

$$S - S_\varepsilon = \frac{d}{dt}(S_0 * (dB * T - B_\varepsilon * T_\varepsilon)) \quad \text{in } \mathcal{B}(X), \qquad (7.49)$$

$$S - S_\varepsilon = \dot{T}_0 * (dB * S_\varepsilon - B_\varepsilon * S_\varepsilon) + \dot{T}_0 * dB * (S - S_\varepsilon) \quad \text{in } \mathcal{B}(Y).$$

Since $\int_0^t |B_\varepsilon(t)|_{\mathcal{B}(Y,X)} d\tau \leq b(t)$, $t > 0$, $\{T_\varepsilon\}_{\varepsilon \in (0,1)} \subset B_p^{1/p,\infty}(J; \mathcal{B}(X,Y))$ is uniformly bounded, $\{dB * T_\varepsilon\}_{\varepsilon \in (0,1)} \subset B_p^{1/p,\infty}(J; \mathcal{B}(X))$ is as well, and therefore $dB * T_\varepsilon - \rho_\varepsilon * dB * T_\varepsilon \to 0$ as $\varepsilon \to 0$ in each $B_p^{\alpha,\infty}(J; \mathcal{B}(X,Y))$, $\alpha p < 1$. By (7.49) this implies $T_\varepsilon \to T$ in $B_p^{\alpha,\infty}(J; \mathcal{B}(X,Y))$, for each $\alpha p < 1$, in particular $T_\varepsilon \to T$ in $L^p(J; \mathcal{B}(X,Y))$. Similarly we also obtain $S_\varepsilon \to S$ in $L^p(J; \mathcal{B}(X) \cap \mathcal{B}(Y))$, where $J = [0, t_0]$, $t_0 > 0$ being arbitrary. Passing to a subsequence $\varepsilon_n \to 0$ if necessary, we obtain a.e. convergence, thereby proving the L^∞-estimates also for $S(t)$, i.e. (i) of Theorem 7.6 follows.

(c) To prove that $S(t)$ is an a-regular resolvent of (7.28) in the sense of Definition 6.2, observe that the first resolvent equation implies $T \in C(\mathbb{R}_+; \mathcal{B}(X))$ and $S \in C(\mathbb{R}_+; \mathcal{B}(Y,X))$; in particular, $S(t)y$ can be extended as a continuous function in X for each $y \in Y$. Since Y is dense in X, the family $\{S(t)\}_{t \geq 0} \subset \mathcal{B}(X)$ is everywhere defined, and $|S(t)|_{\mathcal{B}(X)} \leq C(t_0)$, for all $t \leq t_0$, and so $\{S(t)\}_{t \geq 0}$ is strongly continuous in X by the Banach-Steinhaus theorem. Similarly, the second resolvent equation shows $T \in C(\mathbb{R}_+; \mathcal{B}(Y))$, and so by the same argument $\{T(t)\}_{t \geq 0}$ is strongly continuous in $\mathcal{B}(X,Y)$. This shows that $S(t)$ is a resolvent for (7.28) in the sense of Definition 6.2.

The proofs of the remaining continuity properties (ii) and (iii) do not follow from the L^∞-estimates alone and will be postponed until Section 7.6. $\square$

Let us pause to state some additional regularity properties of $S(t)$ if $B(t)$ has better properties, these follow directly from the integral equations (7.45), (7.46), (7.47), and Theorem 7.6.

Corollary 7.6 *Let the assumptions of Theorem 7.6 be satisfied. Then the resolvent $S(t)$ of (7.28) has the following additional properties.*
(i) If $B \in C(\mathbb{R}_+; \mathcal{B}(Y,X))$ then $S - S_0 \in C(\mathbb{R}_+; \mathcal{B}(X) \cap \mathcal{B}(Y))$ and $T = a * S \in C(\mathbb{R}_+; \mathcal{B}(X,Y))$.
(ii) If $B \in C_0^\alpha(\mathbb{R}_+; \mathcal{B}(Y,X))$ then $S - S_0 \in C_{loc}^\alpha(\mathbb{R}_+; \mathcal{B}(X) \cap \mathcal{B}(Y))$ and $T - T_0 \in C_{loc}^\alpha(\mathbb{R}_+; \mathcal{B}(X,Y))$.

As a consequence of the continuity properties of the resolvent $S(t)$ for (7.28), observe that in Proposition 6.3 for (7.28) the conditions '$f \in W^{1,1}$' in (ii) and (iii) resp. '$g \in W^{1,1}$' in (iv) can be replaced by '$f \in BV \cap C$' resp. '$g \in BV \cap C$'. This follows from the fact that $S * df$ resp. $S * dg$ is now well-defined and continuous, as a short reflection shows.

7.6 Uniform Bounds for the Resolvent

Throughout this subsection we use the same notation as in Sections 7.4 and 7.5. Our aim is to prove the L^∞-estimates for the resolvents $S_\varepsilon(t)$ of the approximating equations (7.28) with dB replaced by B_ε, as stated in Section 7.5. This will be done by a term by term estimate of the Neumann-series corresponding to the solutions of (7.45), (7.46), and (7.47), i.e.

$$T_\varepsilon = \sum_{n=0}^{\infty} (\dot{T}_0 * B_\varepsilon)^{*n} * T_0 \quad \text{in } \mathcal{B}(X, Y), \tag{7.50}$$

$$S_\varepsilon = S_0 + \sum_{n=0}^{\infty} (\dot{S}_0 * B_\varepsilon + B_\varepsilon) * (\dot{T}_0 * B_\varepsilon)^{*n} * T_0 \quad \text{in } \mathcal{B}(X), \tag{7.51}$$

$$S_\varepsilon = \sum_{n=0}^{\infty} (\dot{T}_0 * B_\varepsilon)^{*n} * S_0 \quad \text{in } \mathcal{B}(Y). \tag{7.52}$$

Since the arguments are the same for each of these series we shall concentrate only on the first one, (7.50).

Claim 1 Let
$$R_n(t) = \dot{T}_0 * B_\varepsilon)^{*n} * T_0(t) \quad \text{for } t \in J, \, n \in \mathbb{N}, \tag{7.53}$$
denote the nth summand of the series (7.50) for T_ε. Then

$$R_n(t) = \int_0^t \int_0^{t_n} \cdots \int_0^{t_2} \int_\tau^t \int_\tau^{s_n}$$
$$\cdots \int_\tau^{s_2} \dot{T}_0(t - s_n) B_\varepsilon(\tau_n) \dot{T}_0(s_n - s_{n-1}) B_\varepsilon(\tau_{n-1}) \tag{7.54}$$
$$\cdots B_\varepsilon(\tau_1) T_0(s_1 - \tau) ds_1 \ldots ds_n d\tau_1 \ldots d\tau_n$$

holds, where $t_k = t - \sum_k^n \tau_j$ and $\tau = \sum_1^n \tau_j$.

Proof: Since the integrand is continuous, we may change the order of integration as we please. The induction step from n to $n+1$ is verified as follows.

$$R_{n+1}(t) = \dot{T}_0 * B_\varepsilon * R_n(t) = \int_0^t \dot{T}_0(t - s) \int_0^s B_\varepsilon(\rho) R_n(s - \rho) d\rho ds$$
$$= \int_0^t \int_\rho^t \dot{T}_0(t - s) B_\varepsilon(\rho) R_n(s - \rho) ds d\rho$$
$$= \int_0^t \int_\rho^t \int_0^{s-\rho} \int_0^{s-\rho-\tau_n} \cdots$$

$$\int_0^{s-\rho-\sum_2^n \tau_j} \int_\tau^{s-\rho} \int_\tau^{s_n} \int_\tau^{s_2} \dot{T}_0(t-s)B_\varepsilon(\rho)\dot{T}_0(s-\rho-s_n).$$
$$\ldots\ B_\varepsilon(\tau_1)T_0(s_1-\tau)ds_1 ds_2 \ldots ds_n d\tau_1 \ldots d\tau_n ds d\rho.$$

The transformation $s_j \to s_j + \rho$ and $\tau' = \tau + \rho$ yields

$$R_{n+1}(t) \;=\; \int_0^t \int_\rho^t \int_0^{s-\rho} \ldots \int_0^{s-\rho-\sum_2^n \tau_j} \int_{\tau'}^s \int_{\tau'}^{s_n} \int_{\tau'}^{s_2} \dot{T}_0(t-s)B_\varepsilon(\rho)$$
$$\ldots\ \dot{T}_0(s-s_n)B_\varepsilon(\tau_1)T_0(s_1-\tau')ds_1 \ldots ds_n d\tau_1 \ldots d\tau_n ds d\rho,$$

and commuting the integral over s with those over $\tau_1 \ldots \tau_n$ leads to

$$R_{n+1}(t) = \int_0^t \int_0^{t_{n+1}} \ldots \int_0^{t_2} \int_{\tau'}^t \int_{\tau'}^{s_n} \ldots \int_{\tau'}^{s_2} \{\ldots\}ds_1 \ldots ds_n ds d\tau_1 \ldots d\tau_n d\rho$$

where $t_j = t - \rho - \sum_j^n \tau_k$, i.e (7.54) holds for $n+1$. $\square$

The representation (7.54) of $R_n(t)$ suggests to consider

$$Z_n(t) = \int_\tau^t \int_\tau^{s_n} \ldots \int_\tau^{s_2} \dot{T}_0(t-s_n)B_\varepsilon(\tau_n)\ldots B_\varepsilon(\tau_1)T_0(s_1-\tau)ds_1 \ldots ds_n \quad (7.55)$$

first, where the dependence of Z_n on $\tau_1, \ldots, \tau_n$ has been dropped. To obtain the desired estimates we decompose $Z_n(t)$ according to

Claim 2 We have

$$Z_n(t) = \int_\tau^t \dot{T}_0(t-s)U_n(t,s)ds + \int_\tau^t T_0(t-s)V_n(t,s)ds + T_0(t-\tau)W_n(t), \quad (7.56)$$

where U_n, V_n, W_n are defined inductively by

$$U_1(t,s) = B_\varepsilon(\tau_1)(T_0(s-\tau) - T_0(t-\tau)),$$
$$V_1(t,s) \equiv 0,\ W_1(t) = B_\varepsilon(\tau_1)T_0(t-\tau), \quad (7.57)$$

and

$$W_{n+1}(t) = B_\varepsilon(\tau_{n+1})T_0(t-\tau)W_n(t), \quad (7.58)$$

$$V_{n+1}(t,s) = B_\varepsilon(\tau_{n+1})[\dot{T}_0(t-s)U_n(t,s) + T_0(t-s)V_n(t,s)], \quad (7.59)$$

$$U_{n+1}(t,s) \;=\; B_\varepsilon(\tau_{n+1})\{T_0(s-\tau)W_n(s) - T_0(t-\tau)W_n(t)$$
$$+\ \int_\tau^s (\dot{T}_0(s-r)U_n(s,r) - \dot{T}_0(t-r)U_n(t,r))dr \quad (7.60)$$
$$+\ \int_\tau^s [T_0(s-r)V_n(s,r) - T_0(t-r)V_n(t,r)]dr.\}$$

Proof: For $n = 1$ we have

$$
\begin{aligned}
Z_1(t) &= \int_\tau^t \dot{T}_0(t-s)B_\varepsilon(\tau_1)T_0(s-\tau)ds \\
&= \int_\tau^t \dot{T}_0(t-s)B_\varepsilon(\tau_1)(T_0(s-\tau) - T_0(t-\tau))ds \\
&\quad + (\int_\tau^t \dot{T}_0(t-s)ds)B_\varepsilon(\tau_1)T_0(t-\tau),
\end{aligned}
$$

hence (7.56), and (7.57) hold for $n = 1$. Let the assertions be true for n; then we have

$$
\begin{aligned}
Z_{n+1}(t) &= \int_\tau^t \dot{T}_0(t-s)B_\varepsilon(\tau_{n+1})Z_n(s)ds \\
&= \int_\tau^t \dot{T}_0(t-s)B_\varepsilon(\tau_{n+1}) \int_\tau^s \dot{T}_0(s-r)U_n(s,r)drds \\
&\quad + \int_\tau^t \dot{T}_0(t-s)B_\varepsilon(\tau_{n+1}) \int_\tau^s T_0(s-r)V_n(s,r)drds \\
&\quad + \int_\tau^t \dot{T}_0(t-s)B_\varepsilon(\tau_{n+1})T_0(s-\tau)W_n(s)ds \\
&= \int_\tau^t \dot{T}_0(t-s)B_\varepsilon(\tau_{n+1})\{\int_\tau^s [\dot{T}_0(s-r)U_n(s,r) - \dot{T}_0(t-r)U_n(t,r)]dr \\
&\quad + \int_\tau^s [T_0(s-r)V_n(s,r) - T_0(t-r)V_n(t,r)]dr \\
&\quad + [T_0(s-\tau)W_n(s) - T_0(t-\tau)W_n(t)]\}ds + \int_\tau^t \dot{T}_0(t-s)B_\varepsilon(\tau_{n+1}) \\
&\quad \cdot \{\int_\tau^s [\dot{T}_0(t-r)U_n(t,r) + T_0(t-r)V_n(t,r)]dr + T_0(t-\tau)W_n(t)\}ds \\
&= \int_\tau^t \dot{T}_0(t-s)U_{n+1}(t,s)ds \\
&\quad + \int_\tau^t T_0(t-r)V_{n+1}(t,r)dr + T_0(t-\tau)W_{n+1}(t),
\end{aligned}
$$

hence the assertion also holds for $n + 1$. $\square$

To obtain the desired estimate for $Z_n(t)$ we have to estimate U_n, V_n, W_n separately by induction. This is easy for W_n.

Claim 3 Let $\alpha_n = \prod_1^n |B_\varepsilon(\tau_j)|_{\mathcal{B}(Y,X)}$. Then

$$|W_n(t)| \leq \alpha_n N^n \quad \text{for all } t \geq 0, \tag{7.61}$$

and

$$|W_n(t) - W_n(s)| \leq \alpha_n n N^n \varphi_1(\frac{t-s}{s-\tau}) \qquad \text{for all } \tau < s < t < t_0, \qquad (7.62)$$

where $\varphi_1(a) = \log(1+a)$.

Proof: We have by

$$|W_1(t)| \leq |B_\varepsilon(\tau_1)| \cdot N \leq \alpha_1 N$$

and

$$|W_1(t) - W_1(s)| \leq |B_\varepsilon(\tau_1)||T_0(t-\tau) - T_0(s-\tau)| \leq \alpha_1 N \log(1 + \frac{t-s}{s-\tau}),$$

i.e. (7.61) and (7.62) for $n = 1$. For the induction step we obtain with (7.58)

$$|W_{n+1}(t)| \leq |B_\varepsilon(\tau_{n+1})|N|W_n(t)| \leq \alpha_{n+1} N^{n+1},$$

and

$$
\begin{aligned}
|W_{n+1}(t) - W_{n+1}(s)| \quad &\leq \quad |B_\varepsilon(\tau_{n+1})||T_0(t-\tau) - T_0(s-\tau)||W_n(t)| \\
&+ \quad |B_\varepsilon(\tau_{n+1})||T_0(s-\tau)||W_n(t) - W_n(s)| \\
&\leq \quad |B_\varepsilon(\tau_{n+1})|\alpha_n \log(1 + \frac{t-s}{s-\tau})(N^{n+1} + N n N^n) \\
&= \quad \alpha_{n+1}(n+1)N^{n+1}\varphi_1(\frac{t-s}{s-\tau}). \qquad \square
\end{aligned}
$$

To derive the estimates for U_n and V_n we define functions $\varphi_n(a)$ inductively by

$$\varphi_{n+1}(a) = \int_0^1 \varphi_n(\frac{\sigma}{1-\sigma}) \log(1 + \frac{a}{\sigma})[1 + \log(1 + \frac{\sigma}{a})]\frac{d\sigma}{\sigma}. \qquad (7.63)$$

Some properties of φ_n are collected as

Claim 4 Given $n \in \mathbb{N}$, we have
(i) $0 \leq \varphi_n(a) \leq c(n, \beta)a^\beta$, $a \geq 0$, $\beta \in (0, 1)$;
(ii) $\varphi_n(a) \leq \varphi_{n+1}(a)$ for $a \geq 0$;
(iii) $\varphi_n(a)$ is increasing for $a \geq 0$;
(iv) $\varphi_n(a)/a$ is decreasing for $a \geq 0$.

Proof: By means of the elementary inequality

$$\frac{a}{1+a} \leq \log(1+a) \leq a^\beta/\beta, \qquad a > 0, \qquad (7.64)$$

we obtain for the induction step

$$0 \leq \varphi_{n+1}(a)$$

$$\leq \int_0^1 c(n,\beta+\delta)(\frac{\sigma}{1-\sigma})^{\beta+\delta}[(\frac{a}{\sigma})^\beta \beta^{-1} + (\frac{a}{\sigma})^{\beta+\delta}(\frac{\sigma}{a})^\delta(\beta+\delta)^{-1}\delta^{-1}]\frac{d\sigma}{\sigma}$$

$$= c(n,\beta+\delta) = (\beta^{-1} + (\beta+\delta)^{-1}\delta^{-1})\int_0^1 (1-\sigma)^{-(\beta+\delta)}\sigma^{\delta-1}d\sigma$$

$$\leq c(n+1,\beta)a^\beta,$$

where $\delta > 0$ is such that $\beta + \delta < 1$. Again using (7.64) we get

$$\varphi_2(a) = \int_0^1 \log(1+\frac{\sigma}{1-\sigma})\log(1+\frac{a}{\sigma})[1+\log(1+\frac{\sigma}{a})]\frac{d\sigma}{\sigma}$$

$$\geq \log(1+a)\int_0^1 \frac{\sigma/(1-\sigma)}{1+\sigma/(1-\sigma)}\cdot\frac{d\sigma}{\sigma}$$

$$= \log(1+a) = \varphi_1(a);$$

hence (ii) follows inductively. (iii) and (iv) are implied by $\log(1+x)(1+\log(1+1/x))$ increasing and $x^{-1}\log(1+x)[1+\log(1+1/x)]$ decreasing for $x > 0$. $\quad\square$

By means of the function $\varphi_n(a)$ we obtain the following estimates for U_n and V_n.

Claim 5 Let $C \geq 6N$. Then for each $n \in \mathbb{N}$ we have

$$|U_n(t,s)| \leq C^n \alpha_n \varphi_n(\frac{t-s}{s-\tau}), \quad t > s > \tau; \tag{7.65}$$

$$|V_n(t,s)| \leq C^n \alpha_n (t-s)^{-1}\varphi_{n-1}(\frac{t-s}{s-\tau}), \quad t > s > \tau; \tag{7.66}$$

$$|U_n(t,s) - U_n(\bar{t},s)| \leq C^n \alpha_n \cdot \frac{t-\bar{t}}{\bar{t}-s}\cdot\varphi_n(\frac{\bar{t}-s}{s-\tau})\varphi_0(\frac{t-\bar{t}}{\bar{t}-s}), \quad t > \bar{t} > s > \tau; \tag{7.67}$$

$$|V_n(t,s) - V_n(\bar{t},s)| \leq C^n \alpha_n (\bar{t}-s)^{-1}\varphi_1(\frac{t-\bar{t}}{\bar{t}-s})\varphi_{n-1}(\frac{\bar{t}-s}{s-\tau})\varphi_0(\frac{t-\bar{t}}{\bar{t}-s}), \tag{7.68}$$

for all $t > \bar{t} > s > \tau$, where $\varphi_0(a) = 1 + \log(1+1/a)$, $a > 0$.

Proof: (i) For $n = 1$, (7.66) and (7.68) are trivial; with (7.30) we get

$$|U_1(t,s)| \leq |B_\varepsilon(\tau_1)||T_0(s-\tau) - T_0(t-\tau)| \leq \alpha_1 N \log(1+\frac{t-s}{s-\tau}),$$

and

$$|U_1(t,s) - U_1(\bar{t},s)| \leq \alpha_1|T_0(t-\tau) - T_0(\bar{t}-\tau)| \leq \alpha_1 N \log(1+\frac{t-\bar{t}}{\bar{t}-\tau});$$

with (7.64)

$$\log(1 + \frac{t - \bar{t}}{\bar{t} - \tau}) \leq \frac{t - \bar{t}}{\bar{t} - \tau} = \frac{t - \bar{t}}{\bar{t} - s} \cdot \frac{(\bar{t} - s)/(s - \tau)}{1 + (\bar{t} - s)/(s - \tau)} \leq \frac{t - \bar{t}}{\bar{t} - s} \log(1 + \frac{\bar{t} - s}{s - \tau}),$$

hence the assertion holds for $n = 1$ since $\varphi_0 \geq 1$.

(ii) Let (7.65)$\sim$(7.68) be satisfied for n. Then (7.59) and (7.30) yield

$$
\begin{aligned}
|V_{n+1}(t, s)| &\leq |B_\varepsilon(\tau_{n+1})| \cdot \{N(t - s)^{-1}\alpha_n C^n \varphi_n(\frac{t - s}{s - \tau}) \\
&\quad + NC^n \alpha_n (t - s)^{-1} \varphi_{n-1}(\frac{t - s}{s - \tau})\} \\
&\leq \alpha_{n+1} 2NC^n(t - s)^{-1} \varphi_n(\frac{t - s}{s - \tau})
\end{aligned}
$$

since $\varphi_{n-1}(a) \leq \varphi_n(a)$ by Claim 4 (ii). Furthermore, with $\varphi_0(a) = 1 + \log(1 + 1/a)$ we get

$$
\begin{aligned}
|V_{n+1}(t, s) - V_{n+1}(\bar{t}, s)| &\leq |B_\varepsilon(\tau_{n+1})|\{|(\dot{T}_0(t - s) - \dot{T}_0(\bar{t} - s))U_n(\bar{t}, s)| \\
&\quad + |\dot{T}_0(t - s)| \cdot |U_n(t, s) - U_n(\bar{t}, s)| + |T_0(t - s) - T_0(\bar{t} - s)||V_n(\bar{t}, s)| \\
&\quad + |T_0(t - s)||V_n(t, s) - V_n(\bar{t}, s)|\} \\
&\leq \alpha_{n+1} C^n\{2N\frac{t - \bar{t}}{(t - s)(\bar{t} - s)}\varphi_n(\frac{\bar{t} - s}{s - \tau})\varphi_0(\frac{t - \bar{t}}{\bar{t} - s}) \\
&\quad + N(\bar{t} - s)^{-1}\varphi_1(\frac{t - \bar{t}}{\bar{t} - s})\varphi_{n-1}(\frac{\bar{t} - s}{s - \tau}) \cdot (1 + \varphi_0(\frac{t - \bar{t}}{\bar{t} - s})) \\
&\leq \alpha_{n+1} 4NC^n(\bar{t} - s)^{-1}\varphi_n(\frac{\bar{t} - s}{s - \tau})\varphi_0(\frac{t - \bar{t}}{\bar{t} - s})\varphi_1(\frac{t - \bar{t}}{\bar{t} - s}),
\end{aligned}
$$

hence (7.66) and (7.68) hold for $n + 1$; here we used (7.64) as

$$\frac{t - \bar{t}}{t - s} = \frac{t - \bar{t}}{t - \bar{t} + \bar{t} - s} = \frac{(t - \bar{t})/(\bar{t} - s)}{1 + (t - \bar{t})/(\bar{t} - s)} \leq \log(1 + \frac{t - \bar{t}}{\bar{t} - s}) = \varphi_1(\frac{t - \bar{t}}{\bar{t} - s}),$$

and Claim 4 (ii).

(iii) For $|U_{n+1}(t, s)|$ we obtain

$$
\begin{aligned}
|U_{n+1}(t, s)| &\leq |B_\varepsilon(\tau_{n+1})|\{|T_0(s - \tau) - T_0(t - \tau)||W_n(s)| \\
&\quad + |T_0(t - \tau)||W_n(t) - W_n(s)| \\
&\quad + \int_\tau^s (|\dot{T}_0(s - r) - \dot{T}_0(t - r)||U_n(s, r)| + |\dot{T}_0(t - r)||U_n(s, r) - U_n(t, r)|)dr \\
&\quad + \int_\tau^s (|T_0(s - r) - T_0(t - r)||V_n(s, r)| + |T_0(t - r)||V_n(s, r) - V_n(t, r)|)dr\} \\
&\leq \alpha_{n+1} C^n\{2N\varphi_1(\frac{t - s}{s - \tau}) + 2N\int_\tau^s \frac{t - s}{(s - r)(t - r)} \cdot \varphi_n(\frac{s - r}{r - \tau})\varphi_0(\frac{t - s}{s - r})dr \\
&\quad + 2N\int_\tau^s \varphi_1(\frac{t - s}{s - r})(s - r)^{-1}\varphi_{n-1}(\frac{s - r}{r - \tau})\varphi_0(\frac{t - s}{s - r})dr\}
\end{aligned}
$$

$$\leq \alpha_{n+1} C^n \{ 2N\varphi_1(\frac{t-s}{s-\tau})$$

$$+4N \int_\tau^s (s-r)^{-1}\varphi_1(\frac{t-s}{s-r})\varphi_n(\frac{s-r}{r-\tau})\varphi_0(\frac{t-s}{s-r})dr\},$$

since

$$\frac{t-s}{t-r} \leq \varphi_1(\frac{t-s}{s-r})$$

by (7.64). The transformation

$$r = s - \sigma(s - \tau)$$

now yields

$$\int_\tau^s (s-r)^{-1}\varphi_1(\frac{t-s}{s-r})\varphi_n(\frac{s-r}{r-\tau})\varphi_0(\frac{t-s}{s-r})dr$$

$$= \int_0^1 \varphi_n(\frac{\sigma}{1-\sigma}) \log(1 + \frac{t-s}{s-\tau}\sigma^{-1})[1 + \log(1 + \frac{s-\tau}{t-s}\sigma)]\frac{d\sigma}{\sigma}$$

$$= \varphi_{n+1}(\frac{t-s}{s-\tau}),$$

hence by Claim 4 (ii) we obtain (7.67) for $n+1$.

(iv) Similarly we get

$$|U_{n+1}(t,s) - U_{n+1}(\bar{t},s)|$$

$$\leq |B_\varepsilon(\tau_{n+1})|\{|T_0(t-\tau) - T_0(\bar{t}-\tau)||W_n(t)| + |T_0(\bar{t}-\tau)||W_n(t) - W_n(\bar{t})|$$

$$+ \int_\tau^s (|\dot{T}_0(t-r) - \dot{T}_0(\bar{t}-r)||U_n(\bar{t},r)| + |\dot{T}_0(t-r)||U_n(t,r) - U_n(\bar{t},r)|)dr$$

$$+ \int_\tau^s (|T_0(t-r) - T_0(\bar{t}-r)||V_n(\bar{t},r)| + |T_0(t-r)||V_n(t,r) - V_n(\bar{t},r)|)dr\}$$

$$\leq \alpha_{n+1} C^n \{2N\varphi_1(\frac{t-\bar{t}}{\bar{t}-\tau}) + 2N \int_\tau^s \frac{t-\bar{t}}{(t-r)(\bar{t}-r)}\varphi_n(\frac{\bar{t}-r}{r-\tau})\varphi_0(\frac{\bar{t}-\bar{t}}{\bar{t}-r})dr$$

$$+N \int_\tau^s \varphi_1(\frac{t-\bar{t}}{\bar{t}-r})(\bar{t}-r)^{-1}\varphi_{n-1}(\frac{\bar{t}-r}{r-\tau})[1 + \varphi_0(\frac{t-\bar{t}}{\bar{t}-r})]dr\}$$

$$\leq \alpha_{n+1} C^n \{2N\varphi_1(\frac{t-\bar{t}}{\bar{t}-\tau})$$

$$+4N \int_\tau^s \varphi_1(\frac{t-\bar{t}}{\bar{t}-r})(\bar{t}-r)^{-1}\varphi_n(\frac{\bar{t}-r}{r-\tau})\varphi_0(\frac{t-\bar{t}}{\bar{t}-r})dr\},$$

with $(t-\bar{t})/(t-r) \leq \varphi_1((t-\bar{t})/(\bar{t}-r))$ by (7.64). The first term is okay by the estimate

$$\varphi_1(\frac{t-\bar{t}}{\bar{t}-\tau}) \leq \frac{t-\bar{t}}{\bar{t}-\tau} = \frac{t-\bar{t}}{\bar{t}-s} \cdot \frac{\bar{t}-s}{1+(\bar{t}-s)/(s-\tau)} \leq \frac{t-\bar{t}}{\bar{t}-s}\varphi_1(\frac{\bar{t}-s}{s-\tau}).$$

The transformation $r = \tau + (s - \tau)\sigma$ yields

$$\int_\tau^s \varphi_1(\frac{t-\bar{t}}{\bar{t}-r})(\bar{t}-r)^{-1}\varphi_n(\frac{\bar{t}-r}{r-\tau})\varphi_0(\frac{t-\bar{t}}{\bar{t}-r})dr$$

$$= \int_0^1 \varphi_n(\frac{a+1-\sigma}{\sigma})\varphi_1(\frac{b}{a+1-\sigma})\varphi_0(\frac{b}{a+1-\sigma})\frac{d\sigma}{a+1-\sigma} = q,$$

where $a = (\bar{t}-s)/(s-\tau)$ and $b = (t-\bar{t})/(s-\tau)$. With $\varphi_0(\frac{b}{a+1-\sigma}) \leq \varphi_0(\frac{b}{a})\varphi_0(\frac{a}{1-\sigma})$, (7.64) and Claim 4 (iv) imply

$$q \leq \int_0^1 \frac{b}{a+1-\sigma}\varphi_n(\frac{a+1-\sigma}{\sigma})\varphi_0(\frac{b}{a+1-\sigma})\frac{d\sigma}{a+1-\sigma}$$

$$\leq \frac{b}{a}\int_0^1 \varphi_n(\frac{1-\sigma}{\sigma})\varphi_0(\frac{b}{a+1-\sigma})\frac{a}{a+1-\sigma} \cdot \frac{d\sigma}{1-\sigma}$$

$$\leq \frac{b}{a}\int_0^1 \varphi_n(\frac{1-\sigma}{\sigma})\varphi_1(\frac{a}{1-\sigma})\varphi_0(\frac{b}{a+1-\sigma})\frac{d\sigma}{1-\sigma}$$

$$\leq \frac{b}{a}\varphi_0(\frac{b}{a})\int_0^1 \varphi_n(\frac{1-\sigma}{\sigma})\varphi_1(\frac{a}{1-\sigma})\varphi_0(\frac{a}{1-\sigma})\frac{d\sigma}{1-\sigma}$$

$$= \frac{t-\bar{t}}{\bar{t}-s}\varphi_{n+1}(\frac{\bar{t}-s}{s-\tau})\varphi_0(\frac{t-\bar{t}}{\bar{t}-s}),$$

and from this (7.67) for $n+1$ follows. □

Returning to $R_n(t)$, Claim 5 yields the desired estimate.

Claim 6 For each $n \in \mathbb{N}$ we have with $\eta = \eta(\omega) = \int_0^{t_0} e^{-\omega\tau}db(\tau)$

$$|R_n(t)| \leq C^{n+1}\eta^n e^{\omega t}\int_0^1 \varphi_n(\frac{1-\sigma}{\sigma})\frac{d\sigma}{1-\sigma}. \qquad (7.69)$$

Proof: (7.61), (7.65) and (7.66) yield by means of (7.56)

$$|Z_n(t)| \leq NC^n\alpha_n\{2\int_\tau^t (t-s)^{-1}\varphi_n(\frac{t-s}{s-\tau})ds + 1\}$$

$$\leq C^n N\alpha_n\{1 + 2\int_0^1 \varphi_n(\frac{1-\sigma}{\sigma})\frac{d\sigma}{1-\sigma}\} \leq 3NC^n\alpha_n\int_0^1 \varphi_n(\frac{1-\sigma}{\sigma})\frac{d\sigma}{1-\sigma}$$

and therefore integration over $\tau_1,\ldots,\tau_n$ implies (7.69). □

Thus it remains to derive estimates for the integrals over φ_n in (7.69). For this purpose we consider the singular integral equation

$$\varphi(t) = \log(1+t) + \rho\int_0^1 \varphi(\frac{s}{1-s})\log(1+\frac{t}{s})[1 + \log(1+\frac{s}{t})]\frac{ds}{s}, \quad t > 0. \quad (7.70)$$

Suppose $\varphi \in C(\mathbb{R}_+, \mathbb{R}_+)$ is a solution of (7.70) such that $\varphi(t) \leq c\sqrt{t}$. Then it is easy to see by induction that $\varphi(t) \geq \sum_0^n \rho^j \varphi_j(t)$ for each $n \in \mathbb{N}$, hence

$$\sum_0^\infty \rho^n \int_0^1 \varphi_n\left(\frac{s}{1-s}\right)\frac{ds}{s} = \int_0^1 \sum_0^\infty \rho^n \varphi_n\left(\frac{s}{1-s}\right)\frac{ds}{s} \leq \int_0^1 \varphi\left(\frac{s}{1-s}\right)\frac{ds}{s}$$

$$\leq c\int_0^1 s^{-1/2}(1-s)^{-1/2}ds = c\pi < \infty,$$

i.e. with $\rho = C\eta$ we obtain by (7.69)

$$|T_\varepsilon(t)| \leq \sum_0^\infty |R_n(t)| \leq Ce^{\omega t}\sum_{n=0}^\infty \rho^n \int_0^1 \varphi_n\left(\frac{s}{1-s}\right)\frac{ds}{s} \leq Cc\pi e^{\omega t}, \qquad (7.71)$$

for all $t \geq 0$, $\varepsilon > 0$, provided η is sufficiently small, i.e. ω sufficiently large. This is the uniform estimate for $T_\varepsilon(t)$. Hence it remains to prove

Claim 7 Let $\rho > 0$ be sufficiently small. Then (7.70) has a solution $\varphi \in C(\mathbb{R}_+, \mathbb{R}_+)$ such that

$$\varphi(t) \leq c\sqrt{t} \quad \text{for } t \geq 0 \text{ and some } c > 0. \qquad (7.72)$$

Proof: Let Z denote the Banach space of all functions $\psi \in C(\mathbb{R}_+, \mathbb{R})$ such that (7.72) holds, with norm $||\psi|| = \sup_{t\geq 0} |\psi(t)|t^{-1/2}$ and $L : Z \to Z$ the linear operator

$$(L\psi)(t) = \int_0^1 \psi\left(\frac{s}{1-s}\right)\kappa\left(\frac{t}{s}\right)\frac{ds}{s}, \quad t \geq 0,$$

where $\kappa(a) = \log(1+a)[1 + \log(1+1/a)]$, $a > 0$. Since $\varphi_1(t) = \log(1+t)$ belongs to Z, we only have to show that $L : Z \to Z$ is bounded. For this purpose let $\psi \in Z$ be fixed. Then we have

$$(L\psi)(t) \leq \int_0^1 |\psi\left(\frac{s}{1-s}\right)|\kappa\left(\frac{t}{s}\right)\frac{ds}{s} \leq ||\psi|| \int_0^1 s^{-1/2}(1-s)^{-1/2}\kappa\left(\frac{t}{s}\right)ds;$$

with (7.64) one gets

$$\kappa(a) \leq 2\sqrt{a} + \frac{4}{3}a^{3/4} \cdot 4a^{-1/4} = \frac{22}{3}\sqrt{a}, \quad a > 0,$$

hence

$$\int_{1/2}^1 s^{-1/2}(1-s)^{-1/2}\kappa\left(\frac{t}{s}\right)ds \leq \frac{22}{3}\sqrt{t}\int_{1/2}^1 (1-s)^{-1/2}\frac{ds}{s} = \frac{44}{3}\sqrt{2}\sqrt{t}, \quad t > 0.$$

On the other hand, the change of variables $a = t/s$ yields

$$\int_0^{1/2} s^{-1/2}(1-s)^{-1/2}\kappa\left(\frac{t}{s}\right)ds = \sqrt{t}\int_{2t}^\infty \kappa(a)a^{-3/2}(1-t/a)^{-1/2}da$$

$$\leq \sqrt{t}\sqrt{2}\int_0^\infty \kappa(a)a^{-3/2}da \leq c\sqrt{t}, \quad t > 0,$$

since the last integral is finite. Therefore, $L : Z \to Z$ is bounded. $\square$

Finally, we prove the continuity properties (ii) and (iii) asserted in Theorem 7.6. The arguments for $S(t)$ in $\mathcal{B}(X)$ and $\mathcal{B}(Y)$, resp. for $T(t)$ in $\mathcal{B}(X,Y)$ are again similar, and so we restrict attention to $T(t)$. Continuity of $T(t)$ in $\mathcal{B}(X,Y)$ follows from the Neumann Series (7.50) for $T_\varepsilon(t)$. Since we know already that this series is absolutely and uniformly convergent w.r.t. $t \geq 0$ and $\varepsilon > 0$, it suffices to show that each term $R_{n\varepsilon}(t)$ is $\mathcal{B}(X,Y)$-continuous, uniformly w.r.t. $\varepsilon > 0$. This can be achieved by means of the estimates derived above. In fact, observe first that the decomposition

$$Z_n(t) - Z_n(\bar{t}) = \int_{\bar{t}}^t [\dot{T}_0(t-s)U_n(t,s)ds + T_0(t-s)V_n(t,s)]ds$$

$$+ \int_\tau^{\bar{t}} [(\dot{T}_0(t-s) - \dot{T}_0(\bar{t}-s))U_n(\bar{t},s) + \dot{T}_0(t-s)(U_n(t,s) - U_n(\bar{t},s))]ds$$

$$+ \int_\tau^{\bar{t}} [T_0(t-s) - T_0(\bar{t}-s))V_n(\bar{t},s)] + T_0(t-s)(V_n(t,s) - V_n(\bar{t},s))]ds$$

$$+ (T_0(t-\tau) - T_0(\bar{t}-\tau))W_n(t) + T_0(\bar{t}-\tau)(W_n(t) - W_n(\bar{t}))$$

for $t > \bar{t} > \tau$ implies by Claim 4 and Claim 5

$$|Z_n(t) - Z_n(\bar{t})| \leq 2N\alpha_n C^n\{\int_{\bar{t}}^t (t-s)^{-1}\varphi_n(\frac{t-s}{s-\tau})ds + \varphi_1(\frac{t-\bar{t}}{\bar{t}-\tau})$$

$$+ \int_\tau^{\bar{t}} \frac{t-\bar{t}}{(t-s)(\bar{t}-s)}\varphi_n(\frac{\bar{t}-s}{s-\tau})\varphi_0(\frac{t-\bar{t}}{\bar{t}-s})ds$$

$$+ \int_\tau^{\bar{t}} (\bar{t}-s)^{-1}\varphi_1(\frac{t-\bar{t}}{\bar{t}-s})\varphi_{n-1}(\frac{\bar{t}-s}{s-\tau})\varphi_0(\frac{t-\bar{t}}{\bar{t}-s})ds\}$$

$$\leq 8N\alpha_n C^n\varphi_{n+1}(\frac{t-\bar{t}}{\bar{t}-\tau}).$$

Therefore, by integration we obtain

$$R_{n\varepsilon}(t) - R_{n\varepsilon}(\bar{t}) = \int_{D_t} Z_n(t)d\tau_1\ldots d\tau_n - \int_{D_{\bar{t}}} Z_n(\bar{t})d\tau_1\ldots d\tau_n$$

$$= \int_{D_t\backslash D_{\bar{t}}} Z_n(t)d\tau_1\ldots d\tau_n + \int_{D_{\bar{t}}} (Z_n(t) - Z_n(\bar{t}))d\tau_1\ldots d\tau_n,$$

where $D_t = \{(\tau_1,\ldots,\tau_n) : 0 \leq \tau_j \leq t - \sum_{j+1}^n \tau_j, j = 1,\ldots,n\}$. Since $Z_n(t)$ is bounded by $C^{n+1}\alpha_n \cdot C_n$ for some constant C_n by Claim 6, this yields

$$|R_{n\varepsilon}(t) - R_{n\varepsilon}(\bar{t})| = C_n C^{n+1}\{\int_{D_t\backslash D_{\bar{t}}} \alpha_n d\tau_1\ldots d\tau_n$$

$$+ \int_{D_{\bar{t}-h}} \alpha_n\varphi_{n+1}(\frac{t-\bar{t}}{\bar{t}-\tau})d\tau_1\ldots d\tau_n + \int_{D_{\bar{t}}\backslash D_{\bar{t}-h}} \alpha_n d\tau_1\ldots d\tau_n\}$$

$$\leq C_n C^{n+1}\{b_n(t) - b_n(\bar{t}) + \varphi_{n+1}(\frac{t-\bar{t}}{h})b(t_0)^n + (b_n(\bar{t}) - b_n(\bar{t}-h))\},$$

where $b_n(t) = (db)^{*n-1} * b(t)$. This estimate shows that $R_{n\varepsilon}(t)$ is left-continuous in $\mathcal{B}(X,Y)$ and right-continuous in $\mathcal{B}(X,Y)$ at each point of continuity of $b_n(t)$, uniformly w.r.t. $\varepsilon > 0$. Since $R_{n\varepsilon} \to R_n$ in $L^1_{loc}(\mathbb{R}_+, \mathcal{B}(X,Y))$ this implies that $R_n(t)$ has the same continuity properties and therefore $T(t)$ is left-continuous in $\mathcal{B}(X,Y)$ and right-continuous in $\mathcal{B}(X,Y)$ at each point of continuity of the resolvent kernel r of b, i.e. of the solution of $r = b + db * r$. This completes the proof of Theorem 7.6. □

7.7 Comments

a) Analytic resolvents for equations of nonscalar type have been studied for the first time by Grimmer and Pritchard [144]. Their results have been improved for first order equations with main part, i.e.

$$\dot{u}(t) = Au(t) + \int_0^t dB(t-s)u(s) + f(t), \quad t \in J, \qquad (7.73)$$

where A generates an analytic semigroup in X and $B \in BV_{loc}(\mathbb{R}_+; \mathcal{B}(X_A, X))$, by Da Prato and Iannelli [68], Lunardi [224], Prüss [271], [273], and Tanabe [322].

b) The main result of Section 7.2 is new. It is an extension of the approach presented in Section 3 for equations of salar type.

c) Theorem 7.3 for equations of variational type seems also to be new. It appears to be possible to replace the assumption of uniform measurability by mere weak measurability, however, the proof then cannot be based on Theorems 7.1 and 7.2, it must be carried through directly. For the variational approach to partial differential equations there is an enormous literature; let us mention only the monographs of Lions [215] and Showalter [303].

d) Theorems 7.4 and 7.6 are extensions of results of Prüss [271], [273] where the first order case with main part, i.e. (7.73), was considered. Friedman and Shinbrot [123] is probably the earliest paper which considers (7.73) even in the nonautonomous case, however only solvability for initial values $u_0 \in \mathcal{D}(A^\alpha)$ for some $\alpha > 0$ was obtained under additional regularity assumptions on $B(t)$. In Prüss [270], and Tanabe [320], [321] the nonautonomous resolvent has been constructed. Grimmer and Kappel [140] used perturbation series in the frequency domain to establish existence of the resolvent for (7.73) assuming a certain decay of $\hat{B}(\lambda)$. This decay requires additional regularity of $B(t)$, but implies an improved behaviour of the resolvent. Lunardi [225] contains maximal regularity results for nonautonomous parabolic first order problems.

8 Parabolic Problems in L^p-Spaces

The subject of this section is the L^p-theory for parabolic equations with main part. The first three subsections prepare the approach via sums of commuting linear operators; the two basic results, i.e. a vector-valued Fourier-multiplier theorem and the Dore-Venni theorem, are stated without proof. After a thorough study of Volterra operators in $L^p(\mathbb{R}; X)$, these results are then applied to prove maximal regularity for parabolic equations with main part in $L^p(J; X)$, where $J \subset \mathbb{R}_+$ is a closed interval, X belongs to the class $\mathcal{HT}$ introduced in Section 8.2, and $p \in (1, \infty)$.

8.1 Operators with Bounded Imaginary Powers

Let X be a complex Banach space, and A a closed linear operator in X with dense domain $\mathcal{D}(A)$, such that

(**SL1**) $\mathcal{N}(A) = \{0\}$, $\overline{\mathcal{R}(A)} = X$, $(-\infty, 0) \subset \rho(A)$;

(**SL2**) $|t(t + A)^{-1}| \leq M$ for all $t > 0$, and some $M < \infty$;

are satisfied. Operators satisfying (SL1) and (SL2) will be called *sectorial*, and $\mathcal{S}(X)$ denotes the class of sectorial operators in X. Observe that $A \in \mathcal{S}(X)$ implies $A^{-1} \in \mathcal{S}(X)$, and $cA \in \mathcal{S}(X)$, for each $c > 0$. Also, if $A \in \mathcal{S}(X)$ and $T \in \mathcal{B}(X, Z)$ is invertible then $TAT^{-1} \in \mathcal{S}(Z)$. If $A \in \mathcal{S}(X)$, then $t(t + A)^{-1} \to I$ strongly as $t \to \infty$, and $t(t + A)^{-1} \to 0$ strongly as $t \to 0+$. For any $A \in \mathcal{S}(X)$ one can define *complex powers* A^z, where $z \in \mathbb{C}$ is arbitrary; this will be done as follows.

Assume first that A and A^{-1} are both bounded. Choose a piecewise smooth simple closed path Γ, surrounding the spectrum $\sigma(A)$ of A counterclockwise, which does not intersect the negative real axis $(-\infty, 0]$. Then define A^z by means of the Dunford integral

$$A^z = \frac{1}{2\pi i} \int_\Gamma \lambda^z (\lambda - A)^{-1} d\lambda, \quad z \in \mathbb{C}, \tag{8.1}$$

where $\lambda^z = \exp(z \log \lambda) = \exp(z \log |\lambda| + iz \arg \lambda)$ denotes the principal branch of the multivalued function λ^z. Obviously, $A^z \in \mathcal{B}(X)$ for each $z \in \mathbb{C}$, the mapping $\mathbb{C} \ni z \mapsto A^z \in \mathcal{B}(X)$ is holomorphic, and $\{A^z : z \in \mathbb{C}\} \subset \mathcal{B}(X)$ has the group property

$$A^z A^w = A^{z+w}, \ A^0 = I, \quad \text{for all } z, w \in \mathbb{C},$$

since the Dunford integral induces an algebra homomorphism. For $0 > \text{Re } z > -1$ we may contract the contour Γ to the negative real axis to the result

$$A^z = -\frac{\sin \pi z}{\pi} \int_0^\infty t^z (t + A)^{-1} dt, \quad -1 < \text{Re } z < 0. \tag{8.2}$$

Splitting the integral in (8.2) at $t = 1$ and using the identities

$$t(t + A)^{-1} = I - (t + A)^{-1} A \quad \text{and} \quad \frac{1}{t}(t + A)^{-1} = \frac{1}{t}A^{-1} - (t + A)^{-1} A^{-1},$$

(8.2) becomes

$$A^z = \frac{\sin \pi z}{\pi}\{\frac{1}{z} - \frac{1}{1+z}A^{-1} + \int_0^1 t^{z+1}(t+A)^{-1}A^{-1}dt$$
$$+ \int_1^\infty t^{z-1}(t+A)^{-1}Adt\}. \tag{8.3}$$

This formula not only makes sense for $-1 < \mathrm{Re}\, z < 0$, but by analytic continuation is valid for all $|\mathrm{Re}\, z| < 1$.

Let $A \in \mathcal{S}(X)$ be arbitrary now. Then (8.1) can no longer be used to define A^z since the integral diverges at $\lambda = 0$ and also at $\lambda = \infty$, in general. However, we may use (8.3) to define $A^z x$, $x \in \mathcal{D}(A) \cap \mathcal{R}(A)$, $|\mathrm{Re}\, z| < 1$; the function $f(z) = A^z x$ is still holomorphic in this strip. The complex powers of A defined this way are densely defined and closable, since $\mathcal{D}(A) \cap \mathcal{R}(A)$ is dense in X, A^z commutes with $(t + A)^{-1}$, and is bounded from $\mathcal{D}(A) \cap \mathcal{R}(A)$ equipped with its natural norm $|x| + |Ax| + |A^{-1}x|$ to X. We denote the closure of A^z defined by (8.3) on $\mathcal{D}(A) \cap \mathcal{R}(A)$ again by A^z.

Next we show that A^z still has the group property. For this purpose we introduce the approximations

$$A_\varepsilon = (\varepsilon + A)(1 + \varepsilon A)^{-1}, \quad \varepsilon > 0,$$

of elements $A \in \mathcal{S}(X)$. Obviously, A_ε is bounded and invertible, and

$$\begin{aligned}(t + A_\varepsilon)^{-1} &= (t + (\varepsilon + A)(1 + \varepsilon A)^{-1})^{-1} \\ &= (1 + \varepsilon A)(t + \varepsilon + (1 + \varepsilon t)A)^{-1} \\ &= \frac{1}{1 + \varepsilon t}(1 + \varepsilon A)(\frac{t + \varepsilon}{1 + \varepsilon t} + A)^{-1}, \quad t, \varepsilon > 0,\end{aligned}$$

i.e. $\rho(A_\varepsilon) \supset (-\infty, 0]$, and as $\varepsilon \to 0$, $(t + A_\varepsilon)^{-1} \to (t + A)^{-1}$ in $\mathcal{B}(X)$ for each $t > 0$, $A_\varepsilon x \to Ax$ for each $x \in \mathcal{D}(A)$, $A_\varepsilon^{-1}x \to A^{-1}x$ for each $x \in \mathcal{R}(A)$. Since

$$|t(t + A_\varepsilon)^{-1}| \le \frac{Mt}{t + \varepsilon} + \frac{\varepsilon t(M + 1)}{1 + \varepsilon t} \le 2M + 1, \quad t, \varepsilon > 0,$$

we have $A_\varepsilon \in \mathcal{S}(X)$ for each $\varepsilon > 0$, and there is a constant M for (SL2) which is independent of ε. From (8.3) it is now evident that

$$A_\varepsilon^z x \to_{\varepsilon \to 0} A^z x, \quad |\mathrm{Re}\, z| < 1, \ x \in \mathcal{D}(A) \cap \mathcal{R}(A),$$

holds. The group property for A_ε^z then implies the group property for A^z.

Let us summarize what we have proved so far in

Proposition 8.1 *Suppose* $A \in \mathcal{S}(X)$, *let* A^z *be defined by (8.3) for* $|\mathrm{Re}\, z| < 1$. *Then*
(i) $A^z x$ *is holomorphic on* $|\mathrm{Re}\, z| < 1$, *for each* $x \in \mathcal{D}(A) \cap \mathcal{R}(A)$;
(ii) A^z *is closable for each* $|\mathrm{Re}\, z| < 1$;

(iii) $A^{z+w}x = A^z A^w x$ for $\operatorname{Re} z$, $\operatorname{Re} w$, $\operatorname{Re}(z+w) \in (-1,1)$, $x \in \mathcal{D}(A) \cap \mathcal{R}(A)$;

(iv) $A_\varepsilon = (\varepsilon + A)(1 + \varepsilon A)^{-1}$ is bounded, sectorial, and invertible, for each $\varepsilon > 0$.

(v) $A^z x = \lim_{\varepsilon \to 0} A_\varepsilon^z x$, $x \in \mathcal{D}(A) \cap \mathcal{R}(A)$, $|\operatorname{Re} z| < 1$.

Proposition 8.1 shows that the following definition makes sense.

Definition 8.1 *Suppose $A \in \mathcal{S}(X)$. Then A is said to admit **bounded imaginary powers** if $A^{is} \in \mathcal{B}(X)$ for each $s \in \mathbb{R}$, and there is a constant $C > 0$ such that $|A^{is}| \leq C$ for $|s| \leq 1$. The class of such operators will be denoted by $\mathcal{BIP}(X)$.*

Since by Proposition 8.1, A^{is} has the group property, it is clear that A admits bounded imaginary powers if and only if $\{A^{is} : s \in \mathbb{R}\}$ forms a strongly continuous group of bounded linear operators in X. The growth bound θ_A of this group, i.e.

$$\theta_A = \overline{\lim}_{|s| \to \infty} \frac{1}{|s|} \log |A^{is}| \tag{8.4}$$

will be called the *power angle* of A. This is motivated by the fact that $\sigma(A) \subset \overline{\Sigma(0, \theta_A)}$ holds; see Prüss and Sohr [283].

It is in general not easy to verify that a given $A \in \mathcal{S}(X)$ belongs to $\mathcal{BIP}(X)$, although quite a few classes of operators are known for which the answer is positive. In Section 8.4 it will be shown that Volterra operators in $L^p(J; X)$ admit bounded imaginary powers if the kernels are nice enough. For further results and remarks on the class $\mathcal{BIP}(X)$ see the comments given in Section 8.7.

For a first application consider the fractional power spaces

$$X_\alpha = X_{A^\alpha} = (\mathcal{D}(A^\alpha), |\cdot|_\alpha), \quad |x|_\alpha = |x| + |A^\alpha x|, \quad 0 < \alpha < 1,$$

where $A \in \mathcal{S}(X)$; the embeddings

$$X_A \hookrightarrow X_\beta \hookrightarrow X_\alpha \hookrightarrow X, \quad 1 > \beta > \alpha > 0,$$

are well known. If A belongs to $\mathcal{BIP}(X)$, a characterization of X_α in terms of *complex interpolation spaces* can be derived.

Theorem 8.1 *Suppose $A \in \mathcal{BIP}(X)$. Then*

$$X_\alpha \cong [X, X_A]_\alpha, \quad \alpha \in (0, 1), \tag{8.5}$$

the complex interpolation space between X and $X_A \hookrightarrow X$ of order α.

For a proof we refer to Triebel [328], pp. 103-104, or Yagi [342]. The importance of Theorem 8.1 is twofold. It shows on one hand that X_α is largely independent of A; for instance if $A, B \in \mathcal{BIP}(X)$ are such that $\mathcal{D}(A) = \mathcal{D}(B)$ then $\mathcal{D}(A^\alpha) = \mathcal{D}(B^\alpha)$ for all $\alpha \in (0, 1)$. On the other hand, (8.5) makes the tools of complex interpolation theory available for fractional power spaces and it becomes possible to compute

X_α in many cases; see also Section 8.4 below. For example, the reiteration theorem yields the relation

$$[X_\alpha, X_\beta]_\theta = X_{\alpha(1-\theta)+\theta\beta}, \quad \text{for all } 0 \le \alpha < \beta \le 1, \ \theta \in (0,1).$$

for complex interpolation of fractional power spaces of operators $A \in \mathcal{BIP}(X)$.

8.2 Vector-Valued Multiplier Theorems
One of the most important tools to prove that translation invariant operators on L^p-spaces admit bounded imaginary powers are Fourier multiplier theorems. Let us recall first the Marcienkiewicz multiplier theorem for $L^p(\mathbb{R})$, $1 < p < \infty$.

Marcienkiewicz Multiplier Theorem *Suppose $m \in BV_{loc}(\mathbb{R} \setminus \{0\})$ satisfies*

$$||m|| := |m|_\infty + \sup_{R>0} \frac{1}{2R} \int_{-R}^{R} |\rho||dm(\rho)| < \infty. \tag{8.6}$$

Then there is an operator $T_m \in \mathcal{B}(L^p(\mathbb{R}))$, $1 < p < \infty$, such that

$$(T_m f)^\sim = m(\rho)\tilde{f}(\rho), \quad \text{for } \rho \in \mathbb{R}, \ \tilde{f} \in C_0^\infty(\mathbb{R} \setminus \{0\}), \tag{8.7}$$

and

$$|T_m|_{\mathcal{B}(L^p(\mathbb{R}))} \le C_p||m||, \tag{8.8}$$

where C_p denotes a constant only depending on $p \in (1, \infty)$.

Note that the set of all $f \in L^p(\mathbb{R})$ such that $\tilde{f} \in C_0^\infty(\mathbb{R} \setminus \{0\})$ is dense in $L^p(\mathbb{R})$; therefore (8.7) uniquely determines T_m. For many applications it is enough to check the *Mikhlin multiplier condition*

$$m \in W_{loc}^{1,\infty}(\mathbb{R} \setminus \{0\}), \quad \text{ess sup}\{|m(\rho)| + |\rho m'(\rho)| : \rho \in \mathbb{R} \setminus \{0\}\} < \infty, \tag{8.9}$$

which obviously implies (8.6).

Let us illustrate the connection between multiplier theorems and imaginary powers of translation invariant operators on $L^p(\mathbb{R})$ by means of the following instructive example.

Example 8.1 Let $X = L^p(\mathbb{R})$, $1 \le p < \infty$, $A = d/dt$, $\mathcal{D}(A) = W^{1,p}(\mathbb{R})$; A clearly is the generator of the C_0-group of translations defined by $(T(\tau)f)(t) = f(t+\tau)$, $t, \tau \in \mathbb{R}$. It is not difficult to compute the Fourier multipliers, the *symbols* associated with A, $T(\tau)$, $(\lambda - A)^{-1}$, etc.

$$T(\tau) \longleftrightarrow e^{i\tau\rho}, \ \tau \in \mathbb{R}; \quad (\lambda - A)^{-1} \longleftrightarrow (\lambda - i\rho)^{-1}, \ \text{Re } \lambda \ne 0;$$

$$A \longleftrightarrow i\rho; \quad A_\varepsilon \longleftrightarrow a_\varepsilon(\rho) = \frac{\varepsilon + i\rho}{1 + i\varepsilon\rho}, \ \varepsilon > 0;$$

$$A_\varepsilon^{is} \longleftrightarrow [a_\varepsilon(\rho)]^{is}, \ \varepsilon > 0, \ s \in \mathbb{R}; \quad A^{is} \longleftrightarrow (i\rho)^{is}, \quad s \in \mathbb{R}.$$

Thus the symbol associated with A^{is} is given by $m_s(\rho) = (i\rho)^{is}$. Since $m_s'(\rho) = ism_s(\rho)/\rho$, $\rho \neq 0$, it is not difficult to check the Mikhlin condition (8.9).

$$
\begin{aligned}
|m_s(\rho)| + |\rho m_s'(\rho)| &= (1+|s|)|m_s(\rho)| = (1+|s|)\exp(-s\frac{\pi}{2}\,\text{sign}\,\rho) \\
&\leq (1+|s|)e^{\frac{\pi}{2}|s|}, \quad s \in \mathbb{R}.
\end{aligned}
$$

By the Marcienkiewicz multiplier theorem we therefore obtain $A \in \mathcal{BIP}(L^p(\mathbb{R}))$ and

$$
|A^{is}|_{\mathcal{B}(L^p(\mathbb{R}))} \leq C_p(1+|s|)e^{\frac{\pi}{2}|s|}, \quad s \in \mathbb{R}, \ p \in (1,\infty). \tag{8.10}
$$

On the other hand, $A \notin \mathcal{BIP}(L^1(\mathbb{R}))$, since it is well known that $m(\rho)$ is a multiplier for $L^1(\mathbb{R})$ if and only if $m(\rho) = \widetilde{dk}(\rho)$, $\rho \in \mathbb{R}$, for some function $k \in BV(\mathbb{R})$; in particular, $m(\rho)$ must be uniformly continuous on $\mathbb{R}$, which is not the case for $m_s(\rho)$.   □

Turning to the vector-valued case now, it is obvious that Example 8.1 carries over to $L^p(\mathbb{R};X)$, X a Banach space, whenever the Marcienkiewicz multiplier theorem holds for $L^p(\mathbb{R};X)$. One of the simplest multipliers which satisfy (8.6) or (8.9) is the symbol of the *Hilbert transform*

$$
h(\rho) = -i\,\text{sgn}\,\rho, \quad \rho \neq 0, \tag{8.11}
$$

and therefore the Banach space X must be subject to the property that the Hilbert transform defined by

$$
(Hf)(t) = \lim_{\substack{\varepsilon \to 0 \\ R \to \infty}} \frac{1}{\pi}\int_{\varepsilon \leq |s| \leq R} f(t-s)ds/s, \quad t \in \mathbb{R}, \tag{8.12}
$$

is bounded in $L^p(\mathbb{R};X)$ for some $p \in (1,\infty)$; the limit in (8.12) is to be understood in the L^p-sense. Note that (8.12) exists in $L^p(\mathbb{R};X)$ for any Banach space X, $p \in (1,\infty)$ if $\tilde{f} \in C_0^\infty(\mathbb{R}\setminus\{0\};X)$. The class of Banach spaces X such that the Hilbert transform H is bounded in $L^p(\mathbb{R};X)$ will be denoted by $\mathcal{HT}$.

Even more is true. Suppose $A = d/dt$ belongs to $\mathcal{BIP}(L^p(\mathbb{R};X))$ for some $p \in (1,\infty)$. Then $-A \in \mathcal{BIP}(L^p(\mathbb{R};X))$ as well, since $-A$ is similar to A with similarity $(Rf)(t) = f(-t)$, $t \in \mathbb{R}$, the reflection at 0, which induces an isomorphism on $L^p(\mathbb{R};X)$. Consider now the operator $K = (-i/\sinh\pi)[A^{-i}(-A)^i - \cosh\pi] \in \mathcal{B}(L^p(\mathbb{R};X))$; its symbol is given by

$$
\begin{aligned}
k(\rho) &= -(i/\sinh\pi)[(i\rho)^{-i}(-i\rho)^i - \cosh\pi] \\
&= -(i/\sinh\pi)[\exp(-i\log|\rho| + \frac{\pi}{2}\text{sgn}\rho)\cdot\exp(i\log|\rho| + \frac{\pi}{2}\text{sgn}\rho) - \cosh\pi] \\
&= -i\cdot(\exp(\pi\text{sgn}\rho) - \cosh\pi)/\sinh\pi = -i\text{sgn}\rho, \quad \rho \neq 0,
\end{aligned}
$$

i.e. $k(\rho) = h(\rho)$. Thus if $A = d/dt \in \mathcal{BIP}(L^p(\mathbb{R};X))$ then X must belong to the class $\mathcal{HT}$.

It is surprising that the Marcienkiewicz theorem is valid in the vector-valued case if and only if $A = d/dt \in \mathcal{BIP}(L^p(\mathbb{R};X))$ for some $p \in (1,\infty)$, and if and only if the Banach space X belongs to the class $\mathcal{HT}$.

Theorem 8.2 *Suppose X belongs to the class $\mathcal{HT}$, and that $m \in BV_{loc}(\mathbb{R} \setminus \{0\})$ satisfies (8.6). Then the operator T_m, uniquely defined by*

$$(T_m f)\~(\rho) = m(\rho)\tilde{f}(\rho), \quad \rho \in \mathbb{R}, \ \tilde{f} \in C_0^\infty(\mathbb{R} \setminus \{0\}; X), \tag{8.13}$$

is bounded on $L^p(\mathbb{R};X)$, $1 < p < \infty$, and

$$|T_m|_{\mathcal{B}(L^p(\mathbb{R};X))} \le C_p(X)\|m\|, \tag{8.14}$$

where $C_p(X)$ denotes a constant only depending on p and X.

For the proof we refer to Bourgain [32] and Zimmermann [349]].

The basic reference for the class $\mathcal{HT}$ is the survey article Burkholder [35], where also two other characterizations for the class $\mathcal{HT}$ are given, a probabilistic one, and a geometrical one. To describe the latter, recall that a Banach space X is termed ζ-convex, if there is a function $\zeta : X \times X \to \mathbb{R}$ which is convex in each of its variables and such that $\zeta(0,0) > 0$ and

$$\zeta(x,y) \le |x + y| \quad \text{for all } x, y \in X \text{ with } |x| = |y| = 1.$$

A Banach space X belongs to the class $\mathcal{HT}$ if and only if X is ζ-convex. See also Section 8.7 for comments on further properties of the class $\mathcal{HT}$.

8.3 Sums of Commuting Linear Operators

Let X be a Banach space, A, B closed linear operators in X, and let $A + B$ be defined by

$$(A + B)x = Ax + Bx, \quad x \in \mathcal{D}(A + B) = \mathcal{D}(A) \cap \mathcal{D}(B). \tag{8.15}$$

If $A + B$ is closed and $\rho(A + B) \ne \emptyset$, then for each $\lambda \in \rho(A + B)$, the solution x of

$$(A + B)x = \lambda x + y \tag{8.16}$$

belongs to $\mathcal{D}(A) \cap \mathcal{D}(B)$; consequently it has the regularity induced by A as well as that coming from B, it has *maximal regularity*. On the other hand, if $A + B$ is closable but not closed, and $\lambda \in \rho(\overline{A + B})$ then (8.16) only admits generalized solutions in the sense that there are sequences $(x_n) \subset \mathcal{D}(A) \cap \mathcal{D}(B)$, $x_n \to x$, and $y_n \to y$ such that

$$Ax_n + Bx_n = \lambda x_n + y_n, \quad n \in \mathbb{N}.$$

In general, nothing can be said on $A + B$, it need not even be closable, unless very restrictive assumptions on A and B are imposed.

Two closed linear operators A, B in X are said to *commute* if there are $\lambda_0 \in \rho(A)$, $\mu_0 \in \rho(B)$ such that

$$(\lambda_0 - A)^{-1}(\mu_0 - B)^{-1} = (\mu_0 - B)^{-1}(\lambda_0 - A)^{-1}. \tag{8.17}$$

The *spectral angle* $\phi_A \in [0, \pi]$ of $A \in \mathcal{S}(X)$ is defined by

$$\phi_A = \inf\{\phi : \rho(-A) \supset \Sigma(0, \pi - \phi), \ |\lambda(\lambda + A)^{-1}| \leq C_\phi \ \text{on} \ \Sigma(0, \pi - \phi)\}. \tag{8.18}$$

It is obvious that $\sigma(A) \subset \overline{\Sigma(0, \phi_A)}$ holds, and for $A \in \mathcal{BIP}(X)$ we have the relation $\phi_A \leq \theta_A$; see Prüss and Sohr [283].

We are now in position to state the following fundamental result which is due to Da Prato and Grisvard [64].

Theorem 8.3 *Suppose A, B are closed linear operators in X which commute and satisfy the* **parabolicity condition** *$\phi_A + \phi_B < \pi$. Then*
(i) $A + B$ is closable;
(ii) $L = \overline{A + B} \in \mathcal{S}(X)$ has spectral angle $\phi_L \leq \max\{\phi_A, \phi_B\}$;
(iii) $\mathcal{D}(L) \subset \mathcal{D}_A(1, \infty) \cap \mathcal{D}_B(1, \infty)$;
(iv) If A or B is invertible, then L is also invertible;
(v) $y \in \mathcal{D}_A(\alpha, q)$ [or $y \in \mathcal{D}_B(\alpha, q)$] and $\lambda \in \Sigma(0, \pi - \phi_L)$ then $x = (\lambda + L)^{-1}y \in \mathcal{D}(A) \cap \mathcal{B}(B)$ and $Ax, Bx \in \mathcal{D}_A(\alpha, q)$ [or $y \in \mathcal{D}_B(\alpha, q)$], for any $\alpha \in (0, 1)$ and $q \in [1, \infty]$.

Here $\mathcal{D}_A(\alpha, q) = (X, X_A)_{\alpha, q}$ denote the real interpolation spaces between X and X_A whenever $A \in \mathcal{S}(X)$, and

$$\mathcal{D}_A(1, \infty) = \{x \in X : \ \text{ess sup} \, |tA^2(t + A)^{-2}x| < \infty\};$$

observe $\mathcal{D}(A) \subset \mathcal{D}_A(1, \infty) \subset \mathcal{D}_A(\alpha, q)$, for all $\alpha \in (0, 1)$ and $q \in [1, \infty]$.

There are examples showing that even in a Hilbert space, $A + B$ need not be closed, in the situation of Theorem 8.3; cp. Baillon and Clément [16]. However, if the assumptions on A, B and X are strengthened, one can prove the following result which extends a remarkable recent theorem due to Dore and Venni [101], [102].

Theorem 8.4 *Suppose X belongs to the class $\mathcal{HT}$, and assume $A, B \in \mathcal{BIP}(X)$ commute and satisfy the* **strong parabolicity condition** *$\theta_A + \theta_B < \pi$. Then*
(i) $A + B$ is closed and sectorial;
(ii) $A + B \in \mathcal{BIP}(X)$ with $\theta_{A+B} \leq \max\{\theta_A, \theta_B\}$;
(iii) there is a constant $C > 0$ such that

$$|Ax| + |Bx| \leq C|Ax + Bx|, \quad x \in \mathcal{D}(A) \cap \mathcal{D}(B). \tag{8.19}$$

In particular, if A or B is invertible, then $A + B$ is invertible as well.

For a proof of this result we refer to Prüss and Sohr [283].

Combining Theorem 8.4 with the methods introduced by Da Prato and Grisvard [64] we obtain the following result.

Theorem 8.5 *Suppose* X *belongs to the class* $\mathcal{HT}$, *and assume*
(i) $\omega_A + A, \omega_B + B \in \mathcal{BIP}(X)$, *for some* $\omega_A, \omega_B \in \mathbb{R}$;
(ii) A *and* B *commute;*
(iii) $\theta_{A+\omega_A} + \theta_{B+\omega_B} < \pi$.
Then $A + B$ *is closed with domain* $\mathcal{D}(A + B) = \mathcal{D}(A) \cap \mathcal{D}(B)$ *and* $\sigma(A + B) \subset \sigma(A) + \sigma(B)$. *In particular, if* $\sigma(A) \cap \sigma(-B) = \emptyset$ *then* $A + B$ *is invertible.*

Proof: a) Theorem 8.4 applied to $\omega_A + A$ and $\omega_B + B$ shows that $\omega_A + \omega_B + A + B$ is closed, hence $A + B$ is closed.

Suppose $\lambda \notin \sigma(A) + \sigma(B)$; based on ideas of da Prato and Grisvard [64], we will derive an explicit formula for $(\lambda - A - B)^{-1}$ which then shows that $\lambda \notin \sigma(A+B)$. For this observe first that $\lambda \notin \sigma(A) + \sigma(B)$ is equivalent to $\sigma(A) \cap \sigma(\lambda - B) = \emptyset$. On the other hand, the assumptions on A and B imply $\sigma(A) \subset \overline{\Sigma(-\omega_A, \theta_{\omega_A+A})}$, $\sigma(\lambda - B) \subset \overline{-\Sigma(-\lambda - \omega_B, \theta_{\omega_B+B})}$ and we may choose $\theta_{\omega_B+B} < \theta < \pi - \theta_{\omega_A+A}$ such that

$$|(z + A)^{-1}| \leq \frac{M_1}{|z - \omega_A|} \quad \text{for all } z \in \Sigma(\omega_A, \theta), \tag{8.20}$$

as well as

$$|(z - (\lambda - B))^{-1}| \leq \frac{M_2}{|z - \omega_B - \lambda|} \quad \text{for all } z \in \Sigma(\omega_B + \lambda, \theta). \tag{8.21}$$

w.l.o.g. we may assume $\theta_{\omega_B+B} < \pi/2$ and then also $\theta < \pi/2$.

b) Consider the oriented rays $\Gamma_{-\infty} = -\omega_A - (\infty, R]e^{i\theta}$, $\Gamma_{\infty} = -\omega_A - [R, \infty]e^{-i\theta}$, where $R > \sqrt{2}|\lambda + \omega_A + \omega_B|$ and so large that $\Gamma_{-\infty} \cup \Gamma_{\infty} \subset \Sigma(\omega_B + \lambda, \theta)$. By virtue of (iii) and since $\sigma(A) \cap \sigma(\lambda - B) = \emptyset$ it is apparent that $r_0 = d(\sigma(A), \sigma(\lambda - B))/4 > 0$. Consider the open square Q_R with center at $-\omega_A$ generated by the corners $-\omega_a \pm Re^{\pm i\theta}$; w.l.o.g. $0 \in Q_R$. Choose a grid of mesh size less than r_0 on Q_R, and denote the open squares generated by this grid by Q_j, $1 \leq j \leq N$; w.l.o.g. we may assume $0 \notin \partial Q_j$ for all j. With $\mathcal{J} = \{j : \overline{Q_j} \cap \sigma(A) \neq \emptyset\}$ define $U_0 = \cup_{j \in \mathcal{J}} \overline{Q_j}$, and let $\Gamma_0' = \partial U_0$ with orientation induced by negative orientation of the boundaries of the squares Q_j; observe that $\Gamma_0' \cap \sigma(B) = \emptyset$. Let Γ_R^+ denote that part of ∂Q_R running from $-\omega_A - Re^{i\theta}$ to $-\omega_A Re^{-i\theta}$ which is contained in $\Sigma(\omega_B + \lambda, \theta)$ and $\Gamma_R^- = \partial Q_R \backslash \Gamma_R^+$ with positive orientation; c.p. Figure 8.1.

If $\Gamma_0' \cap \partial Q_R = \emptyset$ we let $\Gamma_0 = \Gamma_R^+$; otherwise Γ_0 will defined as follows. Starting from $-\omega_A - Re^{i\theta}$ move along Γ_R^+ until Γ_0' is reached; then follow Γ_0' until Γ_R^+ is met again. Continuing this way, after finitely many switches between Γ_R^+ and Γ_0' one arrives at $-\omega_A - Re^{i\theta}$; from the curve obtained this way remove all closed connected subarcs, and denote the remaining curve by Γ_1. By construction, $\Gamma_1 \cup \Gamma_R^-$ then is a closed positively oriented simple curve which does not intersect $\sigma(A)$. Therefore by the Jordan curve theorem, for each $\mu \in \sigma(A)$ we have either $n(\Gamma_1 \cup \Gamma_R^-, \mu) = 0$ or $n(\Gamma_1 \cup \Gamma_R^-, \mu) = 1$ according to whether μ is outside or inside of $\Gamma_1 \cup \Gamma_R^-$, here n denotes the winding number.

Let U_1 denote the union of all Q_j such that $\overline{Q_j} \cap \sigma(A) \cap \{\mu : n(\Gamma_1 \cup \Gamma_R^-, \mu) = 1\} \neq \emptyset$, and let $\Gamma_2 = \partial U_1$ with negative orientation; then we have $\Gamma_2 \cap \sigma(A) = \emptyset$ and $n(\Gamma_1 \cup \Gamma_R^- \cup \Gamma_2, \mu) = 0$ for all $\mu \in \sigma(A)$.

Similarly, for each $\mu \in \sigma(\lambda - B) \cap Q_R$ we have either $n(\Gamma_1 \cup \Gamma_R^- \cup \Gamma_2, \mu) = 0$ or $n(\Gamma_1 \cup \Gamma_R^- \cup \Gamma_2, \mu) = 1$. Denoting by U_2 the union of all Q_j such that $\overline{Q_j} \cap \sigma(\lambda - B) \cap \{\mu : n(\Gamma_1 \cup \Gamma_R^- \cup \Gamma_2, \mu) = 0\} \neq \emptyset$, $\Gamma_3 = \partial U_2$ with positive orientation, we obtain $n(\Gamma_1 \cup -\Gamma_R^+ \cup \Gamma_2 \cup \Gamma_3, \mu) = n(\Gamma_1 \cup \Gamma_R^- \cup \Gamma_2, \mu) + n(\Gamma_3; \mu) - 1 = 0$ for all $\mu \in \sigma(\lambda - B)$; Finally, we let $\Gamma = \Gamma_{-\infty} \cup \Gamma_1 \cup \Gamma_\infty \cup \Gamma_2 \cup \Gamma_3$, note that $n(\Gamma_3, \mu) = 0$ for all $\mu \in \sigma(A)$.

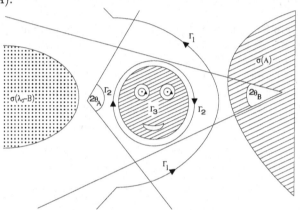

Figure 8.1:   Integration path for (8.22)

c) Define

$$S_\lambda = \frac{1}{2\pi i} \int_\Gamma (z - \lambda + B)^{-1}(z - A)^{-1} dz; \tag{8.22}$$

Since the unbounded parts of Γ, i.e. the rays $\Gamma_\infty, \Gamma_{-\infty}$, are contained in $\Sigma(\omega_B + \lambda, \theta) \cap -\Sigma(\omega_A, \theta)$, with (8.20) and (8.21), the integral in (8.22) is absolutely convergent, and so $S_\lambda \in \mathcal{B}(X)$; note that Γ does not intersect $\sigma(A) \cup \sigma(\lambda - B)$. We claim that $S_\lambda = (\lambda - (A + B))^{-1}$; by closedness of $A + B$ for this it is sufficient to prove

$$S_\lambda(\lambda - (A + B))x = x \quad \text{for each } x \in \mathcal{D}(A) \cap \mathcal{D}(B). \tag{8.23}$$

For $z \in \Gamma$ we have

$$
\begin{aligned}
& (z - \lambda + B)^{-1}(z - A)^{-1}(\lambda x - Ax - Bx) \\
={} & (\lambda - B)(z - \lambda + B)^{-1}(z - A)^{-1}x - (z - \lambda + B)^{-1}A(z - A)^{-1}x \\
={} & -(z - A)^{-1}x + (z - \lambda + B)^{-1}x \\
={} & [-(z - A)^{-1}Ax + (z - (\lambda - B))^{-1}(\lambda - B)x]/z,
\end{aligned}
$$

hence

$$S_\lambda(\lambda - (A+B))x = -\frac{1}{2\pi i}\int_\Gamma (z-A)^{-1}Ax\,dz/z + \frac{1}{2\pi i}\int_\Gamma (z - (\lambda - B))^{-1}(\lambda - B)x\,dz/z.$$

Contracting Γ into $\Gamma_A = \Gamma_1 \cup \Gamma_R^- \cup \Gamma_2 \cup \Gamma_3$ within the sector $-\Sigma(\omega_A, \theta)$ in the first integral of the right hand side of this identity, and Γ into $\Gamma_B = \Gamma_1 \cup (-\Gamma_R^+) \cup \Gamma_2 \cup \Gamma_3$ within $\Sigma(-\omega_A, \pi - \theta)$ in the second one, by Cauchy's theorem we obtain with $n(\Gamma_A, \sigma(A)) = 0$ and $n(\Gamma_B, \sigma(\lambda - B)) = 0$ by construction

$$
\begin{aligned}
S_\lambda(\lambda - (A+B))x &= -\frac{1}{2\pi i} \int_{\Gamma_A} (z - A)^{-1} Ax\, dz/z \\
&+ \frac{1}{2\pi i} \int_{\Gamma_B} (z - (\lambda - B))^{-1}(\lambda - B)x\, dz/z \\
&= [n(\Gamma_A, 0) - n(\Gamma_B, 0)]x = x,
\end{aligned}
$$

since $\Gamma_B = \Gamma_A - \partial Q_R$, i.e. $n(\Gamma_B, 0) = n(\Gamma_A, 0) - 1$; observe that $0 \notin \Gamma_A \cup \Gamma_B$. Thus we have derived (8.23) and the proof is complete. $\square$

Observe that the commutativity assumption in Theorems 8.4 and 8.5 is not as severe as it is in Theorem 8.3, since a posteriori it may be relaxed by introducing a perturbation $C : \mathcal{D}(A) \cap \mathcal{D}(B) \to X$ which is $A + B$-bounded with sufficiently small bound, i.e.

$$
|Cx| \le a|Ax| + b|Bx| + c|x|, \quad x \in \mathcal{D}(A) \cap \mathcal{D}(B),
$$

where a, b are sufficiently small. In the situation of Theorem 8.3 only perturbations of 'lower order' are possible, i.e. $\mathcal{D}(C) \supset \mathcal{D}_A(1, \infty) \cap \mathcal{D}_B(1, \infty)$. We refer also to the comments in Section 8.7 where more remarks are given.

8.4 Volterra Operators in L^p

Let X be a Banach space of class $\mathcal{HT}$, and consider Volterra operators in the spaces $L^p(\mathbb{R}; X)$, $1 < p < \infty$, of the form $b * u$, where $b \in L^1_{loc}(\mathbb{R}_+)$ is of subexponential growth. If b is θ-sectorial and 1-regular, we show that b gives rise to an operator $B \in \mathcal{BIP}(X)$; this will basically be a consequence of Theorem 8.2.

We shall need the following result containing further properties of 1-regular kernels.

Lemma 8.1 *Suppose $b \in L^1_{loc}(\mathbb{R}_+)$ is of subexponential growth and 1-regular. Then*
(i) $\hat{b}(i\rho) := \lim_{\lambda \to i\rho} \hat{b}(\lambda)$ exists for each $\rho \ne 0$;
(ii) $\hat{b}(\lambda) \ne 0$ for each $\operatorname{Re} \lambda \ge 0, \lambda \ne 0$;
(iii) $\hat{b}(i\cdot) \in W^{1,\infty}_{loc}(\mathbb{R} \setminus \{0\})$;
(iv) $|\rho \hat{b}'(i\rho)| \le c|\hat{b}(i\rho)|$ for a.a. $\rho \in \mathbb{R}$;
(v) there is a constant $c > 0$ such that

$$
c|\hat{b}(|\lambda|)| \le |\hat{b}(\lambda)| \le c^{-1}|\hat{b}(|\lambda|)|, \quad \operatorname{Re} \lambda \ge 0, \ \lambda \ne 0;
$$

(iv) $\lim_{r \to \infty} \hat{b}(re^{i\varphi}) = 0$ uniformly for $|\varphi| \le \frac{\pi}{2}$.

Proof: We already observed in Section 3.2 that 1-regular kernels do not admit zeros in $\mathbb{C}_+$. Therefore, by simple connectedness of $\mathbb{C}_+$, there is a holomorphic function $g(\lambda)$ such that $\exp(g(\lambda)) = \hat{b}(\lambda)$ on $\mathbb{C}_+$. 1-regularity of $\hat{b}(\lambda)$ then implies

$$|g'(\lambda)| = |\hat{b}'(\lambda)/\hat{b}(\lambda)| \leq c/|\lambda|, \quad \text{Re } \lambda > 0,$$

i.e. $g'(\lambda)$ is bounded on $\mathbb{C}_+ \setminus \mathcal{B}_\eta(0)$, $\eta > 0$, arbitrary; hence

$$g(i\rho) = \lim_{\lambda \to i\rho} g(\lambda), \quad \rho \neq 0,$$

exists, even uniformly for $|\rho| \geq \eta$. Therefore,

$$\hat{b}(i\rho) = \lim_{\lambda \to i\rho} \hat{b}(\lambda) = \exp(g(i\rho)), \quad \rho \neq 0,$$

also exists, and the convergence is uniform for $0 < \eta \leq |\rho| \leq \eta^{-1}$; this implies (i) and (ii), as well as the boundedness of $\hat{b}(\lambda)$ on compact subsets of $\overline{\mathbb{C}_+} \setminus \{0\}$. The estimate (3.7) of 1-regularity of b then shows that also $\hat{b}'(\lambda)$ is bounded on compact subsets, and so the nontangential limits $\varphi(\rho)$ of $\hat{b}'(\lambda)$ exist a.e. on $\mathbb{R}$, cp. Duren [106], Theorem 1.3. From the identity

$$\hat{b}(\sigma + i\rho) - \hat{b}(\sigma + i\tau) = i \int_\tau^\rho \hat{b}'(\sigma + is)ds, \quad \sigma > 0, \ \rho \cdot \tau > 0$$

we then may conclude

$$\hat{b}(i\rho) - \hat{b}(i\tau) = i \int_\tau^\rho \varphi(s)ds, \quad \rho \cdot \tau > 0,$$

i.e. $\hat{b}(i\cdot) \in W^{1,\infty}_{loc}(\mathbb{R} \setminus \{0\})$ and $\varphi(s) = \hat{b}'(i\rho)$ a.e.; we have proved (iii). (iv) follows from 1-regularity of b. Finally, to prove (v) and (vi) we write

$$g(\lambda) = g(|\lambda|) + i \int_0^{\arg \lambda} g'(|\lambda|e^{i\varphi})|\lambda|e^{i\varphi}d\varphi, \quad \text{Re } \lambda \geq 0, \ \lambda \neq 0,$$

hence by 1-regularity of $b(t)$

$$-c\pi/2 + \text{Re } g(|\lambda|) \leq \text{Re } g(\lambda) \leq \text{Re } g(|\lambda|) + c\pi/2$$

and therefore with $|\hat{b}(\lambda)| = \exp(\text{Re } g(\lambda))$, (v) follows. (vi) is a direct consequence of (v) since $\hat{b}(\lambda) \to 0$ as $\lambda \to \infty$. $\quad \square$

We are now ready for the main result of this subsection. In the sequel, $L^p(\mathbb{R}_+; X)$ is identified with a subspace of $L^p(\mathbb{R}; X)$, extending functions from $L^p(\mathbb{R}_+; X)$ by 0 to $(-\infty, 0)$.

Theorem 8.6 *Suppose X belongs to the class $\mathcal{HT}$, $p \in (1, \infty)$, and let $b \in L^1_{loc}(\mathbb{R}_+)$ be of subexponential growth; assume that b is 1-regular and θ-sectorial, for some $\theta < \pi$. Then there is a unique operator $B \in \mathcal{S}(L^p(\mathbb{R}; X))$ such that*

$$(Bf)^{\sim}(\rho) = \frac{1}{\hat{b}(i\rho)} \tilde{f}(\rho), \quad \rho \in \mathbb{R}, \ \tilde{f} \in C_0^\infty(\mathbb{R} \setminus \{0\}; X). \tag{8.24}$$

Moreover, B has the following properties:
(i) B commutes with the group of translations;
*(ii) $(\mu + B)^{-1} L^p(\mathbb{R}_+; X) \subset L^p(\mathbb{R}_+; X)$ for each $\mu > 0$, i.e. B is **causal**;*
(iii) $B \in \mathcal{BIP}(L^p(\mathbb{R}; X))$, and $\theta_B = \phi_B = \theta_b$, where

$$\theta_b = \sup\{|\arg \hat{b}(\lambda)| : Re\, \lambda > 0\}; \tag{8.25}$$

(iv) $\sigma(B) = \overline{\{1/\hat{b}(i\rho) : \rho \in \mathbb{R} \setminus \{0\}\}}$.

Proof: By Lemma 8.1, $\hat{b}(\lambda)$ admits continuous extension to $\overline{\mathbb{C}}_+ \setminus \{0\}$, and the inequality $|\arg \hat{b}(i\rho)| \leq \theta_b$ holds for all $\rho \neq 0$. Thus for each $\eta > 0$, there is a constant $c(\eta)$ such that

$$c(\eta)|1 + \mu \hat{b}(\lambda)| \geq 1 + |\mu \hat{b}(\lambda)|, \quad \lambda \in \overline{\mathbb{C}}_+ \setminus \{0\}, \ |\arg \mu| \leq \pi - \theta_b - \eta = \vartheta_\eta$$

holds. Define multipliers $r_\mu(\rho)$ by means of

$$r_\mu(\rho) = \frac{\hat{b}(i\rho)}{1 + \mu \hat{b}(i\rho)}, \quad \rho \in \mathbb{R}, \ \rho \neq 0, \ \mu \in \Sigma(0, \vartheta_\eta);$$

then

$$|r_\mu(\rho)| \leq c(\eta) \frac{|\hat{b}(i\rho)|}{1 + |\mu||\hat{b}(i\rho)|} \leq \frac{c(\eta)}{|\mu|}, \quad \rho \in \mathbb{R}, \ \rho \neq 0, \ \mu \in \Sigma(0, \vartheta_\eta),$$

and from

$$r'_\mu(\rho) = \frac{i\hat{b}'(i\rho)}{(1 + \mu \hat{b}(i\rho))^2}$$

using the 1-regularity of $b(t)$ we obtain by Lemma 8.1

$$|\rho r'_\mu(\rho)| \leq |\rho \frac{\hat{b}'(i\rho)}{\hat{b}(i\rho)}| \cdot \frac{c(\eta)^2 |\hat{b}(i\rho)|}{(1 + |\mu||\hat{b}(i\rho)|)^2} \leq \frac{c \cdot c(\eta)^2}{|\mu|}, \quad \rho \in \mathbb{R} \setminus \{0\}, \ \mu \in \Sigma(0, \vartheta_\eta).$$

Therefore, by Theorem 8.2, there are operators $R_\mu \in \mathcal{B}(L^p(\mathbb{R}; X))$, uniquely defined by

$$\widetilde{R_\mu f}(\rho) = r_\mu(\rho) \tilde{f}(\rho), \quad \rho \in \mathbb{R}, \ \tilde{f} \in C_0^\infty(\mathbb{R} \setminus \{0\}; X), \tag{8.26}$$

and they are subject to the estimate

$$|R_\mu| \leq \frac{C(\eta)}{|\mu|}, \quad \mu \in \Sigma(0, \vartheta_\eta). \tag{8.27}$$

The relation

$$r_\mu(\rho) - r_\nu(\rho) = (\nu - \mu)r_\mu(\rho)r_\nu(\rho), \quad \rho \in \mathbb{R}, \ \rho \neq 0, \ \mu, \nu \in \Sigma(0, \vartheta_\eta),$$

then implies the identity

$$R_\mu - R_\nu = (\nu - \mu)R_\mu R_\nu, \quad \mu, \nu \in \Sigma(0, \vartheta_\eta). \tag{8.28}$$

Thus the family $\{R_\mu\}_{\mu \in \Sigma(0, \vartheta_\eta)} \subset \mathcal{B}(L^p(\mathbb{R}; X))$ is a pseudo-resolvent in the sense of Hille and Phillips [180], p.185. According to Theorem 5.8.3 of [180], the kernels $\mathcal{N}(R_\mu)$ and the ranges $\mathcal{R}(R_\mu)$ are independent of μ, and there is a unique closed linear operator B in $L^p(\mathbb{R}; X)$ such that $R_\mu = (\mu + B)^{-1}$ iff the kernel $\mathcal{N}(R_\mu)$ of R_μ is trivial. The operator B (if it exists) is densely defined iff $\mathcal{R}(R_\mu)$ is dense in $L^p(\mathbb{R}; X)$; in fact, B is given by $B = R_\mu^{-1} - \mu I$, i.e. $\mathcal{D}(B) = \mathcal{R}(R_\mu)$.

Since $\hat{b}(i\rho) \neq 0$ for $\rho \neq 0$ there follows $\mu r_\mu(\rho) \to 1$ a.e. as $\mu \to \infty$ and $\mu r'_\mu(\rho) \to 0$ a.e.. Hence for $\tilde{f} \in C_0^\infty(\mathbb{R} \setminus \{0\}; X)$ we obtain $\mu(R_\mu f)^\sim \to \tilde{f}$ and $\mu(tR_\mu f)^\sim \to t\tilde{f}$ in $L^1(\mathbb{R}; X)$, and so $\mu R_\mu f \to f$ and $\mu t R_\mu f \to tf$ in $L^\infty(\mathbb{R}; X)$; this then implies $\mu R_\mu f \to f$ in $L^p(\mathbb{R}; X)$. Since the set of all $f \in L^p(\mathbb{R}; X)$ with $\tilde{f} \in C_0^\infty(\mathbb{R} \setminus \{0\}; X)$ is dense in $L^p(\mathbb{R}; X)$, by (8.27) we obtain $\mu R_\mu \to I$ strongly as $\mu \to \infty$, $\mu \in \Sigma(0, \vartheta_\eta)$. In particular, $\mathcal{N}(R_\mu) = \{0\}$ and there is a closed linear operator B in $L^p(\mathbb{R}; X)$ such that $R_\mu = (\mu + B)^{-1}$, $\mu \in \Sigma(0, \vartheta_\eta)$, thanks to Theorem 5.8.3 of [180]. (8.27) now shows that the spectral angle ϕ_B of B satisfies $\phi_B \leq \theta_b < \pi$ since $\eta > 0$ has been arbitrary. Since $\mu r_\mu(\rho) \to 0$ as $\mu \to 0$ for $\rho \in \mathbb{R} \setminus \{0\}$, a similar reasoning yields $\mu(\mu + B)^{-1} \to 0$ strongly as $\mu \to 0$, $\mu \in \Sigma(0, \vartheta_\eta)$ and so $B(\mu + B)^{-1} \to I$ strongly. In particular $\mathcal{R}(B)$ is dense in X, hence $\mathcal{N}(A) = 0$, and therefore B belongs to $\mathcal{S}(X)$.

To complete the construction of B, by the uniqueness of the Fourier transform and the density of functions f with $\tilde{f} \in C_0^\infty(\mathbb{R} \setminus \{0\}; X)$ in $L^p(\mathbb{R}; X)$, it is clear that (8.24) determines B uniquely. As $0 < \mu \to \infty$, we have $\mu(1 + \mu \hat{b}(i\rho))^{-1} \to \hat{b}(i\rho)^{-1}$ a.e., hence as above $\mu B(\mu + B)^{-1} f \to_{\mu \to \infty} Bf$ for each $\tilde{f} \in C_0^\infty(\mathbb{R} \setminus \{0\}; X)$, i.e. B satisfies (8.24).

To verify (i) it is sufficient to observe the relation $(\widetilde{T_\tau f})(\rho) = e^{i\rho\tau}\tilde{f}(\rho)$, $\rho, \tau \in \mathbb{R}$, valid for each $f \in L^1(\mathbb{R}; X)$.

Next we show (ii), i.e. the causality of B. For this purpose observe that there are functions $b_\mu \in L^1_{loc}(\mathbb{R}_+)$ such that

$$\hat{b}_\mu(\lambda) = \frac{\hat{b}(\lambda)}{1 + \mu \hat{b}(\lambda)}, \quad \operatorname{Re} \lambda > 0, \ \mu > 0.$$

For $f \in C_0^\infty(\mathbb{R}_+; X)$ we therefore obtain

$$\widehat{b_\mu * f}(\lambda) = \hat{b}_\mu(\lambda)\hat{f}(\lambda) = \frac{\hat{b}(\lambda)}{1 + \mu \hat{b}(\lambda)}\hat{f}(\lambda), \quad \operatorname{Re} \lambda > 0.$$

Passing to the horizontal limit $\lambda = \varepsilon + i\rho \to i\rho$ this yields

$$\widehat{b_\mu * f}(i\rho) = r_\mu(\rho)\hat{f}(i\rho), \quad \text{for a.a. } \rho \in \mathbb{R},$$

and so we have the representation $R_\mu f = b_\mu * f$ for all $f \in C_0^\infty(\mathbb{R}_+; X)$, by uniqueness of the Fourier transform. This implies causality.

In the next step we prove

$$|B^{is}| \leq K_p(1 + |s|)e^{|s|\theta_b}, \quad s \in \mathbb{R}, \tag{8.29}$$

i.e. $B \in \mathcal{BIP}(L^p(\mathbb{R}; X))$ and $\theta_B \leq \theta_b$. By Proposition 8.1, it is sufficient to prove (8.29) for $B_\varepsilon = (\varepsilon + B)(1 + \varepsilon B)^{-1}$, uniformly for $1 > \varepsilon > 0$. It is apparent from the preceding parts of this proof that the symbol of B_ε is given by

$$m_\varepsilon(\rho) = (\varepsilon + \hat{b}(i\rho)^{-1})/(1 + \varepsilon\hat{b}(i\rho)^{-1}) = \frac{\varepsilon + z}{1 + \varepsilon z} \circ \varphi(\rho), \quad \rho \neq 0,$$

where $\varphi(\rho) = 1/\hat{b}(i\rho)$. Since $\varphi(\rho)$ is θ_b-sectorial and the function $h_\varepsilon(z) = (\varepsilon+z)(1+\varepsilon z)^{-1}$ preserves any sector $\Sigma(0, \phi)$, $\phi < \pi$, we have $|\arg m_\varepsilon(\rho)| \leq \theta_b$, $\rho \in \mathbb{R} \setminus \{0\}$, $\varepsilon \in (0, 1)$. From this it is immediate that the multiplier $m_\varepsilon(\rho)^{is}$ which corresponds to B_ε^{is} satisfies

$$|m_\varepsilon(\rho)^{is}| \leq e^{|s||\arg m_\varepsilon(\rho)|} \leq e^{|s|\theta_b}, \quad \rho \in \mathbb{R}, \rho \neq 0,$$

and

$$\begin{aligned}
\rho\frac{d}{d\rho}(m_\varepsilon(\rho)^{is}) &= ism_\varepsilon(\rho)^{is} \cdot \frac{\varphi(\rho)h_\varepsilon'(\varphi(\rho))}{h_\varepsilon(\varphi(\rho))} \cdot \frac{\rho\varphi'(\rho)}{\varphi(\rho)} \\
&= ism_\varepsilon(\rho)^{is} \cdot \frac{(1 - \varepsilon^2)\varphi(\rho)}{(1 + \varepsilon\varphi(\rho))(\varepsilon + \varphi(\rho))} \cdot (-i\rho \cdot \frac{\hat{b}'(i\rho)}{\hat{b}(i\rho)})
\end{aligned}$$

then implies with 1-regularity of $b(t)$ by Lemma 8.1

$$|\rho\frac{d}{d\rho}(m_\varepsilon(\rho)^{is})| \leq C|s|e^{|s|\theta_b}, \quad \rho \in \mathbb{R}, \rho \neq 0, s \in \mathbb{R}.$$

Theorem 8.2 therefore yields (8.29) for B_ε, uniformly for $\varepsilon \in (0, 1)$, where K_p depends only on p, X, and b, but not on $s \in \mathbb{R}$.

To prove (iv) observe that the symbol of $(\mu - B)^{-1}$ is given by $r_{-\mu}(\rho)$ for $\mu \in \rho(B)$; therefore $\mu \in \rho(B)$ implies $|\mu - 1/\hat{b}(i\rho)| \geq c > 0$ for each $\rho \in \mathbb{R} \setminus \{0\}$, i.e. $\sigma(B) \supset \{1/\hat{b}(i\rho) : \rho \in \mathbb{R} \setminus \{0\}\}$. Conversely, if $\mu \notin \{1/\hat{b}(i\rho) : \rho \in \mathbb{R} \setminus \{0\}\}$ then $|\mu - 1/\hat{b}(i\rho)| \geq c > 0$ for $\rho \in \mathbb{R} \setminus \{0\}$. Thus $|r_{-\mu}(\rho)| \leq 1/c$ and also $|\rho r'_{-\mu}(\rho)| \leq C$ by 1-regularity of $b(t)$, and so Theorem 8.2 yields $\mu \in \rho(B)$. Finally, we have from (iv) $\theta_b \leq \phi_B \leq \theta_B \leq \theta_b$, hence $\theta_b = \phi_B = \theta_B$. □

Observe that for each component $\mathcal{C}$ of $\rho(B)$ we either have $(\mu - B)^{-1}L^p(\mathbb{R}_+; X) \subset L^p(\mathbb{R}_+; X)$ for all $\mu \in \mathcal{C}$ or for none; this follows from the expansion

$$(\mu - B)^{-1} = \sum_{n=0}^{\infty}(\mu_0 - B)^{-n}(\mu - \mu_0)^n$$

and the connectedness of $\mathcal{C}$. In particular, from the construction of B_ε^{is} in Section 8.1 and (ii) of Theorem 8.6 we obtain $B_\varepsilon^{is} L^p(\mathbb{R}_+; X) \subset L^p(\mathbb{R}_+; X)$, hence B^{is} has the same property since $B_\varepsilon^{is} \to B^{is}$ strongly as $\varepsilon \to 0$.

Of interest in applications is also the domain $\mathcal{D}(B)$ of B. We want to compare $\mathcal{D}(B)$ with the fractional power domains $\mathcal{D}(B_0^\alpha)$, where $B_0 = -(d^2/dt^2)$. It is well known that $B_0 \in \mathcal{S}(X)$, and

$$((\lambda + B_0)^{-1} f)(t) = \frac{1}{2\sqrt{\lambda}} \int_{-\infty}^{\infty} e^{-\sqrt{\lambda}(t-s)} f(s) ds, \quad t \in \mathbb{R}, \ f \in L^p(\mathbb{R}; X);$$

i.e. $\sigma(B_0) \subset [0, \infty)$ and $\phi_{B_0} = 0$. Since the symbol of B_0 is $m_0(\rho) = \rho^2$, by application of Theorem 8.2 it follows as in the proof of Theorem 8.5 that $B_0 \in \mathcal{BIP}(L^p(\mathbb{R}; X))$ and $\theta_{B_0} = 0$. Define then

$$H^{\alpha,p}(\mathbb{R}; X) = \mathcal{D}(B_0^{\alpha/2}), \quad \alpha \in \mathbb{R}_+. \tag{8.30}$$

It is then clear that $H^{2m,p}(\mathbb{R}; X) = W^{2m,p}(\mathbb{R}; X)$, $m \in \mathbb{N}$ and Theorem 8.1 yields

$$H^{2m\alpha,p}(\mathbb{R}; X) = [L^p(\mathbb{R}; X); W^{2m,p}(\mathbb{R}; X)]_\alpha, \quad \alpha \in (0, 1), \ m \in \mathbb{N}.$$

Since the symbol of $B_0^{\alpha/2}$ is given by $|\rho|^\alpha$, we also have the following characterization in terms of Fourier transforms. Let $f \in L^p(\mathbb{R}; X)$; then

$$f \in H^{\alpha,p}(\mathbb{R}; X) \Leftrightarrow \text{there exists } f_\alpha \in L^p(\mathbb{R}; X) \text{ such that } \widetilde{f_\alpha}(\rho) = |\rho|^\alpha \tilde{f}(\rho),$$

for $\rho \in \mathbb{R}$. Here the Fourier transform is meant in the distributional sense.

After these preparations we can now prove

Corollary 8.1 Let the assumptions of Theorem 8.6 hold, let B be defined by (8.24), and let $\alpha, \beta \geq 0$. Then
(i) $\overline{\lim}_{\mu \to \infty} |\hat{b}(\mu)| \mu^\alpha < \infty$ implies $\mathcal{D}(B) \hookrightarrow H^{\alpha,p}(\mathbb{R}; X)$;
(ii) $\underline{\lim}_{\mu \to \infty} |\hat{b}(\mu)| \mu^\beta > 0$ and $\underline{\lim}_{\mu \to 0} |\hat{b}(\mu)| > 0$ imply $H^{\beta,p}(\mathbb{R}; X) \hookrightarrow \mathcal{D}(B)$.

Proof: (i) The goal is to show that $m_\alpha(\rho) = |\rho|^\alpha \cdot \hat{b}(i\rho)(1 + \hat{b}(i\rho))^{-1}$ is a multiplier for $L^p(\mathbb{R}; X)$; this will follow from Theorem 8.2. For $|\rho| \leq 1$, we have $|m_\alpha(\rho)| \leq |r_1(\rho)| \leq C$, while for $|\rho| \geq 1$ by assumption $|m_\alpha(\rho)| \leq C$. The derivative of $m_\alpha(\rho)$ results to

$$\rho m_\alpha'(\rho) = (\rho \hat{b}'(i\rho)/\hat{b}(i\rho)) m_\alpha(\rho)(1 + \hat{b}(i\rho))^{-1} + \alpha m_\alpha(\rho),$$

from which the boundedness of $|m_\alpha(\rho)| + |\rho m_\alpha'(\rho)|$ follows by 1-regularity of b and Lemma 8.1(v), and so Theorem 8.2 applies.

(ii) Here we have to prove that $m_\beta(\rho) = [\hat{b}(i\rho)(1 + |\rho|^\beta)]^{-1}$ is a multiplier for $L^p(\mathbb{R}; X)$. Boundedness of $m_\beta(\rho)$ near $\rho = 0$ is equivalent to $|\hat{b}(i\rho)| \geq c > 0$ for $|\rho| \leq 1$, and near $\rho = \infty$ to $|\hat{b}(i\rho)||\rho|^\beta \geq c > 0$; this then follows by assumption. For the derivative of $m_\beta(\rho)$ we obtain

$$\rho m_\beta'(\rho) = -(i\rho \hat{b}'(i\rho)/\hat{b}(i\rho)) m_\beta(\rho) - \beta |\rho|^\beta (1 + |\rho|^\beta)^{-1} m_\beta(\rho), \quad \rho \neq 0,$$

and so boundedness of $\rho m'_\beta(\rho)$ follows from 1-regularity of b. Theorem 8.2 yields the assertion. $\square$

Observe that by a well known Abelian theorem (cf. Widder [339], Thm. V.I.I.),

$$\lim_{t \to 0} t^{-\alpha} \int_0^t b(s)ds = b_\alpha \cdot \Gamma(\alpha + 1), \quad \text{for some } \alpha \geq 0,$$

implies

$$\lim_{\mu \to \infty} \mu^\alpha \hat{b}(\mu) = b_\alpha,$$

hence we obtain $|\hat{b}(\mu)| \sim \mu^{-\alpha}$ as $\mu \to \infty$, in case $b_\alpha \neq 0$. This observation yields with Corollary 8.1

Corollary 8.2 *Let the assumptions of Theorem 8.5 hold, B be defined by (8.24), and assume in addition*
(i) $\underline{\lim}_{\mu \to 0+} |\hat{b}(\mu)| > 0$;
(ii) $\lim_{t \to 0} t^{-\alpha} \int_0^t b(s)ds \neq 0, \infty$ *exists, for some $\alpha \geq 0$.*
Then $\mathcal{D}(B) = H^{\alpha,p}(\mathbb{R}; X)$.

8.5 Maximal Regularity in L^p

Let X be ζ-convex and consider the Volterra equation with main part

$$u(t) + \int_0^t a(t-s)Au(s)ds = \int_0^t a(t-s)\left(\int_0^s dB(\tau)u(s-\tau)\right)ds + f(t), \quad t \in J, \quad (8.31)$$

where $a \in L^1_{loc}(\mathbb{R}_+)$, A closed linear densely defined, $B \in BV_{loc}(\mathbb{R}_+; \mathcal{B}(X_A, X))$ and $f \in L^p(J; X)$. By means of Theorems 8.4 and 8.6 we can now prove the following result on maximal regularity in $L^p(J; X)$, where $J = [0, T]$. Here $H^{\alpha,p}(J; X) = \{f_{|_J} : f \in H^{\alpha,p}(\mathbb{R}; X)\}$.

Theorem 8.7 *Let X belong to the class $\mathcal{HT}$, $p \in (1, \infty)$, and assume*
(i) $A_{\omega_A} = A + \omega_A \in \mathcal{BIP}(X)$, *for some $\omega_A \geq 0$;*
(ii) $a_{\omega_a} = e^{-\omega_a t}a(t)$ *is 1-regular and θ_a-sectorial, for some $\omega_a, \theta_a \geq 0$;*
(iii) $B \in BV_{loc}(\mathbb{R}_+; \mathcal{B}(X_A, X))$, $B(0) = B(0+) = 0$;
(iv) $\theta_a + \theta_{A+\omega_A} < \pi$;
(v) $\overline{\lim}_{\mu \to \infty} |\hat{a}(\mu)| \mu^\alpha < \infty$, *for some $\alpha \geq 0$.*
Then
(a) for each $f = a * g$ with $g \in L^p(J; X)$ there is a unique a.e. strong solution $u \in H^{\alpha,p}(J; X) \cap L^p(J; X_A)$ of (8.31), and there is a constant $C(T) > 0$ such that

$$|u|_{\alpha,p} + |Au|_p \leq C(T)|g|_p; \quad (8.32)$$

(b) for each $f \in L^p(J; X_A)$ there is a unique a.e. strong solution $u \in L^p(J; X_A)$ of (8.31) with $u - f \in H^{\alpha,p}(J; X)$ and

$$|u|_p + |u - f|_{\alpha,p} + |Au|_p \leq C(T)[|f|_p + |Af|_p]; \quad (8.33)$$

(c) for each $f \in L^p(J;X)$ there is a unique a.e. mild solution $u \in L^p(J;X)$ with $a * u \in L^p(J;X_A) \cap H^{\alpha,p}(J;X)$, and

$$|u|_p + |a * u|_{\alpha,p} + |Aa * u|_p \le C(T)|f|_p. \tag{8.34}$$

Here $|\cdot|_{\alpha,p}$ denotes the norm of $H^{\alpha,p}(J;X)$.

Proof: To prove (a), (8.31) will first be reformulated in the following way. Let $v(t) = e^{-\omega t}u(t)$, $g(t) = e^{-\omega t}f(t)$, $a_\omega(t) = e^{-\omega t}a(t)$, $K(t) = \int_0^t e^{-\omega s}dB(s)(\omega_0 + A)^{-1}$, where $\omega_0 > \omega_A$, is fixed and $\omega > \omega_a$ will be chosen later. Then (8.31) is equivalent to

$$v + a_\omega * Av = a_\omega * dK * (\omega_0 + A)v + a_\omega * g. \tag{8.35}$$

Define an operator $\mathcal{A}$ in $L^p(J;X)$ by means of

$$(\mathcal{A}v)(t) = (\omega_0 + A)v(t), \quad \text{for a.a. } t \in J,$$

with

$$\mathcal{D}(\mathcal{A}) = L^p(J;X_A);$$

then $\mathcal{A} \in \mathcal{BIP}(L^p(J;X))$ is invertible and $\theta_\mathcal{A} \le \theta_{A+\omega_A}$. Define $b \in L^1_{loc}(\mathbb{R}_+)$ as the solution of

$$b = \omega_0 a_\omega * b + a_\omega.$$

Since $|\hat{a}(\lambda)| \to 0$ with Re $\lambda \to \infty$, by the Paley-Wiener lemma $b \in L^1(\mathbb{R}_+)$ provided ω has been chosen large enough, and b is again 1-regular and sectorial, in fact, $\theta_b \le \theta_a + \theta(\omega)$ where $\theta(\omega) \to 0$ as $\omega \to \infty$. Define $\mathcal{B} \in \mathcal{BIP}(L^p(J;X))$ as the restriction of the operator B constructed in Theorem 8.6 to $L^p(J;X)$; this makes sense in virtue of causality. $\mathcal{B}$ is invertible and $\mathcal{B}^{-1}v = b * v$ for all $v \in L^p(J;X)$. Observe also that the constants $C(\eta)$ and K_p are uniform for $\omega \ge \omega_1$, where ω_1 is sufficiently large. Thus we can rewrite (8.35) in abstract form as

$$\mathcal{A}v + \mathcal{B}v = \mathcal{K}\mathcal{A}v + g \tag{8.36}$$

in $Z = L^p(J;X)$. Since $\mathcal{A}$ and $\mathcal{B}$ obviously commute and with ω_2 sufficiently large,

$$\theta_\mathcal{A} + \theta_\mathcal{B} \le \theta_{A+\omega_A} + \theta_a + \theta(\omega) < \pi, \quad \text{for } \omega \ge \omega_2,$$

holds, Theorem 8.4 applies; note that $L^p(J;X)$ belongs to $\mathcal{HT}$ since X has this property and $1 < p < \infty$. Therefore $\mathcal{A} + \mathcal{B}$ is invertible and $\mathcal{L} = (\mathcal{A} + \mathcal{B})^{-1} : Z \to Z_\mathcal{A} \cap Z_\mathcal{B}$ has norm independent of $\omega \ge \max\{\omega_1, \omega_2\}$. Thus (8.36) can be rewritten as

$$w = \mathcal{K}\mathcal{A}\mathcal{L}w + g \tag{8.37}$$

in Z, where $w = (\mathcal{A} + \mathcal{B})v$. From (iii) we obtain

$$|\mathcal{K}|_{\mathcal{B}(Z)} \le \int_0^T |dK(t)|_{\mathcal{B}(X)} = \int_0^T e^{-\omega t}|dB(t)|_{\mathcal{B}(X_A;X)}|(\omega_0 + A)^{-1}|_{\mathcal{B}(X;X_A)} \to 0$$

as $\omega \to \infty$, and since $|\mathcal{AL}|_{\mathcal{B}(Z)}$ is bounded as $\omega \to \infty$, there follows $|\mathcal{KAL}|_{\mathcal{B}(Z)} < 1$ for ω sufficiently large. Then (8.37) admits a unique solution $w \in Z$ for each $g \in Z$, hence (8.36) admits a unique solution $v \in Z_A \cap Z_B$ for each $g \in Z$, i.e. (8.31) admits a unique solution $u \in L^p(J; X_A)$ for each $f \in L^p(J; X)$. Moreover, by Corollary 8.1 we have $Z_B = \mathcal{D}(\mathcal{B}) \hookrightarrow H^{\alpha,p}(J; X)$, hence the solution u of (8.31) also belongs to $H^{\alpha,p}(J; X)$. (8.32) follows from the closed graph theorem. This proves (a).

To prove (b) we reformulate (8.31) with $w = u - f$ as

$$w + a * Aw = a * dB * w + a * (dB * f - Af),$$

and so (b) follows from (a).

To prove (c) we first solve (8.36) with $g = f$ and obtain $v \in \mathcal{D}(\mathcal{B}) \cap L^p(J; X_A)$. Then $u_\omega = \mathcal{B}v + \omega_0 v \in L^p(J; X)$ and $a_\omega * u_\omega = v$. This then shows that $u(t) = e^{\omega t} u_\omega(t)$ is an a.e. mild solution of (8.31) with the desired properties. Further, if u is an a.e. mild solution then $a * u$ is an a.e. strong solution, hence unique by (a), and so Titchmarsh's theorem implies uniqueness of a.e. mild solutions. $\quad\square$

8.6 Strong L^p-Stability on the Halfline

As another application of the approach via sums of linear operators from Section 8.3 we consider (8.31) on the halfline.

Definition 8.2 *Equation (8.31) is called* **strongly L^p-stable**, *if for each* $g \in L^p(\mathbb{R}_+, X)$ *there is a unique a.e. strong solution* $u \in L^p(\mathbb{R}_+, X_A))$ *of (8.31) on* $\mathbb{R}_+$ *with* $f = a * g$.

If (8.31) is strongly L^p-stable then by the closed graph theorem there is a constant $C > 0$ such that the estimate

$$|u|_{L^p(\mathbb{R}_+; X_A)} \leq C |f|_{L^p(\mathbb{R}_+; X_A)}, \quad f \in L^p(\mathbb{R}_+; X_A),$$

holds, i.e. the solution operator $f \mapsto u$ is well-defined and bounded in $L^p(\mathbb{R}_+; X_A)$.

Theorems 8.5 and 8.6 yield the following result which gives sufficient conditions in the parabolic case; necessary conditions will be derived in Section 10.1.

Theorem 8.8 *Let X belong to the class $\mathcal{HT}$, $p \in (1, \infty)$, A a closed linear operator in X with dense domain, $a \in L^1_{loc}(\mathbb{R}_+)$ of subexponential growth, and assume*

(i) $\omega_A + A \in \mathcal{BIP}(X)$, *for some* $\omega_A \geq 0$;
(ii) $a(t)$ *is 1-regular, and* $\lim_{\lambda \to 0+} |\hat{a}(\lambda)| > 0$;
(iii) $\theta_a := \overline{\lim}_{|\lambda| \to \infty} |\arg \hat{a}(\lambda)| < \pi - \theta_{A + \omega_A}$;
(iv) $\sigma(A) \cap \overline{\{-1/\hat{a}(\lambda) : \ \mathrm{Re}\, \lambda > 0\}} = \emptyset$.
Then there is a constant $\beta > 0$ such that equation (8.31) is strongly L^p-stable, for each $B \in BV(\mathbb{R}_+; \mathcal{B}(X_A, X))$ with $\mathrm{Var}\, B|_0^\infty < \beta$.

Proof: The result will be proved by means of Theorem 8.5. Define the operator $\mathcal{A}$ in $Z = L^p(\mathbb{R}_+; X)$ by

$$(\mathcal{A}u)(t) = Au(t), \quad \text{for a.a. } t > 0, \ u \in \mathcal{D}(\mathcal{A}) = L^p(\mathbb{R}; X_A);$$

then by assumption (i), we have $\omega_A + \mathcal{A} \in \mathcal{BIP}(Z)$ and $\theta_{\omega_A + \mathcal{A}} \leq \theta_{\omega_A + A}$. Define $\mathcal{B}_0$ on $L^p(\mathbb{R}; X)$ by

$$(\mathcal{B}_0 u)^{\sim}(\rho) = \frac{1}{\hat{a}(i\rho)} \tilde{f}(\rho), \quad \rho \in \mathbb{R}, \ \tilde{f} \in C_0^\infty(\mathbb{R} \setminus \{0\}; X),$$

and denote by $\mathcal{B}$ the restriction of $\mathcal{B}_0$ to Z. We show by Theorem 8.6 that $\omega + \mathcal{B}_0 \in \mathcal{BIP}(L^p(\mathbb{R}; X))$ for ω sufficiently large.

For this purpose we let $a_\omega \in L^1_{loc}(\mathbb{R}_+)$ be defined by $\widehat{a_\omega}(\lambda) = \hat{a}(\lambda)/(1 + \omega\hat{a}(\lambda))$, and $\theta_\omega = \sup\{|\arg(\lambda)| : \text{Re } \lambda > 0\}$. Since by (ii) and Lemma 8.1, for each $R > 0$ there is $\varepsilon = \varepsilon(R) > 0$ such that $|\hat{a}(\lambda)| \geq \varepsilon$ on $\overline{\mathbb{C}}_+ \cap \overline{B}_R(0)$, for any given $\eta > 0$ there is $\omega(\eta)$ such that

$$\theta_\omega \leq \theta_a + \eta \quad \text{for } \omega \geq \omega(\eta). \tag{8.38}$$

In fact, by definition of θ_a there is $R = R(\eta)$ such that

$$|\arg \widehat{a_\omega}(\lambda)| = |\arg(\hat{a}(\lambda) + \omega|\hat{a}(\lambda)|^2)| \leq |\arg \hat{a}(\lambda)| \leq \theta_a + \eta, \quad \lambda \in \overline{\mathbb{C}}_+, \ |\lambda| \geq R(\eta).$$

Choose $\omega(\eta) \geq (\varepsilon(R(\eta)) \sin \theta_a)^{-1}$, in particular $\omega\varepsilon > 1$; then for $\lambda \in \overline{\mathbb{C}}_+ \cap \overline{B}_R(0)$ we obtain

$$
\begin{aligned}
|\arg(\hat{a}(\lambda) + \omega|\hat{a}(\lambda)|^2)| &\leq |\arg(\hat{a}(\lambda) + \omega\varepsilon|\hat{a}(\lambda)|)| \\
&\leq \arcsin \frac{1}{\omega\varepsilon} \leq \theta_a, \quad \text{if } \omega \geq \omega(\eta),
\end{aligned}
$$

and so (8.38) follows. Thus a_ω is 1-regular and θ_ω-sectorial, hence Theorem 4.6 yields $\omega + \mathcal{B}_0 \in \mathcal{BIP}(Z)$ and $\theta_{\omega + \mathcal{B}_0} \leq \theta_\omega$. Clearly the restriction $\mathcal{B}$ of $\mathcal{B}_0$ to Z also satisfies $\omega + \mathcal{B} \in \mathcal{BIP}(Z)$ and $\theta_{\omega + \mathcal{B}} \leq \theta_\omega$ by causality, and moreover $\mathcal{A}$ and $\mathcal{B}$ commute. By (iii) the assumptions of Theorem 8.5 are satisfied, and so $\mathcal{A} + \mathcal{B}$ is closed with domain $\mathcal{D}(\mathcal{A}) \cap \mathcal{D}(\mathcal{B})$. To prove the invertibility of $\mathcal{A} + \mathcal{B}$ by means of Theorem 8.5, it remains to show $\sigma(\mathcal{A}) \cap \sigma(-\mathcal{B}) = \emptyset$. This, however, follows from assumption (iv), since $\sigma(\mathcal{A}) \subset \sigma(A)$ and

$$\sigma(\mathcal{B}) = \overline{\{1/\hat{a}(\lambda) : \text{Re } \lambda > 0\}},$$

which is implied by Proposition 8.2 below.

For $f = a * g$, $g \in L^p(\mathbb{R}; X)$, (8.31) is therefore equivalent to

$$\mathcal{A}u + \mathcal{B}u = \mathcal{K}u + g$$

where

$$(\mathcal{K}u)(t) = \int_0^t dB(\tau)u(t - \tau), \quad \text{for a.a. } t > 0, \ u \in \mathcal{D}(\mathcal{A}),$$

or to

$$w = \mathcal{K}(\mathcal{A} + \mathcal{B})^{-1}w + g, \quad w = (\mathcal{A} + \mathcal{B})u.$$

If $\text{Var } B|_0^\infty < \beta = |(\mathcal{A} + \mathcal{B})^{-1}|_{\mathcal{B}(Z, Z_A)}^{-1}$ then $|\mathcal{K}(\mathcal{A} + \mathcal{B})^{-1}| < 1$, hence $\mathcal{I} - \mathcal{K}(\mathcal{A} + \mathcal{B})^{-1}$ is invertible in Z, and the result follows. $\square$

Proposition 8.2 *Suppose X belongs to the class $\mathcal{HT}$, $p \in (1,\infty)$, and let $b \in L^1_{loc}(\mathbb{R}_+)$ be of subexponential growth, 1-regular and θ-sectorial, for some $\theta < \pi$. Let $B \in \mathcal{S}(L^p(\mathbb{R}; X))$ be defined by (8.24) in Theorem 8.6 and denote by $\mathcal{B}$ the restriction of B to the subspace $L^p(\mathbb{R}_+; X)$. Then*

$$\sigma(\mathcal{B}) = \overline{\{1/\hat{b}(\lambda) : Re\,\lambda > 0\}}. \tag{8.39}$$

Proof: $\supset$: Suppose $\mu \in \rho(\mathcal{B})$, choose any $x \in X$, $x \neq 0$, and $x^* \in X^*$ such that $< x, x^* > = 1$ and consider $f(t) = e^{-t}x$. Since $f \in L^p(\mathbb{R}_+; X)$, there is a unique solution $u \in L^p(\mathbb{R}_+; X)$ of $\mu u - \mathcal{B}u = f$; therfore $\hat{\varphi}(\lambda) = < \hat{u}(\lambda), x^* >$ is holomorphic on $\mathbb{C}_+$ and satisfies

$$\hat{\varphi}(\lambda)(\mu - 1/\hat{b}(\lambda)) = 1/(1 + \lambda), \quad Re\,\lambda > 0.$$

This implies $\mu \neq 1/\hat{b}(\lambda)$ for all $Re\,\lambda > 0$, i.e. $\{1/\hat{b}(\lambda) : Re\,\lambda > 0\} \subset \sigma(\mathcal{B})$, and so $\supset$ of (8.39) follows from the closedness of $\sigma(\mathcal{B})$.

$\subset$: Conversely, suppose $\text{dist}(\mu; \{1/\hat{b}(\lambda) : Re\,\lambda > 0\}) > 0$; then by the Paley-Wiener lemma, there is $b_\mu \in L^1_{loc}(\mathbb{R}_+)$ such that

$$\hat{b}_\mu(\lambda) = \frac{\hat{b}(\lambda)}{\mu\hat{b}(\lambda) - 1}, \quad Re\,\lambda > 0.$$

Therefore, $u = b_\mu * f$ is the unique solution of $u = \mu b * u - b * f$. On the other hand, $\mu \notin \sigma(B)$ by Theorem 8.6, hence for each $f \in C_0^\infty(\mathbb{R}_+; X)$ there is a unique solution $v \in L^p(\mathbb{R}; X)$ of $\mu v - Bv = f$. As in the proof of Theorem 8.6 then

$$\hat{u}(i\rho) = \widehat{b_\mu}(i\rho)\hat{f}(i\rho) = \tilde{v}(\rho), \quad \rho \in \mathbb{R} \setminus \{0\},$$

hence $u = v$ by uniqueness of the Fourier transform, and so $v \in L^p(\mathbb{R}_+; X)$. This implies $\mu \notin \sigma(\mathcal{B})$. $\square$

It is known that the Mikhlin multiplier theorem (i.e. (8.6) replaced by (8.9)) is valid also for an operator-valued multiplier acting between two Hilbert spaces; see e.g. Bergh and Löfstrom [22], Theorem 6.1.6. By means of this result, Theorem 8.8 can be extended considerably, in case X and Y are Hilbert spaces. Here we prove only one such result for nonscalar equations; cp. Section 7.2.

Theorem 8.9 *Let X, Y be Hilbert spaces with continuous and dense embedding $Y \overset{d}{\hookrightarrow} X$, and let $A \in L^1_{loc}(\mathbb{R}_+; \mathcal{B}(Y, X))$ be of subexponential growth. Assume (7.1) is weakly parabolic and $A(t)$ is weakly 1-regular, let $S(t)$ denote the weak resolvent of (7.1), and consider the solution map T defined by*

$$(Tf)(t) = \frac{d}{dt}\int_0^t S(t - \tau)f(\tau)d\tau, \quad t \geq 0, \; f \in W^{1,1}(\mathbb{R}_+; X). \tag{8.40}$$

Then $T \in \mathcal{B}(L^p(\mathbb{R}_+; X))$, $1 < p < \infty$. If in addition (7.1) is parabolic and $A(t)$ is 1-regular, then $T \in \mathcal{B}(L^p(\mathbb{R}_+; Y))$.

Proof: Let $M(\lambda) = (I - \hat{A}(\lambda))^{-1}$, Re $\lambda > 0$; then

$$(Tf)^\wedge(\lambda) = M(\lambda)\hat{f}(\lambda), \quad \text{Re } \lambda > 0, \ f \in W^{1,1}(\mathbb{R}_+; X).$$

By weak parabolicity of (7.1) and weak 1-regularity of $A(t)$, we obtain as in the proof of Theorem 7.2

$$|M(\lambda)|_{\mathcal{B}(X)} + |\lambda M'(\lambda)|_{\mathcal{B}(X)} \leq C, \quad \text{Re } \lambda > 0.$$

Hence the Mikhlin multiplier theorem for operator-valued multipliers on Hilbert spaces implies that there is a constant $C_p > 0$ such that

$$|e^{-\varepsilon t} Tf|_p \leq C_p |f e^{-\varepsilon t}|_p, \quad f \in W^{1,1}(\mathbb{R}_+; X), \ \varepsilon > 0,$$

with $\varepsilon \to 0$ this implies $T \in \mathcal{B}(L^p(\mathbb{R}_+; X))$. For the second assertion replace X by Y. □

8.7 Comments

a) The theory of complex powers of linear operators was developed independently by Balakrishnan, Kato, and Krasnosel'skii and Sobolevskii. For general expositions of this theory, we refer e.g. to the monographs of Krein [200], Tanabe [319], Triebel [328], and Yosida [347]. Komatsu [199] seems to be the most general treatment which is available presently; there neither $\mathcal{D}(A)$ nor $\mathcal{R}(A)$ need to be dense in X and $\mathcal{N}(A)$ is allowed to be nontrivial.

b) The class $\mathcal{BIP}(X)$ has been introduced in Prüss and Sohr [283]. So far no nontrivial characterization of this class seems to be known, even in the Hilbert space case. On the other hand, some permanence properties have been proved; we only mention the following:
(i) $A \in \mathcal{BIP}(X) \Leftrightarrow A^{-1} \in \mathcal{BIP}(X)$;
(ii) $A \in \mathcal{BIP}(X)$, $T \in \mathcal{B}(X, Z)$ invertible $\Rightarrow TAT^{-1} \in \mathcal{BIP}(Z)$;
(iii) $A \in \mathcal{BIP}(X)$, $0 < \alpha < \pi/\theta_A \Rightarrow A^\alpha \in \mathcal{BIP}(X)$.
(iv) $A \in \mathcal{BIP}(X)$, $c > 0 \Rightarrow cA \in \mathcal{BIP}(X)$;
(v) $A \in \mathcal{BIP}(X)$, $\varepsilon > 0 \Rightarrow \varepsilon + A \in \mathcal{BIP}(X)$.
Here (i)~(iv) are more or less simple consequences of the definition of A^z, and a proof of (v) can be found in Prüss and Sohr [283].

c) The paper of Baillon and Clément [16] contains examples of sectorial operators in a Hilbert space which do not belong to the class $\mathcal{BIP}(X)$. On the other hand, in the Hilbert space case large classes of operators are known which do belong to $\mathcal{BIP}(X)$; we mention the following.
(i) A normal and sectorial $\Rightarrow A \in \mathcal{BIP}(X)$, $\theta_A = \phi_A = \sup_{\lambda \in \sigma(A)} |\arg \lambda|$;
(ii) A generator of a bounded strongly continuous group
$\Rightarrow \pm A \in \mathcal{BIP}(X)$ and $\theta_A = \phi_A = \pi/2$;

(iii) A m-accretive $\Rightarrow A \in \mathcal{BIP}(X)$, $\phi_A \leq \theta_A \leq \pi/2$;

(iv) A regularly m-accretive with angle φ_A, i.e. A m-accretive and

$$(\tan \varphi_A)|Im(Ax,x)| \leq \text{Re}\,(Ax,x) \quad \text{for all } x \in D(A)$$

$\Rightarrow A \in \mathcal{BIP}(X)$, $\phi_A \leq \theta_A \leq \pi/2 - \varphi_A$.

Here (i) and (ii) are implied by the functional calculus for normal operators in Hilbert space -in case (ii), A is similar to a skew-adjoint operator-, while (iii) and (iv) are based on the functional calculus for contractions in Hilbert spaces, the H^∞-calculus; see Foias and Nagy [118].

d) Let $\Omega \subset \mathbb{C}$ be open. A closed linear densely defined operator A in X is said to have an $H^\infty(\Omega)$-*functional calculus* if $\sigma(A) \subset \overline{\Omega}$, and there is a continuous algebra homomorphism $f \mapsto f(A)$ from $H^\infty(\Omega)$ to $\mathcal{B}(X)$, such that $(\lambda - A)^{-1} = f_\lambda(A)$ for each $\lambda \notin \overline{\Omega}$, where $f_\lambda(z) = (\lambda - z)^{-1}$. Obviously, if A is sectorial and admits an $H^\infty(\Sigma(0,\theta))$-functional calculus, then $A \in \mathcal{BIP}(X)$ and $\phi_A \leq \theta_A \leq \theta$; however, the converse of this is in general not true. A admits a $H^\infty(\Sigma(0,\theta))$-functional calculus, $\theta < \pi/2$, if and only if $-A$ generates a bounded analytic semigroup in X and

$$\int_0^\infty |<Ae^{-Ate^{i\phi}}x,x^*>|\,dt < \infty, \quad \text{for all } x \in X, \; x^* \in X^*, \; |\phi| < \theta,$$

is satisfied. The latter was recently proved in Boyadzhiev and deLaubenfels [34]. For other results concerning H^∞-calculus see Boyadzhiev and deLaubenfels [33], Duong [104], [105], McIntosh [238], McIntosh and Yagi [239], and Yagi [343].

e) A Banach space X is said to have the *unconditional martingale difference property* (UMD) if for each $p \in (1,\infty)$ there is a constant C_p such that for any martingale $(f_n)_{n\geq 0} \subset L^p(\Omega,\Sigma,\mu;X)$ and any choice of signs $(\varepsilon_n)_{n\geq 0} \subset \{-1,1\}$ and any $N \in \mathbb{N}$ the following estimate holds.

$$|f_0 + \sum_{n=1}^N \varepsilon_n(f_n - f_{n-1})|_{L^p(\Omega,\Sigma,\mu;X)} \leq C_p|f_N|_{L^p(\Omega,\Sigma,\mu;X)}$$

for the notion of vector-valued martingales see Diestel and Uhl [99]. The class of UMD-spaces was introduced by Burkholder who also established the equivalence to ζ-convexity; Burkholder and McConnell proved the implication UMD$\Rightarrow \mathcal{HT}$, and its converse is due to Bourgain [31]. For these results and others see the survey article by Burkholder [35].

f) It is known that a ζ-convex Banach space is superreflexive, and therefore admits a uniformly convex, uniformly smooth equivalent norm. On the other hand, Bourgain [31] contains an example of a superreflexive Banach lattice which is not ζ-convex.

Every Hilbert space is ζ-convex, and if $(\Omega, \Sigma; \mu)$ is a measure space, $p \in (1, \infty)$ then $L^p(\Omega, \Sigma, \mu; X)$ is ζ-convex if X has this property. Closed subspaces, quotients, duals, finite products of ζ-convex spaces are again ζ-convex. If X, Y are ζ-convex then the complex interpolation spaces $[X, Y]_\alpha$ as well as the real interpolation spaces $(X, Y)_{\alpha,q}$ are again ζ-convex, provided $1 < q < \infty$; cp. Burkholder [35].

g) For a classical proof of the Marcienkiewicz multiplier theorem for $L^p(\mathbb{R})$ we refer to Stein [316]. Its vector-valued version, Theorem 8.2, is due to Bourgain [32]. McConnell [237] and Zimmermann [349] contain extensions of this result to the n-dimensional case, i.e. for $L^p(\mathbb{R}^n; X)$.

h) The Marcienkiewicz multiplier theorem can also be used as a starting point to prove that the Laplacian $-\triangle$ with Dirichlet or Neumann or Robin boundary conditions belongs to $\mathcal{BIP}(L^p(\Omega))$, where $\Omega \subset \mathbb{R}^n$ is open, $\partial\Omega$ sufficiently smooth, and $p \in (1, \infty)$. See Fujiwara [125], Seeley [299] for much more general elliptic differential operators with C^∞- coefficients, and Prüss and Sohr [284] for second order operators with C^α-coefficients. If positivity is present, i.e. if second order operators in divergence form are considered, the transference principle of Coifman and Weiss [56] can be used; see also Clément and Prüss [54]. This approach has been carried out by Duong [104].

The Stokes operator introduced in Section 5.6 is also known to have bounded imaginary powers in $L^p(\Omega; \mathbb{R}^n)$; see e.g. Giga [130] and Giga and Sohr [131].

i) In the situation of Theorem 8.3, Sobolevskii [310] calls the pair (A, B) a *coercive pair* if $A + tB$ with domain $\mathcal{D}(A) \cap \mathcal{D}(B)$ is closed for all $t > 0$ and there is a constant $M > 0$ such that

$$|Ax| + t|Bx| \leq M|Ax + tBx|, \quad \text{for all } x \in \mathcal{D}(A) \cap \mathcal{D}(B), \ t > 0.$$

He then obtains the so-called *mixed derivative theorem*, i.e. there is a constant $C > 0$ such that

$$|A^\beta B^{1-\beta} x| \leq C|(A + B)x|, \quad x \in \mathcal{D}(A) \cap \mathcal{D}(B)$$

for each $\beta \in (0, 1)$. This result is very useful in particular for nonlinear problems, since e.g. in Theorem 8.7 one obtains the additional regularity $u \in H^{\alpha\beta,p}(J; X_{1-\beta})$ for the solution of 8.31.

j) The results of Section 8.4 generalize and improve Theorem 1 in Prüss [278] where $a \in BV(\mathbb{R}_+)$ was assumed. Theorems 8.7 and 8.8 extend Theorem 6 of Prüss and Sohr [283]; see also Clément and Prüss [54], where applications to heat flow with memory are given. For applications of results like Theorems 8.7 and 8.8 to nonlinear problems see Clément and Prüss [55], Giga and Sohr [132], Orlov [265] and Prüss [279].

k) In Sections 8.5 and 8.6 one may apply of course the Da Prato Grisvard result, Theorem 8.3, instead of the Dore-Venni theorem, Theorem 8.4, but then without further restrictions on the inhomogeneity f no BV_0-perturbations can be included. For the case $B \equiv 0$ this approach has been worked out in the paper Clément and Da Prato [52] for the whole line case where the kernel is integrable and completely positive. For an interesting application of Theorem 8.3 to equations of the form $u + Aa * u + Bb * u = f$ see Pugliese [285].

9 Viscoelasticity and Electrodynamics with Memory

The discussion of problems in linear viscoelasticity leading to linear abstract Volterra equations which was begun in Section 5 is continued here. Models for viscoelastic beams and plates are introduced and their well-posedness is studied by means of the results on Volterra equations of scalar type from Chapter I but also by those on equations of nonscalar type from this chapter. In Sections 9.3 and 9.4 two approaches to general linear thermoviscoelasticity based on the results of Sections 6, 7, and 8 are carried through. Under physically reasonable assumptions these yield well-posedness in the variational setting, but also in the strong if the material in question is almost separable. In Sections 9.5 and 9.6 memory effects in isotropic linear electrodynamics are discussed and via the perturbation method well-posedness of the whole space problem as well as of a transmission problem are proved.

9.1 Viscoelastic Beams

Consider a beam $[0, l] \times \Omega$ of length l and uniform cross section $\Omega \subset \mathbb{R}^2$ made of a homogeneous isotropic viscoelastic material. W.l.o.g. we assume $\int_\Omega z \, dy \, dz = \int_\Omega y \, dy \, dz = 0$, i.e. $(0,0)$ is the center of Ω. Let the bending take place only in the (x, z)-plane; then the stress components σ_{12} and σ_{23} do not depend on y. For thin beams, i.e. diam $\Omega = \delta << l$, the normal stresses are in general negligible, hence we assume also $\sigma_{22} = \sigma_{33} = 0$. Concentrating on the components u_1 and u_3 of the displacement, σ_{11} and σ_{13} are the only relevant entries of the stress. As in Section 5.4, the stress-strain relations (5.14) lead to

$$\sigma_{11} = de * \dot{\varepsilon}_{11} \tag{9.1}$$

where de denotes the tensile modulus defined by (5.48), and to

$$\sigma_{13} = 2da * \dot{\varepsilon}_{13},$$

where da denotes the shear modulus. Following a device introduced by Timoshenko the latter will be replaced by

$$\sigma_{13} = 2\kappa da * \dot{\varepsilon}_{13}, \tag{9.2}$$

where $\kappa > 0$ is called the *shear correction coefficient*. The factor κ was introduced to account for the fact that, in general, the shearing strain is not uniform over a cross section of the beam.

Balance of momentum then yields the following equations for u_1 and u_3.

$$\rho_0 \ddot{u}_1 = de * \dot{u}_{1xx} + \kappa da * (\dot{u}_{1zz} + \dot{u}_{3xz}) + \rho_0 b_1, \tag{9.3}$$

$$\rho_0 \ddot{u}_3 = \kappa da * (\dot{u}_{1zx} + \dot{u}_{3xx}) + \rho_0 b_3; \tag{9.4}$$

observe that for reasons of consistency one should have $b_2 \equiv 0$. If the beam is *clamped* at one end, say at $x = 0$, *free* at the other end, the boundary conditions become

$$u_1(t, 0, z) = u_3(t, 0, z) = 0, \tag{9.5}$$

$$\sigma_{11}(t, l, z) = g_1(t, z), \quad \sigma_{13}(t, l, z) = g_3(t, z), \tag{9.6}$$

where g denotes the force acting on the end of the beam; for consistency we assume $g_2 \equiv 0$. With (9.1) and (9.2) boundary conditions (9.6) become

$$de * \dot{u}_{1x}(t, l, z) = g_1(t, z), \quad \kappa da * (\dot{u}_{1z} + \dot{u}_{3x})(t, l, z) = g_3(t, z). \tag{9.7}$$

At the lateral boundary $[0, l] \times \partial\Omega$ we obtain

$$\sigma_{13}(t, x, y, z)n_3(y, z) = h_1(t, x, y, z), \tag{9.8}$$

where h is the force acting on the surface of the beam, and $n = (n_2, n_3)$ denotes the outer normal of Ω. Again for consistency reasons we have to assume $h_2 \equiv h_3 \equiv 0$. With (9.2) this yields the boundary condition

$$n_3 \kappa da * (\dot{u}_{1z} + \dot{u}_{3x}) = h_1 \quad \text{on } \mathbb{R} \times [0, l] \times \partial\Omega. \tag{9.9}$$

To derive the Timoshenko beam model we introduce new dependent variables, namely

$$w(t, x) = \frac{1}{\alpha} \int_\Omega u_3(t, x, z) dy \, dz, \tag{9.10}$$

the *mean displacement* of the beam,

$$\phi(t, x) = \frac{1}{\beta} \int_\Omega z \, u_1(t, x, z) dy \, dz, \tag{9.11}$$

the *mean angle of rotation* of a cross section of the beam, and

$$\psi(t, x) = \frac{1}{\alpha} \int_\Omega u_{1z}(t, x, z) dy \, dz, \tag{9.12}$$

where

$$\alpha = \int_\Omega dy \, dz = \text{mes } (\Omega), \quad \beta = \int_\Omega z^2 dy \, dz.$$

Taking the mean value of (9.4) over Ω and the first moment w.r. to z of (9.3) we obtain with an integration by parts and by (9.8)

$$\begin{aligned} \rho_0 \ddot{w} &= \kappa da * (\ddot{w}_{xx} + \dot{\psi}_x) + f_s \\ \rho_0 \ddot{\phi} &= de * \phi_{xx} - (\alpha\kappa/\beta)da * (\dot{w}_x + \dot{\psi}) + f_b, \end{aligned} \tag{9.13}$$

where

$$f_s = \frac{\rho_0}{\alpha} \int_\Omega b_3 dy \, dz, \quad f_b = \frac{\rho_0}{\beta} \int_\Omega z b_1 dy \, dz + \frac{1}{\beta} \int_{\partial\Omega} z h_1 d\omega.$$

Boundary conditions (9.5) and (9.6) yield

$$w(t,0) = \phi(t,0) = 0, \quad t \in \mathbb{R} \tag{9.14}$$

and

$$de * \dot{\phi}_x(t,l) = g_b(t), \quad \kappa da * (\dot{w}_x + \dot{\phi})(t,l) = g_s(t), \tag{9.15}$$

where

$$g_b(t) = \frac{1}{\beta} \int_\Omega z g_1(t,z) dy\, dz \quad , \quad g_s(t) = \frac{1}{\alpha} \int_\Omega g_3(t,z) dy\, dz,$$

are the bending moment resp. shear force applied at the free end $x = l$ of the beam.

To obtain a complete set of equations, a relation between ϕ and ψ has to be added. If in addition $\int_\Omega z^3 dy\, dz = 0$, e.g. if Ω is symmetric w.r. to the (x,y)-plane, Taylor expansion of u_1 yields

$$\psi = \phi + O(\delta^2),$$

i.e. for thin beams the approximation $\psi \approx \phi$ is reasonable. The resulting model for thin beams is called viscoelastic *Timoshenko beam*, in honour of the pioneering work of Timoshenko in this field.

$$\begin{aligned}
\rho_0 \ddot{w} &= \kappa da * (\dot{w}_{xx} + \dot{\phi}_x) + f_s \\
\rho_0 \ddot{\phi} &= de * \dot{\phi}_{xx} - \kappa \gamma da * (\dot{w}_x + \dot{\phi}) + f_b,
\end{aligned} \tag{9.16}$$

where $\gamma = \alpha/\beta$, and the boundary conditions in the homogeneous case become

$$w = 0, \quad \phi = 0 \ \text{ at } x = 0$$
$$\phi_x = 0, \quad w_x + \phi = 0 \quad \text{ at } x = l. \tag{9.17}$$

Other boundary conditions are possible, we mention here only the *supported end*

$$w = 0, \quad \phi_x = 0$$

and the *guided end*

$$\phi = 0, \quad \phi + w_x = 0.$$

Of interest is also the limiting case $\kappa \to \infty$, which gives the so called (viscoelastic) *Rayleigh beam*. As $\kappa \to \infty$, (9.16) formally yields $\phi = -w_x$ and elimination of the terms containing κ leads to

$$\gamma \rho_0 \ddot{w} - \rho_0 \ddot{w}_{xx} + de * \dot{w}_{xxxx} = f \tag{9.18}$$

with $f = \gamma f_s + f_{bx}$. The boundary conditions in the homogeneous case then become for the clamped end

$$w = 0, \quad w_x = 0,$$

for the supported end

$$w = 0, \quad w_{xx} = 0,$$

for the guided end

$$w_x = 0, \ -\rho_0 \ddot{w}_x + de * \dot{w}_{xxx} = 0, \quad (\text{i.e. } w_{xxx} = 0),$$

and for the free end

$$w_{xx} = 0, \quad -\rho_0 \ddot{w}_x + de * \dot{w}_{xxx} = 0.$$

Since γ is of the order δ^{-2}, in many papers the term $\rho_0 \ddot{w}_{xx}$ in (9.18) is neglected; this leads to the viscoelastic *Euler-Bernoulli beam*

$$\gamma \rho_0 \ddot{w} + de * \dot{w}_{xxxx} = f. \tag{9.19}$$

In this situation the boundary conditions for the clamped, supported and guided end are the same as before, while for the free end they become

$$w_{xx} = 0, \quad w_{xxx} = 0.$$

To begin the discussion of the well-posedness of the three beam models described above, it is clear that the *Euler-Bernoulli beam* with any of the boundary conditions at either end considered here leads to Volterra equations of scalar type of the form (1.1) in $X = L^2[0, l]$ and with a selfadjoint operator A which is negative semidefinite. Thus results of Chapter I are applicable, and keeping in mind that the tensile modulus $de(t)$ can be expected to be at least of positive type, well-posedness follows e.g. from Corollary 1.2. Moreover, the discussion of the one-dimensional problems in Section 5.4 apply for the *Euler-Bernoulli beam* as well.

The situation is a little different for the *Rayleigh beam*. Excluding free ends for the moment, one may proceed as follows. (9.18) can be written as (w.l.o.g. $\rho_0 = 1$)

$$(\gamma + B)w = e * Aw + g \tag{9.20}$$

in $X = L^2[0, l]$, where $A = -(d/dx)^4$ with $\mathcal{D}(A) \subset W^{4,2}[0, l]$ including the involved boundary conditions, and $B = -(d/dx)^2$ with $\mathcal{D}(A) \subset \mathcal{D}(B) \subset W^{2,2}[0, l]$ such that B is positive semidefinite; A is negative semidefinite as before. With $v = (\gamma + B)^{1/2}w$, (9.20) can be rewritten as

$$v = e * A_B v + h, \tag{9.21}$$

where $A_B = (\gamma + B)^{-1/2}A(\gamma + B)^{-1/2}$ is again negative semidefinite. In this way (9.18) has been reduced to a scalar Volterra equation in $L^2[0, l]$; note that w then lives in its 'natural space' $W^{1,2}[0, l]$. This approach does not work if one end of the beam is free since then one of the corresponding boundary conditions contains the kernel de. In this case one may still use a variational approach as in Section 6.7, where $H = W^{1,2}[0, l]$, $V = W^{2,2}[0, l]$, modulo low order boundary conditions. The resulting resolvent will be subject to the energy inequality $|S(t)|_{\mathcal{B}(H)} \leq 1$.

Turning to the viscoelastic *Timoshenko beam*, we observe first that (9.16) is of the form (1.1) in $X = L^2[0, l] \times L^2[0, l]$, provided the material is *synchronous*; cp. Section 5.4 (iii). The operator A is defined by the matrix

$$A = \begin{pmatrix} \kappa \rho_0^{-1} \partial_x^2, & \kappa \rho_0^{-1} \partial_x \\ -\kappa \gamma \rho_0^{-1} \partial_x, & \beta_e \partial_x^2 - \kappa \gamma \rho_0^{-1} \end{pmatrix},$$

where $\beta_e = 9\beta/[\rho_0(3\beta+1)]$, β from Section 5.4 (iii), and the domain of A, $\mathcal{D}(A) \subset W^{2,2}[0, l] \times W^{2,2}[0, l]$ contains the boundary conditions in question. With any of the boundary conditions discussed above at either end, A is negative semidefinite, and so results of Chapter I are applicable.

In case the material is not synchronous then (9.16) will not be of scalar type. In this situation we may apply the results of Section 6, in particular Theorems 6.4. and 6.5. In fact, with

$$A(t) = \begin{pmatrix} \kappa \rho_0^{-1} a(t) \partial_x^2, & \kappa \rho_0^{-1} a(t) \partial_x \\ -\kappa \gamma \rho_0^{-1} a(t) \partial_x, & e(t) \rho_0^{-1} \partial_x^2 - \kappa \gamma \rho_0^{-1} a(t) \end{pmatrix}$$

in $X = L^2[0, l] \times L^2[0, l]$, $Y \equiv \mathcal{D}(A(t)) \subset W^{2,2}[0, l] \times W^{2,2}[0, l]$ incorporating the boundary conditions in question, (9.16) can be rewritten as (6.1). It is easy to check that $dA(t)$ is of positive type in X, $\widehat{dA}(\lambda)$ is negative semidefinite for each $\lambda > 0$, provided only da and de are of positive type. Therefore Theorem 6.4 yields a weak resolvent for (9.16) in X, which is even an a-regular resolvent by Corollary 6.6 if in addition there is $k \in BV_{loc}(\mathbb{R}_+)$ with $k(0+) > 0$ such that $b = dk * a$; such materials will be called *almost synchronous*. We may apply Corollary 6.7 instead if both ends are either guided or supported to obtain a resolvent for arbitrary isotropic and homogeneous materials. For example, if the end at $x = 0$ is supported and that at $x = l$ is guided we have

$$Y = \{(w, \phi) \in W^{2,2}([0, l]; \mathbb{R}^2) : w(0) = \phi_x(0) = \phi(l) = w_x(l) = 0\},$$

and it is easy to check that

$$Z = \{(w, \phi) \in W^{4,2}([0, l]; \mathbb{R}^2) \cap Y : w_{xx}(0) = \phi_{xxx}(0) = \phi_{xx}(l) = w_{xxx}(l) = 0\}$$

satisfies the assumptions of Corollary 6.7.

On the other hand, if a free or a clamped end are involved it is not clear whether such a space Z exists, and so in the not almost synchronous case no more regularity of the weak resolvent can be obtained by the results of Section 6.6. But still the variational approach of Section 6.7 can be applied, which gives some more regularity, at least. We leave the developement of this to the interested reader.

9.2 Viscoelastic Plates

Similarly to the derivation in Section 9.1, models for thin viscoelastic plates may be obtained. We shall not carry this through here, but simply state the resulting models, and refer to Lagnese and Lions [204] for more details.

Let the plate be represented by a bounded domain $\Omega \subset \mathbb{R}^2$ located in the (x, y)-plane. In the *Midlin-Timoshenko plate* model there are three dependent variables, the mean normal displacement w, and the mean angles of rotation ψ, ϕ of the cross sections $x =$const, $y =$ const. Assuming again that the material is homogeneous and isotropic with mass density ρ_0 and material functions $a(t)$ and $b(t)$, the equations are

$$\rho_0 \ddot{w} = \kappa da * [\dot{w}_{xx} + \dot{w}_{yy} + \dot{\psi}_x + \dot{\phi}_y] + f_w$$
$$\rho_0 \ddot{\psi} = da * [\dot{\psi}_{xx} + \dot{\psi}_{yy}] + dc * [\dot{\psi}_{xx} + \dot{\phi}_{xy}] - \gamma\kappa da * [w_x + \psi] + f_\psi \quad (9.22)$$
$$\rho_0 \ddot{\phi} = da * [\dot{\phi}_{xx} + \dot{\phi}_{yy}] + dc * [\dot{\phi}_{yy} + \dot{\psi}_{xy}] - \gamma\kappa da * [w_y + \phi] + f_\phi$$

for $t \in \mathbb{R}$, $(x, y) \in \Omega \subset \mathbb{R}^2$. Here κ denotes the shear correction coefficient, $\gamma = 6/\delta^2$, $\delta > 0$ the thickness of the plate, and the modulus $c(t)$ is defined by the relation

$$\widehat{dc}(\lambda) = \frac{9\widehat{da}(\lambda)\widehat{db}(\lambda)}{3\widehat{db}(\lambda) + 4\widehat{da}(\lambda)}, \quad \lambda > 0. \quad (9.23)$$

Observe that as in Section 5.4 (iii), $c(t)$ exists and is a Bernstein function if $a(t)$ and $b(t)$ are such; in the synchronous case $b(t) = \beta a(t)$ we have $c(t) = [9\beta/(3\beta+4)]a(t)$, i.e then (9.22) involves only one kernel, hence is of scalar type. The boundary conditions on the part $\Gamma_c \subset \partial\Omega$ where the plate is clamped are

$$w = \psi = \phi = 0, \quad t \in \mathbb{R}, \ (x, y) \in \Gamma_c, \quad (9.24)$$

on the free part $\Gamma_f \subset \partial\Omega$ we have

$$\kappa da * [\partial\dot{w}/\partial n + n_x\dot{\psi} + n_y\dot{\phi}] = g_w$$
$$da * [n_x(\dot{\psi}_x - \dot{\phi}_y) + n_y(\dot{\psi}_y + \dot{\phi}_x)] + dc * [n_x(\dot{\psi}_x + \dot{\phi}_y)] = g_\psi \quad (9.25)$$
$$da * [n_y(\dot{\phi}_y - \dot{\psi}_x) + n_x(\dot{\psi}_y + \dot{\phi}_x)] + dc * [n_y(\dot{\psi}_x + \dot{\phi}_y)] = g_\phi,$$

where $n = (n_x, n_y)$ denotes the outer normal of Ω at $(x, y) \in \partial\Omega$. Another boundary condition of interest on $\Gamma_s \subset \partial\Omega$, the simply supported part, is

$$w = 0$$
$$da * [n_x(\dot{\psi}_x - \dot{\phi}_y) + n_y(\dot{\psi}_y + \dot{\phi}_x)] + dc * [n_x(\dot{\psi}_x + \dot{\phi}_y)] = 0 \quad (9.26)$$
$$da * [n_y(\dot{\phi}_y - \dot{\psi}_x) + n_x(\dot{\psi}_y + \dot{\phi}_x)] + dc * [n_y(\dot{\psi}_x + \dot{\phi}_y)] = 0.$$

In the limiting case $\kappa \to \infty$ one obtains the viscoelastic *Kirchhoff plate* model, i.e. $\psi = -w_x$, $\phi = -w_y$, and

$$\gamma\rho_0\ddot{w} - \rho_0\Delta\ddot{w} + (da + dc) * \Delta^2\dot{w} = f, \quad t \in \mathbb{R}, \ (x, y) \in \Omega. \quad (9.27)$$

The boundary conditions become on the clamped part $\mathbb{R} \times \Gamma_c$

$$w = \partial w/\partial n = 0,$$

on the supported part $\mathbb{R} \times \Gamma_s$

$$w = 0$$
$$(da + dc) * \Delta \dot{w} - 2da * (n_y^2 \dot{w}_{xx} + n_x^2 \dot{w}_{yy} - 2n_x n_y \dot{w}_{xy}) = 0,$$

and on the free part $\mathbb{R} \times \Gamma_f$

$$(da + dc) * \Delta \dot{w} - 2da * (n_y^2 \dot{w}_{xx} + n_x^2 \dot{w}_{yy} - 2n_x n_y \dot{w}_{xy}) = g$$
$$-\rho_0 \partial \ddot{w}/\partial n + (da + dc) * \partial(\Delta \dot{w})/\partial n + dc * (\dot{w}_{xxy} - \dot{w}_{yyx})(n_x - n_y) = h.$$

Observe that besides the dimension of the underlying domain, there is a fundamental difference between the beam models discussed in Section 9.1 and the corresponding plate models. Namely both kernels da, dc characteristic for plates not only enter the equations but also the boundary conditions mentioned here, except for an entirely clamped plate. As a consequence, for nonsynchronous materials so far only the variational approach of Sections 6.7 and 7.3 can be used to establish well-posedness and existence of the resolvent, unless the plate is entirely clamped.

In the synchronous case the Midlin-Timoshenko plate is of scalar type, and the Kirchhoff plate is so if the boundary $\partial\Omega$ contains no free part, i.e. $\Gamma_f = \emptyset$. Since $\beta_e = 9\beta/(3\beta+4) \leq 3$ it is not difficult to show that the corresponding operators A of these models are selfadjoint and negative semidefinite in $X = L^2(\Omega)^3$ for (9.22) and $X = L^2(\Omega)$ for (9.27). Therefore the results of Chapter I apply again.

If the plate is entirely clamped we obtain similarly to the case of beams a weak resolvent by means of Theorem 6.4, which is also an a-regular resolvent if the material is almost synchronous, thanks to Corollary 6.6.

In the general case the variational approach can be used if both, $da(t)$ and $dc(t)$ are of positive type, i.e. e.g. if $a(t)$ and $b(t)$ are Bernstein functions. However, we shall not do this here since Section 9.4 is entirely devoted to the variational approach to the more complicated problem of three-dimensional thermoviscoelasticity, and we want to avoid too many repetitions.

9.3 Thermoviscoelasticity: Strong Approach
Consider a 3-dimensional body which is represented by a bounded open domain $\Omega \subset \mathbb{R}^3$ with smooth boundary $\partial\Omega$, and let $\rho_0 \in C(\overline{\Omega})$, $\rho_0 > 0$, be its density of mass. According to Section 5.3, linear thermoviscoelasticity of this body is governed by the equations

$$\rho_0(x)\dot{v}(t,x) = \text{div } \mathcal{S}(t,x) + \rho_0(x)g(t,x)$$
$$\dot{\varepsilon}(t,x) = -\text{div } q(t,x) + r(t,x)$$
$$\mathcal{S}(t,x) = \int_0^\infty d\mathcal{A}(\tau,x)\dot{\mathcal{E}}(t-\tau,x) - \int_0^\infty d\mathcal{B}(\tau,x)\dot{\Theta}(t-\tau,x) \quad (9.28)$$
$$\varepsilon(t,x) = \int_0^\infty dm(\tau,x)\Theta(t-\tau,x) + \int_0^\infty d\mathcal{B}(\tau,x):\dot{\mathcal{E}}(t-\tau,x)$$
$$q(t,x) = -\int_0^\infty d\mathcal{C}(\tau,x)\nabla\Theta(t-\tau,x),$$

for $t \in \mathbb{R}$, $x \in \Omega$, where $v, \varepsilon, \mathcal{S}, q, \theta, g, r$ have the same meaning as in Section 5. For the formulation of the boundary conditions, let $\Gamma_d, \Gamma_s, \Gamma_b, \Gamma_f \subset \partial\Omega$ be closed, such that $\overset{\circ}{\Gamma}_d \cap \overset{\circ}{\Gamma}_s = \emptyset$, $\overset{\circ}{\Gamma}_b \cap \overset{\circ}{\Gamma}_f = \emptyset$, $\Gamma_d \cup \Gamma_s = \Gamma_b \cup \Gamma_f = \partial\Omega$, and let $n(x)$ denote the outer normal of Ω at $x \in \partial\Omega$. Then

$$
\begin{aligned}
v(t, x) &= v_d(t, x), & t \in \mathbb{R}, \ x \in \overset{\circ}{\Gamma}_d, \\
\mathcal{S}(t, x)n(x) &= g_s(t, x), & t \in \mathbb{R}, \ x \in \overset{\circ}{\Gamma}_s, \\
\Theta(t, x) &= \Theta_b(t, x), & t \in \mathbb{R}, \ x \in \overset{\circ}{\Gamma}_b, \\
-q(t, x) \cdot n(x) &= q_f(t, x), & t \in \mathbb{R}, \ x \in \overset{\circ}{\Gamma}_f,
\end{aligned}
\tag{9.29}
$$

are the natural boundary conditions for (9.28). In the sequel we shall always assume that these are homogeneous, i.e. $v_d \equiv g_s \equiv \Theta_b \equiv q_f \equiv 0$; as usual, this can be achieved by suitable transformations of the variables $v, \varepsilon, \mathcal{S}, q, \Theta, g$ and the forcing functions g and r, if the inhomogeneities in (9.29) are sufficiently smooth.

To obtain a reformulation of (9.28) and (9.29) as an abstract Volterra equation of the form (6.1), i.e.

$$
u(t) = f(t) + \int_0^t A(t - \tau)u(\tau)d\tau, \quad t \geq 0,
\tag{9.30}
$$

where $A \in BV_{loc}(\mathbb{R}_+; \mathcal{B}(Y, X))$, $Y \overset{d}{\hookrightarrow} X$, the spaces X, Y and the variable u must be chosen carefully. We intend to work in an L^2-setting, and choose

$$
X = L^2(\Omega; \mathbb{R}^3) \times L^2(\Omega; \text{Sym}\{3\}) \times L^2(\Omega) \times L^2(\Omega; \mathbb{R}^3);
$$

the variable u will be

$$
u = (b_m * \dot{v}, \mathcal{S}, b_e * \dot{\Theta}, q),
$$

where the kernels $b_m, b_e \in L^1_{loc}(\mathbb{R}_+)$ are still to our disposal. Next we introduce two operators T_m, T_e which take care of the differentiations w.r.t. space involved in (9.28) as well as of the boundary conditions (9.29).

$$
T_e \Theta = -\nabla \Theta, \quad \Theta \in D(T_e) = W^{1,2}_{\Gamma_b}(\Omega),
$$

$$
T_m v = \frac{1}{2}(\nabla v + (\nabla v)^T), \quad v \in D(T_m) = W^{1,2}_{\Gamma_d}(\Omega; \mathbb{R}^3).
\tag{9.31}
$$

Observe that $T_e : \mathcal{D}(T_e) \subset L^2(\Omega) \to L^2(\Omega; \mathbb{R}^3)$ is linear, closed, densely defined, and the range $\mathcal{R}(T_e)$ is closed in $L^2(\Omega; \mathbb{R}^3)$ by Poincaré's inequality. Similarly, $T_m : \mathcal{D}(T_m) \subset L^2(\Omega; \mathbb{R}^3) \to L^2(\Omega; \text{Sym}\{3\})$ is linear, closed, densely defined, with range $\mathcal{R}(T_m)$ closed in $L^2(\Omega; \text{Sym}\{3\})$ by Korn's second inequality. Also of interest here are the adjoints of T_e, T_m; they are given by

$$
\begin{aligned}
T_e^* q &= \text{div } q, \qquad T_m^* \mathcal{E} = -\text{div } \mathcal{E} \\
q \in D(T_e^*) &= \{q \in L^2(\Omega; \mathbb{R}^3) : \text{div } q \in L^2(\Omega), q \cdot n = 0 \text{ on } \Gamma_f\} \\
\mathcal{E} \in D(T_m^*) &= \{\mathcal{E} \in L^2(\Omega; \text{Sym}\{3\}) : \text{div } \mathcal{E} \in L^2(\Omega; \mathbb{R}^3), \mathcal{E}n = 0 \text{ on } \Gamma_s\}.
\end{aligned}
\tag{9.32}
$$

Here the divergence as well as the boundary conditions appearing in $D(T_e^*)$ and $D(T_m^*)$ have to be understood in the weak sense. With these notations and eliminating ε from (9.28) the initial value problem for (9.28), (9.29) can be reformulated in the following way. Let $\mathcal{A}_0 \in L_{loc}^1(\mathbb{R}_+; L^\infty(\Omega; \mathcal{B}(\mathrm{Sym}\{3\})))$, $\mathcal{C}_0, \mathcal{B}_1, \mathcal{B}_2 \in L_{loc}^1(\mathbb{R}_+; L^\infty(\Omega; \mathrm{Sym}\{3\}))$, and $l \in L_{loc}^1(\mathbb{R}_+; L^\infty(\Omega))$ be such that the relations

$$\mathcal{A} = b_m * \mathcal{A}_0, \ \mathcal{C} = b_e * \mathcal{C}_0, \ b_e = l * dm, \ \mathcal{B} = 1 * b_e * \mathcal{B}_1$$
$$\mathcal{B} * b_e = m * b_m * \mathcal{B}_2 \tag{9.33}$$

are valid. Then (9.28), (9.29) becomes (9.30), where

$$A(t) = \begin{pmatrix} 0 & -b_m(t)\rho_0^{-1}T_m^* & 0 & 0 \\ \mathcal{A}_0(t)T_m & 0 & -\mathcal{B}_1(t) & 0 \\ -\mathcal{B}_2(t) : T_m & 0 & 0 & -l(t)T_e^* \\ 0 & 0 & \mathcal{C}_0(t)T_e & 0 \end{pmatrix}, \tag{9.34}$$

and the space Y is given by $Y = D(T_m) \times D(T_m^*) \times D(T_e) \times D(T_e^*)$ equipped with its natural norm; obviously, $A \in L_{loc}^1(\mathbb{R}_+; \mathcal{B}(Y, X))$. Relations (9.33) can be fulfilled in fairly general situations, it only prevails restrictions on the behaviour of m near $t = 0$ and smoothness of the coupling kernel $\mathcal{B}(t)$. In fact, suppose $m(\cdot, x)$ is a creep function for each $x \in \Omega$ with uniformly positive instantaneous heat capacity $m(0+, x) \geq m_0 > 0$, and let $\mathcal{B} \in W_{loc}^{1,1}(\mathbb{R}_+; L^\infty(\Omega; \mathrm{Sym}\{3\}))$, $\mathcal{B}(0) = \mathcal{B}(0+) = 0$, $\dot{\mathcal{B}} \in BV_{loc}(\mathbb{R}_+; L^\infty(\Omega; \mathrm{Sym}\{3\}))$. Then any functions $b_m, b_e \in L_{loc}^1(\mathbb{R}_+)$ such that

$$b_m * k_m = 1 = b_e * k_e$$

for some $k_m, k_e \in L_{loc}^1(\mathbb{R}_+)$ will do; we have

$$\mathcal{A}_0 = k_m * d\mathcal{A}, \ \mathcal{C}_0 = k_e * d\mathcal{C}, \ \mathcal{B}_1 = k_e * d\dot{\mathcal{B}}, \ \mathcal{B}_2 = l * k_m * d\dot{\mathcal{B}}$$

where $l \in L_{loc}^1(\mathbb{R}_+; L^\infty(\Omega))$ is the solution of $l * dm = b_e$ which exists by $m_0 > 0$.

Having reformulated (9.28), (9.29) as a Volterra equation of nonscalar type (9.30), we are now in position to apply the results obtained in Sections 6, 7, and 8. Of course, one cannot hope for well-posedness of (9.30) with $A(t)$ given by (9.34) without further hypotheses on the material functions, in particular on $\mathcal{A}(t)$, $\mathcal{C}(t)$, and $m(t)$. In the remainder of this subsection we mainly concentrate on the parabolic case and apply some of the results of Section 7 and 8.

(i) Let us first consider only viscoelasticity, i.e. the system for $b_m * \dot{v}$ and $\mathcal{S}$, assuming $\mathcal{B}(t) \equiv 0$. Suppose $\mathcal{A}(t, x)$ is of the form

$$\mathcal{A}(t, x) = a(t)\mathcal{A}_1(x) + \int_0^t a(t - \tau)d\mathcal{A}_2(\tau, x), \quad t \geq 0, \ x \in \Omega, \tag{9.35}$$

where $a(t)$ is a Bernstein Function, $\mathcal{A}_1 \in L^\infty(\Omega; \mathcal{B}(\mathrm{Sym}\{3\}))$ is uniformly positive definite and $\mathcal{A}_2 \in BV_{loc}(\mathbb{R}_+; L^\infty(\Omega, \mathcal{B}(\mathrm{Sym}\{3\})))$, $\mathcal{A}_2(0, x) = \mathcal{A}_2(0+, x) = 0$ a.e. in Ω. Materials for which $\mathcal{A}(t, x) = a(t)\mathcal{A}_1(x)$, $a, \mathcal{A}_1$ as above, will be called (viscoelastically) *separable*, while those which are subject to (9.35) are termed (viscoelastically) *almost separable*. Define $b_m \in L^1_{loc}(\mathbb{R}_+)$ by $\hat{b}_m(\lambda) = \sqrt{\hat{a}(\lambda)}$. $b_m(t)$ exists and is completely monotonic; cp. Lemma 4.2. The equation for the purely viscoelastic problem then has the main part $b_m(t)A_1$, where

$$A_1 = \begin{pmatrix} 0 & -\rho_0^{-1}T_m^* \\ \mathcal{A}_1 T_m & 0 \end{pmatrix} \tag{9.36}$$

is skew-adjoint w.r.t. the inner product on $X_1 = L^2(\Omega;\mathbb{R}^3) \times L^2(\Omega;\mathrm{Sym}\{3\})$ defined by

$$((u_1,u_2),(v_1,v_2))_{X_1} = \int_\Omega u_1(x) \cdot \overline{v_1(x)}\rho_0(x)dx + \int_\Omega \mathcal{A}_1^{-1}(x)u_2(x) : \overline{v_2(x)}dx,$$

and we choose $Y_1 = D(A_1) = D(T_m) \times D(T_m^*)$. Thus A_1 generates a unitary group in X_1, and therefore admits bounded imaginary powers which satisfy $|A_1^{is}| \leq e^{|s|\pi/2}$, $s \in \mathbb{R}$. Therefore in case $\mathcal{A}_2 \equiv 0$ this problem is always well-posed, and it is also for $\mathcal{A}_2 \neq 0$ provided $\dot{\mathcal{A}}_2 \in BV_{loc}(\Omega; L^\infty(\Omega; \mathcal{B}(\mathrm{Sym}\{3\})))$, by Theorem 6.1. The problem is parabolic if in addition $|\arg \hat{a}(\lambda)| \leq \Theta_a < \pi$ holds for Re $\lambda > 0$; then without further assumptions on $\mathcal{A}_2(t,x)$ we obtain the well-posedness by Theorem 7.6 as well as maximal regularity of type C^α and L^p by Theorems 7.4 and 8.7.

(ii) Similarly one considers the purely thermodynamical problem for $b_e * \dot{\Theta}$ and q in case $\mathcal{B}(t) \equiv 0$. Materials which are subject to

$$m(t,x) = m(t)\sigma_0(x), \quad C(t,x) = c(t)\mathcal{C}_1(x) \tag{9.37}$$

with m, c Bernstein functions, $\sigma_0 \in L^\infty(\Omega)$, $\sigma_0(x) \geq \sigma_{00} > 0$ a.e., and $\mathcal{C}_1 \in L^\infty(\Omega;\mathrm{Sym}\{3\})$ uniformly positive definite, will be called (thermodynamically) *separable*. Choosing $l \in L^1_{loc}(\mathbb{R}_+)$ such that $\hat{l}(\lambda) = \sqrt{\hat{c}(\lambda)/\widehat{dm}(\lambda)}$, $b_e = l * dm$, as in Section 5.5, the function $l(t)$ is completely monotonic, and the thermodynamical problem becomes of scalar type, with kernel $l(t)A_2$, where

$$A_2 = \begin{pmatrix} 0 & -\sigma_0^{-1}T_e^* \\ \mathcal{C}_1 T_e & 0 \end{pmatrix}.$$

Similar to (i), A_2 is skew-adjoint w.r.t. a suitable inner product on $X_2 = L^2(\Omega) \times L^2(\Omega;\mathbb{R}^3)$, and the conclusions are analogous to (i), also for the (thermodynamically) *almost separable* case

$$m(t,x) = m(t)\sigma_0(x) + \int_0^t m(t-\tau)d\sigma_1(\tau,x)$$

$$C(t,x) = c(t)\mathcal{C}_1(x) + \int_0^t c(t-\tau)d\mathcal{C}_2(\tau,x) \tag{9.38}$$

with $m, c, \sigma_0, \mathcal{C}_1$ as above, $\sigma_1 \in BV_{loc}(\mathbb{R}_+; L^\infty(\Omega))$, $\sigma_1(0,x) = \sigma_1(0+,x) = 0$, and $\mathcal{C}_2 \in BV_{loc}(\mathbb{R}_+; L^\infty(\Omega;\mathrm{Sym}\{3\}))$ such that $\mathcal{C}_2(0,x) = \mathcal{C}_2(0+,x) = 0$. Observe that the problem is parabolic if $|\arg(\hat{c}(\lambda)/\widehat{dm}(\lambda))| \leq \Theta_l < \pi$ for Re $\lambda > 0$.

(iii) Finally, consider the thermoviscoelastic case $\mathcal{B}(t) \not\equiv 0$, when the material is almost separable, both viscoelastically and thermodynamically. The coupling involving $\mathcal{B}(t)$ will be considered as a perturbation, therefore we must have $\mathcal{B}_2 = l * d\mathcal{B}_3$, $\mathcal{B}_3(0) = \mathcal{B}_3(0+) = 0$. If e.g. $\dot{\mathcal{B}} \in BV_{loc}(\mathbb{R}_+; L^\infty(\Omega; \mathrm{Sym}\{3\}))$ this factorization follows, hence the results of Sections 7 and 8 apply to the result that the full thermoviscoelastic problem is well-posed and enjoys maximal regularity of type C^α and L^p, provided the problem is parabolic, viscoelastically but also thermodynamically. In the hyperbolic case Theorem 6.1 implies well-posedness if e.g. $\ddot{\mathcal{B}} \in BV_{loc}(\mathbb{R}_+; L^\infty(\Omega; \mathrm{Sym}\{3\}))$.

9.4 Thermoviscoelasticity: Variational Approach

Consider again problem (9.28), (9.29) where $v_d \equiv 0 \equiv \Theta_b$. We want to rewrite it in variational form. For this purpose assume again $m(\cdot, x) \in \mathcal{CF}$ for each $x \in \overline{\Omega}$, $\sigma_0(x) = m(0+, x) > 0$. Let $V = W^{1,2}_{\Gamma_d}(\Omega; \mathbb{C}^3) \times W^{1,2}_{\Gamma_b}(\Omega)$; then $V \overset{d}{\hookrightarrow} H \overset{d}{\hookrightarrow} V^*$, where $H = L^2(\Omega; \mathbb{C}^3) \times L^2(\Omega)$ is equipped with the inner product

$$((v_1; \Theta_1), (v_2, \Theta_2))_H = \int_\Omega \rho_0(x) v_1(x) \cdot \overline{v_2(x)} dx + \int_\Omega \sigma_0(x) \Theta_1(x) \overline{\Theta_2(x)} dx.$$

We define a bounded sesquilinear form on V by means of

$$\beta(t, (v_1, \Theta_1), (v_2, \Theta_2)) = \int_\Omega \nabla v_1(x) : \overline{\mathcal{A}(t, x) \nabla v_2(x)} dx$$

$$+ \int_\Omega \nabla v_1(x) : \overline{\dot{\mathcal{B}}(t, x) \Theta_2(x)} dx + \int_\Omega \Theta_1(x) \overline{\dot{\mathcal{B}}(t, x) : \nabla v_2(x)} dx \qquad (9.39)$$

$$+ \int_\Omega \Theta_1(x) \overline{m_1(t, x) \Theta_2(x)} dx + \int_\Omega \nabla \Theta_1 \cdot \overline{\mathcal{C}(t, x) \nabla \Theta_2(x)} dx.$$

With $u = (v, \Theta)$ there follows the variational formulation of (9.28), (9.29) by an integration by parts.

$$(w, u(t))_H + \int_0^t \beta(t - s, w, u(s)) ds = \; <w, f(t)>, \quad t \geq 0, \qquad (9.40)$$

for each $w \in V$, where $f(t) \in \overline{V}^*$ contains $g(t, x)$, $r(t, x)$, $g_s(t, x)$, $q_f(t, x)$ as well as the histories of $v(t, x)$ and $\Theta(t, x)$. Since $\beta(t, \cdot, \cdot)$ is bounded, there is a family of bounded linear operators $\{A(t)\}_{t \geq 0} \subset \mathcal{B}(V, \overline{V}^*)$ such that

$$\beta(t, w, u) = \; - <w, A(t)u>, \quad \text{for all } u, w \in V, \; t \geq 0. \qquad (9.41)$$

There follows $A \in BV_{loc}(\mathbb{R}_+; \mathcal{B}(V, \overline{V}^*))$, provided $m_1 \in BV_{loc}(\mathbb{R}_+; L^\infty(\Omega))$, $\dot{\mathcal{B}}, \mathcal{C} \in BV_{loc}(\mathbb{R}_+; L^\infty(\Omega; \mathrm{Sym}\{3\}))$, and $\mathcal{A} \in BV_{loc}(\mathbb{R}_+; L^\infty(\Omega; \mathcal{B}(\mathrm{Sym}\{3\})))$. Thus the variational formulation of (9.28), (9.29) becomes

$$u(t) = \int_0^t A(t - \tau) u(\tau) d\tau + f(t), \quad t \geq 0; \qquad (9.42)$$

in $X = \overline{V}^*$, $Y = V$.

In the case of homogeneous isotropic materials the form β simplifies to

$$
\begin{aligned}
\beta(t, u_1, u_2) \;=\; & 2a(t)[(\mathcal{E}_{v_1} : \mathcal{E}_{v_2}) - (\,\text{div } v_1, \ \text{div } v_2)/3] \\
& + \ b(t)(\,\text{div } v_1, \ \text{div } v_2) + c(t)(\nabla\Theta_1, \nabla\Theta_2) + m_1(t)(\Theta_1, \Theta_2) \\
& + \ (db * d\alpha)(t)[(\,\text{div } v_1, \Theta_2) - (\Theta_1, \ \text{div } v_2)], \quad t > 0
\end{aligned}
$$

where the moduli $a(t)$, $b(t)$, $c(t)$, $\alpha(t)$, and $m(t) = \sigma_0 + 1 * m_1$ are as in Section 5.3.

Let us concentrate on the hyperbolic case, and apply the positive type methods presented in Section 6.7. For this purpose consider the inequalities

$$
\text{Re } q \cdot \overline{\widehat{dC}(\lambda, x)q} \geq 0 \ \text{ for all } q \in \mathbb{C}^3, \ x \in \Omega, \ \text{Re } \lambda > 0, \tag{9.43}
$$

and

$$
\text{Re } [\mathcal{E} : \overline{\widehat{dA}(\lambda, x)\mathcal{E}} - \mathcal{E} : \overline{\widehat{dB}(\lambda, x)\vartheta} + \vartheta\overline{d\widehat{B}(\lambda, x) : \mathcal{E}} + \vartheta\overline{\widehat{dm_1}(\lambda, x)\vartheta}] \geq 0, \tag{9.44}
$$

for all symmetric $\mathcal{E} \in \mathbb{C}^{3\times3}$, $\vartheta \in \mathbb{C}$, $x \in \Omega$, $\text{Re } \lambda > 0$. Physically (9.43), (9.44) are related to the second law of thermodynamics; cp. Christensen [42]. These inequalities will be termed *dissipation inequalities* in the sequel, in analogy to those of pure viscoelasticity. In the homogeneous isotropic case (cp. Section 5.3), they reduce to the requirements that dc is of positive type and

$$
\text{Re } [2\widehat{da}(|\mathcal{E}|^2 - \frac{1}{3}|\text{tr}\mathcal{E}|^2) + \widehat{db}|\text{tr } \mathcal{E}|^2 + 2i\lambda\widehat{db}\widehat{d\alpha}\text{Im}(\text{tr } \mathcal{E}\overline{\vartheta}) + \widehat{dm_1}|\vartheta|^2] \geq 0,
$$

for all symmetric $\mathcal{E} \in \mathbb{C}^{3\times3}$, $\vartheta \in \mathbb{C}$, $\text{Re } \lambda > 0$. But this in turn is equivalent to da, db, dc, dm_1 of positive type, and existence of a constant $\delta \leq 1$ such that

$$
[\text{Im } (\lambda\widehat{db}(\lambda)\widehat{d\alpha}(\lambda))]^2 \leq \delta\text{Re } \widehat{db}(\lambda) \cdot \text{Re } \widehat{dm_1}(\lambda), \quad \text{Re } \lambda > 0; \tag{9.45}
$$

observe that (9.45) holds for classical thermoelasticity given by (5.41). The dissipation inequalities imply

$$
\text{Re } < u, \widehat{dA}(\lambda)u > = -\text{Re } \widehat{d\beta}(\lambda, u, u) \leq 0, \quad u \in Y, \ \text{Re } \lambda > 0, \tag{9.46}
$$

i.e. the part of $\widehat{dA}(\lambda)$ in H, $\widehat{dA}(\lambda)|_H$ is dissipative for each $\text{Re } \lambda > 0$, and the closure in H of $\widehat{dA}(\lambda)|_H$ is even m-dissipative. However, this is not enough for Theorem 6.4 to apply, neither in H nor in $\overline{V}^*$. One reason for this lies in the weakness of the dissipation inequalities. But even if these are strengthened, Theorem 6.4 still is of no use here, since in general, $\widehat{dA}(\lambda)$ will not be dissipative in $\overline{V}^*$ and $D(\widehat{dA}(\lambda)|_H)$ will depend on λ.

On the other hand, let the dissipation inequalities (9.43) and (9.44) be replaced by the *strong dissipation inequalities*

$$
\text{Re } q \cdot \overline{\widehat{dC}(\lambda, x)q} \geq \gamma(\text{Re}\frac{1}{\lambda+\eta})|q|^2, \quad \text{Re } \lambda > 0, \ q \in \mathbb{C}^3, \ x \in \Omega, \tag{9.47}
$$

and

$$\text{Re} \, [\mathcal{E} : \overline{d\mathcal{A}(\lambda, x)\mathcal{E}} + \mathcal{E} : \overline{d\dot{\mathcal{B}}(\lambda, x)\vartheta} + \overline{\vartheta d\dot{\mathcal{B}}(\lambda, x)} : \mathcal{E} + \overline{\vartheta d m_1(\lambda, x)\vartheta}]$$
$$\geq \gamma (\text{Re} \, \frac{1}{\lambda + \eta}) |\mathcal{E}|^2, \qquad (9.48)$$
$$\text{for Re} \, \lambda > 0, \quad x \in \Omega, \ \mathcal{E} \in \mathbb{C}^{3 \times 3} \text{ symmetric}, \ \vartheta \in \mathbb{C},$$

where $\gamma > 0$ is a constant, and assume one more degree of regularity for the coefficients $\mathcal{A}, \dot{\mathcal{B}}, \mathcal{C}, m_1$. Then in case $\eta = 0$, resp. $\eta > 0$, the assumptions of Theorem 6.5, resp. Corollary 6.7, are fulfilled, provided in addition $\overset{\circ}{\Gamma}_d \neq \emptyset$ and $\overset{\circ}{\Gamma}_b \neq \emptyset$ by the inequalities of Korn and Poincaré, and we obtain a resolvent $S(t)$ which enjoys properties $(a) \sim (c)$ of Theorem 6.5, resp. Corollary 6.7, in particular is leaving H invariant and satisfies $|S(t)|_H \leq 1$.

Observe that in the homogeneous, isotropic case this means

$$\sigma_0 > 0, \ a_0 = b_0 = c_0 = \alpha_0 = 0,$$
$$a_1(0+), \ b_1(0+), \ c_1(0+), \ m_1(0+), \ \alpha_1(0+) < \infty,$$

whenever a, b, c, m, α are creep functions (cp. Section 5.3). The strong dissipation inequalities hold if a_1, b_1, c_1, m_1 are strongly positive (or merely of positive type if we allow for an exponential shift) and (9.45) holds with $\delta < 1$.

Similarly, one can apply Theorem 7.3 in the parabolic case; the translation of conditions (PV1)$\sim$(PV4) into parabolic dissipation inequalities is left to the reader.

9.5 Electrodynamics with Memory

As is well known, the macroscopic electrodynamic fields in a medium at rest in the considered inertial system, are governed by *Maxwell's equations*

$$\mathcal{B}_t + \text{curl} \, \mathcal{E} = 0, \qquad \text{div} \, \mathcal{B} = 0,$$
$$\mathcal{D}_t - \text{curl} \, \mathcal{H} + J = 0, \qquad \text{div} \, \mathcal{D} = \rho, \qquad (9.49)$$

where $\mathcal{E}$ denotes the electric field, $\mathcal{H}$ the magnetic field, $\mathcal{B}$ the magnetic induction, $\mathcal{D}$ the electric induction, J the free current, and ρ the free charge. Clearly (9.49) is underdetermined and constitutive equations have to be supplemented, which specify the electrodynamic properties of the medium in question.

For the simplest medium - the vacuum - these relations are given by

$$\mathcal{B} = \mu_0 \mathcal{H}, \quad \mathcal{D} = \varepsilon_0 \mathcal{E}, \quad J = 0, \qquad (9.50)$$

where $\varepsilon_0, \mu_0 > 0$ are well known fundamental physical constants which are connected with the speed of light in vacuum c_0 by the relation

$$\varepsilon_0 \mu_0 = c_0^{-2}.$$

The next simplest medium is the rigid linear isotropic dieletric which Maxwell defined by

$$\mathcal{B} = \mu \mathcal{H}, \quad \mathcal{D} = \varepsilon \mathcal{E}, \quad J = 0, \tag{9.51}$$

where ε, μ are material constants ($\varepsilon \geq \varepsilon_0 > 0, \mu > 0$). Soon it was realized that such simple constitutive relations do not account for the dielectric losses observed, when the medium is placed in a rapidly varying electromagnetic field. This defect cannot be removed by the introduction of a conductivity $\sigma > 0$, i.e. replacing the last equation in (9.51) by a relation of the form $J = \sigma \mathcal{E}$ (Ohm's law), i.e

$$\mathcal{B} = \mu \mathcal{H}, \quad \mathcal{D} = \varepsilon \mathcal{E}, \quad J = \sigma \mathcal{E}. \tag{9.52}$$

More precisely, if one considers periodic fields $\mathcal{E}(t) = \mathcal{E}_0 e^{i\omega t}$, $\mathcal{D}(t) = \mathcal{D}_0 e^{i\omega t}$, one still observes relations of the form $\mathcal{D}_0 = \hat{\varepsilon} \mathcal{E}_0$ for each frequency $\omega > 0$, however, $\hat{\varepsilon}$ will be complex in general, and depend on ω. This phenomenon is known as *electromagnetic dispersion* in the physical literature.

To account for dispersion in a rigid linear isotropic medium the following constitutive relations have been proposed

$$\mathcal{B} = d\mu * \mathcal{H}, \quad \mathcal{D} = d\varepsilon * \mathcal{E}, \quad J = d\sigma * \mathcal{E}, \tag{9.53}$$

where $\mu, \varepsilon, \sigma \in BV(\mathbb{R}_+)$ are given material functions. Observe that the quantities $\mu(\infty)$, $\varepsilon(\infty)$, $\sigma(\infty)$ correspond to the constants appearing in (9.52) in the static (time-dependent) case. Not much seems to be known about general material functions; however, from the discussion in Kapitel IX of Landau and Lifschitz [205] one can guess that $\varepsilon(t)$ should be (at least) a bounded creep function and that $\mu(0+) = \mu_0$, $\varepsilon(0+) = \varepsilon_0$ are the fundamental constants of the vacuum introduced above. It is also reasonable to expect that $\sigma(t)$ is a bounded creep function as well. In the paramagnetic case one has $\mu(\infty) > \mu_0$, while $\mu(\infty) < \mu_0$ holds for diamagnetic media; therefore in the former case $\mu(t)$ can be considered a bounded creep function, while in the latter $\mu_0 - \mu(t)$ should be such.

The continuity equation for the charge density ρ which follows from Maxwell's equation (9.49) reads

$$\rho_t + \operatorname{div} J = 0. \tag{9.54}$$

Convolving (9.54) with $d\varepsilon$, (9.53) and (9.49) then imply

$$d\varepsilon * \rho_t + d\sigma * \rho = 0; \tag{9.55}$$

considering the history of $\rho(t, x)$ up to $t = 0$ as known, it is clear that (9.55) determines $\rho(t, x)$ for all $t \in \mathbb{R}$, and so in the sequel $\rho(t, x)$ will be considered as known.

Assuming the constitutive laws (9.53), Maxwell's equations can be decoupled as usual. Taking the curl of the first equation of (9.49), with the identity

$$\operatorname{curl} \operatorname{curl} \mathcal{E} = \operatorname{grad} \operatorname{div} \mathcal{E} - \Delta \mathcal{E},$$

by div $\mathcal{D} = \rho$ and with (9.53) one obtains e.g. for $\mathcal{D}$

$$d\varepsilon * d\mu * \mathcal{D}_{tt} + d\mu * d\sigma * \mathcal{D}_t \;=\; \Delta\mathcal{D} - \nabla\rho$$
$$\operatorname{div}\mathcal{D} \;=\; \rho. \tag{9.56}$$

Analogously, applying curl to the third equation of (9.49) leads to the following equation e.g. for $\mathcal{B}$

$$d\varepsilon * d\mu * \mathcal{B}_{tt} + d\mu * d\sigma * \mathcal{B}_t \;=\; \Delta\mathcal{B}$$
$$\operatorname{div}\mathcal{B} \;=\; 0. \tag{9.57}$$

Note that the relations (9.53) are invertible; therefore $\mathcal{H}$ satisfies again (9.57) while $\mathcal{E}$ solves (9.56) with ρ replaced by $d\nu * \rho$, where $d\nu * d\varepsilon = de_0$; $e_0(t)$ as before denotes Heaviside's function. Thus as in the case of the vacuum, once one of the fields $\mathcal{E}, \mathcal{D}, \mathcal{H}$, or $\mathcal{B}$ is known then Maxwell's equations and the constitutive equations (9.53) determine the others by pointwise integration resp. convolution.

We are now in position to apply the results on scalar Volterra equations obtained in Chapter I, say to (9.57). For simplicity we consider (9.57) on all of $\mathbb{R}^3$. Let $X = L^2_\sigma(\mathbb{R}^3)$ and denote by A_2 the Stokes operator introduced in Section 5.6. With $b \in BV_{loc}(\mathbb{R}_+)$ defined by

$$\hat{b}(\lambda) \;=\; (\lambda^2 \widehat{d\varepsilon}(\lambda)\widehat{d\mu}(\lambda) + \lambda\widehat{d\mu}(\lambda)\widehat{d\sigma}(\lambda))^{-1}$$
$$=\; (\lambda\widehat{d\mu}(\lambda))^{-1}(\lambda\widehat{d\varepsilon}(\lambda))^{-1}(1 + \widehat{d\sigma}(\lambda)/\lambda\widehat{d\varepsilon}(\lambda))^{-1}, \quad \operatorname{Re}\lambda > 0,$$

the initial value problem for (9.57) is then equivalent to

$$u(t) = f(t) + \int_0^t b(t - \tau)A_2 u(\tau)d\tau, \quad t > 0, \tag{9.58}$$

where $u(t)$ means magnetic field or magnetic induction and f is given. Now we can claim the following.

(i) Assume $\varepsilon(t) = \varepsilon_0 + \int_0^t \varepsilon_1(s)ds$, $\mu(t) = \mu_0 \pm \int_0^t \mu_1(s)ds$ where $\varepsilon_1, \mu_1 \in BV_{loc}(\mathbb{R}_+)$, and $\sigma \in BV_{loc}(\mathbb{R}_+)$. Then $b(t)$ is of the form $b(t) = tc_0^2 + t * r(t)$, with $r \in BV_{loc}(\mathbb{R}_+)$, hence Theorem 1.2 applies and shows that (9.58) is well-posed. The resolvent $S(t)$ for (9.58) is of the form $S(t) = \operatorname{Co}(c_0 t) + S_1(t)$ where $\operatorname{Co}(\tau)$ denotes the bounded cosine family generated by the Stokes operator A_2, and $S_1(t)$ is a perturbation which, however, may grow exponentially.

(ii) A bounded resolvent can be obtained by means of Corollary 1.2. Suppose ε, μ, σ are bounded creep functions with $\varepsilon(0+) = \varepsilon_0$, $\mu(0+) = \mu_0$ and such that μ_1 and $\sigma_1(t)$ are of positive type, and assume in addition

$$\operatorname{Re}\hat{\varepsilon_1}(i\rho) + \operatorname{Im}\hat{\sigma_1}(i\rho)/\rho \geq 0 \quad \text{for} \quad \rho > 0. \tag{9.59}$$

Then $db(t)$ is of positive type, hence Corollary 1.2 yields boundedness of the resolvent $S(t)$.

(iii) Stronger assertions can be obtained by the subordination principle of Section 4. For this assume that ε, μ, σ are bounded Bernstein functions with

$\varepsilon(0+) = \varepsilon_0$, $\mu(0+) = \mu_0$ and such that in addition $-\dot{\varepsilon}_1 - \sigma_1$ is completely mono-tonic on $(0, \infty)$; note that the latter implies (9.59). Then $\varepsilon(t) + \int_0^t \sigma(s)ds$ is a Bernstein function, hence by Lemma 4.3 there is a Bernstein function k such that

$$\begin{aligned}
\hat{k}(\lambda) &= \hat{\mu}(\lambda)[\hat{\mu}(\lambda)/(\hat{\varepsilon}(\lambda) + \hat{\sigma}(\lambda)/\lambda)]^{-1/2} \\
&= [\hat{\mu}(\lambda)(\hat{\varepsilon}(\lambda) + \hat{\sigma}(\lambda)/\lambda)]^{1/2}, \quad \text{Re } \lambda > 0.
\end{aligned}$$

From this there follows $b = c * c$ for some completely monotonic function $c \in L^1_{loc}(\mathbb{R}_+)$, hence by Theorem 4.3 there is a bounded resolvent $S(t)$ for (9.58) which by Theorem 4.4 has the form

$$S(t) = C_0(t/\kappa)e^{-\alpha t/\kappa} + S_0(t), \quad t > 0,$$

where $\kappa = k_0 = \sqrt{\varepsilon_0 \mu_0} = c_0^{-1}$, and $\alpha = k_\infty + k_1(0+) = k_1(0+)$, i.e.

$$\alpha = \frac{1}{2}\mu_1(0+)\sqrt{\frac{\varepsilon_0}{\mu_0}} + \frac{1}{2}(\varepsilon_1(0+) + \sigma_0)\sqrt{\frac{\mu_0}{\varepsilon_0}}.$$

Thus the (maximum) wave speed is c_0, the speed of light in vacuum, and the resolvent in question decomposes into the resolvent of the vacuum, exponentially damped with attenuation α and a dispersive part $S_0(t)$ which is smooth. Observe that we always have $\alpha > 0$, unless $\mu_1(0+) = \varepsilon_1(0+) = \sigma_0 = 0$, i.e. $\mu(t) \equiv \mu_0$, $\varepsilon(t) \equiv \varepsilon_0$ and $\sigma_0 = 0$.

9.6 A Transmission Problem for Media with Memory

Problems in electrodynamics involving media with memory can in general not be formulated as Volterra equations of scalar type, even if the medium in question is homogeneous and isotropic. We want to discuss briefly a transmission problem for such media in the framework of the perturbation results for nonscalar equations obtained in Section 6.3.

Suppose $\Omega_1 \subset \mathbb{R}^3$ is a domain with smooth boundary Γ and let $\Omega_2 = \mathbb{R}^3 \setminus \Omega_1$; $\Omega_i, i = 1, 2$ be filled with a homogeneous isotropic medium with material functions $\varepsilon_i, \mu_i, \sigma_i$ which satisfy the assumptions (i) of Section 9.5, i.e. $\varepsilon_i, \mu_i \in W^{1,1}_{loc}(\mathbb{R}_+)$, $\varepsilon_i(0+) = \varepsilon_0, \mu_i(0+) = \mu_0$, and $\dot{\varepsilon}_i, \dot{\mu}_i, \sigma_i \in BV_{loc}(\mathbb{R}_+)$. As before, let $\mathcal{B}_i, \mathcal{H}_i, \mathcal{D}_i, \mathcal{E}_i$ denote the magnetic induction, magnetic field, electrical induction, electric field, respectively, in domain Ω_i. Then we have Maxwell's equations in each domain Ω_i,

$$\begin{aligned}
\mathcal{B}_{it} + \text{curl } \mathcal{E}_i = 0, \quad &\text{div } \mathcal{B}_i = 0, \\
\mathcal{D}_{it} - \text{curl } \mathcal{H}_i + \mathcal{J}_i = 0, \quad &\text{div } \mathcal{D}_i = \rho_i,
\end{aligned} \tag{9.60}$$

and the material relations

$$\mathcal{B}_i = d\mu_i * \mathcal{H}_i, \quad \mathcal{D}_i = d\varepsilon_i * \mathcal{E}_i, \quad \mathcal{J}_i = d\sigma_i * \mathcal{E}_i, \tag{9.61}$$

for $i = 1, 2$. At the interface Γ we have to supplement transmission conditions. If $n(x)$ denotes the outer normal at $x \in \partial\Omega_1 = \Gamma$ of Ω_1 these can be stated as

$$\begin{aligned}
n \cdot \mathcal{B}_1 = n \cdot \mathcal{B}_2, \quad &n \times \mathcal{E}_1 = n \times \mathcal{E}_2 \\
n \cdot (\mathcal{D}_2 - \mathcal{D}_1) = \rho_s, \quad &n \times (\mathcal{H}_2 - \mathcal{H}_1) = \mathcal{J}_s
\end{aligned} \tag{9.62}$$

on Γ, where ρ_s denotes the density of surface charges, and $\mathcal{J}_s$ the density of surface currents; observe that $\mathcal{J}_s \cdot n = 0$.

The initial value problem for (9.60)~(9.62) will now be reformulated as a Volterra equation of the form (6.24), i.e. as

$$\dot{u} = Au + B_1 * u + dB_0 * u + f, \quad u(0) = u_0, \tag{9.63}$$

where $u = (\mathcal{H}_1, \mathcal{E}_1, \mathcal{H}_2, \mathcal{E}_2)$ lives in $X = L^2(\Omega_1; \mathbb{R}^3)^2 \times L^2(\Omega_2; \mathbb{R}^3)^2$ with inner product

$$(u, v)_X = \int_{\Omega_1} (\mu_0 u_1 \cdot \bar{v}_1 + \varepsilon_0 u_2 \cdot \bar{v}_2) dx + \int_{\Omega_2} (\mu_0 u_3 \cdot \bar{v}_3 + \varepsilon_0 u_4 \cdot \bar{v}_4) dx.$$

We assume that there are no surface currents, i.e. $\mathcal{J}_s \equiv 0$. Define an operator A_0 in X by

$$A_0 u = (-\mu_0^{-1} \text{curl } u_2, \; \varepsilon_0^{-1} \text{curl } u_1, \; -\mu_0^{-1} \text{curl } u_4, \; \varepsilon_0^{-1} \text{curl } u_3),$$

with

$$D(A_0) = \{u \in X : u_1, u_2 \in H^{1,2}(\Omega_1; \mathbb{R}^3), u_3, u_4 \in H^{1,2}(\Omega_2; \mathbb{R}^3),$$
$$n \times (u_2 - u_4) = n \times (u_1 - u_3) = 0 \text{ on } \Gamma\}.$$

It then follows by integration by parts that $A_0^* \supset -A_0$, hence A_0 is skew-symmetric, in particular closable. One can also show $\mathcal{R}(I - A_0) = X$, and therefore the closure A of A_0 is skew-adjoint. Thus A generates a unitary C_0-group $T(t)$ in X.

In the sequel it is of advantage to invert the material relations for $\mathcal{B}_i$ and $\mathcal{D}_i$, i.e.

$$\mathcal{H}_i = \mu_0^{-1} \mathcal{B}_i + \nu_i * \mathcal{B}_i, \quad \mathcal{E}_i = \varepsilon_0^{-1} \mathcal{D}_i + \eta_i * \mathcal{D}_i \tag{9.64}$$

where $\nu_i, \eta_i \in BV_{loc}(\mathbb{R}_+)$ by the assumptions on ε_i on μ_i. Let the Hilbert space Y be defined by

$$Y = \{u \in X : \text{curl } u = (\text{curl } u_i)_{i=1}^4 \in X\}$$

with inner product

$$(u, v)_Y = (u, v)_X + (\text{curl } u, \text{curl } v)_X;$$

then $D(A_0) \subset Y$ and $|u|_Y \leq C(|A_0 u| + |u|), u \in D(A_0)$, for some constant $C > 0$. We can now define $B_1 \in BV_{loc}(\mathbb{R}_+; \mathcal{B}(Y, X))$ by means of

$$B_1(t)u = (-\nu_1(t)\text{curl } u_2, \; \eta_1(t)\text{curl } u_1, \; -\nu_2(t)\text{curl } u_4, \; \eta_2(t)\text{curl } u_3)$$

and $B_0 \in BV_{loc}(\mathbb{R}_+; \mathcal{B}(Y, X))$ by

$$B_0(t)u = (0, \; \varepsilon_0^{-1}\sigma_1(t)u_2 - (\eta_1 * \sigma_1)(t)u_2, \; 0, \; -\varepsilon_0^{-1}\sigma_2(t)u_4 - (\eta_2 * \sigma_2)(t)u_4).$$

Then (9.60)~(9.62) is equivalent to (9.63), where $u_0 = (\mathcal{H}_1(0), \mathcal{E}_1(0), \mathcal{H}_2(0), \mathcal{E}_2(0))$ and f contains the history of $u(t), t \leq 0$. We may apply now the perturbation theorem, Theorem 6.1, in the form of Corollary 6.2, thereby proving the well-posedness of the problem in question for the L^2-setting.

The resolvent $S(t)$ for (9.63) obtained this way is of at most exponential growth provided $\varepsilon_i, \mu_i, \sigma_i$ have this property, this is the most which can be said, in general. However, if in addition $\dot{\varepsilon}_i, \dot{\mu}_i, d\sigma_i$ are of positive type, $i = 1, 2$, then $|S(t)|_{\mathcal{B}(X)} \le 1$. In fact, (9.60)~(9.62) yield the identity

$$\frac{1}{2}\frac{d}{dt}\{\int_{\Omega_1} (\varepsilon_0|\mathcal{E}_1|^2 + \mu_0|\mathcal{H}_1|^2)dx + \int_{\Omega_2} (\varepsilon_0|\mathcal{E}_2|^2 + \mu_0|\mathcal{H}_2|^2)dx\}$$

$$= -\int_{\Omega_1} (\dot{\varepsilon}_1 * \mathcal{E}_1 \cdot \mathcal{E}_1 + \dot{\mu}_1 * \mathcal{H}_1 \cdot \mathcal{H}_1 + d\sigma_1 * \mathcal{E}_1 \cdot \mathcal{E}_1)dx$$

$$- \int_{\Omega_2} (\dot{\varepsilon}_2 * \mathcal{E}_2 \cdot \mathcal{E}_2 + \dot{\mu}_2 * \mathcal{H}_2 \cdot \mathcal{H}_2 + d\sigma_2 * \mathcal{E}_2 \cdot \mathcal{E}_2)dx,$$

hence after integration $|u(t)|_X \le |u_0|_X$, $t \ge 0$. In this situation one can also apply Theorem 6.4 instead of the perturbation result.

9.7 Comments

a) For the classical theory of elastic beams and plates we refer e.g. to Landau and Lifschitz [207] and Timoshenko and Woinowsky-Krieger [325]. The recent monograph Lagnese and Lions [204] contains a very readable introduction into classical plate models and their history. For viscoelastic beams and plates see e.g. Bland [24], Flügge [117], Lagnese [203], and Lagnese and Lions [204].

b) The derivation of the viscoelastic beam models in Section 9.1 is different from the usual one, where the correspondence principle of viscoelasticity is applied to the classical models of purely elastic beams. We think that the derivation presented here is somewhat more rigorous, and gives more insight where approximations come in. In particular, it seems that neglecting the term $\rho_0\ddot{w}_{xx}$ in (9.18) when passing to the Euler-Bernoulli beam causes the trouble with robustness of the latter; see (c) below.

c) There is a recent interest in well-posedness, stability, stabilization and destabilization by boundary feedback laws, and robustness w.r.t. delays in boundary feedback for viscoelastic beams and plates; cf. Desch and Wheeler [94], Desch, Hannsgen, and Wheeler [88], Desch, Hannsgen, Renardy, and Wheeler [87], Grimmer, Lenczewski, and Schappacher [141], Hannsgen and Wheeler [HaWh89], Hannsgen, Renardy, and Wheeler [166], Kim and Renardy [197], Lagnese [203], Lagnese and Lions [204], Leugering [210], [211]. In particular, in [87] it is shown that boundary feedback with delay of the Euler-Bernoulli beam may lead to a loss of well-posedness; on the contrary for the Timoshenko beam this cannot happen [141]. It is not clear whether the Rayleigh beam also has this defect, but from (b) above one might expect that the answer is no.

d) Poincaré's inequality is standard in the modern theory of partial differential equations and can be found in any decent textbook on this subject. For Korn's inequalities we refer e.g. to Marsden and Hughes [234] or Leis [208]. The approach to thermoviscoelasticity given in Section 9.3 seems to be new. It heavily depends on the almost separability of the material in question, and it is unclear at present whether positive type methods like Theorem 6.4 are can be used here.

e) The variational approach to viscoelasticity is originally due to Dafermos [75], [76]; see als Adali [2], Bouc, Geymonat, Jean, and Nayroles [30]. The methods used in these papers are based on perturbation techniques, their results use stronger assumptions than those of Section 9.4.

f) For the formulation of Maxwell's equations (9.49) and the transmission conditions (9.62) see any reasonable textbook on electrodynamics, e.g. Landau and Lifschitz [206], [205], Grant and Phillips [135], Simonyi [305].

g) It seems that already Maxwell knew about the phenomenon of dispersion. Hopkinson [183] was the first to use a constitutive relation of the form

$$\mathcal{D}(t) = \varepsilon_0 \mathcal{E}(t) + (\varepsilon_1 * \mathcal{E})(t), \quad t \in \mathbb{R},$$

while he kept the relations $\mathcal{B} = \mu \mathcal{H}$ and $\mathcal{J} = \sigma \mathcal{E}$. By means of these laws he was able to fit his data on the residual charge of a Leyden jar, choosing $\varepsilon_1(t)$ of the form

$$\varepsilon_1(t) = \sum_j a_j \exp(-\alpha_j t), \quad t > 0,$$

where $a_j, \alpha_j \geq 0$. In Volterra [331] the mathematical study of memory effects in electrodynamics was continued.

h) For more recent works on electrodynamics with memory see Toupin and Rivlin [326] and Bloom [25]. The constitutive laws for general time invariant linear media are according to Toupin and Rivlin [326] given by

$$\mathcal{D} = \sum_0^p a_j \mathcal{E}^{(j)} + \sum_0^p c_j \mathcal{B}^{(j)} + \varphi_1 * \mathcal{E} + \varphi_2 * \mathcal{B}$$

$$\mathcal{H} = \sum_0^p d_j \mathcal{E}^{(j)} + \sum_0^p b_j \mathcal{B}^{(j)} + \psi_1 * \mathcal{E} + \psi_2 * \mathcal{B},$$

where a_j, b_j, c_j, d_j are matrices and φ_i, ψ_i are integrable matrix-valued functions.

Chapter III

Equations on the Line

10 Integrability of Resolvents

The connections between stability properties of Volterra equations and integrability of the corresponding resolvent are discussed in Section 10.1; this discussion motivates the study of integrability of resolvents. For the classes of equations of scalar type introduced in Sections 2, 3, and 4 a complete characterization of integrability of $S(t)$ in terms of spectral conditions is derived. For nonscalar parabolic problems sufficient conditions are presented in a fairly general setting, while for nonscalar hyperbolic problems the analysis is valid in Hilbert spaces only, as counterexamples show.

10.1 Stability on the Halfline

Consider the equation of scalar type

$$u(t) = f(t) + \int_0^t a(t - \tau)Au(\tau)d\tau, \quad t \in \mathbb{R}_+, \tag{10.1}$$

on the halfline $\mathbb{R}_+$, where $a \in L^1_{loc}(\mathbb{R}_+)$ is Laplace transformable and A is a closed linear operator in X with dense domain $\mathcal{D}(A)$. In Section 8.6 we introduced the concept of *strong L^p-stability* of (10.1) and already obtained sufficient conditions. In general, however, this concept is too strong to be meaningful; below we show that strong L^p-stability of (10.1) for some $p \in [1, \infty]$ already implies that (10.1) is parabolic in the sense of Definition 3.1. The following concepts are more appropriate in hyperbolic situations.

Definition 10.1 *(i) Equation (10.1) is called L^p-stable (R), if for each $g \in L^p(\mathbb{R}_+; X)$ there is a unique a.e. mild solution $u \in L^p(\mathbb{R}_+; X)$ of (10.1) with $f = a * g$.*
(ii) Equation (10.1) is called L^p-stable (S) if for each $f \in W^{1,p}_{loc}(\mathbb{R}_+; X)$, $\dot{f} \in L^p(\mathbb{R}_+; X)$ there is a unique a.e. mild solution $u \in L^p(\mathbb{R}_+; X)$ of (10.1).

If (10.1) is L^p-stable (R) then for each $f \in L^p(\mathbb{R}_+; X_A)$ there is a unique a.e. mild solution $u \in L^p(\mathbb{R}_+; X)$ of (10.1); in fact, $u = f + v$, where v denotes the a.e. mild solution of $v = a * Af + a * Av$. Moreover, $g \in L^p(\mathbb{R}_+; X_A)$ or $f, Af \in L^p(\mathbb{R}_+; X_A)$ imply $u \in L^p(\mathbb{R}_+; X_A)$. Similarly, if (10.1) is L^p-stable (S) then $f \in W^{1,p}(\mathbb{R}_+; X_A)$ implies $u \in L^p(\mathbb{R}_+; X_A)$ is an a.e. strong solution of (10.1).

To obtain sufficient conditions for L^p-stability in terms of the resolvents $S(t)$ and $R(t)$ associated with (10.1) we have to introduce several notions of integrability of operator families.

Definition 10.2 *Let $\{T(t)\}_{t \geq 0} \subset \mathcal{B}(X, Z)$ be a strongly measurable family of operators, i.e. $T(\cdot)x$ is Bochner-measurable in Z, for each $x \in X$. Then $T(t)$ is called*
*(i) **strongly integrable**, if $T(\cdot)x \in L^1(\mathbb{R}_+; Z)$ for each $x \in X$;*
*(ii) **integrable** if there is $\varphi \in L^1(\mathbb{R}_+)$ such that $|T(t)| \leq \varphi(t)$ a.e. on $\mathbb{R}_+$;*
*(iii) **uniformly integrable**, if $T(\cdot) \in L^1(\mathbb{R}_+; \mathcal{B}(X, Z))$.*

Obviously, every uniformly integrable operator family $T(t)$ is integrable, however not conversely, unless $T(\cdot)$ is Bochner-measurable in $\mathcal{B}(X, Z)$. Similarly, every integrable family $T(t)$ is also strongly integrable, but the converse is not true in general.

Suppose now (10.1) admits an integrable and bounded resolvent $S(t)$ and let $f \in W^{1,p}_{loc}(\mathbb{R}_+; X)$, such that $\dot{f} \in L^p(\mathbb{R}_+; X)$, $1 \le p \le \infty$. Then the mild solution of (10.1) is given by the variation of parameters formula

$$u(t) = S(t)f(0) + \int_0^t S(t - \tau)\dot{f}(\tau)d\tau, \quad t \in \mathbb{R}_+, \tag{10.2}$$

and therefore we obtain $u \in L^p(\mathbb{R}_+; X)$; observe that $W^{1,p}_{loc}(\mathbb{R}_+; X) \hookrightarrow C(\mathbb{R}_+; X)$. Thus (10.1) is L^p-stable (S) for each $p \in [1, \infty]$.

Even more is true. If 10.1 is L^1-stable(S), choosing $f(t) \equiv x$ we obtain $S(\cdot)x \in L^1(\mathbb{R}_+; X)$ for each $x \in X$, i.e. $S(t)$ is strongly integrable. Conversely, if $S(t)$ is strongly integrable and $\dot{f} \in L^1(\mathbb{R}_+; X)$ is a simple function, say $\dot{f} = \sum \chi_{A_i} x_i$, then

$$u = Sf(0) + \sum \chi_{A_i} * (Sx_i)$$

is well-defined, and

$$|u|_1 \le |Sf(0)|_1 + \sum |Sx_i|_1 \mathrm{mes}(A_i) \le M(|f(0)| + |\dot{f}|_1),$$

where $M = \sup\{|Sx|_1 : |x| \le 1\}$. Since simple functions are dense in $L^1(\mathbb{R}_+; X)$, this estimate shows that (10.1) is L^1-stable(S). Thus L^1-stability is equivalent to strong integrability of $S(t)$.

Similarly, suppose (10.1) admits an integrable integral resolvent $R(t)$; cp. Section 1.6. Then the mild solution of (10.1) for $f = a * g$, $g \in L^p(\mathbb{R}_+; X_A)$ is given by

$$u(t) = \int_0^t R(t - \tau)g(\tau)d\tau, \quad t \in \mathbb{R}_+, \tag{10.3}$$

and so $u \in L^p(\mathbb{R}_+; X)$. Thus (10.1) is L^p-stable (R) for each $p \in [1, \infty]$.

Next let us derive necessary conditions for L^p-stability in terms of Laplace transforms. For this purpose we let as before

$$H(\lambda) = \frac{1}{\lambda}(I - \hat{a}(\lambda)A)^{-1}, \tag{10.4}$$

and

$$K(\lambda) = \hat{a}(\lambda)(I - \hat{a}(\lambda)A)^{-1}, \tag{10.5}$$

wherever these operators are defined.

Now suppose (10.1) is L^p-stable (R) for some $p \in [1, \infty]$. By $p' \in [1, \infty]$ we denote the conjugate exponent defined by $1/p + 1/p' = 1$. Let $f = a * g$, $g = e^{-\mu t}x$, where $x \in X$ and $\mu \in \mathbb{C}_+$ are fixed. Then there is a unique solution $u_\mu(\cdot; x) \in L^p(\mathbb{R}_+; X)$, and by the closed graph theorem there is a constant $C > 0$ such that

$$|u_\mu(\cdot; x)|_p \le C|g|_p = C(p\mathrm{Re}\,\mu)^{-1/p}|x|. \tag{10.6}$$

Define an operator family $\{U_\mu(t)\}_{t\geq 0} \subset \mathcal{B}(X)$ by means of

$$U_\mu(t)x = \int_0^t u_\mu(s;x)ds, \quad t \geq 0, \ x \in X;$$

then $U_\mu(t)$ is strongly continuous and $|U_\mu(t)| \leq c(\mu)t^{1/p'}$, $t \geq 0$. By uniqueness of the solutions of (10.1) we have $Au_\mu(t;x) = u_\mu(t;Ax)$ for a.e. $t \in \mathbb{R}_+$, and all $x \in D(A)$, hence $U_\mu(t)$ commutes with A for each $t \geq 0$. Taking Laplace transforms, (10.1) implies

$$(I - \hat{a}(\lambda)A)\hat{U}_\mu(\lambda)x = \hat{a}(\lambda)[\lambda(\lambda+\mu)]^{-1}x, \quad x \in X,$$

for Re λ sufficiently large. Since $\hat{U}_\mu(\lambda)$ commutes with A this identity shows that $I - \hat{a}(\lambda)A$ is bijective for all sufficiently large Re λ for which $\hat{a}(\lambda) \neq 0$ and

$$\hat{a}(\lambda)(I - \hat{a}(\lambda)A)^{-1} = \lambda(\lambda+\mu)\hat{U}_\mu(\lambda).$$

This identity shows as in Section 2.1 that $\hat{a}(\lambda)$ admits a meromorphic extension to $\mathbb{C}_+$ and $\hat{a}(\lambda)$ can only vanish if A is bounded; $1/\hat{a}(\lambda) \in \rho(A)$ on $\mathbb{C}_+$. This shows that $H(\lambda)$ and $K(\lambda)$ defined by (10.4), (10.5), respectively, exist on $\mathbb{C}_+$ and are holomorphic in $\mathcal{B}(X, X_A)$. Moreover, with (10.6) we obtain

$$|\hat{u}_\mu(\lambda;x)| \leq |u_\mu|_p(p'\text{Re }\lambda)^{-1/p'} \leq C'(\text{Re }\mu)^{-1/p}(\text{Re }\lambda)^{-1/p'}|x|, \quad \text{Re }\lambda, \ \mu > 0,$$

and this implies with $\hat{u}_\mu(\lambda;x)(\lambda+\mu) = K(\lambda)x$

$$|K(\lambda)| \leq C'(\text{Re }\mu)^{-1/p}(\text{Re }\lambda)^{-1/p'}|\lambda+\mu|, \quad \text{Re }\lambda, \ \text{Re }\mu > 0.$$

Choosing $\mu = \bar{\lambda}$ this inequality results in $|K(\lambda)| \leq 2C'$, Re $\lambda > 0$, i.e. $K(\lambda)$ is uniformly bounded in $\mathcal{B}(X)$ on $\mathbb{C}_+$. In case $p = 1$ we even see that $K(\lambda)$ admits a strongly continuous and uniformly bounded extension to $\overline{\mathbb{C}}_+$, since $\hat{u}_\mu(\lambda;x)$ is continuous on $\overline{\mathbb{C}}_+$.

If (10.1) is even strongly L^p-stable we obtain by similar reasoning the stronger estimate

$$|AK(\lambda)| + |K(\lambda)| \leq C, \quad \text{Re }\lambda > 0,$$

which implies in particular $|\lambda H(\lambda)| \leq C$ on $\mathbb{C}_+$, i.e. Equation (10.1) is parabolic in the sense of Definition 3.1.

Let us study the behaviour of $K(\lambda)$ on the imaginary axis in more detail. Suppose $K(\lambda)$ is strongly continuous and bounded on $\overline{\mathbb{C}}_+$, and consider a point $i\rho$, $\rho \in \mathbb{R}$, let $(\lambda_n) \subset \mathbb{C}_+$, $\lambda_n \to i\rho$, be such that $1/\hat{a}(\lambda_n) = z_n \to z \in \mathbb{C}$. Then $z \in \rho(A)$ and $K(\lambda_n) \to (z - A)^{-1}$ in $\mathcal{B}(X)$; if $z_n \to \infty$ then, by the boundedness of $K(\lambda_n)$, $K(\lambda_n) \to 0$ strongly. Therefore we see that $\lim_{\lambda \to i\rho} \hat{a}(\lambda) = \hat{a}(i\rho) \in \mathbb{C} \cup \{\infty\}$ exists for each $\rho \in \mathbb{R}$, i.e. $\hat{a}(\lambda)$ admits continuous extension to $\overline{\mathbb{C}}_+$ and the representation (10.1) holds also on $\overline{\mathbb{C}}_+$. Observe that in case $\hat{a}(\lambda) = \infty$ for some $\lambda \in \overline{\mathbb{C}}_+$ then A is necessarily invertible.

We summarize what we have proved so far in

Proposition 10.1 *(i) If (10.1) is strongly L^p-stable then (10.1) is parabolic in the sense of Definition 3.1.*
(ii) If (10.1) admits an integrable integral resolvent $R(t)$ then (10.1) is L^p-stable (R) for each $p \in [1, \infty]$.
(iii) If (10.1) is L^p-stable (R) for some $p \in [1, \infty]$, then $\hat{a}(\lambda)$ admits meromorphic extension to $\mathbb{C}_+$, $1/\hat{a}(\lambda) \in \rho(A)$, and $K(\lambda)$ defined by (10.5) is uniformly bounded on $\mathbb{C}_+$; $\hat{a}(\lambda)$ does not vanish on $\mathbb{C}_+$ in case A is unbounded, and can only have poles if A is invertible.
(iv) If (10.1) is L^1-stable (R) then $\hat{a}(\lambda)$ admits continuous extension in $\mathbb{C} \cup \{\infty\}$ to $\overline{\mathbb{C}}_+$, $1/\hat{a}(\lambda) \in \rho(A)$, and $K(\lambda)$ defined by (10.5) is strongly continuous on $\overline{\mathbb{C}}_+$; $\hat{a}(\lambda) = \infty$ for some $\lambda \in \overline{\mathbb{C}}_+$ implies $0 \in \rho(A)$.

On the other hand, if (10.1) is L^p-stable (S) then a reasoning similar to the one above yields $|H(\lambda)| \le M$ on $\mathbb{C}_+$, and a strongly continuous extension of $H(\lambda)$ to $\overline{\mathbb{C}}_+$ in case $p = 1$.

Next we consider the behaviour of $H(\lambda)$ on $i\mathbb{R}$ in case $p = 1$. If $0 \in \sigma(A)$ then $|z(z - A)^{-1}| \ge 1$ for all $z \in \rho(A)$, hence

$$M \ge \frac{1}{|\lambda|}|(I - \hat{a}(\lambda)A)^{-1}| \ge 1/|\lambda|,$$

a contradiction. Therefore, boundedness of $H(\lambda)$ near zero implies already $0 \in \rho(A)$. Let $(\lambda_n) \subset \mathbb{C}_+$, $\lambda_n \to i\rho \ne 0$, be such that $\hat{a}(\lambda_n) \to z \in \mathbb{C} \cup \{\infty\}$; then $H(\lambda_n) \to H(i\rho) = (I - zA)^{-1}/i\rho$ in $\mathcal{B}(X)$ in case $z \ne 0, \infty$, $H(\lambda_n) \to H(i\rho) = 1/i\rho$ strongly in case $z = 0$, and $H(\lambda_n) \to H(i\rho) = 0$ in $\mathcal{B}(X)$ in case $z = \infty$. This shows that $\hat{a}(\lambda)$ admits continuous extension in $\mathbb{C} \cup \{\infty\}$ to $\overline{\mathbb{C}}_+ \setminus \{0\}$. Similarly, for $\lambda_n \to 0$ we obtain in case $z \in \mathbb{C}$

$$0 = \lim_{n \to \infty} \lambda_n H(\lambda_n)x = \lim_{n \to \infty} (I - \hat{a}(\lambda_n)A)^{-1}x = (I - zA)^{-1}x, \quad x \in X,$$

which is impossible, hence $\hat{a}(\lambda) \to \hat{a}(0) = \infty$ as $\lambda \to 0$. But then

$$-A^{-1}x \lim_{\lambda \to 0} (\lambda\hat{a}(\lambda))^{-1} = \lim_{\lambda \to 0} (\lambda\hat{a}(\lambda))^{-1}(1/\hat{a}(\lambda) - A)^{-1}x = \lim_{\lambda \to 0} H(\lambda)x = H(0)x,$$

i.e. $\lambda\hat{a}(\lambda) \to a(\infty)$ as $\lambda \to 0$ exists and is nonzero but possibly infinity. Let us summarize these results on L^p-stability (S) in

Proposition 10.2 *(i) If (10.1) admits an integrable and bounded resolvent $S(t)$ then (10.1) is L^p-stable (S) for each $p \in [1, \infty]$.*
(ii) If (10.1) is L^p-stable (S) for some $p \in [1, \infty]$ then $\hat{a}(\lambda)$ admits meromorphic extension to $\mathbb{C}_+$, $1/\hat{a}(\lambda) \in \rho(A)$, and $H(\lambda)$ defined by (10.4) is uniformly bounded on $\mathbb{C}_+$; A is invertible, and $\hat{a}(\lambda)$ does not vanish on $\mathbb{C}_+$ in case A is unbounded.
(iii) If (10.1) is L^1-stable (S) then $\hat{a}(\lambda)$ admits continuous extension to $\overline{\mathbb{C}}_+$, $1/\hat{a}(\lambda) \in \rho(A)$, and $H(\lambda)$ is strongly continuous on $\overline{\mathbb{C}}_+$; $\hat{a}(0) = \infty$ and the limit $\lim_{\lambda \to 0} \lambda\hat{a}(\lambda) = a(\infty)$ exists in $\mathbb{C} \cup \{\infty\}$ and is nonzero.
(iv) $S(t)$ is strongly integrable iff (10.1) is L^1-stable(S).

Note that we have the relations

$$\int_0^\infty R(t)dt = K(0) = (1/\hat{a}(0) - A)^{-1}, \qquad (10.7)$$

and

$$\int_0^\infty S(t)dt = H(0) = -A^{-1}/a(\infty), \qquad (10.8)$$

whenever R resp. S are strongly integrable.

In general, it is not true that the necessary conditions on $H(\lambda)$ resp. $K(\lambda)$ obtained in Proposition 10.2 and 10.1 are also sufficient for the integrability of $S(t)$ resp. $R(t)$, even if their existence is a priori known. There are several counterexamples in the case $\hat{a}(\lambda) = 1/\lambda$, i.e. the case of first order differential equations $\dot{u} = Au + g$; then

$$H(\lambda) = K(\lambda) = (\lambda - A)^{-1}, \quad \text{i.e. } R(t) = S(t) = e^{At}.$$

The next subsections deal with the sufficiency of the transform conditions for integrability under various extra assumptions frequently met in applications. At first only equations of scalar type are considered for which the theory is more complete, but later, extensions to nonscalar equations will also be discussed. As before, we concentrate on $S(t)$ which is the more important quantity.

10.2 Parabolic Equations of Scalar Type
Before necessary and sufficient conditions for uniform integrability of resolvents for parabolic equations of scalar type are derived we have to recall the notion of a locally analytic function on the right halfplane.

Definition 10.3 *Let $\mathbb{C}_+^\infty = \overline{\mathbb{C}}_+ \cup \{\infty\}$ and $\varphi : \mathbb{C}_+^\infty \to \mathbb{C}$. Then*
*(i) φ is called **locally analytic** at a point $\lambda_0 \in \overline{\mathbb{C}}_+$ if there are $r > 0$, n functions $k_j \in BV(\mathbb{R}_+)$, and a holomorphic function $\psi : \Omega \to \mathbb{C}$, where $\Omega \subset \mathbb{C}^{n+1}$ is an open neighbourhood of $(\lambda_0, \widehat{dk}_1(\lambda_0), \ldots, \widehat{dk}_n(\lambda_0)) \in \Omega$, such that*

$$\varphi(\lambda) = \psi(\lambda, \widehat{dk}_1(\lambda), \ldots, \widehat{dk}_n(\lambda)), \quad \lambda \in B_r(\lambda_0) \cap \overline{\mathbb{C}}_+. \qquad (10.9)$$

*(ii) φ is called **locally analytic** at ∞ if there are $R > 0$, n functions $a_j \in L^1(\mathbb{R}_+)$, m functions $k_j \in BV(\mathbb{R}_+)$ and a holomorphic function $\psi : \Omega \to \mathbb{C}$, where $\Omega \subset \mathbb{C}^{n+m+1}$ is open with $0 \in \Omega$, and*

$$\varphi(\lambda) = \psi(\frac{1}{\lambda}, \hat{a}_1(\lambda), \ldots, \hat{a}_n(\lambda), \hat{k}_1(\lambda), \ldots, \hat{k}_m(\lambda)), \quad \lambda \in \mathbb{C}_+^\infty \setminus B_R(0). \qquad (10.10)$$

*(iii) φ is called **locally analytic** on $D \subset \mathbb{C}_+^\infty$ if it is locally analytic at every point of D.*

Observe that a function φ is locally analytic on $\mathbb{C}_+$ if and only if is holomorphic on $\mathbb{C}_+$. The number n and the functions k_j in Definition 10.3 (i) may vary from point to point, hence there is a lot of flexibility in proving that a given function is locally analytic on $\mathbb{C}_+^\infty$. Note that locally analytic functions are continuous. The reason for our interest in locally analytic functions is the following result.

Lemma 10.1 *Suppose φ is locally analytic on $\mathbb{C}_+^\infty$. Then there is a function $b \in L^1(\mathbb{R}_+)$ such that*

$$\varphi(\lambda) = \varphi(\infty) + \hat{b}(\lambda) \quad \text{for all } \lambda \in \mathbb{C}_+^\infty. \tag{10.11}$$

Lemma 10.1 is an extension of the general Wiener-Levy theorem for the Laplace transform, which is essentially due to Gelfand, Raikhov and Shilov [128]. The proof relies on several results in complex function theory of several variables and cannot be given here; cp. Section 10.7 for further remarks.

It is not difficult to show that in the definition of local analyticity at $\lambda_0 \in \overline{\mathbb{C}}_+$, the measures dk_j can be replaced by L^1-functions and the explicit dependence on λ can be dropped, by passing to a different holomorphic function ψ_0. In fact, with $\hat{a}_0(\lambda) = (\lambda + 1)^{-1}$ we may write

$$
\begin{aligned}
\psi(\lambda, \widehat{dk}_1, \ldots, \widehat{dk}_n) &= \psi(\frac{1}{\hat{a}_0} - 1, \frac{1}{\hat{a}_0} \cdot \hat{a}_0 \widehat{dk}_1, \ldots, \frac{1}{\hat{a}_0} \cdot \hat{a}_0 \widehat{dk}_n) \\
&= \psi_0(\hat{a}_0, \hat{a}_1, \ldots, \hat{a}_n),
\end{aligned}
$$

where $a_j = a_0 * dk_j \in L^1(\mathbb{R}_+)$, and ψ_0 is analytic in a neighborhood of $(\hat{a}_0(\lambda_0)$, $\hat{a}_1(\lambda_0), \ldots, \hat{a}_n(\lambda_0))$; observe that $\hat{a}_0(\lambda_0) \neq 0$ for $\lambda_0 \in \overline{\mathbb{C}}_+$. Similarly, at ∞ the explicit dependence of ψ on $1/\lambda$, $\hat{k}_1, \ldots, \hat{k}_m$ can be dropped since $1/\lambda = \hat{a}_0/(1 - \hat{a}_0)$, and $\hat{k}_j(\lambda) = \widehat{dk}_j(\lambda)/\lambda$.

Another useful remark is the following. If φ is locally analytic on an open set $E \subset i\mathbb{R}$ and $\chi \in C_0^\infty(\mathbb{R})$ with supp $\chi \subset E$ then $\varphi(i\rho) \cdot \chi(\rho)$ is the Fourier transform of a function in $L^1(\mathbb{R})$. This can be proved similar to the Paley-Wiener lemma by a localization procedure and the general Wiener-Levy theorem for the Fourier transform.

Suppose now that (10.1) admits a strongly integrable resolvent $S(t)$; then for each $\lambda_0 \in \overline{\mathbb{C}}_+$ for which $\hat{a}(\lambda_0) \neq 0$ there are $x \in D(A)$, $x^* \in X^*$ such that $< x, x^* > = 1$ and $\lambda_0 < H(\lambda_0)x, x^* > \neq < x, x^* > = 1$. As in Section 2.1 this implies the representation

$$
\begin{aligned}
\frac{1}{\lambda \hat{a}(\lambda)} &= \frac{< H(\lambda)Ax, x^* >}{\lambda < H(\lambda)x, x^* > -1} \\
&= g(\lambda, \hat{a}_1(\lambda), \hat{a}_2(\lambda)), \quad \lambda \in B_r(\lambda_0) \cap \overline{\mathbb{C}}_+,
\end{aligned}
$$

where $a_1(t) = < S(t)Ax, x^* >$, $a_2(t) = < S(t)x, x^* >$ belong to $L^1(\mathbb{R}_+)$ by assumption, and the function $g(\lambda, z_1, z_2) = z_1/(\lambda z_2 - 1)$ is analytic on a neighborhood of $(\lambda_0, \hat{a}_1(\lambda_0), \hat{a}_2(\lambda_0))$. This shows that the function $1/\lambda \hat{a}(\lambda)$ is locally analytic on $\overline{\mathbb{C}}_+ \setminus \hat{a}^{-1}(0)$. It follows from Proposition 10.2 that $\hat{a}(\lambda) \neq 0$ on $\overline{\mathbb{C}}_+$ if A is unbounded and the resolvent $S(t)$ is integrable, since then $H(\lambda)$ is $\mathcal{B}(X)$-continuous on $\overline{\mathbb{C}}_+$.

We can now prove the following characterization of uniform integrability of analytic resolvents.

Theorem 10.1 *Suppose $S(t)$ is an analytic resolvent for (10.1), and let A be unbounded. Then $S(t)$ is uniformly integrable if and only if the following conditions hold.*
(I1) $1/\hat{a}(\lambda)$ *admits a continuous extension in $\mathbb{C}$ to $\overline{\mathbb{C}}_+ \setminus \{0\}$ which is holomorphic on $\mathbb{C}_+$ and $1/\hat{a}(\lambda) \in \rho(A)$ for all $\lambda \in \overline{\mathbb{C}}_+$, $\lambda \neq 0$;*
(I2) $\lambda\hat{a}(\lambda) \to a(\infty) \neq 0$ *as $\lambda \to 0$, and $0 \in \rho(A)$;*
(I3) $1/\lambda\hat{a}(\lambda)$ *is locally analytic on $i\mathbb{R}$.*

Proof: The necessity of the conditions is clear from Proposition 10.2 and the remarks in front of this theorem. To prove sufficiency recall from Theorem 2.1 that $H(\lambda)$ as well as $1/\hat{a}(\lambda)$ admit analytic extension to a sector $\Sigma(\omega_0, \theta_0 + \pi/2)$ and that $|H(\lambda)| \leq M(\omega, \theta)/|\lambda - \omega|$ holds on $\Sigma(\omega, \theta + \pi/2)$, $\omega > \omega_0$, $\theta < \theta_0$. From Cauchy's theorem we also obtain

$$|H'(\lambda)||\lambda - \omega|^2 + |H''(\lambda)||\lambda - \omega|^3 \leq M(\omega, \theta), \quad \lambda \in \Sigma(\omega, \theta + \pi/2).$$

(I1) and (I2) then show that $H(\lambda)$ is $\mathcal{B}(X)$-continuous on $\overline{\mathbb{C}}_+ \cup \Sigma((\omega, \theta + \pi/2)$ and bounded, and analytic in the interior of this set. Therefore, by Cauchy's theorem and (2.10), we obtain the representation

$$S(t) = \frac{1}{2\pi i} \int_\Gamma H(\lambda) e^{\lambda t} d\lambda, \quad t \in \mathbb{R}_+,$$

where Γ consists of the two halfrays $\pm i(N + 3) + e^{\pm i(\frac{\pi}{2} + \theta)}\mathbb{R}_+$ and the interval $[-i(N + 3), i(N + 3)]$, with $N > \omega \cot\theta$.

Choose a $C_0^\infty(\mathbb{R})$-function $\tilde{\varphi}(\rho)$ such that $\tilde{\varphi}(\rho) = 1$ for $|\rho| \leq N + 1$, $\tilde{\varphi}(\rho) = 0$ for $|\rho| \geq N + 2$, $0 \leq \tilde{\varphi} \leq 1$ elsewhere. Then $S(t)$ decomposes according to

$$
\begin{aligned}
S(t) &= \frac{1}{2\pi} \int_{-\infty}^\infty H(i\rho)\tilde{\varphi}(\rho)e^{i\rho t}d\rho + \frac{1}{2\pi i} \int_\Gamma H(\lambda)\psi(\lambda)e^{\lambda t}d\lambda \\
&= S_1(t) + S_2(t), \quad t \in \mathbb{R}_+,
\end{aligned}
$$

where $\psi(i\rho) = 1 - \tilde{\varphi}(\rho)$, $\rho \in \mathbb{R}$, $\psi(\lambda) = 1$ for $\lambda \notin i\mathbb{R}$. Integrating by parts twice in the integral defining $S_2(t)$ we obtain an estimate of the form $|S_2(t)| \leq C(1+t^2)^{-1}$, $t \geq 0$, hence $S_2 \in L^1(\mathbb{R}_+; \mathcal{B}(X))$. Although $\tilde{\varphi} \in C_0^\infty(\mathbb{R})$, this trick cannot be applied directly to $S_1(t)$, since we only know that $H(i\rho)$ is $\mathcal{B}(X)$-continuous on $\mathbb{R}$, in particular at $\rho = 0$. To circumvent this difficult we use the Paley-Wiener theorem and local analyticity of $1/\lambda\hat{a}(\lambda)$ on $i\mathbb{R}$. For this purpose approximate the function $b(\rho) = (i\rho\hat{a}(i\rho))^{-1}$ by some $b_\varepsilon \in C_0^\infty(\mathbb{R})$ uniformly on $[-(N+5), N+5]$, and consider the operator-valued function $H_\varepsilon(\rho)$ defined by

$$H_\varepsilon(\rho) = \tilde{\varphi}(\rho)b_\varepsilon(\rho)(b_\varepsilon(\rho)i\rho - A)^{-1}, \quad \rho \in \mathbb{R},$$

where $\varepsilon \geq 0$ is small enough. Since $H_\varepsilon(\rho)$ has compact support and is of class C^∞ there follows $H_\varepsilon(\rho) = \tilde{S}_{1\varepsilon}(\rho)$ for some $S_{1\varepsilon} \in L^1(\mathbb{R}; \mathcal{B}(X))$. We have the identity

$$
\begin{aligned}
\tilde{\varphi}(\rho)H(i\rho) &= H_\varepsilon(\rho) + \tilde{\varphi}(\rho)(b(\rho) - b_\varepsilon(\rho))(b_\varepsilon(\rho)i\rho - A)^{-1} \\
&\quad - i\rho(b(\rho) - b_\varepsilon(\rho))(b_\varepsilon(\rho)i\rho - A)^{-1}\tilde{\varphi}(\rho)H(i\rho), \quad \rho \in \mathbb{R},
\end{aligned}
$$

hence with $\tilde{\varphi}_1(\rho) = \tilde{\varphi}(\rho \frac{N+2}{N+4})$

$$
\tilde{\varphi}(\rho)H(i\rho) = H_\varepsilon(\rho) + \tilde{\varphi}(\rho)K_\varepsilon(\rho) - i\rho K_\varepsilon(\rho)\tilde{\varphi}(\rho)H(i\rho), \quad \rho \in \mathbb{R}, \tag{10.12}
$$

where

$$
K_\varepsilon(\rho) = \tilde{\varphi}_1(\rho)(b(\rho) - b_\varepsilon(\rho))(b_\varepsilon(\rho)i\rho - A)^{-1}, \quad \rho \in \mathbb{R}.
$$

Since $1/\lambda \hat{a}(\lambda)$ is locally analytic on $i\mathbb{R}$, there follows existence of a function $R_\varepsilon \in W^{1,1}(\mathbb{R}; \mathcal{B}(X))$ such that $\tilde{R}_\varepsilon(\rho) = K_\varepsilon(\rho)$, $\rho \in \mathbb{R}$. Thus consider the convolution equation corresponding to (10.12), i.e.

$$
S_1 = (S_{1\varepsilon} + \varphi * R_\varepsilon) + \dot{R}_\varepsilon * S_1 \tag{10.13}
$$

and apply the Paley-Wiener lemma to obtain $S_1 \in L^1(\mathbb{R}; \mathcal{B}(X))$; the Paley-Wiener condition is satisfied since

$$
I - \tilde{\dot{R}}_\varepsilon(\rho) = \begin{cases} I & \text{for } \rho \notin M := \operatorname{supp} \tilde{\varphi}_1 \\ \{(\tilde{\varphi}_1(\rho)b(\rho) + (1 - \tilde{\varphi}_1(\rho))b_\varepsilon(\rho))i\rho - A\}(b_\varepsilon(\rho)i\rho - A)^{-1}, & \rho \in M, \end{cases}
$$

is invertible for all $\rho \in \mathbb{R}$, provided $\varepsilon > 0$ is chosen small enough. Therefore $\tilde{\varphi}(\rho)H(i\rho)$ is the Fourier transform of some function $S_1 \in L^1(\mathbb{R}; \mathcal{B}(X))$, and so the resolvent for (10.1) is uniformly integrable. $\square$

For an application of Theorem 10.1 let us take up Examples 2.2 again. So let A generate a bounded analytic semigroup in X and let $a \in L^1_{loc}(\mathbb{R}_+)$ be completely monotonic. Then by Corollary 2.4 there is an analytic resolvent for (10.1) which in fact is also bounded. (I1) is trivially satisfied, and (I2) holds if $a(\infty) = \lim_{\lambda \to 0} \lambda \hat{a}(\lambda) = \lim_{t \to \infty} a(t) > 0$ and A is invertible. Since $1/\hat{a}(\lambda)$ is a Bernstein function we obtain

$$
\frac{1}{\lambda \hat{a}(\lambda)} = k_0 + \frac{k_\infty}{\lambda} + \hat{k}_1(\lambda), \quad \operatorname{Re}\lambda > 0,
$$

for some $k_0, k_\infty \geq 0$, $k_1 \in L^1_{loc}(\mathbb{R}_+)$ completely monotonic, thanks to Proposition 4.6. $a(\infty) > 0$ then implies $k_\infty = 0$, $k_1 \in L^1(\mathbb{R}_+)$ and so $1/\lambda \hat{a}(\lambda)$ is locally analytic on $\mathbb{C}_+^\infty$, i.e. (I3) holds. We have proved

Corollary 10.1 *Suppose A generates an analytic C_0-semigroup, bounded on some sector $\Sigma(0, \theta)$, and is invertible, let $a \in L^1_{loc}(\mathbb{R}_+)$ be completely monotonic and such that $a(\infty) = \lim_{t \to \infty} a(t) > 0$. Then (10.1) admits an uniformly integrable analytic resolvent.*

Another simple computation shows that the resolvent $S(t)$ from Example 2.1 is uniformly integrable iff $\beta \geq 1$.

By the same method of proof, only a slight modification is needed, one obtains a characterization of uniform integrability of resolvents for parabolic equations with 2-regular kernels.

Theorem 10.2 *Suppose (10.1) is parabolic in the sense of Definition 3.1 and let $a(t)$ be 2-regular. Then $S(t)$ is uniformly integrable if and only if $1/\lambda \hat{a}(\lambda)$ is locally analytic at $\lambda = 0$ and $0 \in \rho(A)$.*

Proof: We only have to prove sufficiency. From Definition 3.1 and 2-regularity there follows by Lemma 8.1 that $H(\lambda)$ admits a $\mathcal{B}(X)$-continuous extension to $\mathbb{C}_+^\infty \setminus \{0\}$, $H(\infty) = 0$. Since $0 \in \rho(A)$ by assumption and $\lim_{\lambda \to 0} \lambda \hat{a}(\lambda) = a(\infty) \neq 0$ by local analyticity of $1/\lambda \hat{a}(\lambda)$, we see that $H(\lambda)$ is also continuous at $\lambda = 0$. Therefore we have $|H(\lambda)| \leq M/(1 + |\lambda|)$, Re $\lambda \geq 0$, as well as $|\lambda^2 H'(\lambda)| + |\lambda^3 H''(\lambda)| \leq M$, Re $\lambda > 0$, as in the proof of Theorem 3.1, since $a(t)$ is by assumption 2-regular. For fixed $x \in X$, $x^* \in X^*$, the L^2-theory of the Laplace transform therefore yields the representation

$$< S(t)x, x^* > = \frac{1}{2\pi} \int_{-\infty}^{\infty} < H(i\rho)x, x^* > e^{i\rho t} d\rho, \quad \text{a.a. } t \in \mathbb{R}_+. \qquad (10.14)$$

With $\tilde{\varphi}$ as in the proof of Theorem 10.1, $N > 0$ being arbitrarily fixed, after two integrations by parts (10.14) becomes

$$
\begin{aligned}
< S(t)x, x^* > \ = \ & \frac{1}{2\pi} \int_{-\infty}^{\infty} < \tilde{\varphi}(\rho)H(i\rho)x, x^* > e^{i\rho t} d\rho \\
& - \frac{1}{2\pi t^2} \int_{-\infty}^{\infty} < [(1 - \tilde{\varphi}(i\rho))H(i\rho)]'' x, x^* > e^{i\rho t} d\rho
\end{aligned}
$$

for a.a. $t \in \mathbb{R}_+$. Since both integrals in this formula exist absolutely we may drop x, x^* to the result

$$
\begin{aligned}
S(t) \ = \ & \frac{1}{2\pi} \int_{-\infty}^{\infty} \tilde{\varphi}(\rho)H(i\rho)e^{i\rho t} d\rho - \frac{1}{2\pi t^2} \int_{-\infty}^{\infty} [(1 - \tilde{\varphi}(i\rho))H(i\rho)]'' e^{i\rho t} d\rho \\
= \ & S_1(t) + S_2(t), \quad t \in \mathbb{R}_+.
\end{aligned}
$$

Since $S(t)$ is already known to be bounded by Theorem 3.1 and $S_1(t)$ obviously is, $S_2(t)$ is bounded, but also $|S_2(t)| \leq C/t^2$ holds, i.e. $S_2 \in L^1(\mathbb{R}; \mathcal{B}(X))$. On the other hand, $S_1(t)$ can be treated as in the proof of Theorem 10.1, once it is known that $1/\lambda \hat{a}(\lambda)$ is locally analytic on $i\mathbb{R}$. By assumption this is true for $\lambda = 0$, and for $\lambda \in i\mathbb{R}$, $\lambda \neq 0$, this follows from Lemma 10.2 given below. The proof is complete. $\square$

The following lemma contains conditions sufficient for local analyticity of the function $(\lambda \hat{a}(\lambda)))^{-1}$.

Lemma 10.2 *Suppose $a(t)$ is 1-regular and let $\varphi(\lambda) = 1/(\lambda^k \hat{a}(\lambda))$ for some $k \in \mathbb{N}_0$. Then*
(i) $\varphi(\lambda)$ is locally analytic on $\overline{\mathbb{C}}_+ \setminus \{0\}$;
(ii) if $\lim_{|\lambda| \to 0} \varphi(\lambda) = \varphi(0) \in \mathbb{C}$ exists and $\varphi'(i\cdot) \in L^1(-1,1)$, then $\varphi(\lambda)$ is locally analytic at $\lambda = 0$;
(ii) if $\lim_{|\lambda| \to \infty} \varphi(\lambda) = \varphi(\infty) \in \mathbb{C}$ exists and $\varphi'(i\cdot) \in L^1(\mathbb{R} \setminus [-1,1])$ then $\varphi(\lambda)$ is locally analytic at $\lambda = \infty$.

Proof: (i) From Lemma 8.1 (iv) and (v) there follows the existence of a constant $c > 0$ and $N \in \mathbb{N}$ such that $|\hat{a}(\lambda)| \geq c|\lambda|^{-N}$ as $|\lambda| \to \infty$ and $|\hat{a}(\lambda)| \geq c|\lambda|^N$ as $|\lambda| \to 0$. The function

$$\psi(\lambda) = \frac{\lambda^{N+1}}{(1+\lambda)^{2N+2}\hat{a}(\lambda)}, \qquad \mathrm{Re}\,\lambda > 0,$$

is therefore bounded on $\mathbb{C}_+$, and by 1-regularity of $a(t)$, ψ' belongs to $H^1(\mathbb{C}_+)$. Hardy's inequality (cp. Duren [106]) then yields $b \in L^1(\mathbb{R}_+)$ such that $\hat{b} = \psi$; but this leads to

$$\varphi(\lambda) = \frac{1}{\lambda^k \hat{a}(\lambda)} = \hat{b}(\lambda) \cdot \frac{(1+\lambda)^{2N+2}}{\lambda^{N+1+k}}, \qquad \lambda \in \overline{\mathbb{C}}_+ \setminus \{0\},$$

i.e. $\varphi(\lambda)$ is locally analytic on $\overline{\mathbb{C}}_+ \setminus \{0\}$.
(ii) For the second assertion consider correspondingly

$$\psi(\lambda) = \frac{1}{(\lambda+1)^{N+1}\lambda^k \hat{a}(\lambda)}, \qquad \mathrm{Re}\,\lambda > 0;$$

then again $\psi(\lambda)$ is bounded on $\mathbb{C}_+$, and ψ' belongs to $H^1(\mathbb{C}_+)$, for $|\psi'(\lambda)| \leq c/|\lambda|^2$ as $|\lambda| \to \infty$, and

$$\int_{-1}^{1} |\psi'(i\rho)|d\rho \leq C + C \int_{-1}^{1} |\varphi'(i\rho)|d\rho < \infty.$$

Thus $\psi(\lambda) = \hat{b}(\lambda)$ for some $b \in L^1(\mathbb{R}_+)$ by Hardy's inequality again, and so

$$\varphi(\lambda) = \frac{1}{\lambda^k \hat{a}(\lambda)} = \hat{b}(\lambda)(1+\lambda)^{N+1}, \qquad \lambda \in \overline{\mathbb{C}}_+,$$

i.e. $\varphi(\lambda)$ is locally analytic on $\overline{\mathbb{C}}_+$.
(iii) is proved similarly by taking

$$\psi(\lambda) = \left(\frac{\lambda}{1+\lambda}\right)^{N+k+1} \cdot \frac{1}{\lambda^k \hat{a}(\lambda)}, \qquad \mathrm{Re}\,\lambda > 0. \quad \square$$

In the situation of Corollary 3.2, $1/\lambda\hat{a}(\lambda)$ is locally analytic at zero iff $a(\infty) > 0$; in fact, we may write

$$\frac{1}{\lambda\hat{a}(\lambda)} = \frac{1}{\lambda\hat{a}_1(\lambda) + \widehat{da_2}(\lambda)} = f(\lambda, \hat{a}_1(\lambda), \widehat{da_2}(\lambda)), \qquad \mathrm{Re}\,\lambda \geq 0,$$

where $a_1 \in L^1(\mathbb{R}_+)$, $a_2 \in BV(\mathbb{R}_+)$, and $f(\lambda, z, w)$ is analytic near $\lambda_0 = 0$, $z_0 = \hat{a}_1(0)$, $w_0 = \widehat{da_2}(0) = a(\infty) > 0$. Therefore, $a(\infty) > 0$ and $0 \in \rho(A)$ imply with Theorem 10.2 that the resolvent $S(t)$ obtained in Corollary 3.2 is uniformly integrable.

Similarly, in the situation of Corollary 3.3 where $a(t) = a_0 + \int_0^t a_1(s)ds$, $a_0 \geq 0$, a_1 2-monotone, we have $a(\infty) = a_0 + \int_0^\infty a_1(t)dt > 0$, unless $a(t) \equiv 0$. Setting

$$\varphi(t) = \int_0^t a_1(s)ds, \quad \psi(t) = \int_0^t sa_1(s)ds, \ t > 0,$$

with Propositions 3.8 and 3.9 we get

$$\int_{-1}^1 |\frac{d}{d\rho}(i\rho\hat{a}(i\rho))^{-1}|d\rho = \int_{-1}^1 \frac{|\hat{a}_1'(i\rho)|}{|a_0 + \hat{a}_1(i\rho)|^2}d\rho \leq C \int_0^1 \psi(1/\rho)\varphi(1/\rho)^{-2}d\rho$$

$$= C \int_1^\infty \psi(t)\varphi(t)^{-2}dt/t^2 \leq C(\psi(1)\varphi(1)^{-2} + \int_1^\infty \psi'(t)\varphi(t)^{-2}dt/t)$$

$$= C(\psi(1)\varphi(1)^{-2} + \int_1^\infty \varphi'(t)\varphi(t)^{-2}dt) \leq C(\psi(1)\varphi(1)^{-2} + \varphi(1)^{-1}) < \infty.$$

Therefore, $1/\lambda\hat{a}(\lambda)$ is locally analytic at $\lambda = 0$, by Lemma 10.2. Theorem 10.2 then shows that the resolvent $S(t)$ obtained in Corollary 3.3 is uniformly integrable if and only if $0 \in \rho(A)$.

10.3 Subordinated Resolvents
Suppose $b \in L^1_{loc}(\mathbb{R}_+)$ satisfies $\int_0^\infty |b(t)|e^{-\beta t}dt < \infty$ for some $\beta \geq 0$, let A be closed linear unbounded and densely defined, and assume there is a resolvent $S_b(t)$ for

$$v(t) = g(t) + \int_0^t b(t-s)Av(s)ds, \quad t \in \mathbb{R}_+, \tag{10.15}$$

such that

$$\omega_b = \overline{\lim}_{t\to\infty}t^{-1}\log|S_b(t)| < \infty.$$

Let $c(t)$ be a completely positive function with associated creep function

$$k(t) = \kappa + \omega t + \int_0^t k_1(s)ds, \quad t > 0, \tag{10.16}$$

where $\kappa, \omega \geq 0$, $k_1(t) \geq 0$ nonincreasing with $\lim_{t\to\infty} k_1(t) = 0$, and let $\alpha = \omega + k_1(0+)$. According to Theorem 4.1 there is a resolvent $S_a(t)$ for (10.1), where $a \in L^1_{loc}(\mathbb{R}_+)$ defined by

$$\hat{a}(\lambda) = \hat{b}(1/\hat{c}(\lambda)), \quad \text{Re } \lambda \text{ sufficiently large}, \tag{10.17}$$

is also of exponential growth, by Proposition 4.7. The growth bound of $S_a(t)$ is not greater than ω_a, where

$$\omega_a = \hat{c}^{-1}(1/\omega_b) \quad \text{if } \omega_b > 1/\omega, \quad \omega_a = 0 \text{ otherwise.} \tag{10.18}$$

In this subsection we want to discuss uniform and strong integrability of $S_a(t)$. Our approach is based on the Paley-Wiener lemma and depends heavily on the structure theorem, Theorem 4.4, for subordinated resolvents; therefore we must also assume that the function k_1 belongs to $W_{loc}^{1,1}(0,\infty)$. The decomposition (4.57) shows that $S_a(t)$ can only be expected to be $\mathcal{B}(X)$-measurable if $\alpha + 1/\kappa = \infty$, unless we restrict the class of kernels b and operators A, so that $S_b(t)$ is $\mathcal{B}(X)$-measurable. This indicates that the two cases $\alpha + 1/\kappa = \infty$ and $\alpha + 1/\kappa < \infty$ should be treated separately. We do not impose any extra condition on $b(t)$, except for a mild regularity assumption for $\hat{b}(z)$ at $z = \infty$.

We begin with the case $1/\kappa + \alpha = \infty$.

Theorem 10.3 *Suppose the function* $g(z) = z(\hat{b}(\eta + z) - \hat{b}(z))/\hat{b}(z)$ *is locally analytic at* ∞, *for some* $\eta > \max\{\omega_b, \beta\}$, *and let* $1/\kappa + \alpha = \infty$. *Then* $S_a(t)$ *is uniformly integrable if and only if Conditions (I1), (I2), (I3) of Theorem 10.1 are satisfied.*

Proof: The necessity of (I1), (I2), (I3) is already clear. To prove the sufficiency, fix any $\eta > \max\{\omega_b, \beta\}$ such that $g(z)$ is locally analytic at ∞. To apply the Paley-Wiener lemma Theorem 0.7 we have to have an object $S_\eta \in L^1(\mathbb{R}_+; \mathcal{B}(X))$ to start with. Define $H_\eta(\lambda)$ by means of

$$
\begin{aligned}
H_\eta(\lambda) &= z^{-1}(I - \hat{b}(z)A)^{-1} \circ (\eta + 1/\hat{c}(\lambda)) \\
&= \int_0^\infty S_b(\tau)e^{-\eta\tau}e^{-\tau/\hat{c}(\lambda)}d\tau, \quad \text{Re } \lambda \geq 0;
\end{aligned}
\tag{10.19}
$$

then $H_\eta(\lambda) = \hat{S}_\eta(\lambda)$, Re $\lambda \geq 0$, where

$$
S_\eta(t) = \int_0^\infty S_b(\tau)e^{-\eta\tau}w_t(t,\tau)d\tau, \quad t \geq 0,
\tag{10.20}
$$

belongs to $L^1(\mathbb{R}_+; \mathcal{B}(X))$ since $\alpha + 1/\kappa = \infty$ by assumption; thanks to Propositions 4.9 and 4.10. Next one verifies easily the identity

$$
H(\lambda) = \varphi_\eta(\lambda)H_\eta(\lambda) - \psi_\eta(\lambda)H_\eta(\lambda)H(\lambda), \quad \text{Re } \lambda > 0,
\tag{10.21}
$$

where

$$
\varphi_\eta(\lambda) = (\eta + 1/\hat{c}(\lambda))\hat{b}(\eta + 1/\hat{c}(\lambda))/\lambda\hat{a}(\lambda), \quad \text{Re } \lambda > 0,
$$

and

$$
\psi_\eta(\lambda) = (\eta + 1/\hat{c}(\lambda))(\hat{b}(\eta + 1/\hat{c}(\lambda)) - \hat{b}(1/\hat{c}(\lambda))/\hat{b}(1/\hat{c}(\lambda)), \quad \text{Re } \lambda > 0.
$$

Then the Paley-Wiener condition is implied by (I1), (I2) since

$$
\begin{aligned}
I + \psi_\eta H_\eta &= [(I - \hat{b}(\eta + 1/\hat{c})A) + \hat{b}(\eta + 1/\hat{c})/\hat{b}(1/\hat{c}) - I] \cdot (I - \hat{b}(\eta + 1/\hat{c})A)^{-1} \\
&= \hat{b}(\eta + 1/\hat{c})[1/\hat{a} - A](I - \hat{b}(\eta + 1/\hat{c})A)^{-1}
\end{aligned}
$$

is invertible for each Re $\lambda \geq 0$ and

$$(I + \psi_\eta H_\eta)^{-1} = (1/\hat{b}(\eta + 1/\hat{c}) - A)(1/\hat{a} - A)^{-1};$$

observe that $\hat{b}(\eta + 1/\hat{c})$ is well-defined and nonzero by the choice of η. Thus to apply Theorem 0.7 it remains to show that φ_η and ψ_η are Laplace transforms of integrable functions on the halfline; this will be achieved by showing that these functions are locally analytic on $\mathbb{C}_+^\infty$ and then applying Lemma 10.1.

To prove this observe that $\hat{b}(\eta + 1/\hat{c}) = \hat{a}_\eta$ for some function $a_\eta \in L^1(\mathbb{R}_+)$, by Proposition 4.7 and the choice of η. The function $1/\hat{c}(\lambda)$ is a Bernstein function which is locally analytic on $\overline{\mathbb{C}}_+$ by decomposition (4.9). Together with (I3) these facts imply that $\varphi_\eta(\lambda)$ and $\psi_\eta(\lambda)$ are locally analytic on $\overline{\mathbb{C}}_+$.

The difficult part is to show that $\psi_\eta(\lambda)$ is locally analytic at ∞, and here we need the hypothesis about $g(z)$. By assumption $g(z)$ is locally analytic at infinity, hence $g_\eta(z) = g(z) + \eta g(z)/z$ is as well, and so by the remarks following Lemma 10.1

$$g_\eta(z) = \psi(\hat{a}_1(z), \ldots, \hat{a}_n(z)), \quad \text{Re } z \geq 0, \ |z| > R,$$

for some ψ analytic near 0, $a_j \in L^1(\mathbb{R}_+)$. But then by Proposition 4.7

$$\begin{aligned} \psi_\eta(\lambda) &= g_\eta(1/\hat{c}(\lambda)) = \psi(\hat{a}_1(1/\hat{c}(\lambda)), \ldots, \hat{a}_n(1/\hat{c}(\lambda))) \\ &= \psi(\hat{b}_1(\lambda), \ldots, \hat{b}_n(\lambda)), \quad \text{Re } \lambda \geq 0, \ |\lambda| \geq r \end{aligned}$$

with $b_j \in L^1(\mathbb{R}_+)$ and r sufficiently large; hence $\psi_\eta(\lambda)$ is locally analytic at ∞.

Finally, to show that $\varphi_\eta(\lambda)$ is locally analytic at ∞, write

$$\varphi_\eta(\lambda) = \frac{1}{\lambda}\psi_\eta(\lambda) + \frac{\eta}{\lambda} + \frac{1}{\lambda\hat{c}(\lambda)}, \quad \text{Re } \lambda > 0,$$

and observe that $1/\lambda\hat{c}(\lambda)$ is locally analytic at ∞ by (4.9) again.   □

As an example where $g(z) = z(\hat{b}(\eta + z) - \hat{b}(z))/\hat{b}(z)$ is locally analytic at ∞, choose $\hat{b}(z) = z^{-\gamma}$, i.e. $b(t) = t^{\gamma-1}/\Gamma(\gamma)$, where $\gamma > 0$. Then

$$g(z) = z((\frac{z}{\eta + z})^\gamma - 1) = z((1 - \frac{\eta}{\eta + z})^\gamma - 1),$$

which is even analytic at ∞; in particular the cases of Theorem 4.2 and 4.3, i.e. $b(t) = 1$, resp. $b(t) = t$ are covered.

Next we consider the case $1/\kappa + \alpha < \infty$.

Theorem 10.4 *Suppose the function $g(z, w) = w^{-1}(\hat{b}(z)/\hat{b}(z + zw) - 1)$ is analytic at $(\infty, 0)$, let $1/\kappa + \alpha < \infty$ and assume $\alpha > \omega_b$. Then $S_a(t)$ is integrable if and only if (I1), (I2), (I3) of Theorem 10.1 are satisfied. $S_a(t)$ is uniformly integrable if and only if $S_a(t)$ is integrable and $S_b(t)$ is $\mathcal{B}(X)$-measurable on $\mathbb{R}_+$.*

Proof: The second assertion is clear from the structure theorem, Theorem 4.4; we therefore concentrate on the sufficiency part of the first assertion. To apply again a Paley-Wiener argument, we start with the 'bad' part of $S(t)$, which by (4.57) is $S_1(t) = S_a(t) - S_0(t)$. $S_1(t)$ is integrable since $\alpha > \omega_b$ by assumption, and its Laplace transform is given by

$$H_1(\lambda) = \hat{S}_1(\lambda) = \kappa(\kappa\lambda + \alpha)^{-1}(I - \hat{b}(\kappa\lambda + \alpha)A)^{-1}, \qquad \mathrm{Re}\,\lambda \geq 0; \qquad (10.22)$$

w.o.l.g. we may assume $\alpha > 0$ in the sequel, since otherwise $S_0(t) \equiv 0$, and the assertion is then trivial. A simple computation yields the following identity for $H_0(\lambda) = \hat{S}_0(\lambda)$.

$$H_0(\lambda) = \varphi_1(\lambda)H_1(\lambda) + \varphi_0(\lambda)H_1(\lambda)^2 + \varphi_0(\lambda)H_1(\lambda)H_0(\lambda), \qquad \mathrm{Re}\,\lambda \geq 0, \quad (10.23)$$

where

$$\varphi_1(\lambda) = (1 + \alpha/\kappa\lambda)\hat{b}(\alpha + \kappa\lambda)/\hat{a}(\lambda) - 1, \qquad \mathrm{Re}\,\lambda \geq 0,$$

and

$$\varphi_0(\lambda) = (\lambda + \alpha/\kappa)(1 - \hat{b}(\alpha + \kappa\lambda)/\hat{a}(\lambda)), \qquad \mathrm{Re}\,\lambda \geq 0.$$

As in the proof of Theorem 10.3 it is easy to check by (I1), (I2) that the Paley-Wiener condition is satisfied, and so by Theorem 0.7 we obtain $S_0 \in L^1(\mathbb{R}_+; \mathcal{B}(X))$ once we have shown that $\varphi_0(\lambda)H_1(\lambda)$ and $\varphi_1(\lambda)H_1(\lambda)$ are Laplace transforms of functions from $L^1(\mathbb{R}_+; \mathcal{B}(X))$.

This will be achieved by a local analyticity argument, as in the proof of Theorem 10.3. Since $\hat{b}(z)$ is meromorphic for $\mathrm{Re}\,z > \omega_b$ with poles which only may cluster at the vertical line $\mathrm{Re}\,z = \omega_b$, $\hat{b}(\alpha + \kappa\lambda)$ is holomorphic on $\overline{\mathbb{C}}_+$, except for finitely many poles $\lambda_1, \ldots, \lambda_n$, say; therefore, by (I3), $\varphi_1(\lambda)$ and $\varphi_0(\lambda)$ are locally analytic on $\overline{\mathbb{C}}_+ \setminus \{\lambda_1, \ldots, \lambda_n\}$. Since $g(z, w)$ is analytic at $(\infty, 0)$ we also see that

$$\varphi_0(\lambda) = -\frac{1}{\kappa}\hat{k}_1(\lambda)g(\alpha + \kappa\lambda, \frac{\hat{k}_1(\lambda)}{\alpha + \kappa\lambda}), \qquad \mathrm{Re}\,\lambda > 0,$$

is locally analytic at ∞, and so

$$\varphi_1(\lambda) = -\frac{1}{\lambda}\varphi_0(\lambda) - \frac{\alpha}{\kappa\lambda}, \qquad \mathrm{Re}\,\lambda > 0,$$

is as well, and $\varphi_1(\infty) = \varphi_0(\infty) = 0$. Thus if $\hat{b}(z)$ has no poles in $\mathrm{Re}\,\lambda \geq \alpha$, $\varphi_j(\lambda) = \hat{b}_j(\lambda)$, $\mathrm{Re}\,\lambda \geq 0$, for some functions $b_j \in L^1(\mathbb{R}_+)$ by Lemma 10.1, and so $\varphi_j(\lambda)H_1(\lambda)$ is the Laplace transform of an element of $L^1(\mathbb{R}_+; \mathcal{B}(X))$. In case $\hat{b}(z)$ has poles, then $H_1(\lambda)$ has zeros at these points of the same order, and so the poles 'drop out', and one obtains the same conclusion as before. The details of this pole cancelling are left to the reader. $\square$

Observe that for $\hat{b}(z) = z^{-\gamma}$, $\gamma > 0$, one obtains

$$g(z, w) = \frac{1}{w}((1 + w)^\alpha - 1),$$

and so g obviously is analytic at $(\infty, 0)$; this shows that again the cases of Theorem 4.2 and 4.3 are covered. Due to their importance in applications we discuss the cases $b(t) = 1$ and $b(t) = t$ now separately.

So let A generate a C_0-semigroup in X with growth bound $\omega_0(A)$, and let $b(t) = 1$, while $c(t)$ and $k(t)$ are as above. Then Theorems 10.3 and 10.4 apply whenever $\alpha > \omega_0(A)$, hence in this case only conditions (I1), (I2), (I3) from Theorem 10.1 have to be checked. For this purpose observe that

$$1/\lambda\hat{c}(\lambda) = \kappa + \hat{k}_1 + k_\infty/\lambda \to 1/c(\infty) \neq \infty, \qquad \text{as } \lambda \to 0,$$

if and only if $k_\infty = 0$ and $k_1 \in L^1(\mathbb{R}_+)$, i.e. $k(t)$ is bounded. If this is the case then $1/\lambda\hat{c}(\lambda)$ obviously is locally analytic on $\overline{\mathbb{C}}_+$, so (I3) holds, and $1/\hat{c}(\lambda)$ admits continuous extension in $\mathbb{C}$ to $\overline{\mathbb{C}}_+ \setminus \{0\}$.

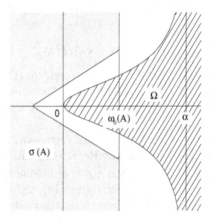

Figure 10.1: Illustration of Corollary 10.2

Therefore Theorems 10.3 and 10.4 imply

Corollary 10.2 *Let A generate a C_0-semigroup in X with growth bound $\omega_0(A)$, let $c(t)$ be a completely positive function with associated creep function $k(t) \in W^{2,1}_{loc}(0, \infty)$ and characteristic numbers κ and α. Define*

$$\Omega = \{1/\hat{c}(\lambda) : \lambda \in \overline{\mathbb{C}}_+, \ \lambda \neq 0\},$$

and assume in addition $\alpha > \omega_0(A)$. Then the resolvent $S(t)$ of (10.1) with $a(t) = c(t)$ is integrable if and only if

$$0 \in \rho(A), \quad k(\infty) < \infty, \quad \Omega \cap \sigma(A) = \emptyset. \tag{10.24}$$

$S(t)$ is uniformly integrable iff in addition $\kappa = 0$ or $\alpha = \infty$ or e^{At} is $\mathcal{B}(X)$-continuous for $t > 0$.

Consequently, if $\omega_0(A) < 0$ then $S(t)$ will be integrable iff $k(\infty) < \infty$; in the limiting case $\omega_0(A) = 0$, $S(t)$ is integrable iff A is invertible and $k(\infty) < \infty$.

Observe that the requirement $\alpha > \omega_0(A)$ is trivially satisfied in case $\omega_0(A) < 0$, and for $\omega_0(A) = 0$ it simply means $k_1 \not\equiv 0$.

For the second order subordination $b(t) = t$, suppose A generates a cosine family $\mathrm{Co}(t)$ in X with growth bound $\omega_1(A)$, and according to the applications dicussed in Sections 5 and 9 let $a(t)$ be a creep function with $\log a_1(t)$ convex; cp. Theorem 4.3 (ii). Then the characteristic numbers of the creep function $k(t)$ associated with the Bernstein function $\varphi(\lambda) = 1/\sqrt{\hat{a}(\lambda)}$ are $\kappa = 1/\sqrt{a_\infty + a_1(0+)}$ if $a_0 = 0$, $\kappa = 0$ if $a_0 > 0$, $\omega = 0$, $\alpha = -\dot{a}_1(0+)\kappa^3/2$; note that $k \in W_{loc}^{2,1}(0, \infty)$, by convexity of $a_1(t)$. To verify (I1)~(I3) observe that $\lambda\hat{a}(\lambda) \to a(\infty) \neq 0$ as $\lambda \to 0$ is always satisfied; $a(\infty)$ is finite in the 'fluid case' infinite otherwise. Local analyticity of $1/\lambda\hat{a}(\lambda)$ on $\overline{\mathbb{C}}_+$ is obtained as in the remarks following the proof of Lemma 10.2, referring to the situation of Corollary 3.3. Therefore Theorems 10.3 and 10.4 yield the following result.

Corollary 10.3 *Let A generate a cosine family $\mathrm{Co}(t)$ in X with growth bound $\omega_1(A)$, let A be unbounded, $a(t)$ be a creep function with $\log a_1(t)$ convex, and let κ and α denote the characteristic numbers of the Bernstein function $\varphi(\lambda) = 1/\sqrt{\hat{a}(\lambda)}$. Define*

$$\Omega = \{1/\hat{a}(\lambda) : \lambda \in \overline{\mathbb{C}}_+\},$$

and assume in addition $\alpha > \omega_1(A)$. Then the resolvent $S(t)$ of (10.1) is integrable if and only if

$$0 \in \rho(A) \quad and \quad \Omega \cap \sigma(A) = \emptyset. \tag{10.25}$$

$S(t)$ is uniformly integrable iff in addition $\kappa = 0$ or $\alpha = \infty$.

Observe that the cosine family generated by A is never $\mathcal{B}(X)$-measurable, in case A is unbounded, and $\omega_1(A) \geq 0$. In case $\omega_1(A) = 0$ the situation is particularly nice. In fact, unless $a(t) \equiv a_\infty t$, the restriction $\alpha > \omega_1(A)$ is always satisfied, and by Proposition 1.5 we have $\sigma(A) \subset (-\infty, 0]$. On the other hand, $1/\hat{a}(\lambda) \in (-\infty, 0]$ iff $\lambda = i\rho$, for some $\rho \in \mathbb{R}$, and $\lambda\hat{a}(\lambda) \in i\mathbb{R}$; but this is impossible since it would require

$$\mathrm{Re}\,(i\rho\hat{a}(i\rho)) = a_0 + \mathrm{Re}\,a_1(i\rho) \leq 0, \quad -\rho\mathrm{Im}\,(i\rho\hat{a}(i\rho)) = a_\infty - \rho\mathrm{Im}\,a_1(i\rho) = 0.$$

Therefore in case $\omega_1(A) = 0$, the resolvent $S(t)$ is integrable iff $0 \in \rho(A)$ and $a(t) \not\equiv a_\infty t$. The latter excludes only the 'purely elastic' case where $S(t) = \mathrm{Co}(t/\kappa)$, which is never integrable. This result exhibits a rather striking phenomenon of dynamics with memory of this type, e.g. in viscoelasticity problems involving only one kernel of the type considered in Corollary 10.3. The presence of a whatsoever small memory, i.e. $a(t) \not\equiv a_\infty t$, stabilizes all frequencies except 0. Stability is obtained if in addition the stationary problem is well-posed, i.e. if A is invertible. When considering specific examples, the latter is usually satisfied if the domain under consideration is bounded, but is also valid for several classes of unbounded domains.

Let us point out the sharp difference between the 'fluid case' $a_\infty = 0$, $a_1 \in$ $L^1(\mathbb{R}_+)$, and that of a solid $a_\infty > 0$. If $S(t)$ is integrable then

$$\int_0^\infty S(t)dt = H(0) = -A^{-1}/a(\infty),$$

hence we obtain $H(0) = 0$ in the 'solid case', while $H(0) \neq 0$ for 'fluids'. For further remarks see Section 10.6, and for applications of the foregoing results cp. Section 13.

10.4 Strong Integrability in Hilbert Spaces
Let X be a Hilbert space and suppose (10.1) admits an exponentially bounded resolvent $S(t)$ in X. If the kernel $a(t)$ is in addition 1-regular, strong integrability of the resolvent can be characterized in terms of the Laplace transform $H(\lambda)$ of $S(t)$. The result reads as follows.

Theorem 10.5 *Suppose (10.1) admits a resolvent $S(t)$ with finite growth bound $\omega_0(S) < \infty$ in the Hilbert space X, and let $a(t)$ be 1-regular. Then $S(t)$ is strongly integrable if and only if (I1), (I2), (I3) of Theorem 10.1 hold and $H(\lambda)$ is uniformly bounded on $\bar{\mathbb{C}}_+$.*

Proof: Since strong integrability of $S(t)$ implies L^1-stability (S) of (10.1), the necessity part follows from Proposition 10.2 and the remarks in front of Theorem 10.1.

To prove sufficiency, choose a cut-off function $\tilde{\varphi}(\rho)$, i.e. $\tilde{\varphi}(\rho) = 1$ for $|\rho| \leq 1$, $\tilde{\varphi}(\rho) = 0$ for $|\rho| \geq 1$, and $\tilde{\varphi} \in C_0^\infty(\mathbb{R})$. Then as in the proof of Theorem 10.1 there is $S_0 \in L^1(\mathbb{R}; \mathcal{B}(X))$ such that

$$\tilde{S}_0(\rho) = \tilde{\varphi}(\rho)H(i\rho), \quad \rho \in \mathbb{R}.$$

Therefore it remains to show that for each $x \in X$ the function $(1 - \tilde{\varphi}(\rho))H(i\rho)x$ is the Fourier transform of a function in $L^1(\mathbb{R}; X)$. So let $x \in X$ be fixed and put $\chi(\rho) = 1 - \tilde{\varphi}(\rho)$. Choose $\omega > \omega_0(S)$; then $f(t) = e^{-\omega t}e_0(t)S(t)x$ belongs to $L^2(\mathbb{R}; X)$, hence by Parseval's theorem, $\hat{f}(\rho) = H(i\rho + \omega)x$ is square integrable. The identity

$$H(i\rho)x = H(i\rho + \omega)x + \frac{\omega}{\omega + i\rho}H(i\rho)x + h(\rho)H(i\rho)(H(\omega + i\rho)x - \frac{x}{\omega + i\rho}),$$

where

$$h(\rho) = i\rho(\hat{a}(i\rho)/\hat{a}(\omega + i\rho) - 1),$$

then implies $\chi(\cdot)H(i\cdot)x \in L^2(\mathbb{R}; X)$ as well since $H(i\rho)$ is bounded by assumption, provided $h(\rho)$ is bounded for $|\rho| \geq 1$. To prove the latter we employ 1-regularity of $a(t)$. As in the proof of Lemma 8.1 there is a holomorphic function $g(\lambda)$, Re $\lambda > 0$, such that $\exp(g(\lambda)) = \hat{a}(\lambda)$, and $|g'(\lambda)| \leq c/|\lambda|$, Re $\lambda > 0$. Then

$$h(\rho) = i\rho \int_0^\omega \frac{d}{ds}\exp(g(i\rho) - g(i\rho + s))ds$$

$$= -i\rho \int_0^\omega \exp(-\int_0^s g'(i\rho + \tau)d\tau)g'(i\rho + s)ds,$$

hence for $|\rho| \geq 1$ we obtain

$$
\begin{aligned}
|h(\rho)| &\leq \int_0^\omega \exp\left(\int_0^s |g'(i\rho + \tau)|d\tau\right)|\rho||g'(i\rho + s)ds \\
&\leq c\omega \exp(c\omega) < \infty.
\end{aligned}
$$

Thus $\chi(\cdot)H(i\cdot)x \in L^2(\mathbb{R}; X)$ and so by Parseval's theorem there is a function $u \in L^2(\mathbb{R}; X)$ with $\tilde{u}(\rho) = \chi(\rho)H(i\rho)x$ for a.a. $\rho \in \mathbb{R}$.

On the other hand,

$$
\begin{aligned}
\tilde{u}'(\rho) &= \chi'(\rho)H(i\rho)x + i\chi(\rho)H'(i\rho)x \\
&= \chi'(\rho)H(i\rho)x + i\chi(\rho)\left(-H(i\rho)x/i\rho + H(i\rho)i\rho\frac{\hat{a}'(i\rho)}{\hat{a}(i\rho)}(H(i\rho)x - \frac{x}{i\rho})\right)
\end{aligned}
$$

also belongs to $L^2(\mathbb{R}; X)$ since $H(\lambda)$ is bounded, $\chi(\rho) = 0$ for $|\rho| \leq 1$, and a is 1-regular; see Lemma 8.1. Therefore, by Parseval's theorem there is $v \in L^2(\mathbb{R}; X)$ such that $\tilde{v}(\rho) = \tilde{u}'(\rho)$ for a.a $\rho \in \mathbb{R}$. But this implies $tu(t) = -v(t)$, hence

$$
|u|_1 = \int_{-1}^1 |u(t)|dt + \int_{|t|\geq 1} |t||u(t)|\frac{dt}{|t|} \leq \sqrt{2}(|u|_2 + |tu|_2) < \infty.
$$

The proof is complete. ☐

Since $D(A^*)$ is dense in X, under the assumptions of Theorem 10.5, $S^*(t)$ is the resolvent for (10.1) with A replaced by A^*, and $a(t)$ replaced by $\overline{a(t)}$. Theorem 10.5 then shows in case $\omega_0(S) < \infty$ and $a(t)$ 1-regular that with $S(t)$ also $S^*(t)$ is strongly integrable. This observation together with the vector-valued Riesz-Thorin interpolation theorem yields

Corollary 10.4 *Suppose (10.1) admits a resolvent $S(t)$ with finite growth bound $\omega_0(S) < \infty$ in the Hilbert space X, and let $a(t)$ be 1-regular. If $S(t)$ is strongly integrable then (10.1) is L^p-stable (S) for each $p \in [1, \infty]$.*

Proof: We already know that strong integrability of $S(t)$ implies L^1-stability (S) of (10.1). Since the remark in front of Corollary 10.2 yields also strong integrability of $S^*(t)$ we obtain for $u(t) = (S * f)(t)$ the estimate

$$
|(u(t), x^*)| \leq \int_0^t |(f(\tau), S^*(t - \tau)x^*)|d\tau \leq |f|_\infty \cdot |S^*(\cdot)x^*|_1
$$

for each $f \in L^\infty(\mathbb{R}_+; X)$, $x^* \in X$, $t \in \mathbb{R}_+$. Thus there is a constant $C > 0$ such that $|u|_\infty \leq |f|_\infty$, i.e. (10.1) L^∞-stable (S). Interpolation then yields L^p-stability (S). ☐

Let us briefy discuss strong integrability of the resolvent $S(t)$ obtained in Corollary 1.2.

Corollary 10.5 *Let X be a Hilbert space, $a \in BV_{loc}(\mathbb{R}_+)$ be 1-regular and such that $da(t)$ is of positive type. Then the resolvent $S(t)$ of (10.1) is strongly integrable for every negative definite selfadjoint operator A in X if and only if the following conditions hold.*
(i) $\mathrm{Re}\ \widehat{da}(i\rho) > 0$ for all $\rho \in \mathbb{R}$, $\rho \neq 0$;
(ii) $\overline{\lim}_{|\rho| \to \infty}\ \mathrm{Im}\ \widehat{da}(i\rho)/[|\rho|\ \mathrm{Re}\ \widehat{da}(i\rho)] < \infty$
(iii) $1/\widehat{da}(\lambda)$ is locally analytic at $\lambda = 0$.

Observe that (i) is equivalent to $1/\hat{a}(i\rho) \notin (-\infty, 0)$ for all $\rho \in \mathbb{R}$, $\rho \neq 0$, (iii) concerns existence and continuity of $H(\lambda)$ near 0, and (ii) corresponds to boundedness of $H(\lambda)$ on $\overline{\mathbb{C}_+}$. The last condition is satisfied e.g. if $a \in BV(\mathbb{R}_+)$, $a(\infty) \neq 0$. (iii) also holds in case $a(t)$ is a creep function with $a_1(t)$ convex, as the remarks following the proof of Lemma 10.2 show; note that in this case 1-regularity of $a(t)$ follows and $da(t)$ is already of positive type, by Proposition 3.3. One should compare Corollary 10.5 with Corollary 10.3. If in addition $a_1 = a_2 + a_3$ where a_2 is 2-monotone and such that $\int_1^\infty a_2(t)dt/t < \infty$ and a_3 is 3-monotone then it has been shown in Carr and Hannsgen [38] that under the assumptions of Corollary 10.5, $S(t)$ is even integrable; see Section 10.7 for further remarks and discussions.

Next we discuss strong integrability of the resolvent for hyperbolic problems of variational type introduced in Section 6.7. So let V, H be Hilbert spaces such that $V \overset{d}{\hookrightarrow} H$ and consider

$$(w, u(t)) + \int_0^t \alpha(t - s, w, u(s))ds = <w, f(t)>, \quad t \in \mathbb{R}_+,\ w \in V, \qquad (10.26)$$

where $\alpha : \mathbb{R}_+ \times V \times V \to \mathbb{C}$ is such that $\alpha(t, \cdot, \cdot)$ is a sesquilinear form on V for each $t > 0$, $\alpha(0, \cdot, \cdot) = 0$, and let $\alpha(t, \cdot, \cdot)$ be represented by $A(t) \in \mathcal{B}(V, \overline{V}^*)$ by means of

$$\alpha(t, u, v) = - <u, A(t)v>, \quad t > 0,\ u, v \in V. \qquad (10.27)$$

Since Corollary 10.3 is the special case $\alpha(t, u, v) = a(t)(A^{1/2}u, A^{1/2}v)$ in this setting, it becomes apparent that for strong integrability of the resolvent $S(t)$ for (10.26) frequency domain conditions reflecting (i) and (ii) of Corollary 10.3 will be needed, as well as some regularity assumptions on the form $\alpha(t, u, v)$ which replace (iii). We first concentrate on the *solid* case which corresponds to $A(t) \to A_\infty \neq 0$ as $t \to \infty$.

Theorem 10.6 *Let $\alpha : \mathbb{R}_+ \times V \times V \to \mathbb{C}$ satisfy (V1) and (V2) of Theorem 6.5 with $\alpha_1(t)$ bounded, and let α be η-coercive for some $\eta > 0$. In addition assume*
(V4) $\alpha(\cdot, u, v) \in W_{loc}^{2,1}(0, \infty)$, $\alpha(0, u, v) = 0$, *for each $u, v \in V$, and*

$$|\ddot{\alpha}(t, u, v) - \ddot{\alpha}(s, u, v)| \leq (\alpha_2(s) - \alpha_2(t))||u||\ ||v||, \quad \text{for all } t > s > 0,\ u, v \in V,$$

where $\alpha_2(t)$ is nonincreasing and such that $-\int_0^\infty t d\alpha_2(t) < \infty$;
(V5) $\overline{\lim}_{|\rho| \to \infty} |\ \mathrm{Re}\ \hat{\alpha}(i\rho, u, u)/[\rho\ \mathrm{Im}\ \hat{\alpha}(i\rho, u, u)]| \leq \phi_\infty < \infty$ *for each $u \in V$;*

(S) *For each* $u, v \in V$, $\alpha_\infty(u, v) = \lim_{t \to \infty} \dot{\alpha}(t, u, v)$ *exists, and with some constant* $\gamma_\infty > 0$

$$Re \, \alpha_\infty(u, u) \geq \gamma_\infty ||u||^2, \quad for \, all \, u \in V.$$

Then the resolvent $S(t)$ for (10.26) and its integral $R(t) = (1 * S)(t)$ satisfy
(i) $v \in V \Rightarrow S(\cdot)x \in L^1(\mathbb{R}_+; V)$, $\dot{S}(\cdot)v \in L^1(\mathbb{R}_+; H)$, $\ddot{S}(\cdot)v \in L^1(\mathbb{R}_+; \overline{V}^*)$;
(ii) $x \in H \Rightarrow S(\cdot)x \in L^1(\mathbb{R}_+; H)$, $\dot{S}(\cdot)x \in L^1(\mathbb{R}_+; \overline{V}^*)$, $R(\cdot)x \in L^1(\mathbb{R}_+; V)$;
(iii) $u \in \overline{V}^* \Rightarrow S(\cdot)u \in L^1(\mathbb{R}_+; \overline{V}^*)$, $R(\cdot)u \in L^1(\mathbb{R}_+; H)$, and $e^{-\eta t} * R(\cdot)u \in L^1(\mathbb{R}_+; V)$.

Observe that η-coercivity together with (S), (V2), and α_1 bounded imply coercivity of α, hence Theorem 6.5 applies, and consequently the exponent 1 in (i), (ii), (iii) of Theorem 10.6 can be replaced by ∞, hence by $p \in [1, \infty]$ by interpolation. Since the adjoint form $\alpha^*(t, u, v) = \overline{\alpha(t, u, v)}$ also satisfies the assumptions of Theorem 10.6, the argument in the proof of Corollary 10.2 may be applied again, and so (10.26) is L^p-stable (S) for each $p \in [1, \infty]$.

Proof of Theorem 10.6: The argument is basically the same as in the proof of Theorem 10.5, therefore we only give an outline here, emphasizing the necessary modifications.

(a) We first show how to obtain the boundedness of $H(\lambda)$ in $\mathcal{B}(\overline{V}^*)$, $\mathcal{B}(H)$, $\mathcal{B}(V)$, of $H(\lambda)/\lambda$ in $\mathcal{B}(\overline{V}^*, H)$, $\mathcal{B}(H, V)$, and of $H(\lambda)/(\lambda(\lambda+\eta))$ in $\mathcal{B}(\overline{V}^*, V)$ which are necessary for (i), (ii), (iii) to be valid. (10.26) implies the identity

$$\lambda |H(\lambda)x|^2 + \hat{\dot{\alpha}}(\bar{\lambda}, H(\lambda)x, H(\lambda)x) = \, < H(\lambda)x, x >, \tag{10.28}$$

valid for $Re \, \lambda > 0$ and $x \in \overline{V}^*$. Taking real parts in (10.28), by η-coerciveness we obtain

$$\eta\gamma ||H(\lambda)x|| \leq ((\eta + \sigma)^2 + \rho^2)||x||_*, \quad Re \, \lambda > 0, \, x \in \overline{V}^*, \tag{10.29}$$

where $\lambda = \sigma + i\rho$. We have

$$-\lambda^2 \widehat{\dot{A}}'(\lambda) = \lambda^2 \widehat{t\dot{A}}(\lambda) = \widehat{td\ddot{A}}(\lambda) + 2\widehat{d\dot{A}}(\lambda) - \dot{A}(0), \tag{10.30}$$

hence $|\lambda^2 \widehat{\dot{A}}'(\lambda)|_{\mathcal{B}(V, \overline{V}^*)} \leq M$ since $-\int_0^\infty td\alpha_2(t) < \infty$ and $\alpha_1(t)$ is bounded. With

$$H'(\lambda) = -H(\lambda)(I - \widehat{\dot{A}}'(\lambda))H(\lambda), \quad Re \, \lambda > 0, \tag{10.31}$$

(10.29) and (10.30) then imply the existence of $H(i\rho) = \lim_{\lambda \to i\rho} H(\lambda)$ in $\mathcal{B}(\overline{V}^*, V)$ for each $\rho \neq 0$; observe that $\widehat{\dot{A}}(\lambda) = \widehat{d\dot{A}}(\lambda)/\lambda$ admits continuous extension to $\overline{\mathbb{C}}_+ \setminus \{0\}$, by boundedness of $\alpha_1(t)$. η-coerciveness and (10.28) for $\lambda = i\rho$ then yield

$$\eta\gamma ||H(i\rho)(\eta + i\rho)^{-1}x||^2 \leq |\, < H(i\rho)x, x > |, \quad \rho \neq 0, \, x \in \overline{V}^*, \tag{10.32}$$

while decomposing (10.28) into real and imaginary parts, (V5) implies

$$|\rho| |H(i\rho)x|^2 \le c(1 + |\rho|)| < H(i\rho)x, x > |, \quad \rho \ne 0, \ x \in \overline{V}^*. \tag{10.33}$$

Finally (S) gives

$$||H(i\rho)x|| \le c|\rho| \, ||x||_*, \quad \rho \ne 0, \ x \in \overline{V}^*; \tag{10.34}$$

in particular $H(0) = \lim_{\lambda \to 0} H(\lambda) = 0$. From (10.32), (10.33), (10.34) one easily obtains the boundedness of $|H(i\rho)|_{\mathcal{B}(H)}$ etc., and then by the maximum principle also on $\overline{\mathbb{C}}_+$.

(b) As in the proof of Theorem 10.5 consider the identity

$$\begin{aligned} H(i\rho)x \ = \ & H(i\rho + \eta)x + \eta H(i\rho)H(i\rho + \eta)x + H(i\rho)(\widehat{A}(i\rho) \tag{10.35} \\ & -\widehat{A}(i\rho + \eta))H(i\rho + \eta)x \end{aligned}$$

where $x \in \overline{V}^*$, $\rho \in \mathbb{R}$. (V4) and

$$(i\rho + \eta)^2(\widehat{A}(i\rho) - \widehat{A}(i\rho + \eta)) = \eta^2 \widehat{A}(i\rho) + \eta \dot{A}(0) + 2\eta \widehat{\ddot{A}}(i\rho) + \widehat{d\ddot{A}}(i\rho) - \widehat{d\ddot{A}}(i\rho + \eta),$$

imply

$$|(i\rho + \eta)^2(\widehat{A}(i\rho) - \widehat{A}(i\rho + \eta))|_{\mathcal{B}(V,\overline{V}^*)} \le C(1 + 1/|\rho|), \quad \rho \ne 0,$$

for $-\int_0^\infty t\,d\alpha_2(t) < \infty$ and α_1 is bounded. Since e.g. $e^{-\eta t}S(t)x \in L^2(\mathbb{R}_+; V)$ for $x \in V$, by Parseval's theorem, we obtain $H(i\rho + \eta)x \in L^2(\mathbb{R}_+; V)$, hence $H(i\rho)x \in L^2(\mathbb{R}_+; V)$ by (10.35) and (a), and so by Parseval's theorem also $S(t)x \in L^2(\mathbb{R}_+; V)$. By similar arguments the assertions of the theorem follow with exponent 2 instead of 1.

(c) To obtain e.g. $tS(t)x \in L^2(\mathbb{R}_+; V)$ consider (10.31) for $\lambda = i\rho$. Then for $x \in V$ one deduces from (10.31)

$$||H'(i\rho)x|| \le |H(i\rho)|_{\mathcal{B}(V)}||H(i\rho)x|| + |H(i\rho)/\rho^2|_{\mathcal{B}(\overline{V}^*,V)}M||H(i\rho)x||,$$

hence $H'(i\rho)x \in L^2(\mathbb{R}_+; V)$, by the boundedness properties of $H(i\rho)$, and by (10.34). Parseval's theorem gives $tS(t)x \in L^2(\mathbb{R}_+; V)$ for each $x \in V$, which together with (b) yields the strong integrability of $S(t)$ in V. The remaining assertions follow by similar arguments. $\square$

By the same type of arguments one can prove the following result, in which the case of a *fluid*, i.e. $\int_0^\infty |\dot{A}(t)|_{\mathcal{B}(V,\overline{V}^*)}dt < \infty$ is considered.

Corollary 10.6 *Let the assumptions of Theorem 10.6 be satisfied, with the exception that (S) is replaced by*
(F) $|\dot{a}(t, u, v)| \le \alpha_3(t)||u|| \, ||v||$, $t > 0$, $u, v \in V$, *where* $\alpha_3 \in L^1(\mathbb{R}_+)$.
Then the assertions of Theorem 10.6 for $S(t)$, $\dot{S}(t)$, $\ddot{S}(t)$ remain valid.

Observe that in the fluid case one cannot expect integrability of $R(t)$ since

$$\int_0^\infty S(t)dt = H(0) = -A(\infty)^{-1} \neq 0.$$

Here the difference between the solid and the fluid case becomes apparent, again.

10.5 Nonscalar Parabolic Problems

Next we consider integrability properties of the resolvent $S(t)$ for the nonscalar parabolic problem

$$u(t) = f(t) + \int_0^t A(t-\tau)u(\tau)d\tau, \quad t \geq 0, \tag{10.36}$$

in the Banach space X, where $A \in L^1_{loc}(\mathbb{R}_+; \mathcal{B}(Y,X))$ is of subexponential growth, and $Y \overset{d}{\hookrightarrow} X$. If (10.36) is (weakly) parabolic and $A(t)$ is (weakly) 1-regular, see Definition 7.3, then by Theorem 7.2, the (weak) resolvent $S(t)$ for (10.36) exists and is bounded in $(\mathcal{B}(X))$ $\mathcal{B}(X) \cap \mathcal{B}(Y)$. If $S(t)$ is integrable in X, by continuity in $\mathcal{B}(X)$ it is also uniformly integrable, and so $H(\lambda) = \hat{S}(\lambda)$ is $\mathcal{B}(X)$-continuous on $\overline{\mathbb{C}}_+$. In particular, $\lambda H(\lambda) \to 0$ as $\lambda \to 0$, hence $\lambda \hat{A}(\lambda)$ is invertible near $\lambda = 0$ and

$$H(0) = \lim_{\lambda \to 0}(\lambda - \lambda \hat{A}(\lambda))^{-1} = -\lim_{\lambda \to 0}(\lambda \hat{A}(\lambda))^{-1} \tag{10.37}$$

exists in $\mathcal{B}(X)$. Similarly, if $S(t)$ is integrable in Y then (10.37) also holds in $\mathcal{B}(Y)$. For the sufficiency of (10.37) stronger assumptions on the regularity of $A(t)$ and on the behavior of $\lambda \hat{A}(\lambda)$ near $\lambda = 0$ are needed.

Theorem 10.7 *Suppose (10.36) is weakly parabolic, $A(t)$ weakly 2-regular, and let $S(t)$ denote the weak reolvent for (10.36). In addition assume that there is $\eta > 0$ such that $\lambda \hat{A}(\lambda)$ is invertible in X for $\lambda \in \mathbb{C}_+ \cap B_\eta(0)$ and $F(\lambda) = (\lambda \hat{A}(\lambda))^{-1} \in \mathcal{B}(X,Y)$ satisfies*

$$\lim_{\lambda \to 0} F(\lambda) = F(0) \quad \text{exists in } \mathcal{B}(X), \tag{10.38}$$

$$\int_{-\eta}^\eta \{|F'(i\rho)|_{\mathcal{B}(X)} + |F(i\rho) - F(0)|_{\mathcal{B}(X)}/|\rho|\}d\rho + \int_{-\eta}^\eta |\rho||dF'(i\rho)|_{\mathcal{B}(X)} < \infty. \tag{10.39}$$

Then $S \in L^1(\mathbb{R}_+; \mathcal{B}(X))$. If (10.36) is parabolic, $A(t)$ 2-regular, and (10.38), (10.39) hold also in $\mathcal{B}(Y)$ then $S \in L^1(\mathbb{R}_+; \mathcal{B}(Y))$.

Proof: (a) Suppose $A(t)$ is weakly 2-regular. We first show that $\hat{A}(\lambda)$ admits a $\mathcal{B}(Y,X)$-continuous extension to $\overline{\mathbb{C}}_+ \setminus \{0\}$ which is such that (10.39) makes sense. For this purpose observe first that

$$\varphi_0(r) = \sup\{|\hat{A}(\lambda)| : \lambda \in \mathbb{C}_+, |\lambda| \geq r\}, \quad r > 0,$$

is finite, nonincreasing, and $\lim_{r \to \infty} \varphi(r) = 0$; this follows from 1-regularity by means of Gronwall's inequality. By 2-regularity we then also have with

$$\varphi_k(r) = \sup\{|\hat{A}^{(k)}(\lambda)| : \lambda \in \mathbb{C}_+, |\lambda| \geq r\}, \quad r > 0, \ k = 1, 2,$$

the estimates

$$\varphi_k(r) \leq (1 + \varphi_0(r))/r^k, \quad r > 0, \ k = 1, 2.$$

Therefore $\hat{A}(\lambda)$ and $\hat{A}'(\lambda)$ are uniformly Lipschitz on $\mathbb{C}_+ \setminus B_\varepsilon(0)$ for each $\varepsilon > 0$, hence admit extension to $\overline{\mathbb{C}}_+ \setminus \{0\}$, and

$$\hat{A}(i\cdot) \in C_0^1(\mathbb{R} \setminus \{0\}; \mathcal{B}(Y, X)), \quad \hat{A}'(i\cdot) \in BV_{loc}(\mathbb{R} \setminus \{0\}; \mathcal{B}(Y, X)),$$

as well as

$$|\hat{A}(i\rho)| \leq \varphi_0(|\rho|), \ |\hat{A}'(i\rho)| \leq \varphi_1(|\rho|), \ |d\hat{A}'(i\rho)| \leq \varphi_2(|\rho|)d\rho, \quad \rho \neq 0.$$

Therefore, by (10.38) we see that $F(i\rho)$ is of class C^1 for $0 < |\rho| \leq \eta$ and $F'(i\cdot) \in BV_{loc}([-\eta, 0) \cup (0, \eta]; \mathcal{B}(X))$; thus (10.39) makes sense.

(b) From (a) and weak parabolicity it now follows easily that $H(\lambda)$ admits continuous extension in $\mathcal{B}(X)$ to $\overline{\mathbb{C}}_+ \setminus \{0\}$; with

$$H'(\lambda) = H(\lambda)(\hat{A}(\lambda) + \lambda\hat{A}'(\lambda) - I)H(\lambda)$$

and

$$\begin{aligned} H''(\lambda) &= H(\lambda)(2\hat{A}'(\lambda) + \lambda\hat{A}''(\lambda))H(\lambda) \\ &+ 2H(\lambda)(\hat{A}(\lambda) + \lambda\hat{A}'(\lambda) - I)H(\lambda)(\hat{A}(\lambda) + \lambda\hat{A}'(\lambda) - I)H(\lambda), \end{aligned}$$

there follows $H(i\cdot) \in C^1(\mathbb{R} \setminus \{0\}; \mathcal{B}(X))$, $H'(i\cdot) \in BV_{loc}(\mathbb{R} \setminus \{0\}; \mathcal{B}(X))$, as well as the estimates

$$|\rho||H(i\rho)| + |\rho|^2|H'(i\rho)| \leq M, \quad |\rho|^3|dH'(i\rho)| \leq Md\rho, \ \rho \neq 0.$$

Furthermore, (10.38) shows the existence of $\lim_{\lambda \to 0} H(\lambda) = -F(0)$, for

$$H(\lambda) = (\lambda - \lambda\hat{A}(\lambda)))^{-1} = -F(\lambda)(I - \lambda F(\lambda))^{-1}, \quad |\lambda| < \eta;$$

in particular, $H(\lambda)$ is bounded in $\mathcal{B}(X)$ near $\lambda = 0$. Finally, (10.39) implies

$$\int_{-\delta}^{\delta} \{|H'(i\rho)|_{\mathcal{B}(X)} + |H(i\rho) - H(0)|_{\mathcal{B}(X)}/|\rho|\}d\rho + \int_{-\delta}^{\delta} |\rho||dH'(i\rho)| < \infty,$$

where $0 < \delta \leq \eta$ is chosen small enough, e.g. such that $\delta|F(\lambda)| \leq 1/2$ for $|\lambda| \leq \eta$; observe that by 2-regularity and (10.38), $F(\lambda)$, $\lambda F'(\lambda)$, $\lambda^2 F''(\lambda)$ are bounded for $\lambda \in \mathbb{C}_+ \cap B_\eta(0)$.

(c) To prove integrability of $S(t)$ in $\mathcal{B}(X)$ we begin with the representation

$$tS(t) = \frac{1}{2\pi} \int_{-\infty}^{\infty} H'(i\rho)e^{i\rho t}d\rho, \quad t > 0,$$

which follows from the complex inversion formula of the Laplace-transform; the integral is absolutely covergent by the properties of $H'(i\rho)$ derived in (b). Integrating by parts for $|\rho| \geq 1/t$ this formula becomes

$$
\begin{aligned}
S(t) &= \frac{1}{2\pi t} \int_{-1/t}^{1/t} H'(i\rho)(e^{i\rho t} - 1)d\rho + \frac{1}{2\pi ti}(H(i/t) - H(-i/t)) \\
&\quad + \frac{1}{2i\pi t^2}(H'(-i/t)e^{-i} - H'(i/t)e^{i}) - \frac{1}{2i\pi t^2}\int_{|\rho|>1/t} e^{i\rho t}dH'(i\rho) \\
&= S_1(t) + S_2(t) + S_3(t) + S_4(t), \quad t \geq 1.
\end{aligned}
$$

We estimate the corresponding integrals separately.

$$
\begin{aligned}
\pi \int_1^\infty |S_1(t)|_{\mathcal{B}(X)} dt &\leq \int_0^1 \int_{-s}^s |H'(i\rho)|_{\mathcal{B}(X)}|\sin(\rho/2s)|d\rho ds/s \\
&= \int_{-1}^1 \left(\int_{|\rho|}^1 |\sin(|\rho|/2s)|ds/s\right)|H'(i\rho)|_{\mathcal{B}(X)}d\rho \\
&= \int_{-1}^1 |H'(i\rho)|_{\mathcal{B}(X)}\left(\int_{|\rho|/2}^{1/2} |\sin\tau|d\tau/\tau\right)d\rho \\
&\leq c\int_{-1}^1 |H'(i\rho)|_{\mathcal{B}(X)}d\rho < \infty.
\end{aligned}
$$

$$
\begin{aligned}
2\pi \int_1^\infty |S_2(t)|_{\mathcal{B}(X)}dt &= \int_0^1 |H(is) - H(-is)|_{\mathcal{B}(X)}ds/s \\
&\leq \int_{-1}^1 |H(is) - H(0)|_{\mathcal{B}(X)}ds/|s| < \infty.
\end{aligned}
$$

$$
\begin{aligned}
2\pi \int_1^\infty |S_3(t)|_{\mathcal{B}(X)}dt &\leq \int_0^1 (|H'(-is)|_{\mathcal{B}(X)} + |H'(is)|_{\mathcal{B}(X)})ds \\
&= \int_{-1}^1 |H'(is)|_{\mathcal{B}(X)}ds < \infty.
\end{aligned}
$$

$$
\begin{aligned}
2\pi \int_1^\infty |S_4(t)|_{\mathcal{B}(X)}dt &\leq \int_0^1 \left(\int_{|\rho|\geq s} |dH'(i\rho)|_{\mathcal{B}(X)}\right)ds \\
&\leq 2M \int_0^1 \left(\int_1^\infty d\rho/\rho^3\right)ds + \int_{-1}^1 \left(\int_0^{|\rho|} ds\right)|dH'(i\rho)|_{\mathcal{B}(X)} \\
&= M + \int_{-1}^1 |\rho||dH'(i\rho)|_{\mathcal{B}(X)} < \infty.
\end{aligned}
$$

Therefore $S \in L^1([1,\infty); \mathcal{B}(X))$, and since $S(t)$ is also bounded in $\mathcal{B}(X)$ we see that $S(t)$ is uniformly integrable in X. The second assertion follows by repeating the above arguments in the space Y. $\square$

Let us discuss (10.38) and (10.39) for equations of variational type

$$(w, u(t)) + \int_0^t \alpha(t - s, w, u(s))ds = <w, f(t)>, \quad t \geq 0, \ w \in V, \tag{10.40}$$

where $V \overset{d}{\hookrightarrow} H$ are Hilbert spaces, and $\alpha : \mathbb{R}_+ \times V \times V \to \mathbb{C}$ is a uniformly measurable and locally integrable bounded sesquilinear form on V satisfiying (PV1), (PV2), (PV3) with $k = 2$; see Section 7.3. Then $\hat{A}(\lambda) \in \mathcal{B}(V, \overline{V}^*)$ is invertible and by (7.22) we have

$$|(\lambda\hat{A}(\lambda))^{-1}|_{\mathcal{B}(\overline{V}^*, V)} \leq \frac{1}{\gamma\varphi(\bar{\lambda})|\lambda|}, \quad \operatorname{Re}\lambda > 0. \tag{10.41}$$

Suppose $|\lambda|\varphi(\bar{\lambda}) \to \infty$ as $\lambda \to 0$; then by (10.41), $F(\lambda) \to 0$ in $\mathcal{B}(\overline{V}^*, V)$ as $\lambda \to 0$, i.e. (10.38) holds with $F(0) = 0$. From

$$F'(\lambda) = -F(\lambda)(\lambda\hat{A}(\lambda))'F(\lambda)$$

and

$$F''(\lambda) = 2F(\lambda)(\lambda\hat{A}(\lambda))'F(\lambda)(\lambda\hat{A}(\lambda))'F(\lambda) - F(\lambda)(\lambda\hat{A}(\lambda))''F(\lambda)$$

as well as (PV4) with $k = 2$, we then obtain

$$|F'(\lambda)|_{\mathcal{B}(\overline{V}^*, V)} \leq C(\frac{1}{\gamma\varphi(\bar{\lambda})|\lambda|})^2\varphi(\bar{\lambda}) \leq \frac{C}{|\lambda|^2\varphi(\bar{\lambda})}, \quad \operatorname{Re}\lambda > 0,$$

and similarly

$$|F''(\lambda)|_{\mathcal{B}(\overline{V}^*, V)} \leq \frac{C}{|\lambda|^3\varphi(\bar{\lambda})}, \quad \operatorname{Re}\lambda > 0.$$

Therefore, if $\varphi(\lambda)$ admits continuous extension to $\overline{\mathbb{C}}_+ \setminus \{0\}$, (10.39) is implied by the condition $\int_{-\eta}^{\eta}(|\rho|^2\varphi(i\rho))^{-1}d\rho < \infty$, for some $\eta > 0$.

Corollary 10.7 *Let the assumptions of Theorem 7.3 (ii) with $k = 2$ be satisfied and assume $\varphi(\lambda)$ admits continuous extension to $\overline{\mathbb{C}}_+ \setminus \{0\}$ such that*

$$\lim_{\lambda \to 0} |\lambda|\varphi(\bar{\lambda}) = \infty \quad and \quad \int_{-\eta}^{\eta} \frac{d\rho}{|\rho|^2\varphi(i\rho)} < \infty \quad for \ some \ \eta > 0. \tag{10.42}$$

Then the resolvent $S(t)$ for (10.40) is uniformly integrable in $\overline{V}^*, V$ and H.

On the other hand, suppose $|\lambda|\varphi(\bar{\lambda}) \to \varphi(0) \neq 0$ as $\lambda \to 0$ and

$$\lambda\hat{\alpha}(\bar{\lambda}, u, v) \to_{\lambda \to 0} \alpha_\infty(u, v) = <u, A_\infty v> \quad \text{uniformly for } ||u||, ||v|| \leq 1.$$

Then $\lambda\hat{A}(\lambda) \to A_\infty$ in $\mathcal{B}(V, \overline{V}^*)$, where A_∞ is invertible. Therefore $F(\lambda) = (\lambda\hat{A}(\lambda))^{-1} \to A_\infty^{-1}$ in $\mathcal{B}(\overline{V}^*, V)$ as $\lambda \to 0$, and so (10.38) follows. Since $F(\lambda)$ is bounded in $\mathcal{B}(\overline{V}^*, V)$ it is easy to see that (10.39) is implied by

$$|\lambda\hat{\alpha}(u, v) - \alpha_\infty(u, v)| + |\lambda(\lambda\hat{\alpha})'(\lambda, u, v)| + |\lambda^2(\lambda\hat{\alpha})''(\lambda, u, v)| \leq \psi(\lambda)||u|| \ ||v|| \tag{10.43}$$

for Re $\lambda > 0$, $u, v \in V$ where $\psi(\lambda)$ admits continuous extension to $\overline{\mathbb{C}}_+ \setminus \{0\}$ such that $\int_{-\eta}^{\eta} \psi(i\rho)d\rho/|\rho| < \infty$.

Corollary 10.8 *Let the assumptions of Theorem 7.3 (ii) with $k = 2$ be satisfied and suppose $|\lambda|\varphi(\bar{\lambda}) \to \varphi(0) \neq 0$ and $\lambda\hat{a}(\lambda, u, v) \to \alpha_\infty(u, v)$ uniformly for $||u||, ||v|| \leq 1$ as $\lambda \to 0$, and let (10.43) hold, with a function $\psi \in C(\overline{\mathbb{C}}_+ \setminus \{0\})$ such that $\int_{-\eta}^{\eta} \psi(i\rho)d\rho/|\rho| < \infty$ for some $\eta > 0$. Then the resolvent $S(t)$ for (10.40) is uniformly integrable in $\overline{V}^*, V$ and H.*

10.6 Comments

a) L^2-stability has been considered in infinite dimensions for the first time by Miller and Wheeler [246]; although their definition is slightly different from ours the underlying ideas are similar. To be more precise, as regards to stability, Miller and Wheeler study the second order problem

$$\ddot{x}(t) = c\dot{x}(t) + (A + \gamma)x(t) + \int_0^t [b(t - \tau)Ax(\tau) + g(t - \tau)x(\tau)]d\tau + f(t),$$

$$x(0) = x_0, \quad \dot{x}(0) = v_0, \quad t \geq 0, \tag{10.44}$$

in a Hilbert space X, where A is a selfadjoint, negative definite operator in X such that $(\mu - A)^{-1}$ is compact for each $\mu \in \rho(A)$, $c, \gamma \in \mathbb{R}$, and $b, g \in C^1(\mathbb{R}_+) \cap L^1(\mathbb{R}_+)$. (10.44) is then called L^2-stable if it admits a mild solution $x \in L^2(\mathbb{R}_+; X)$ for each $x_0, v_0 \in X$ and $f \in L^1(\mathbb{R}_+; X)$. It is then proved that the spectral conditions

$$1 + \hat{b}(\lambda) \neq 0, \quad (\lambda^2 - c\lambda - \gamma - \hat{g}(\lambda))/(1 + \hat{b}(\lambda)) \in \rho(A), \quad \text{for all Re } \lambda \geq 0, \tag{10.45}$$

imply L^2-stability, provided in addition $c + b(0) < 0$ and $\dot{b} \in L^1(\mathbb{R}_+)$. By means of the subordination principle and perturbation techniques which are discussed in Section 12.1, it can be shown that a stronger version of this result is true in a general Banach space for operators A generating a strongly continuous cosine family, $g \in L^1(\mathbb{R}_+)$, and $b \in W^{1,1}(\mathbb{R}_+)$, provided the spectral condition (10.45) holds, and $\omega_0(A) < -(c + b(0))$.

b) The concept of local analyticity is originally due to Gelfand, Raikov, and Shilov [128], Paragraph 13, and is given in the context of a general commutative Banach algebra. The abstract version of Lemma 10.1 and a proof can be found there. This result is a considerable extension of the classical Wiener-Lévy theorem, and is a very powerful tool in modern transform theory; the results in Sections 10.2 and 10.3 are examples of this. The definition of a locally analytic function given here is a slight but nevertheless quite useful extension of the original concept, due to Jordan, Staffans, and Wheeler [189]. They study Volterra equations in finite dimensions in weighted spaces on the halfline and on the line. Based on their Proposition 2.3 (the result corresponding to Lemma 10.1) they were able to unify most results on integrability of resolvents available at that time, and to extend and strengthen many of them. In this connection we refer also to the monograph byGripenberg, Londen, and Staffans [156].

c) The first result on integrability of resolvents for parabolic problems appears already in the paper of Friedman and Shinbrot [123]. In the second part of this work it is proved that the resolvent $S(t)$ of the equation of scalar type (10.1) belongs to $L^p(\mathbb{R}_+; \mathcal{B}(X))$, $p \geq 2$, if the kernel a satisfies

$$a \in W^{1,1}_{loc}(\mathbb{R}+), \ a(0) > 0, \ \dot{a} \in L^1(\mathbb{R}_+),$$

the closed, linear, and densely defined operator A generates an analytic C_0-semigroup in X which is bounded in some sector $\Sigma(0, \theta)$, A is invertible, and the spectral condition

$$\hat{a}(\lambda) \neq 0, \ 1/\hat{a}(\lambda) \in \rho(A), \quad \text{for all Re } \lambda \geq 0,$$

holds. If in addition $t\dot{a}(t)$ is integrable then the uniform integrability of the resolvent is obtained.

Obviously, the assumptions on a and A stated above imply Conditions (I1), (I2), (I3) of Theorem 10.1, however, in general the kernel need not be 2-regular, and therefore neither Theorem 10.2 nor Theorem 10.1 can be applied. But by means of parabolic perturbation results, we show in Section 12 that the moment condition on $\dot{a}(t)$ is not needed for uniform integrability of $S(t)$. This was observed first in Prüss [271]; see also [275]. Theorems 10.1, 10.2, 10.7 are new.

d) Theorems 10.3 and 10.4 on integrability of subordinate resolvents are unifications and extensions of results for the special cases $b(t) = 1$, and $b(t) = t$, i.e. the first and second order case, obtained in Prüss [274]. These special cases are stated here as Corollaries 10.2 and 10.3. By means of the methods presented in Section 10.3 it is also possible to discuss integrability of operator families related to $S(t)$, like $\dot{S}(t)$, $\ddot{S}(t)$, $1 * S(t)$, and so on. For the second order case this has been done for the first two derivatives of $S(t)$ in Prüss [277], and for $(1+t)S(t)$, $1 * S(t)$, and $(1+t)^2 \dot{S}(t)$ even with a weight ρ in Hannsgen and Wheeler [171].

e) In analogy to Theorems 10.3 and 10.4 one obtains a characterization of integrability of the integral resolvent $R(t)$ of subordinated equations. If the assumptions of Theorems 10.3 or 10.4 hold, $R(t)$ is integrable iff the following conditions are satisfied.
(IR1) $1/\hat{a}(\lambda)$ admits a continuous extension in $\mathbb{C}$ to $\overline{\mathbb{C}}_+$, holomorphic on $\mathbb{C}_+$;
(IR2) $1/\hat{a}(\lambda) \in \rho(A)$ for all $\lambda \in \overline{\mathbb{C}}_+$;
(IR3) $1/\hat{a}(\lambda)$ is locally analytic on $i\mathbb{R}$.
Note that these conditions differ from the corresponding ones for $S(t)$ only at $\lambda = 0$.

f) The papers Hannsgen [164], and Carr and Hannsgen [38], contain the first integrability results for resolvents $S(t)$ of equations of scalar type (10.1) in a Hilbert space with negative definite operator A, where the kernels $a(t)$ are -in our terminology- creep functions with $a_1(t)$ convex. Assuming in addition $a_1 = a_2 + a_3$, where a_2 is 3-monotone and a_3 2-monotone and such that $\int_1^\infty a_3(t) dt/t < \infty$, in these papers it has been shown that the scalar functions $s(t; \mu)$ (cp. Section 1.3) are integrable, uniformly w.r.t. $\mu \geq \mu_0 > 0$, i.e.

$$\int_0^\infty \sup_{\mu \geq \mu_0} |s(t; \mu)| dt < \infty, \qquad \text{for each } \mu_0 > 0.$$

(In case $a(t)$ is even a Bernstein function Hannsgen and Wheeler [169] gave later a different argument which is based on the decomposition of $s(t; \mu)$ into a completely monotonic part which is infinitely smooth at $t = 0$, and an exponentially decreasing part.) Via the functional calculus for selfadjoint operators such estimates imply integrability of $S(t)$, rather than strong integrability, i.e. as proved in Section 10.4; see Corollary 10.3. Although the analysis is beautiful, these estimates are of a very technical nature and therefore are not reproduced here. Later they have been extended by Carr and Hannsgen [39] and Noren [260] to obtain also integrability of $\dot{S}(t)$ and $\ddot{S}(t)$. These estimates do not seem to carry over to nonscalar equations, however, as we have shown in Corollary 10.2, strong integrability of the resolvent is enough to obtain L^p-stability. Theorem 10.6 can therefore be considered as the natural extension of the Carr-Hannsgen result to nonscalar problems; cp. Prüss [281]. This paper contains also the straightforward application of Theorem 10.6 and its corollary to nonisotropic hyperbolic viscoelasticity.

g) A topic very much related to stability of (10.1) and integrability of its associated resolvent $S(t)$ is the question of *stabilizability* of (10.1) via feedback controls. For this subject the reader is referred e.g. to the papers Da Prato and Lunardi [74], Desch and Miller [89], [90], Jeong [188], [187], and Leugering [209], [211], [212].

11 Limiting Equations

The solvability behaviour of Volterra equations on the line will be studied in this section. The central concept which corresponds to well-posedness is that of admissibility of homogeneous spaces of functions on the line. Necessary conditions for admissibility are derived and some consequences of this property are studied. Thereafter the connections between the equations on the line and on the halfline are studied and the notion of limiting equation is justified.

11.1 Homogeneous Spaces

Let X, Y be Banach spaces such that $Y \overset{d}{\hookrightarrow} X$, $\{A(t)\}_{t \geq 0} \subset \mathcal{B}(Y, X)$ measurable, and consider the Volterra equation on the line

$$u(t) = g(t) + \int_0^\infty A(\tau)u(t - \tau)d\tau, \quad t \in \mathbb{R}, \tag{11.1}$$

where $g \in L^1_{loc}(\mathbb{R}; X)$. This section is devoted to a study of the solvability behaviour of (11.1) in various spaces of functions on the line.

In general, (11.1) is more difficult to handle than its local version (6.1), however, in contrast to (6.1) it enjoys the property of *translation invariance*, i.e. if u is a solution of (11.1) in the sense of Definition 11.2 below, then $(T_h u)(t) = u(t + h)$, $t \in \mathbb{R}$, is again a solution of (11.1), with g replaced by $T_h g$, for each $h \in \mathbb{R}$. To be able to exploit this property, we consider the solvability behaviour of (11.1) only in spaces of locally integrable functions on $\mathbb{R}$ which are translation invariant, more precisely, homogeneous in the sense of

Definition 11.1 *A Banach space $\mathcal{H}(X)$ of locally integrable X-valued functions is said to be **homogeneous** if*
(i) $\mathcal{H}(X) \hookrightarrow L^1_{loc}(\mathbb{R}; X)$, i.e. for each $T > 0$ there is a constant $C(T) > 0$ such that

$$\int_{-T}^T |f(t)|dt \leq C(T)|f|_{\mathcal{H}(X)} \quad \text{for each } f \in \mathcal{H}(X);$$

(ii) $\{T_h : h \in \mathbb{R}\} \subset \mathcal{B}(\mathcal{H}(X))$ is strongly continuous and uniformly bounded;
(iii) $\mathcal{B}(X) \hookrightarrow \mathcal{B}(\mathcal{H}(X))$, i.e. $f \in \mathcal{H}(X)$, $K \in \mathcal{B}(X)$ imply $Kf \in \mathcal{H}(X)$;
are satisfied.

Examples of homogeneous spaces are $L^p(X) := L^p(\mathbb{R}; X)$, $1 \leq p < \infty$, the space of bounded uniformly continuous X-valued functions $C_{ub}(X) := C_{ub}(\mathbb{R}; X)$, the space of almost periodic functions $AP(X)$, the space of continuous ω-periodic functions $P_\omega(X)$, etc. $L^\infty(X) := L^\infty(\mathbb{R}; X)$ and $C_b(X) := C_b(\mathbb{R}; X)$ are not homogeneous since the group of translations is not strongly continuous in these spaces.

The largest space we will consider is $BM(X)$, the space of functions *bounded in mean*, defined by

$$BM(X) = \{f \in L^1_{loc}(\mathbb{R}; X) : |f|_1^b < \infty\},$$

where the norm $| \cdot |_1^b$ in $BM(X)$ is given by

$$|f|_1^b = \sup\{\int_t^{t+1} |f(s)|ds : t \in \mathbb{R}\}.$$

$BM(X)$ is not homogeneous, however, its closed subspace $BM^0(X)$, defined by

$$BM^0(X) = \{f \in BM(X) : |T_h f - f|_1^b \to 0 \quad \text{as } h \to 0\},$$

has this property; in fact, it is the largest homogeneous space, as (a) of the next proposition shows.

If $\mathcal{H}(X)$ is homogeneous, the generator of the translation group will be denoted by $D_\mathcal{H}$, or simply by $D = d/dt$ if there is no danger of confusion. We then may define

$$\mathcal{H}^m(X) = \mathcal{D}(D_\mathcal{H}^m), \quad |f|_{\mathcal{H}^m(X)} := \sum_{l=0}^m |D^l f|_{\mathcal{H}(X)}, \quad m \in \mathbb{N},$$

and

$$\mathcal{H}^\infty(X) = \cap_{m \geq 1} \mathcal{H}^m(X);$$

obviously $\mathcal{H}^m(X)$ are again Banach spaces, and $\mathcal{H}^\infty(X)$ is a Fréchet space with the seminorms $p_m(f) = |D^m f|_{\mathcal{H}(X)}$, $m \in \mathbb{N}_0$. Of interest also is the space $\mathcal{H}^a(X)$ of entire vectors w.r.t. the translation group, i.e.

$$\mathcal{H}^a(X) = \{f \in \mathcal{H}^\infty(X) : \overline{\lim}_{n\to\infty} |D^n f|_{\mathcal{H}(X)}^{1/n} < \infty\}.$$

This is precisely the space of all functions $f \in \mathcal{H}(X)$ such that $T_h f$ extends to an entire function on $\mathbb{C}$ of exponential growth. Since $\{T_h\}$ is a C_0-group on $\mathcal{H}(X)$ it is well-known that $\mathcal{H}^a(X)$ is dense in $\mathcal{H}(X)$; this is the second assertion of the next proposition.

Proposition 11.1 *Let $\mathcal{H}(X)$ be homogeneous. Then*
(i) $\mathcal{H}(X) \hookrightarrow BM^0(X)$;
(ii) $\mathcal{H}^a(X)$ is dense in $\mathcal{H}(X)$;
(iii) $K \in BV(\mathbb{R}; \mathcal{B}(X))$ induces via $Tf := dK * f$ an operator $T \in \mathcal{B}(\mathcal{H}(X))$, there is a constant $C > 0$ such that $|T| \leq C\mathrm{Var}\, K|_{-\infty}^\infty$.

Proof: (i) For $f \in \mathcal{H}(X)$ we have by (i) and (ii) of Definition 11.1

$$\int_t^{t+1} |f(s)|ds = \int_0^1 |(T_t f)(s)|ds \leq C(1)|T_t f|_{\mathcal{H}(X)} \leq C(1)M|f|_{\mathcal{H}(X)}, \quad t \in \mathbb{R},$$

where $M = \sup\{|T_t|_{\mathcal{B}(\mathcal{H}(X))} : t \in \mathbb{R}\}$.
(ii) is standard in semigroup theory; see e.g. Davies [77], p. 30.
(iii) follows from the representation

$$Tf = dK * f = \int_{-\infty}^\infty dK(\tau)f(\cdot - \tau) = \int_{-\infty}^\infty dK(\tau)T_{-\tau}f$$

$$= \lim_{\delta \to 0} \sum_{i=1}^N (K(\tau_i) - K(\tau_{i-1}))T_{-\tau_i}f,$$

where $\delta = \sup\{\tau_i - \tau_{i-1} : i = 1, \dots, N\}$ denotes the size of the decomposition $-\infty < \tau_0 < \dots < \tau_N < \infty$; observe that the function $\tau \mapsto T_\tau f \in \mathcal{H}(X)$ is uniformly continuous on $\mathbb{R}$. □

In particular, any integrable function $F : \mathbb{R} \to \mathcal{B}(X)$ induces by Proposition 11.1 (iii) via $Tf := F * f$ an operator $T \in \mathcal{B}(\mathcal{H}(X))$, for every homogeneous space $\mathcal{H}(X)$.

Functions $f \in BM(X)$ are of polynomial growth, since

$$\int_{-\infty}^{\infty} |f(t)| \frac{dt}{1+t^2} \leq \sum_{j=0}^{\infty} \frac{1}{1+j^2} \left(\int_0^1 |T_j f(t)| dt + \int_{-1}^0 |T_{-j} f(t)| dt \right) \leq c |f|_1^b,$$

for some constant $c > 0$, independent of f. Therefore, $\sigma(f)$, the spectrum of f, introduced in Section 0.5 via the Fourier-Carleman transform is well-defined. This gives rise to the family $\mathcal{H}_\Lambda(X)$ of subspaces of a homogeneous space $\mathcal{H}(X)$, defined by

$$\mathcal{H}_\Lambda(X) = \{f \in \mathcal{H}(X) : \sigma(f) \subset \Lambda\}.$$

Theorem 0.8 shows that $\mathcal{H}_\Lambda(X)$ is closed, whenever $\Lambda \subset \mathbb{R}$ is closed. For the spaces $C_{ub\Lambda}(X)$ we use the abbreviation $\Lambda(X)$, i.e.

$$\Lambda(X) = \{f \in C_{ub}(X) : \sigma(f) \subset \Lambda\}.$$

If $\mathcal{H}(X)$ is homogeneous and $\Lambda \subset \mathbb{R}$ is closed, then Proposition 0.4 implies that $\mathcal{H}_\Lambda(X)$ is again homogeneous. This gives a fine structure of subspaces suitable to describe the solvability behaviour of (11.1) in terms of the spectrum of the inhomogeneity g.

If $\mathcal{H}(X)$ is homogeneous, it is not difficult to verify $\mathcal{H}^a(X) \subset \mathcal{H}^\sigma(X)$, where

$$\mathcal{H}^\sigma(X) = \{f \in \mathcal{H}(X) : \sigma(f) \text{ compact}\};$$

therefore the space $\mathcal{H}^\sigma(X)$ is also dense in $\mathcal{H}(X)$.

Next given $b \in \mathcal{W}(\mathbb{R}) = \{b \in L^1(\mathbb{R}) : \hat{b}(\rho) \neq 0 \text{ for all } \rho \in \mathbb{R}\} \cup \{\delta_0\}$, the *Wiener class*, we define

$$\mathcal{H}^b(X) = \{f \in \mathcal{H}(X) : f = b * g \quad \text{for some } g \in \mathcal{H}(X)\},$$

i.e. $\mathcal{H}^b(X)$ is the range of the convolution operator on $\mathcal{H}(X)$ induced by b. By local analyticity of $1/\tilde{b}$ on $\mathbb{R}$ and by Proposition 0.6, there follows $\mathcal{H}^\sigma(X) \subset \mathcal{H}^b(X)$, for each $b \in \mathcal{W}(\mathbb{R})$, hence these spaces are also dense in $\mathcal{H}(X)$. It should be noted that for the kernels $b_\beta(t) = e_0(t)e^{-t}t^{\beta-1}/\Gamma(\beta)$, one obtains $\mathcal{H}^{b_\beta}(X) = \mathcal{H}^\beta(X) = \mathcal{D}(D^\beta)$, for all $\beta > 0$; in fact, it is easy to verify the relation $(I+D)^{-\beta}f = b_\beta * f$.

Let us summarize this discussion in

Proposition 11.2 *Let $\Lambda \subset \mathbb{R}$ be closed, $\mathcal{H}(X)$ be homogeneous, $b \in \mathcal{W}(\mathbb{R})$, and let $\mathcal{H}_\Lambda(X)$, $\mathcal{H}^a(X)$, $\mathcal{H}^\sigma(X)$, $\mathcal{H}^b(X)$ be defined as above. Then*
(i) $\mathcal{H}_\Lambda(X)$ is a homogeneous subspace of $\mathcal{H}(X)$;

(ii) $\mathcal{H}^a(X) \subset \mathcal{H}^\sigma(X) \subset \mathcal{H}^b(X)$;

(iii) $\mathcal{H}^\sigma(X)$ and $\mathcal{H}^b(X)$ are dense in $\mathcal{H}(X)$.

11.2 Admissibility

Let's turn attention to (11.1) now. If $A \in L^1(\mathbb{R}_+; \mathcal{B}(Y, X))$ and $u \in BM^0(Y)$ then $A * u$ is well-defined, belongs to $BM^0(X)$, and $|A * u|_1^b \le |A|_1 |u|_1^b$. For many applications, however, integrability of $A(t)$ is too stringent, but then the convolution $A * u$ over $\mathbb{R}$ is in general not defined. For this reason we restrict attention to kernels $A(t)$ of the form

$$A(t) = A_0(t) + A_1(t), \quad t \ge 0, \quad \text{where} \tag{11.2}$$
$$A_0 \in L^1(\mathbb{R}_+; \mathcal{B}(Y, X)) \quad \text{and } A_1 \in BV(\mathbb{R}_+; \mathcal{B}(Y, X)),$$

and consider the differentiated version of (11.1):

$$\dot{u}(t) = \int_0^\infty A_0(\tau)\dot{u}(t - \tau)d\tau + \int_0^\infty dA_1(\tau)u(t - \tau) + f(t), \quad t \in \mathbb{R}. \tag{11.3}$$

In the sequel we always assume (11.2) and confine our study to (11.3). Remarks on (11.1) with $A \in L^1(\mathbb{R}_+; \mathcal{B}(Y, X))$ and on the second order case where (11.1) is differentiated twice will be given in Section 11.7.

Definition 11.2 *Let $f \in BM^0(X)$. Then*
(a) $u \in BM^0(X)$ is called **strong solution** *of (11.3) if $u \in BM^0(Y)$, $u - A_0 * u \in BM^1(X)$, and $(u - A_0 * u)^{\cdot} = dA_1 * u + f$ a.e. on $\mathbb{R}$.*
(b) $u \in BM^0(X)$ is called **mild solution** *of (11.3) if there are $f_n \in BM^0(X)$, $f_n \to f$ in $BM^0(X)$, and strong solutions $u_n \in BM^0(Y)$ of (11.3) with f replaced by f_n, such that $u_n \to u$ in $BM^0(X)$.*

The solvability behaviour of (11.3) is described in terms of admissibility of homogeneous spaces which we introduce next.

Definition 11.3 *Let $b \in \mathcal{W}(\mathbb{R})$ be given. A homogeneous space $\mathcal{H}(X)$ is called* **b-admissible** *for equation (11.3), if for each $f \in \mathcal{H}^b(X)$ there is a unique strong solution $u \in \mathcal{H}(X) \cap BM^0(Y)$ of (11.3), and $(f_n) \subset \mathcal{H}^b(X)$, $f_n \to 0$ in $\mathcal{H}(X)$, implies $u_n \to 0$ in $\mathcal{H}(X)$. $\mathcal{H}(X)$ is called* **0-admissible** *if it is δ_0-admissible.*

Thus the kernel $b \in \mathcal{W}(\mathbb{R})$ measures the amount of regularity in time of an inhomogeneity f needed to allow for a strong solution. Suppose the homogeneous space $\mathcal{H}(X)$ is b-admissible. Then we may define the *solution operator* $G : \mathcal{H}^b(X) \to \mathcal{H}(X) \cap BM^0(Y)$ by means of $Gf = u$, where $u \in \mathcal{H}(X) \cap BM^0(Y)$ denotes the unique solution of (11.3) with $f \in \mathcal{H}^b(X)$. By definition of b-admissibility, G admits a unique bounded linear extension to $G \in \mathcal{B}(\mathcal{H}(X))$ since $\mathcal{H}^b(X)$ is dense in $\mathcal{H}(X)$ by Proposition 11.1. Moreover, the closed graph theorem implies also $b * G \in \mathcal{B}(\mathcal{H}(X), BM^0(Y))$, where we deliberately use the notation $b * G$ for the operator $G(b * f)$. Therefore, Gf is a mild solution of (11.3), for each $f \in \mathcal{H}(X)$.

Translation invariance and uniqueness of strong solutions for $f \in \mathcal{H}^b(X)$ then imply that G commutes with the group of translations, hence also with its generator D, in particular $G \in \mathcal{B}(\mathcal{H}^m(X))$ for each $m \geq 0$, as well as $b * G \in \mathcal{B}(\mathcal{H}^m(X), BM^m(Y))$. Observe also the relation

$$G(\varphi * f) = \varphi * Gf, \quad f \in \mathcal{H}(X), \ \varphi \in L^1(\mathbb{R}), \tag{11.4}$$

which follows again from translation invariance and Proposition 11.1. These observations are collected in

Proposition 11.3 *Suppose the homogeneous space $\mathcal{H}(X)$ is b-admissible, and let the solution operator G be defined as above. Then*
(i) $G \in \mathcal{B}(\mathcal{H}^m(X))$, and $b * G \in \mathcal{B}(\mathcal{H}^m(X), BM^m(Y))$ for each $m \geq 0$;
(ii) G commutes with the translation group, i.e. $T_h G = G T_h$ for all $h \in \mathbb{R}$;
(iii) G commutes with D;
(iv) $G(\varphi * f) = \varphi * Gf$ for all $f \in \mathcal{H}(X)$, $\varphi \in L^1(\mathbb{R})$.

Next we draw a connection between integrability properties of the resolvent $S(t)$ for the local version (6.1) of (11.1) and admissibility.

Proposition 11.4 *Suppose (6.1) admits a resolvent $S(t)$, which is integrable in $\mathcal{B}(X)$ and such that $b * S$ is integrable in $\mathcal{B}(X, Y)$, for some $b \in \mathcal{W}(\mathbb{R}_+)$. Then each homogeneous space $\mathcal{H}(X)$ is b-admissible.*

Proof: We will show the representation

$$(Gf)(t) = \int_0^\infty S(\tau) f(t - \tau) d\tau, \quad t \in \mathbb{R}. \tag{11.5}$$

Since $S(t)$ is integrable in $\mathcal{B}(X)$ by assumption, Proposition 11.1 shows $G \in \mathcal{B}(\mathcal{H}(X))$ for each homogeneous space. On the other hand, since $b * S$ is integrable in $\mathcal{B}(X, Y)$ we obtain $b * G \in \mathcal{B}(\mathcal{H}(X), BM^0(Y))$. From the first resolvent equation (6.2) we obtain with $R = b * S$

$$(R - A_0 * R)^{\cdot} = b + dA_1 * R, \quad t \in \mathbb{R},$$

hence $u = b * Gg = Gb * g = R * g$ with $g \in \mathcal{H}(X)$ is a strong solution of (11.3) for $f = b * g$. Conversely, if u is a strong solution of (11.3), by the second resolvent equation (6.3),

$$(S - S * A_0)^{\cdot} = \delta_0 + S * dA_1,$$

we obtain $u = S * f$, hence uniqueness. $\quad \square$

To obtain conditions necessary for admissibility, suppose $\Lambda(X)$ is b-admissible. For $\rho \in \Lambda$, the function $f(t) = e^{i\rho t} x$, $x \in X$, belongs to $\Lambda^a(X)$, hence there is a unique strong solution $u \in \Lambda(X) \cap BM^0(Y)$ of (11.3). Translation invariance, i.e. Proposition 11.3 then yields

$$\dot{u} = DGf = G\dot{f} = i\rho Gf = i\rho u,$$

hence $u(t) = e^{i\rho t}y$ for some unique $y \in Y$. (11.3) now implies

$$(i\rho - i\rho\hat{A}_0(i\rho) - \widehat{dA}_1(i\rho))y = x,$$

hence the operators $i\rho - \hat{A}_0(i\rho) - \widehat{dA}_1(i\rho) = i\rho(I - \hat{A}(i\rho)) \in \mathcal{B}(Y,X)$ are invertible for each $\rho \in \Lambda$. Moreover, boundedness of G in $\Lambda(X)$, and of $b * G$ from $\Lambda(X)$ to $BM^0(Y)$ yield the estimates

$$|y|_X = |u|_{\Lambda(X)} \leq |G|_{\mathcal{B}(\Lambda(X))}|f|_{\Lambda(X)} \leq c|x|_X,$$
$$|\tilde{b}(\rho)||y|_Y = |b * u|_{BM^0(Y)} \leq |b * G|_{\mathcal{B}(\Lambda(X),BM^0(Y))}|f|_{\Lambda(X)} \leq c|x|_X.$$

With the notation

$$H(\lambda) = \frac{1}{\lambda}(I - \hat{A}(\lambda))^{-1} = (\lambda - \lambda\hat{A}_0(\lambda) - \widehat{dA}_1(\lambda))^{-1}, \tag{11.6}$$

which was used before, this implies

$$|H(i\rho)|_{\mathcal{B}(X)} + |\tilde{b}(\rho)||H(i\rho)|_{\mathcal{B}(X,Y)} \leq M, \quad \rho \in \Lambda.$$

Defining the *real spectrum* of (11.3) by

$$\Lambda_0 = \{\rho \in \mathbb{R} : i\rho - i\rho\hat{A}_0(i\rho) - \widehat{dA}_1(i\rho) \in \mathcal{B}(Y,X) \text{ is not invertible}\}, \tag{11.7}$$

we have proved

Proposition 11.5 *Suppose $\Lambda(X)$ is b-admissible, $\Lambda \subset \mathbb{R}$ closed. Then $\Lambda \cap \Lambda_0 = \emptyset$, and there is a constant $M > 0$ such that*

$$|H(i\rho)|_{\mathcal{B}(X)} + |\tilde{b}(\rho)||H(i\rho)|_{\mathcal{B}(X,Y)} \leq M, \quad \text{for all } \rho \in \Lambda, \tag{11.8}$$

where $H(\lambda)$ is defined by (11.6).

The converse of Proposition 11.5 is in general false, even in the semigroup case $A(t) \equiv A$; see Section 12.7 for further remarks in the semigroup case. However, for large classes of equations or closed sets $\Lambda \subset \mathbb{R}$ characterization of admissibility in terms of the spectral condition $\Lambda \cap \Lambda_0 = \emptyset$ and uniform bounds on $H(i\rho)$ like (11.8) is possible. The next subsection and all of Section 12 deal with this question.

11.3 Λ-Kernels for Compact Λ
A main tool in the sequel will be the following

Lemma 11.1 *Suppose $\varphi \in L^1(\mathbb{R})$ is such that $\tilde{\varphi} \in C_0^\infty(\mathbb{R})$ and supp $\tilde{\varphi} \cap \Lambda_0 = \emptyset$. Then there is $G_\varphi \in W^{\infty,1}(\mathbb{R}; \mathcal{B}(X,Y))$ such that*

$$\dot{G}_\varphi = A_0 * \dot{G}_\varphi + dA_1 * G_\varphi + \varphi = \dot{G}_\varphi * A_0 + G_\varphi * dA_1 + \varphi. \tag{11.9}$$

Moreover, G_φ admits extension to an entire function $G_\varphi(z)$ with values in $\mathcal{B}(X,Y)$ of exponential growth.

Proof: Since $\tilde{\varphi}$ has compact support and supp $\tilde{\varphi} \cap \Lambda_0 = \emptyset$, $H(i\rho)$ is bounded in $\mathcal{B}(X, Y)$ on supp $\tilde{\varphi}$, and therefore $G_\varphi(z)$ defined by

$$G_\varphi(z) = \frac{1}{2\pi} \int_{-\infty}^{\infty} H(i\rho)\tilde{\varphi}(\rho)e^{i\rho z}d\rho, \quad z \in \mathbb{C}, \tag{11.10}$$

is an entire function with values in $\mathcal{B}(X, Y)$. Clearly G_φ and all its derivatives are bounded on $\mathbb{R}$ and of exponential growth on $\mathbb{C}$, and it is easily verified that (11.9) holds. Thus it remains to prove $G_\varphi^{(n)} \in L^1(\mathbb{R}; \mathcal{B}(X, Y))$ for each $n \in \mathbb{N}_0$. Replacing $\tilde{\varphi}$ by $\tilde{\varphi} \cdot (i\rho)^n$, i.e. φ by $\varphi^{(n)}$, it is enough to show $G_\varphi \in L^1(\mathbb{R}; \mathcal{B}(X, Y))$ since $G_\varphi^{(n)} = G_{\varphi^{(n)}}$.

(i) Suppose first that in addition to (11.2) we have

$$\int_0^\infty |A_0(t)|_{\mathcal{B}(Y,X)}(1 + t^2)dt < \infty, \quad \text{and} \quad \int_0^\infty |dA_1(t)|_{\mathcal{B}(Y,X)}(1 + t^2)dt < \infty.$$

Then $i\rho\hat{A}(i\rho)$ is twice continuously differentiable on $\mathbb{R}$, hence $\Phi(\rho) = H(i\rho)\tilde{\varphi}(\rho)$ is so as well. Integrating by parts twice, (11.10) becomes

$$G_\varphi(t) = -\frac{1}{2\pi t^2} \int_{-\infty}^{\infty} \Phi''(\rho)e^{i\rho t}d\rho,$$

and so we obtain $|G_\varphi(t)|_{\mathcal{B}(X,Y)} \le C(1 + t^2)^{-1}$, i.e. $G_\varphi \in L^1(\mathbb{R}; \mathcal{B}(X, Y))$.

(ii) To remove the moment conditions, consider $A_m(t)$ given by

$$A_m(t) = A_{0m}(t) + A_{1m}(t), \ t > 0, \ \text{where}$$

$$A_{0m}(t) = A_0(t) \ \text{for} \ t \le m, \ A_{0m}(t) = 0 \ \text{for} \ t > m,$$

$$A_{1m}(t) = A_1(t) \ \text{for} \ t \le m, \ A_{1m}(t) = A_1(m) \ \text{for} \ t > m.$$

Then $\lambda\hat{A}_m(\lambda) \to \lambda\hat{A}(\lambda)$ in $\mathcal{B}(Y, X)$ as $m \to \infty$, uniformly on compact subsets of $\overline{\mathbb{C}}_+$, and A_{0m}, dA_{1m} have moments of any order. The relation

$$i\rho(I - \hat{A}_m(i\rho)) = i\rho(I - \hat{A}(i\rho))(I - H(i\rho)i\rho(\hat{A}_m(i\rho) - \hat{A}(i\rho))) \tag{11.11}$$

therefore shows that $H_m(i\rho) = (i\rho - i\rho\hat{A}_m(i\rho))^{-1}$ exists on a neighborhood U of supp $\tilde{\varphi}$ and $|H_m(i\rho)|_{\mathcal{B}(X,Y)} \le M < \infty$ on supp $\tilde{\varphi}$, provided m is large enough. (11.11) then yields

$$H(i\rho)\tilde{\varphi}(\rho) = H_m(i\rho)\tilde{\varphi}(\rho) + H(i\rho)\tilde{\varphi}(\rho)i\rho[\hat{A}(i\rho) - \hat{A}_m(i\rho)]H_m(i\rho)\tilde{\varphi}_1(\rho), \tag{11.12}$$

where $\tilde{\varphi}_1 \in C_0^\infty$ is such that $\tilde{\varphi}_1 = 1$ on supp $\tilde{\varphi}$, $0 \le \tilde{\varphi}_1 \le 1$, and supp $\tilde{\varphi}_1 \subset U$. By step (i) of this proof there are $G_m, G_{jm} \in L^1(\mathbb{R}; \mathcal{B}(X, Y))$ such that $\tilde{G}_m = \tilde{\varphi}H_m$, $\tilde{G}_{jm} = (i\rho)^{1-j}\tilde{\varphi}_1 H_m$. Let

$$K = (A_0 - A_{0m}) * G_{0m} + (dA_1 - dA_{1m}) * G_{1m},$$

which belongs to $L^1(\mathbb{R}; \mathcal{B}(X))$; then (11.12) becomes the convolution equation

$$G_\varphi = G_m + G_\varphi * K. \qquad (11.13)$$

Since

$$I - \widetilde{K} = I - i\rho(\widetilde{A} - \widetilde{A}_m)H_m\tilde{\varphi}_1 = (i\rho - i\rho\widetilde{A}_m(1 - \tilde{\varphi}_1) - i\rho\widetilde{A}\tilde{\varphi}_1)H_m$$

is invertible on $\mathbb{R}$ by the choice of $\tilde{\varphi}_1$, if m is large enough, by Theorem 0.6 there is a resolvent kernel $R \in L^1(\mathbb{R}; \mathcal{B}(X))$ for (11.13), and so the representation

$$G_\varphi = G_m + G_m * R$$

shows $G_\varphi \in L^1(\mathbb{R}; \mathcal{B}(X, Y))$. The proof is complete. $\quad\square$

By means of Lemma 11.1 we can now prove the converse of Proposition 11.5 for compact Λ.

Theorem 11.1 *Suppose $\Lambda \subset \mathbb{R}$ is compact, $\Lambda \cap \Lambda_0 = \emptyset$, and let $\mathcal{H}(X)$ be homogeneous. Then $\mathcal{H}_\Lambda(X)$ is 0-admissible and there is $G_\Lambda \in W^{\infty,1}(\mathbb{R}; \mathcal{B}(X, Y))$ such that the solution operator G for $\mathcal{H}_\Lambda(X)$ is represented by*

$$(Gf)(t) = \int_{-\infty}^\infty G_\Lambda(\tau)f(t - \tau)d\tau, \quad t \in \mathbb{R}, \ f \in \mathcal{H}_\Lambda(X). \qquad (11.14)$$

$G_\Lambda(t)$ extends to an entire $\mathcal{B}(X, Y)$-valued function of exponential growth.

Proof: Since $\Lambda \subset \mathbb{R}$ is compact and $\Lambda \cap \Lambda_0 = \emptyset$, there is a neighborhood $\Lambda_{3\varepsilon} = \Lambda + \overline{B_{3\varepsilon}}(0)$ of Λ such that $\Lambda_{3\varepsilon} \cap \Lambda_0 = \emptyset$; observe that Λ_0 is closed. Choose $\varphi_0 \in L^1(\mathbb{R})$ such that $\tilde{\varphi}_0 \in C_0^\infty(\mathbb{R})$, $\tilde{\varphi}_0 \equiv 1$ on Λ_ε, $\tilde{\varphi}_0 \equiv 0$ on $\mathbb{R} \setminus \Lambda_{2\varepsilon}$, $0 \leq \tilde{\varphi}_0 \leq 1$, and let $G_\Lambda \in W^{\infty,1}(\mathbb{R}; \mathcal{B}(X, Y))$ denote the function G_φ from Lemma 11.1 with $\varphi = \varphi_0$. If $f \in \mathcal{H}_\Lambda(X)$, then $\sigma(f) \subset \Lambda$ hence $\varphi_0 * f = f$ by Proposition 0.6, and so $u = G_\Lambda * f$ is a strong solution of (11.3), by (11.9). On the other hand, if $u \in \mathcal{H}_\Lambda(X)$ is a strong solution of (11.3) then (11.9) yields $G_\Lambda * f = \varphi_0 * u = u$ by Proposition 0.6 again, hence uniqueness. Since $G_\Lambda \in L^1(\mathbb{R}; \mathcal{B}(X, Y))$, we finally obtain boundedness of the solution operator G in $\mathcal{H}_\Lambda(X)$, as well as from $\mathcal{H}_\Lambda(X)$ to $BM^0(Y)$, and so $\mathcal{H}_\Lambda(X)$ is 0-admissible. $\quad\square$

A kernel G_Λ representing the solution operator G of a homogeneous space $\mathcal{H}_\Lambda(X)$ via (11.14) will be called a Λ-*kernel* in the sequel. Observe that 0-admissible implies b-admissible for any $b \in \mathcal{W}(\mathbb{R})$. Lemma 11.1 also yields dense solvability and uniqueness in $\mathcal{H}_\Lambda(X)$ whenever $\Lambda \cap \Lambda_0 = \emptyset$, as we show next.

Proposition 11.6 *Let $\mathcal{H}(X)$ be homogeneous and suppose $\Lambda \cap \Lambda_0 = \emptyset$. Then*
(i) $\sigma(u) \subset \Lambda_0$ for each strong solution of (11.3) with $f = 0$;
(ii) for each $f \in \mathcal{H}_\Lambda(X)$ there is at most one strong solution $u \in \mathcal{H}_\Lambda(X)$;

(iii) for each $f \in \mathcal{H}_\Lambda^\sigma(X)$ there is a strong solution $u \in \mathcal{H}_\Lambda(X)$;
(iv) if $f \in \mathcal{H}_\Lambda(X)$, and $u \in \mathcal{H}_\Lambda(X)$ is a strong solution, then $\sigma(u) = \sigma(f)$.

Proof: (i) Since $\sigma(u) = \operatorname{supp} \tilde{D}_u$, where D_u denotes the distribution generated by $u \in BM^0(Y) \subset BM^0(X)$, in virtue of Proposition 0.5, we have to show

$$[\tilde{D}_u, \tilde{\varphi}] = [D_u, \check{\tilde{\varphi}}] = (\varphi * u)(0) = 0$$

for each $\tilde{\varphi} \in C_0^\infty(\mathbb{R})$ with $\operatorname{supp} \tilde{\varphi} \cap \Lambda_0 = \emptyset$. But given such φ, let G_φ denote the kernel from Lemma 11.1; (11.9) and (11.3) then imply $\varphi * u = 0$, in particular $(\varphi * u)(0) = 0$.

(ii) follows from (i), since $\sigma(u) \subset \Lambda_0 \cap \Lambda = \emptyset$ implies $u = 0$, by Proposition 0.5.

(iii) If $f \in \mathcal{H}_\Lambda^\sigma(X)$ then $\sigma(f) \subset \Lambda$ is compact and does not intersect Λ_0. Applying Theorem 11.1 to the set $\sigma(f)$ we obtain a strong solution $u \in \mathcal{H}_{\sigma(f)}(X) \subset \mathcal{H}_\Lambda(X)$ of (11.3).

(iv) Suppose $u \in \mathcal{H}_\Lambda(X)$ is a strong solution of (11.3), and let $\tilde{\varphi} \in C_0^\infty(\mathbb{R})$ be such that $\operatorname{supp} \tilde{\varphi} \cap \sigma(u) = \emptyset$. Then $\sigma(\varphi * u) = \emptyset$, hence $\varphi * u = 0$ by Proposition 0.5, and so (11.3) implies also $\varphi * f = 0$. But this implies as above $[\tilde{D}_f, \tilde{\varphi}] = 0$, i.e. $\sigma(f) \subset \sigma(u)$.

Conversely, let $\tilde{\varphi} \in C_0^\infty(\mathbb{R})$ satisfy $\operatorname{supp} \tilde{\varphi} \cap \sigma(f) = \emptyset$. Then $\varphi * f = 0$, hence $\varphi * u$ is a strong solution of (11.3) with $f = 0$. Since $\sigma(\varphi * u) \subset \operatorname{supp} \tilde{\varphi}$, but also $\sigma(\varphi * u) \subset \Lambda_0$ by (i) of this proposition, we obtain $\sigma(\varphi * u) = \emptyset$, hence $\varphi * u = 0$ by Proposition 0.5 again; this implies $\sigma(u) = \sigma(f)$. $\square$

A reasoning similar to that of the last step in the proof of Proposition 11.6 yields

Corollary 11.1 *Suppose $\mathcal{H}(X)$ is a homogeneous space which is b-admissible, and let G denote the solution operator. Then*

$$\sigma(Gf) = \sigma(f) \quad \text{for each } f \in \mathcal{H}(X).$$

Proof: Consider first $f \in \mathcal{H}^b(X)$ and let $u = Gf$. Then as in (iv) of the proof of Proposition 11.6 we have $\sigma(f) \subset \sigma(u)$. Conversely, let $\tilde{\varphi} \in C_0^\infty(\mathbb{R})$ satisfy supp $\widetilde{\varphi} \cap \sigma(f) = \emptyset$, i.e. $\varphi * f = 0$ by Proposition 0.6. Then $\varphi * u$ belongs to $\mathcal{H}(X)$ and is a strong solution of (11.3) with $f = 0$, hence by admissibility $\varphi * u = 0$. With $[\widetilde{D_u}, \tilde{\varphi}] = (\varphi * u)(0)$ this then implies $\sigma(u) = \sigma(f)$, i.e. we have $\sigma(Gf) = \sigma(f)$ for each $f \in \mathcal{H}^b(X)$.

For the general case consider $\tilde{\varphi} \in C_0^\infty(\mathbb{R})$ such that $0 \leq \tilde{\varphi} \leq 1$, $\tilde{\varphi}(\rho) = 1$ for $|\rho| \leq 1$, $\tilde{\varphi}(\rho) = 0$ for $|\rho| \geq 2$, and let $\varphi_n(t) = n\varphi(nt)$. φ_n is then an approximation of the identity, i.e.

$$\varphi_n * f = \int_{-\infty}^\infty \varphi_n(\tau) T_{-\tau} f d\tau = \int_{-\infty}^\infty \varphi(\tau) T_{-\tau/n} f d\tau \to f \quad \text{in } \mathcal{H}(X)$$

since the translation group is strongly continuous in $\mathcal{H}(X)$. But $\sigma(\varphi_n * f) \subset \sigma(f) \cap \operatorname{supp} \tilde{\varphi}_n \subset \sigma(f) \cap \bar{B}_{2n}(0)$ is compact, i.e. $\varphi_n * f \in \mathcal{H}^\sigma(X)$. From Proposition

11.3 we obtain $\varphi_n * Gf = G\varphi_n * f$, hence these are strong solutions, and so $\sigma(\varphi_n * Gf) = \sigma(\varphi_n * f)$. But with Theorem 0.8 we obtain

$$\sigma(Gf) = \cap_{m\geq 1}\overline{\cup_{n\geq m}\sigma(\varphi_n * Gf)} = \cap_{m\geq 1}\overline{\cup_{n\geq m}\sigma(\varphi_n * f)} = \sigma(f). \quad \square$$

From Proposition 11.6 it is evident, that the main part in proving admissibility of $\mathcal{H}_\Lambda(X)$, where $\Lambda \cap \Lambda_0 = \emptyset$, consists in establishing a bound M such that

$$|u|_{\mathcal{H}(X)} \leq M|f|_{\mathcal{H}(X)} \quad \text{and} \quad |b * u|_{BM^\circ(Y)} \leq M|f|_{\mathcal{H}(X)}$$

holds for each $f \in \mathcal{H}_\Lambda(X)$. This question is studied for unbounded sets $\Lambda \subset \mathbb{R}$ in Section 12. Here we discuss next further consequences of admissibility.

11.4 Almost Periodic Solutions

Recall that a function $f \in C_b(X)$ is called *almost periodic* (a.p.) if $\{T_\tau f : \tau \in \mathbb{R}\}$ is relatively compact in $C_b(X)$. Since almost periodic functions are uniformly continuous, the space $AP(X)$ of all a.p. functions is a closed translation invariant subspace of $C_{ub}(X)$, which is also homogeneous. Note that periodic functions are characterized among a.p. functions by the stronger property that $\{T_\tau f : \tau \in \mathbb{R}\}$ is compact in $C_b(X)$.

The *Bohr transform* given by

$$\alpha(\rho, f) := \lim_{N\to\infty} N^{-1} \int_0^N e^{-i\rho t} f(t)dt, \quad \rho \in \mathbb{R}, \ f \in AP(X),$$

is well-defined and continuous from $AP(X)$ to X. For $f \in AP(X)$ its exponent set

$$\exp(f) = \{\rho \in \mathbb{R} : \alpha(\rho, f) \neq 0\}$$

is at most countable, and $\sigma(f) = \overline{\exp(f)}$. This can be proved by means of Bochner's approximation theorem, which states that, given a fixed countable subset $\{\rho_j\}_1^\infty$, there are convergence factors $\gamma_{nj} \in \mathbb{R}$ with $\gamma_{nj} \to 1$ as $n \to \infty$ such that the trigonometric polynomials

$$B_n f = \sum_{j=1}^n \gamma_{nj}\alpha(\rho_j, f)e^{i\rho_j t}, \quad n \in \mathbb{N},$$

converge to f uniformly on $\mathbb{R}$, provided $\exp(f) \subset \{\rho_j\}_1^\infty$. By virtue of this result, it is also clear that $f \in AP(X)$ is uniquely determined by its Bohr transform.

All of these results and many others can be found in the monograph of Amerio and Prouse [9].

Recall also that a function $f \in C_{ub}(X)$ is called *asymptotically almost periodic* (a.a.p.) (to the right) if $\{T_\tau f|_{[0,\infty)}\}_{\tau\geq 0} \subset C_b(\mathbb{R}_+; X)$ is relatively compact. It can be shown (see e.g. Fréchet [119]) that then there is an a.p. function f_a such that $f_0(t) = f(t) - f_a(t) \to 0$ as $t \to \infty$, i.e. $f_0 \in C_0^+(X)$, where

$$C_0^+(X) = \{f \in C_{ub}(X) : f(t) \to 0 \quad \text{as } t \to \infty\}$$

is another closed subspace of $C_{ub}(X)$. The a.p. function f_a is necessarily unique, since

$$\alpha(\rho, f_a) = \alpha(\rho, f), \quad \rho \in \mathbb{R},$$

and the Bohr transform is unique. Moreover, $|f_a|_\infty \leq |f|_\infty$, and therefore we have the direct sum decomposition of the space $AAP(X)$ of a.a.p. functions

$$AAP^+(X) = AP(X) \oplus C_0^+(X).$$

The space

$$C_l^+(X) = \{f \in C_{ub}(X) : \lim_{t \to \infty} f(t) = f(\infty) \text{ exists }\}$$

is a closed subspace of $AAP^+(X)$. It is easy to see that $C_0^+(X)$, $C_l^+(X)$, and $AAP^+(X)$ are homogeneous.

After these preparations we are ready to prove the following result on a.p. and a.a.p. solutions of (11.3).

Theorem 11.2 *Suppose $\Lambda(X)$ is b-admissible, where $\Lambda \subset \mathbb{R}$ is closed, and let G denote the corresponding solution operator. Then*
(i) $f \in AP(X)$, $\exp(f) \subset \Lambda$, imply $Gf \in AP(X)$, $\exp(Gf) = \exp(f)$, and

$$\alpha(\rho, Gf) = H(i\rho)\alpha(\rho, f), \quad \rho \in \exp(f). \tag{11.15}$$

(ii) $f \in AAP^+(X)$, $\sigma(f) \subset \Lambda$, imply $Gf \in AAP^+(X)$, $(Gf)_0 = Gf_0$, $(Gf)_a = Gf_a$ and also (11.15) holds;
(iii) $0 \in \Lambda$, $f \in C_l^+(X)$, $\sigma(f) \subset \Lambda$ imply $Gf \in C_l^+(X)$ and

$$(Gf)(\infty) = H(0)f(\infty). \tag{11.16}$$

Proof: (i) Suppose $\Lambda(X)$ is b-admissible. Then if $\rho \in \Lambda$, as in the discussion before Proposition 11.5 we obtain $G(xe^{i\rho t}) = H(i\rho)xe^{i\rho t}$; note that $\Lambda \cap \Lambda_0 = \emptyset$. Hence if $f(t) = \sum_{n=1}^N f_n e^{i\rho_n t}$ is a trigonometric polynomial with $\exp(f) = \{\rho_n\}_1^N \subset \Lambda$ we obtain by linearity $(Gf)(t) = \sum_{n=1}^N H(i\rho_n)f_n e^{i\rho_n t}$, i.e. (i) of Theorem 11.2 holds for trigonometric polynomials. The general case then follows by Bochner's approximation theorem mentioned above since G is bounded and the Bohr transform is continuous.

(ii) If $f = f_a + f_0 \in AAP^+(X) \cap \Lambda(X)$, then $T_{\tau_n} f \to f_a$ uniformly on compact subsets of $\mathbb{R}$, for some sequence $\tau_n \to \infty$. Hence by Theorem 0.8 we obtain $\sigma(f_a) \subset \sigma(f) \subset \Lambda$, and so $\sigma(f_0) = \sigma(f - f_a) \subset \sigma(f) \cup \sigma(f_a) \subset \sigma(f)$ as well. Since $\Lambda(X)$ is b-admissible we therefore have $Gf = Gf_a + Gf_0$, where $Gf_a \in AP(X)$ by step (i) of this proof, and (11.15) holds again. Thus it remains to show $Gf_0 \in C_0^+(X)$, i.e. $(Gf_0)(t) \to 0$ as $t \to \infty$. By Proposition 11.2 it is enough to assume $\sigma(f_0)$ compact, i.e. w.l.o.g. Λ compact. Let $G_\Lambda \in W^{\infty,1}(\mathbb{R}; \mathcal{B}(X,Y))$ denote a Λ-kernel for (11.3) which exists by Theorem 11.1. Then $Gf_0 = G_\Lambda * f_0$ and since $G_\Lambda \in L^1(\mathbb{R}; \mathcal{B}(X,Y))$, $(Gf_0)(t) \to 0$ as $t \to \infty$ follows from $f_0(t) \to 0$ as $t \to \infty$.

(iii) This is a consequence of (ii) with $f_a(t) \equiv f(\infty)$. $\square$

Observe that in case $f = b * g$, with $g \in \Lambda(X) \cap AP(X)$ and $\Lambda(X)$ is b-admissible then $u \in BM^0(Y)$ is a.p. in the sense of Stepanov in Y, i.e. $\{T_t u : t \in \mathbb{R}\} \subset BM^0(Y)$ is relatively compact.

If $f \in C_{ub}(X)$ is ω-periodic, then $\exp(f) \subset (2\pi/\omega)\mathbb{Z}$ and

$$\alpha(2\pi n/\omega, f) = f_n = \frac{1}{\omega} \int_0^\omega e^{-2\pi i n t/\omega} f(t)\,dt, \quad n \in \mathbb{Z},$$

are the Fourier coefficients of f. Then with $\Lambda = (2\pi/\omega)\mathbb{Z}$ we have $\Lambda(X) = P_\omega(X)$, and Theorem 11.2 yields the following

Corollary 11.2 *Suppose $\Lambda = (2\pi/\omega)\mathbb{Z}$ and $\Lambda(X)$ is b-admissible. Then for every $f \in P_\omega(X)$ there is a unique mild solution $u \in P_\omega(X)$ of (11.3) and for their Fourier coefficients we have*

$$u_n = H(2\pi i n/\omega)f_n, \quad n \in \mathbb{Z}. \tag{11.17}$$

In passing we note the following result for a.p. solutions, without assuming any admissibility.

Proposition 11.7 *Let $f \in AP(X)$ and suppose $u \in BM^0(Y)$ is a strong solution of (11.3) which is a.p. in Y in the sense of Stepanov. Then*

$$(i\rho - i\rho\hat{A}_0(i\rho) - \widehat{dA_1}(i\rho))\alpha(\rho, u) = \alpha(\rho, f) \quad \text{for all } \rho \in \mathbb{R}. \tag{11.18}$$

In particular, the condition

$$\alpha(\rho, f) \in \mathcal{R}(i\rho - i\rho\hat{A}_0(i\rho) - \widehat{dA_1}(i\rho)), \quad \rho \in \mathbb{R},$$

is necessary for existence of u.

This follows from direct computation on the Fourier coefficients of u by using (11.3).

Another concept of almost periodicity was introduced by Eberlein [107] and is defined as follows. A function $f \in C_b(X)$ is called *weakly almost periodic* (w.a.p.) in the sense of Eberlein, if $\{T_\tau f : \tau \in \mathbb{R}\}$ is weakly relatively compact in $C_b(X)$. The space $W(X)$ of all w.a.p. functions can be shown to be a closed translation invariant subspace of $C_{ub}(X)$ which is also homogeneous. The inclusion $AP(X) \subset W(X)$ is obvious. According to the decomposition theorem of Jacobs, Glicksberg, Deleeuw (see e.g. Krengel [201], Sect. 2.4),

$$W(X) = AP(X) \oplus W_0(X),$$

where the homogeneous space $W_0(X) \subset W(X)$ is given by

$$W_0(X) = \{f \in W(X) \ : \ w - \lim_{n\to\infty} T_{\tau_n} f = w - \lim_{n\to\infty} T_{\sigma_n} f = 0,$$

$$\text{for some sequences } \tau_n \to \infty, \ \sigma_n \to -\infty\}.$$

Similarly, a function $f \in C_{ub}(X)$ is called *weakly asymptotically almost periodic* (w.a.a.p.) if $\{T_\tau f|_{[0,\infty)}\}_{\tau \geq 0} \subset C_b(\mathbb{R}_+; X)$ is weakly relatively compact. The space of all w.a.a.p. functions is also a closed translation invariant subspace of $C_{ub}(X)$ which will be denoted by $W^+(X)$. There is again a decomposition following from the Jacobs-Glicksberg-Deleeuw theorem.

$$W^+(X) = AP(X) \oplus W_0^+(X),$$

where $W_0^+(X)$ is definded according to

$$W_0^+(X) = \{f \in W^+(X) \quad : \quad w - \lim_{n \to \infty} T_{\tau_n} f|_{[0,\infty)} = 0 \text{ in } C_b(\mathbb{R}_+; X),$$

$$\text{for some sequence } \tau_n \to \infty\};$$

both spaces $W^+(X)$ and $W_0^+(X)$ are easily seen to be homogeneous and clearly $W_0(X) \subset W_0^+(X)$.

For $f \in W^+(X)$ the Bohr transform of f is still well-defined and with $f = f_a + f_0$, where $f_a \in AP(X)$, $f_0 \in W_0^+(X)$, we have

$$\alpha(\rho, f) = \lim_{N \to \infty} N^{-1} \int_0^N e^{-i\rho t} f(t) dt = \alpha(\rho, f_a), \quad \rho \in \mathbb{R}. \tag{11.19}$$

This is implied by the general mean ergodic theorem for bounded C_0-semigroups; see e.g. Hille and Phillips [180], Chap. XVIII, and Ruess and Summers [298]. Furthermore, as in the case of $AAP^+(X)$ we have the relation

$$\sigma(f) = \sigma(f_a) \cup \sigma(f_0), \quad f \in W^+(X). \tag{11.20}$$

In fact, the inclusion $\subset$ is obvious by Proposition 0.4; the converse follows from Proposition 0.7 and (11.19), since $\rho \notin \sigma(f)$ implies $\alpha(\rho, f) = 0$, hence $\alpha(\rho, f_a) = 0$, which in turn yields $\rho \notin \exp(f_a)$, i.e. $\sigma(f) \supset \exp(f_a)$. Closedness of $\sigma(f)$ then gives $\sigma(f_a) = \overline{\exp(f_a)} \subset \sigma(f)$, but also $\sigma(f_0) = \sigma(f - f_a) \subset \sigma(f) \cup \sigma(f_a) \subset \sigma(f)$, by Proposition 0.4.

These arguments show that the second part of Theorem 11.2 remains valid if $AAP^+(X)$ is replaced by $W^+(X)$ and by $W(X)$, thanks to Proposition 11.1 and to homogeneity of the spaces $W_0^+(X)$ and by $W_0(X)$.

Corollary 11.3 *Suppose $\Lambda(X)$ is b-admissible for some $b \in \mathcal{W}(\mathbb{R})$, where $\Lambda \subset \mathbb{R}$ is closed, and let G denote the corresponding solution operator. Then $f \in W^+(X)$ and $\sigma(f) \subset \Lambda$ imply $Gf \in W^+(X)$, $(Gf)_a = Gf_a$, $(Gf)_0 = Gf_0$, and (11.15) holds. The same assertions are true for $W(X)$ instead of $W^+(X)$.*

11.5 Nonresonant Problems
Consider (6.1) in differentiated form, i.e.

$$\dot{u}(t) = \int_0^t A_1(\tau) \dot{u}(t - \tau) d\tau + \int_0^t dA_2(\tau) u(t - \tau) + f(t), \quad t \in \mathbb{R}_+,$$

$$u(0) = u_0, \tag{11.21}$$

as well as its limiting equation

$$\dot{v}(t) = \int_0^\infty A_1(\tau)\dot{v}(t-\tau)d\tau + \int_0^\infty dA_2(\tau)v(t-\tau) + f(t), \quad t \in \mathbb{R}. \tag{11.22}$$

The purpose of this and the next subsection is the study of the relations between the solutions of (11.21) and (11.22). For this we need the *(complex) spectrum* of (11.21).

$$\Sigma_0 = \{\lambda \in \overline{\mathbb{C}}_+ : \lambda - \lambda\hat{A}_1(\lambda) - \widehat{dA}_2(\lambda) \in \mathcal{B}(Y, X) \quad \text{is not invertible}\}. \tag{11.23}$$

A standing assumption will be the well-posedness of (11.21), i.e. we take for granted the existence of an exponentially bounded weak resolvent $S(t)$ for (11.21).

Concerning (11.22) suppose the (real) spectrum Λ_0 of (11.22) satisfies $\Lambda_0 = \emptyset$ and there is an $\mathbb{R}$-kernel $G(t)$ for (11.22); recall that $G : \mathbb{R} \to \mathcal{B}(X)$ satisfies at least

(G1) $G(\cdot)x \in L^1(\mathbb{R}_+; X)$ *for each* $x \in X$;

(G2) $\tilde{G}(\rho) = H(i\rho)$ *for each* $\rho \in \mathbb{R}$.

Here as before we use the notation

$$H(\lambda) = (\lambda - \lambda\hat{A}_1(\lambda) - \widehat{dA}_2(\lambda))^{-1}, \quad \lambda \notin \Sigma_0. \tag{11.24}$$

To draw the connection between $S(t)$ and $G(t)$ we need in addition compactness of Σ_0 and a uniform bound on $H(\lambda)$ in $\mathcal{B}(X)$ away from Σ_0. Let $\Gamma_0 \subset \mathbb{C}_+$ denote some Jordan curve surrounding Σ_0 counter-clockwise, and define

$$S_0(z) = \frac{1}{2\pi i} \int_{\Gamma_0} e^{\lambda z} H(\lambda) d\lambda, \quad z \in \mathbb{C}; \tag{11.25}$$

clearly, $S_0 : \mathbb{C} \to \mathcal{B}(X, Y)$ is an entire function of exponential growth, and

$$|S_0(t)|_{\mathcal{B}(X,Y)} \leq M e^{\eta t/2}, \quad t \leq 0, \tag{11.26}$$

where $\eta = \text{dist}(\Sigma_0, i\,\mathbb{R}) = \inf \text{Re}\,\Sigma_0$. S_0 satisfies the equations

$$\dot{S}_0(t) = \int_0^\infty A_1(\tau)\dot{S}_0(t-\tau)d\tau + \int_0^\infty dA_2(\tau)S_0(t-\tau), \quad t \in \mathbb{R}, \tag{11.27}$$

and

$$\dot{S}_0(t) = \int_0^\infty \dot{S}_0(t-\tau)A_1(\tau)d\tau + \int_0^\infty S_0(t-\tau)dA_2(\tau), \quad t \in \mathbb{R}, \tag{11.28}$$

as a simple computation shows.

Let $c(t) = e^{-t}e_0(t)$, and define $S(t) \equiv 0$ for $t < 0$; then for $\omega > \omega_0(S)$, the growth bound of $S(t)$, the inversion formula (0.9) for the vector-valued Laplace transform yields

$$(c * S)(t) = \lim_{N \to \infty} \frac{1}{2\pi i} \int_{\omega-iN}^{\omega+iN} \hat{c}(\lambda)H(\lambda)e^{\lambda t}d\lambda, \quad t \in \mathbb{R}. \tag{11.29}$$

Similarly, the inversion formula of the Fourier transform gives

$$(c * G)(t) = \lim_{N \to \infty} \frac{1}{2\pi i} \int_{-N}^{N} \hat{c}(i\rho) H(i\rho) e^{i\rho t} d\rho, \quad t \in \mathbb{R}; \qquad (11.30)$$

(11.29) and (11.30) have to be understood in the strong sense in X. By means of Cauchy's theorem we then obtain the identity

$$(c * S)(t) = (c * S_0)(t) + (c * G)(t), \quad t \in \mathbb{R},$$

which after a differentation becomes

$$S(t) = S_0(t) + G(t), \quad t \in \mathbb{R}. \qquad (11.31)$$

From (11.31) further properties of the $\mathbb{R}$-kernel $G(t)$ can be deduced.

(G3) $G(\cdot)x$ *is continuous on* $\mathbb{R} \setminus \{0\}$, *for each* $x \in X$;

(G4) *The limits* $G(0+) = \lim_{t \to 0+} G(t)$ *and* $G(0-) = \lim_{t \to 0-} G(t)$ *exist strongly, and the jump relation* $G(0+) - G(0-) = I$ *is satisfied.*

Moreover, $G(t)$ also fulfills the second resolvent equation on the line, i.e.

$$G(t)y = \int_{-\infty}^{t} G(\tau) A(t - \tau) y d\tau + e_0(t)y, \quad t \in \mathbb{R} \setminus \{0\}, \ y \in Y, \qquad (11.32)$$

which follows from (G1), (G2) and (11.26), (11.31). Note that $G(t) = -S_0(t)$ for $t < 0$, hence $G(t)$ behaves very nicely for negative t. On the other hand, for $t > 0$, $G(t)$ enjoys the same regularity properties as $S(t)$, in particular if $S(t)$ is a resolvent, the first resolvent equation on the line is valid too, i.e.

$$G(t)y = \int_{-\infty}^{t} A(t - \tau) G(\tau) y d\tau + e_0(t)y, \quad t \in \mathbb{R} \setminus \{0\}, \ y \in Y. \qquad (11.33)$$

However, this way it is not possible to obtain more information about the asymptotic behaviour of $G(t)$, like boundedness as $t \to \infty$, or $\lim_{t \to \infty} G(t)x = 0$ for each $x \in X$. Here the relation $\tilde{G}(\rho) = H(i\rho)$ must be employed.

Let us summarize the above discussion.

Theorem 11.3 *Suppose* $A(t)$ *is of the form (11.2), let* Σ_0 *be defined by (11.23) and* $H(\lambda)$ *by (11.24); assume in addition*
(i) $\Sigma_0 \cap i\mathbb{R} = \emptyset$, Σ_0 *is compact and*

$$\overline{\lim}_{|\lambda| \to \infty} |H(\lambda)|_{\mathcal{B}(X)} < \infty; \qquad (11.34)$$

(ii) there is a weak resolvent $S(t)$ *for (11.21) with growth bound* $\omega_0(S) < \infty$; *set* $S(t) = 0$ *for* $t < 0$;
(iii) there is an $\mathbb{R}$-kernel $G(t)$ *for (11.22) satisfying (G1) and (G2).*

Then $S_0(t) = S(t) - G(t)$, $t \in \mathbb{R}$, *belongs to* $C^\infty(\mathbb{R}; \mathcal{B}(X,Y))$, *and satisfies (11.26), (11.27) and (11.28);* $S_0(t)$ *admits extension to an entire function of exponential growth. In particular, if* $\Sigma_0 = \emptyset$, *then* $S(t)$ *is strongly integrable.*

Observe that Theorem 11.3 yields further results on integrability of resolvents, whenever $\Sigma_0 = \emptyset$, $H(\lambda)$ is bounded on $\mathbb{C}_+$, once there is a weak resolvent for (11.21) of subexponential growth, and there is an $\mathbb{R}$-kernel for (11.22) satisfying (G1) and (G2). For the latter consult Section 12.

A remark about compactness of Λ_0 seems to be necessary. In hyperbolic situations one cannot expect that Λ_0 is compact, in general, hence assumption (i) of Theorem 11.3 is a serious restriction. But even in the parabolic case there are examples showing that Λ_0 need not be compact.

Example 11.1 Let A_0 be a selfadjoint negative definite unbounded linear operator in the Hilbert space X, $Y = X_{A_0}$, and consider the problem

$$\dot{u}(t) = A_0 u(t) + \beta A_0 u(t-1) + f(t), \quad t \in \mathbb{R},$$

where $\beta \in \mathbb{R}$. This problem is clearly of the form (11.22) with $A_1(t) \equiv 0$ and $A_2(t) = A_0 e_0(t) + \beta A_0 e_0(t-1)$. We obtain by a simple computation

$$\Sigma_0 = \{\lambda \in \overline{\mathbb{C}}_+ : \lambda/(1 + \beta e^{-\lambda}) \in \sigma(A_0) \cup \{\infty\}\}.$$

For $|\beta| < 1$ we have $1 + \beta e^{-\lambda} \neq 0$ on $\overline{\mathbb{C}}_+$ and $\varphi(\lambda) = \lambda(1 - \beta e^{-\lambda})^{-1}$ stays within the sector $\Sigma(0, \pi/2 + \arcsin|\beta|)$, hence $\Sigma_0 = \emptyset$. But for $\beta = -1$ say, zeros of $1 - e^{-\lambda}$ pop up at $\lambda = 2\pi i n$, $n \in \mathbb{Z}$, and so $\Lambda_0 = 2\pi\mathbb{Z}$, in particular Σ_0 is no longer compact. If β decreases below -1 then these zeros are shifted to the right by the amount $\log|\beta|$. These are poles for $\varphi(\lambda)$, hence they are limit points of sequences $(\lambda_n) \subset \mathbb{C}_+$ such that $\varphi(\lambda_n) \in \sigma(A_0)$, i.e. the structure of Σ_0 becomes quite complicated, although Λ_0 can be even empty, and $|H(i\rho)| + |A_0 H(i\rho)|(1 + |\rho|)^{-1}$ bounded on $\mathbb{R}$. $\square$

Observe that the resolvent $S(t)$ for Example 11.1 was shown to exists in Example 1.1. From the explicit formula given there one can see that $S(t)$ is uniformly integrable provided $|\beta| < 1$.

11.6 Asymptotic Equivalence

Suppose $S(t)$ is a weak resolvent for (11.21) with finite growth bound $\omega_0(S)$, let Σ_0 be compact, assume (11.34) holds, and let $G(t)$ be an $\mathbb{R}$-kernel for (11.22) which satisfies (G1) and (G2), hence also (G3) and (G4) as shown in Section 11.5. Assume in addition that $G(t)$ is also subject to

(G5) $\lim_{|t|\to\infty} G(t)x = 0$ *for each* $x \in X$,

this is the case in all of those results in Section 12 claiming the existence of an $\mathbb{R}$-kernel. Last but not least let $C_{ub}(X)$ be b-admissible for some $b \in \mathcal{W}(\mathbb{R})$.

We want to consider the relations between the solutions $u(t)$ and $v(t)$ of (11.21) and (11.22). So let $u_0 \in X$ and $f \in C_{ub}(X)$; then

$$u(t) = S(t)u_0 + \int_0^t S(t - \tau)f(\tau)d\tau, \quad t \geq 0, \tag{11.35}$$

and

$$v(t) = \int_{-\infty}^{\infty} G(t - \tau)f(\tau)d\tau, \quad t \in \mathbb{R}. \tag{11.36}$$

Subtracting these equations and using the decomposition (11.31) we obtain

$$u(t) - v(t) = G(t)u_0 - \int_t^{\infty} G(\tau)f(0)d\tau - \int_{-\infty}^0 G(t - \tau)[f(\tau) - f(0)]d\tau + w(t), \tag{11.37}$$

where

$$w(t) = S_0(t)u_0 + \int_0^{\infty} S_0(t - \tau)f(\tau)d\tau, \quad t \in \mathbb{R}. \tag{11.38}$$

By (G5), $G(t)u_0 \to 0$, and $\int_t^{\infty} G(\tau)f(0)d\tau \to 0$ as $t \to \infty$ by (G1); since $C_0^+(X)$ is also b-admissable there follows

$$\int_{-\infty}^0 G(t - \tau)[f(\tau) - f(0)]d\tau = Gf_-(t) \to 0 \quad \text{as } t \to \infty,$$

with $f_-(t) = (f(t) - f(0))(1 - e_0(t))$. Thus if we can show $w(t) \to 0$ as $t \to \infty$, then the solutions $u(t)$ of (11.21) and $v(t)$ of (11.22) are asymptotic to each other. Since $S_0(t)$ is exponentially growing when nontrivial, it is not at all clear whether $w \in C_0(\mathbb{R}_+; X)$, and in general this will not be true. However, if $u(t)$ stays bounded as $t \to \infty$, then it is, we prove below, even $w(t) \equiv 0$.

This is the main result of this subsection, saying that for bounded solutions (11.21) and (11.22) are *asymptotically equivalent*.

Theorem 11.4 *Suppose $A(t)$ is of the form (11.2), let Σ_0 be defined by (11.23) and $H(\lambda)$ by (11.24); assume in addition*
(i) $\Sigma_0 \cap i\mathbb{R} = \emptyset$, Σ_0 is compact, and

$$\overline{\lim}_{|\lambda| \to \infty} |H(\lambda)|_{\mathcal{B}(X)} < \infty;$$

(ii) there is a weak resolvent $S(t)$ for (11.21) with growth bound $\omega_0(S) < \infty$;
(iii) the space $C_{ub}(X)$ is b-admissible for (11.22), for some $b \in \mathcal{W}(\mathbb{R})$;
(iv) there is an $\mathbb{R}$-kernel $G(t)$ for (11.22) satisfying (G1), (G2), (G5).
    Let $u_0 \in X$, $f \in C_{ub}(X)$, and let $u(t)$ and $v(t)$ be given by (11.35) and (11.36); assume $u(t)$ is bounded on $\mathbb{R}_+$. Then $u(t) - v(t) \to 0$ as $t \to \infty$. In particular,
(a) $f \in AAP^+(X)$ implies $u \in AAP^+(X)$, and

$$\alpha(\rho, u) = H(i\rho)\alpha(\rho, f), \quad \rho \in \mathbb{R}; \tag{11.39}$$

(b) $f \in W^+(X)$ *implies* $u \in W^+(X)$, *and (11.39);*
(c) $f \in C_l^+(X)$ *implies* $u \in C_l^+(X)$, *and* $u(\infty) = H(0)f(\infty)$.

Proof: Assertions (a), (b) and (c) follow from Theorem 11.2 and Corollary 11.3, once $u(t) - v(t) \to 0$ as $t \to \infty$ has been proved. This in turn is a consequence of $w(t) \equiv 0$, as explained above.

So let $u(t)$ be bounded and let $w(t)$ be given by (11.38). Then (11.37) shows that $w(t)$ is also bounded, hence $w(t) \equiv 0$ if we can show $\sigma(w) = \emptyset$. The Fourier-Carleman transform $\hat{w}(\lambda)$ is defined for Re $\lambda \neq 0$; however, (11.26) and (11.38) yield

$$|w(t)| \leq Me^{\eta t}(|u_0| + 2|f|_\infty/\eta), \quad t \leq 0,$$

hence $\hat{w}(\lambda)$ admits a holomorphic extension from the left halfplane Re $\lambda < 0$ to the halfplane Re $\lambda < \eta$. For $0 < $ Re $\lambda < \eta$, (11.38) leads to a representation for this extension, namely

$$
\begin{aligned}
\hat{w}(\lambda) &= -\int_{-\infty}^0 e^{-\lambda t}[S_0(t)u_0 + \int_0^\infty S_0(t-\tau)f(\tau)d\tau]dt \\
&= \hat{S}_0(\lambda)u_0 - \int_0^\infty (\int_{-\infty}^0 S_0(t-\tau)e^{-\lambda(t-\tau)}dt)f(\tau)e^{-\lambda t}d\tau \\
&= \hat{S}_0(\lambda)u_0 + \hat{S}_0(\lambda)\hat{f}(\lambda) + \int_0^\infty (\int_0^\tau S_0(t)e^{-\lambda t}dt)f(\tau)d\tau,
\end{aligned}
$$

for all $0 < $ Re $\lambda < \eta$; observe that all integrals are absolutely convergent.

On the other hand, for Re $\lambda > \omega = \sup\{$ Re $\mu : \mu \in \Gamma_0\}$ (cp. (11.25)) we obtain by (11.38)

$$
\begin{aligned}
\hat{w}(\lambda) &= \int_0^\infty e^{-\lambda t}[S_0(t)u_0 + \int_0^\infty S_0(t-\tau)f(\tau)d\tau]dt \\
&= \hat{S}_0(\lambda)u_0 + \int_0^\infty (\int_0^\infty e^{-\lambda(t-\tau)}S_0(t-\tau)d\tau)f(\tau)e^{-\lambda\tau}dt \\
&= \hat{S}_0(\lambda)u_0 + \hat{S}_0(\lambda)\hat{f}(\lambda) + \int_0^\infty (\int_0^\tau S_0(t-\tau)e^{-\lambda t}dt)f(\tau)d\tau,
\end{aligned}
$$

for all $0 < $ Re $\lambda < \eta$, where again all integrals are absolutely convergent. The last term in this representation is holomorphic in Re $\lambda > 0$, $\hat{f}(\lambda)$ is so as the Laplace transform of a bounded function, and the Fourier Carleman transform $\hat{S}_0(\lambda)$ of $S_0(t)$ is holomorphic outside the Jordan curve Γ_0; in fact,

$$\hat{S}_0(\lambda) = \frac{1}{2\pi i}\int_{\Gamma_0} H(\mu)d\mu/(\lambda - \mu), \quad \lambda \text{ outside of } \Gamma_0.$$

Therefore the extension of $\hat{w}(\lambda)$ to the halfplane Re $\lambda < \eta$ coincides with $\hat{w}(\lambda)$ in the strip $0 < $ Re $\lambda < \eta$, i.e. $\hat{w}(\lambda)$ is entire. This implies $\sigma(w) = 0$, hence $w(t) \equiv 0$, by Proposition 0.5. $\square$

Let us briefly discuss the question when $u(t)$ will be bounded. From the decomposition $S = S_0 + G$ and from (11.35) it is clear that $u(t)$ is bounded iff

$$z(t) = S_0(t)u_0 + \int_0^t S_0(t - \tau)f(\tau)d\tau, \quad t \geq 0, \tag{11.40}$$

is bounded. Therefore a necessary condition for boundedness of $u(t)$ is that

$$\hat{z}(\lambda) = \hat{S}_0(\lambda)(u_0 + \hat{f}(\lambda)) \tag{11.41}$$

is holomorphic in the right halfplane, i.e. $u_0 + \hat{f}(\lambda)$ must 'cancel' the singularities of $\hat{S}_0(\lambda)$.

In case $\Sigma_0 = \{\lambda_1, \ldots, \lambda_N\}$ finite we obtain even necessary and sufficient conditions for boundedness of $u(t)$. For this purpose consider the Laurent expansion of $H(\lambda)$ at $\lambda = \lambda_j$, i.e.

$$H(\lambda) = \sum_{m=1}^{\infty} H_{j,m}(\lambda - \lambda_j)^{-m} + H_j(\lambda). \tag{11.42}$$

Then by residue calculus $S_0(t)$ becomes

$$S_0(t) = \sum_{j=1}^{N} e^{\lambda_j t}(\sum_{m=1}^{\infty} H_{j,m}\frac{t^{m-1}}{(m-1)!}),$$

hence

$$z(t) = \sum_{j=1}^{N} e^{\lambda_j t} \sum_{l=1}^{\infty} \frac{t^{l-1}}{(l-1)!} z_{j,l}(t),$$

where

$$z_{j,l}(t) = \sum_{m=l}^{\infty} H_{j,m}[\int_0^t e^{-\lambda_j \tau}\frac{(-\tau)^{m-l}}{(m-l)!} f(\tau)d\tau + \delta_{l,m}u_0],$$

$\delta_{l,m}$ Kronecker's delta. In case $z(t)$ is bounded we must necessarily have $z_{j,l}(\infty) = 0$ for all j, l, i.e.

$$\sum_{m=l}^{\infty} H_{j,m}[\frac{1}{(m-l)!}\hat{f}^{(m-l)}(\lambda_j) + \delta_{l,m}u_0] = 0 \quad \text{for all } l, j.$$

If all singularities of $H(\lambda)$ are simple poles these conditions reduce to

$$H_{j,1}(u_0 + \hat{f}(\lambda_j)) = 0, \quad j = 1, \ldots, N,$$

and in both cases they are necessary *and* sufficient.

Proposition 11.8 *In addition to the assumptions of Theorem 11.4, suppose $\Sigma_0 = \{\lambda_1, \ldots, \lambda_N\}$, and let $H_{j,m}$ denote the coefficients in the Laurent expansion of $H(\lambda)$ at $\lambda = \lambda_j$ in $\mathcal{B}(X)$.*

Then, given $u_0 \in X$, $f \in C_{ub}(X)$, $u(t)$ defined by (11.35) is bounded on $\mathbb{R}_+$ if and only if the compatibility conditions

$$\sum_{m=0}^{\infty} H_{j,m+l}[\frac{1}{m!}\hat{f}^{(m)}(\lambda_j) + \delta_{0,m}u_0] = 0 \quad \text{for all } l \in \mathbb{N}, j \in \{1,\ldots,N\}$$

are satisfied, where $\delta_{l,m} = 1$ for $l = m$, $\delta_{l,m} = 0$ otherwise.

The proof of the sufficiency part follows by a direct estimation of $z(t)$, where $z_{j,l}(t)$ can be replaced by $z_{j,l}(t) - z_{j,l}(\infty)$, since the compatilibity condition is equivalent to $z_{j,l}(\infty) = 0$. The details are left to the reader.

11.7 Comments
a) The concept of homogeneous spaces as defined in Section 11.1 is the direct extension of the definition given in Katznelson [194] to the vector-valued case. The spaces $\Lambda(X)$ have been employed in Prüss [271], [275] before.

b) Admissibility of function spaces is a familiar topic in the theory of Volterra equations in finite dimensions; see e.g. the monographs of Miller [242], Corduneanu [60], and Gripenberg, Londen and Staffans [156]. In infinite dimensions, this concept has been introduced for Volterra equations with main part A, a closed linear operator in X with dense domain, and $Y = X_A$ in Prüss [271]; see also Prüss [275].

c) The results in Section 11.2, 11.3, and the first part of Section 11.4 are extensions of Prüss [271], [275], as are Theorems 11.3 and 11.4; they have been published recently in Prüss [280]. For the results on weakly almost periodic solutions see Prüss and Ruess [282].

d) In Staffans [314], 2π-periodic L^2-solutions of

$$da * \dot{u} + db * u + dc * Au = f$$

in case X is a Hilbert space are studied, where a, b, c are of bounded variation on $\mathbb{R}$. In our terminology, this paper contains the full characterization of admissability of $L^2_\Lambda(X)$, where $\Lambda = \mathbb{Z}$, in terms of boundedness properties of $H(i\rho) = (i\rho\widetilde{da}(\rho) + \widetilde{db}(\rho) + \widetilde{dc}(\rho)A)^{-1}$ on $\mathbb{Z}$, like Proposition 11.5. In Da Prato and Lunardi [71] periodic solutions of $\dot{u} = Au + B * u + f$ are studied, where A generates an analytic semigroup in X and $B \in L^1(\mathbb{R}_+; \mathcal{B}(X_A, X))$ subject to very strong regularity assumptions.

e) For the concept of scalar-valued almost periodic functions see the classical monograph of Bohr [27]; the definition given in Section 11.4, however, is due to Bochner who also extended this notion to the vector-valued case. The decomposition of a.a.p. functions was first obtained by Fréchet [119]. As a general reference for $AP(X)$, we refer to Amerio and Prouse [9] and to Levitan and Zhikov [214].

The notion of weakly almost periodic function was introduced by Eberlein [107]. It should be observed that the definition due to Eberlein is different from that given in Amerio and Prouse. However, it seems that Eberlein's concept is more appropriate for evolutionary problems, as recent results about w.a.p. solutions and ergodicity of nonlinear evolution equations show; see Ruess and Summers [294], [295], [296], [297], [298] and the survey paper [293]. The paper of Rosenblatt, Ruess and Sentilles [292] contains very interesting results about properties of W_0^+.

f) Not much seems to be known about the structure of the solution set of the homogeneous equation (11.3) with $f = 0$. In the simplest case $\Lambda_0 = \{\rho_1, \ldots, \rho_n\}$, the relation $\sigma(u) \subset \Lambda_0$ implies via Proposition 0.5 that u(t) is of the form

$$u(t) = \sum_{k=1}^{n} a_k(t) \exp(i\rho_k), \quad t \in \mathbb{R},$$

where the $a_k(t)$ are X-valued polynomials. If in addition u belongs to any homogeneous space then the a_j are constant, i.e. $u(t)$ is a trigonometric polynomial with coefficients $a_j \in \mathcal{N}(i\rho_j(I - \widehat{A}(i\rho_j)))$. Much further study is necessary to extend the results which are known in finite dimensions to the infinite dimensional case. The same remark applies to problems with resonance, i.e. $\Lambda \cap \Lambda_0 \neq \emptyset$. For the latter the only reference is Miller and Wheeler [245] where for equations of scalar type in a Hilbert space with negative definite operator results on existence of an $\mathbb{R}$-kernel $G(t)$ are proved, when $\Lambda \cap \Lambda_0$ consists of finitely many points and certain moment conditions hold.

g) A theory similar to that presented here can be developed for equations of n-order, i.e. for

$$u^{(k)} - A_0 * u^{(n)} = \sum_{k=1}^{n} dA_k * u^{(n-k)} + f, \tag{11.43}$$

where $A_0 \in L^1(\mathbb{R}_+; \mathcal{B}(Y, X))$, and $A_k \in BV(\mathbb{R}_+; \mathcal{B}(Y, X))$, $k = 1, \ldots, n$, and $n \in \mathbb{N}_0$ is arbitrary. We have restricted the exposition to $n = 1$ since this case is most frequent in applications. The main difference consists in the behaviour of $H(i\rho)$ at $\rho = 0$, as the formal representation of the Fourier transform of the solution of (11.43) shows.

$$\widetilde{u}(\rho) = H(i\rho)(i\rho)^{1-k}\widetilde{f}, \quad \rho \in \mathbb{R}. \tag{11.44}$$

Note that $0 \notin \Lambda_0$ for (11.43) iff $\widehat{dA_n}(0) = A_n(\infty) \in \mathcal{B}(Y, X)$ is invertible.

It is also possible to reduce a second order problem to a first order one, convolving the second order equation with $e^{-t}e_0(t)$ and integrating by parts. In Section 13 we shall apply this technique to the Timoshenko beam model.

h) If (11.3) is of scalar type then the approach via sums of commuting linear operators is applicable, in particular the results of Da Prato and Grisvard [64], and Dore and Venni [101], Prüss and Sohr [283] are available; cp. Section 8. This approach has been carried through in Clément and Da Prato [52] via Theorem 8.3 and its consequences for (11.43) with $A_0(t) = c(t)A$, where $c \in L^1(\mathbb{R}_+)$ is completely positive and A generates an analytic semigroup. In Clément and Prüss [54] Theorem 8.4 has been applied to linear heat conduction with memory; see also the survey papers of Clément [44], [45]. Another application of the sum method for equations on the line is contained in Prüss [279]. In each of these papers it turns out that $\Lambda_0 = \emptyset$.

12 Admissibility of Function Spaces

This section is concerned with sufficiency of the frequency domain conditions derived in Proposition 11.5 for admissibility of evolutionary integral equations on the line as well as on the existence of Λ-kernels and their properties. The main results cover subordinated equations, hyperbolic problems in Hilbert spaces, and parabolic problems in arbitrary spaces. The discussion includes also perturbation problems and maximal regularity on the line for parabolic problems.

12.1 Perturbations: Hyperbolic Case

For $\Lambda \subset \mathbb{R}$ unbounded, admissibility of $\mathcal{H}_\Lambda(X)$ in the hyperbolic case is in general quite difficult to prove. There are not many tools available for this at present, the Paley-Wiener lemma, subordination and Parseval's theorem in the Hilbert space case are the almost only ones. The main difficulty is the lack of decay properties of $H(i\rho)$ as $|\rho| \to \infty$ which is in contrast to the parabolic case, and for this reason also multiplier theory does not work well, except for $\mathcal{H}(X) = L^2(\mathbb{R}; X)$, when X and Y are Hilbert spaces.

In this section we consider again the problem

$$\dot{u} = A_0 * \dot{u} + dA_1 * u + f, \tag{12.1}$$

on the line, where

$$A_0 \in L^1(\mathbb{R}_+; \mathcal{B}(Y, X)), \quad A_1 \in BV(\mathbb{R}_+; \mathcal{B}(Y, X)), \tag{12.2}$$

and $A(t) = A_0(t) + A_1(t)$, as in Section 11. The spectrum of (12.1) will be denoted by Λ_0, as before.

We begin the discussion here with an application of the Paley-Wiener lemma to the perturbed equation

$$\dot{u} = A_0 * \dot{u} + dA_1 * u + B_0 * u + b * B_1 * u + f, \tag{12.3}$$

where the unperturbed equation (12.1) has some admissibility properties at infinity.

Theorem 12.1 *Suppose the kernels A_0, A_1 satisfy (12.2), let $B_0 \in L^1(\mathbb{R}_+; \mathcal{B}(X))$, $B_1 \in L^1(\mathbb{R}_+; \mathcal{B}(Y, X))$, and $b \in \mathcal{W}(\mathbb{R})$. Let $\Lambda \subset \mathbb{R}$ be closed, $\mathcal{H}(X)$ homogeneous, and assume*
(i) there is $R > 0$ such that $\mathcal{H}_{\Lambda_R}(X)$ is b-admissible for (12.1), where $\Lambda_R = \{\rho \in \Lambda : |\rho| \geq R\}$;
(ii) the spectrum Λ_1 of (12.3) satisfies $\Lambda \cap \Lambda_1 = \emptyset$.
Then $\mathcal{H}_\Lambda(X) = BM_\Lambda^0(X)$ is b-admissible for (12.3) if $\mathcal{H}(X) = BM^0(X)$. Moreover, if (12.1) admits a Λ_R-kernel G_R, integrable in $\mathcal{B}(X)$, $b * G_R$ integrable in $\mathcal{B}(X, Y)$, then there is a Λ-kernel G_Λ for (12.3) such that $G_\Lambda - G_R \in L^1(\mathbb{R}; \mathcal{B}(X))$, $b * (G_\Lambda - G_R) \in L^1(\mathbb{R}; \mathcal{B}(X, Y))$. In this case $\mathcal{H}_\Lambda(X)$ is b-admissible, for any homogeneous space $\mathcal{H}(X)$.

Proof: Choose $N \geq R + 1$ and let $G_N \in \cap_{m \geq 0} W^{m,1}(\mathbb{R}; \mathcal{B}(X,Y))$ be a kernel for $\Lambda_N = \Lambda \cap \bar{B}_{N+2}$ for (12.3), according to Theorem 11.1; observe that this theorem applies by assumption (ii). Choose a cut-off function $\tilde{\varphi} \in C_0^\infty(\mathbb{R})$, $0 \leq \tilde{\varphi} \leq 1$, such that $\tilde{\varphi} = 1$ for $|\rho| \leq N + 1$, $\tilde{\varphi} = 0$ for $|\rho| \geq N + 2$ and decompose any $f \in \mathcal{H}_\Lambda(X)$ as

$$f = \varphi * f + (f - \varphi * f) = f_1 + f_2.$$

Then $\sigma(f_1) \subset \Lambda_N$ and $\sigma(f_2) \subset \Lambda \cap \{\rho \in \mathbb{R} : |\rho| \geq N + 1\} = \Lambda_\infty$, by Proposition 0.6. By virtue of Theorem 11.1, the admissibility of $\mathcal{H}_\Lambda(X)$ is this way reduced to admissibility of $\mathcal{H}_{\Lambda_\infty}(X)$; we may therefore assume $\Lambda_N = \emptyset$, where $N \geq R + 1$ is sufficiently large, specified later.

Let G_0 denote the solution operator for (12.1) in $\mathcal{H}_\Lambda(X) = BM_\Lambda^0(X)$ which exists and belongs to $\mathcal{B}\mathcal{H}_\Lambda(X)$, with $R_0 = b * G_0 \in \mathcal{B}(BM_\Lambda^0(X), BM_\Lambda^0(Y))$, by assumption (i). Then (12.3) may be rewritten as the fixed point equation

$$u = G_0 f + G_0 B_0 * u + G_0 B_1 * b * u \tag{12.4}$$

or equivalently with $u = G_0 v$,

$$v = f + B_0 * G_0 v + B_1 * R_0 v. \tag{12.5}$$

Thus the solution operators G and $b * G$ for (12.3) are given by

$$G = G_0(I - B_0 * G_0 - B_1 * R_0)^{-1}, \quad R = R_0(I - B_0 * G_0 - B_1 * R_0)^{-1},$$

and to prove b-admissibility of $\mathcal{H}_\Lambda(X)$ it is therefore sufficient to show

$$|B_0 * G_0 + B_1 * R_0| < 1.$$

For this purpose consider the Fejer approximations

$$B_j^N(t) = \frac{1}{2\pi} \int_{-N}^{N} (1 - |\rho|/N)\hat{B}_j(i\rho)e^{i\rho t} d\rho, \quad t \in \mathbb{R}, \ j = 0, 1;$$

since $B_0 \in L^1(\mathbb{R}; \mathcal{B}(X))$ we have $B_0^N \to B_0$ in $L^1(\mathbb{R}; \mathcal{B}(X))$, and similarly $B_1^N \to B_1$ in $L^1(\mathbb{R}; \mathcal{B}(Y,X))$. Now for $f \in BM_\Lambda^0(X)$ we have

$$\sigma(B_0^N * f) \subset \sigma(f) \cap \mathrm{supp}\, \tilde{B}_0^N \subset \sigma(f) \cap \bar{B}_N(0) \subset \Lambda_N = \emptyset,$$

hence $B_0^N * f = 0$, and similarly $B_1^N * g = 0$ for $g \in BM_\Lambda^0(Y)$. This implies

$$|B_0 * G_0 + B_1 * R_0| \leq |B_0 - B_0^N|_1 |G_0| + |B_1 - B_1^N|_1 |R_0| < 1,$$

provided N is sufficiently large. Note that the norms of G_0 and R_0 do not depend on $N \geq R + 1$.

If there exists a Λ_R-kernel G_R which is integrable in $\mathcal{B}(X)$, $b * G_R$ integrable in $\mathcal{B}(X,Y)$, then $K = B_0 * G_R + B_1 * b * G_R$ belongs to $L^1(\mathbb{R}; \mathcal{B}(X))$. By the

Paley-Wiener lemma, Theorem 0.6, there is a resolvent kernel $R \in L^1(\mathbb{R}; \mathcal{B}(X))$ for K. It follows readily from (12.4) that then

$$G_\Lambda = G_R + G_R * R$$

is a Λ-kernel for (12.4) with the properties asserted in Theorem 12.1.   □

Unfortunately, there are two shortcomings of Theorem 12.1 which limit its applicability. The first one is only conceptual, and could be removed, however, the second is of a more principal nature.

For the first defect, recall that b-admissibility of the perturbed equation is only asserted in case $\mathcal{H}(X) = BM^0(X)$. The reason for this is the fact that we do not know whether $B_1 * b * G_0$ leaves invariant $\mathcal{H}_\Lambda(X)$ whenever this space is b-admissible for the unperturbed equation. This is due to our definition of a solution; if a strong solution in $\mathcal{H}(X)$ would also require $u \in \mathcal{H}(Y)$ this defect would not occur. However, the problem is that there is no 'natural' map $X \mapsto \mathcal{H}(X)$, so when considering a homogeneous space $\mathcal{H}(X)$, the 'corresponding' space $\mathcal{H}(Y)$ is not at all clear to exist. If one restricts attention to such homogeneous spaces as $L^p(X)$ there would be no problem. Note that in case $B_1 = 0$ the result is true for any $\mathcal{H}(X)$.

The second shortcoming of Theorem 12.1 is the assumption of integrability of G_R and $b * G_R$, rather than their strong integrability. The reason for this is that it is not clear whether $B_0 * G_R$ and $B_1 * b * G_R$ are uniformly integrable which is needed for the Paley-Wiener lemma. This can be easily shown in case G_R and $b * G_R$ are integrable rather than strongly integrable.

12.2 Subordinated Equations
Next consider the equation of scalar type

$$\dot{u} = a_0 * A\dot{u} + da_1 * Au + f, \tag{12.6}$$

where $a_0 \in L^1(\mathbb{R}_+)$, $a_1 \in BV(\mathbb{R}_+)$ are subordinate to a completely positive function $c(t)$ according to $a(t) = a_0(t) + a_1(t)$, and

$$\hat{a}(\lambda) = \hat{b}(1/\hat{c}(\lambda)), \quad \lambda > 0, \tag{12.7}$$

where $\int_0^\infty |b(t)| e^{-\beta t} dt < \infty$ for some $\beta \in \mathbb{R}$. Here the creep function $k(t)$ associated with $c(t)$ is given by

$$k(t) = \kappa + \omega t + \int_0^t k_1(\tau) d\tau, \quad t > 0. \tag{12.8}$$

Let A be a closed linear operator in X with dense domain and let $S_b(t)$ denote the resolvent of

$$v(t) = g(t) + \int_0^t b(t - \tau) Av(\tau) d\tau, \quad t > 0, \tag{12.9}$$

and suppose
$$\omega_b = \overline{\lim}_{t\to\infty} t^{-1} \log |S_b(t)| < \infty.$$

According to Theorem 4.1, there is a resolvent $S_a(t)$ for the equation subordinate to (12.9), and the growth bound for $S_a(t)$ is not greater than ω_a defined by

$$\omega_a = \hat{c}^{-1}(1/\omega_b) \text{ if } \omega_b > 1/\omega, \quad \omega_a = 0 \text{ otherwise.} \tag{12.10}$$

We set again $\alpha = \omega + k_1(0+)$, and as in Section 10.3 assume $k_1 \in W^{1,1}(0, \infty)$, to be able to apply Proposition 4.9 and the structure theorem for subordinated resolvents, Theorem 4.4. Observe that if $\alpha > \omega_b$ then the spectrum Λ_0 of (12.6) is compact; in fact $1/\hat{b}(z) \in \rho(A)$ for Re $z > \omega_b$ by Theorem 1.3' and Re $1/\hat{c}(i\rho) \to \alpha$ as $|\rho| \to \infty$, hence $1/\hat{a}(i\rho) \in \rho(A)$ for $|\rho| \geq R$, say.

Theorem 12.2 *Assume either of the following.*
(i) $1/\kappa + \alpha = \infty$; the function $g(z) = z(\hat{b}(\eta + z)/\hat{b}(z) - 1)$ is locally analytic at ∞, for some $\eta > \max\{\beta, \omega_b\}$;
(ii) $1/\kappa + \alpha < \infty$, $\alpha > \omega_b$; the function $g(z, w) = (\hat{b}(z)/\hat{b}(z + zw) - 1)/w$ is analytic at $(\infty, 0)$.
If $\Lambda \subset \mathbb{R}$ is closed such that $\Lambda \cap \Lambda_0 = \emptyset$, then there is a Λ-kernel $G_\Lambda(t)$ for (12.6). G_Λ is of the form

$$G_\Lambda(t) = S_b(t/\kappa)e^{-\alpha t/\kappa}e_0(t) + G_0(t), \quad t \in \mathbb{R}, \tag{12.11}$$

where $G_0 \in L^1(\mathbb{R}; \mathcal{B}(X)) \cap L^\infty(\mathbb{R}; \mathcal{B}(X))$ is $\mathcal{B}(X)$-continuous on $\mathbb{R} \setminus \{0\}$, even on $\mathbb{R}$ in case (ii), $\lim_{|t|\to\infty} |G_\Lambda(t)|_{\mathcal{B}(X)} = 0$, and the jump relation

$$\lim_{t\to 0+} G_\Lambda(t)x - \lim_{t\to 0-} G_\Lambda(t)x = x, \quad \text{for all } x \in X, \tag{12.12}$$

is satisfied. Moreover, if $r_\mu \in L^1(\mathbb{R})$ is such that

$$\tilde{r}_\mu(\rho) = \frac{\hat{a}(i\rho)}{1 + \mu\hat{a}(i\rho)}, \quad \rho \in \mathbb{R}, \tag{12.13}$$

then $\mathcal{H}_\Lambda(X)$ is r_μ-admissible, for any homogeneous space $\mathcal{H}(X)$.

Proof: As in the proof of Theorem 12.1 we may assume $\Lambda \cap B_R(0) = \emptyset$ where R can be chosen arbitrarily large, thanks to Theorem 11.1. Since $\Lambda_0 \subset \mathbb{R}$ is compact as mentioned before, choose R so large that $\Lambda_0 \subset \bar{B}_{R-5}(0)$, and then w.l.o.g. $\Lambda = \mathbb{R} \setminus B_R(0)$. Choose a cut-off function $\chi_0 \in C^\infty(\mathbb{R})$ such that $\chi_0 = 1$ for $|\rho| \geq R - 1$, $\chi_0 = 0$ for $|\rho| \leq R - 2$, $0 \leq \chi \leq 1$ elsewhere; then $\chi_0 = 1 - \tilde{\psi}_0$, where $\psi_0 \in L^1(\mathbb{R})$. Choose another cut-off function $\chi_1 \in C^\infty(\mathbb{R})$ such that $\chi_1 = 1$ on $|\rho| \geq R - 3$, $\chi_1 = 0$ for $|\rho| \leq R - 4$, $0 \leq \chi_1 \leq 1$ elsewhere; we have $\chi_1 = 1 - \tilde{\psi}_1$ for some $\psi_1 \in L^1(\mathbb{R})$.

$G_\Lambda(t)$ will be constructed as in the proofs of Theorems 10.3 and 10.4 via the Paley-Wiener lemma applied to the convolution equations

$$\chi_0 H = \chi_0 \varphi_\eta H_\eta - (\chi_1 \psi_\eta H_\eta)\chi_0 H, \tag{12.14}$$

corresponding to (10.21) in case (i), respectively,

$$\chi_0 H_0 = \chi_0(\varphi_1 + \varphi_0 H_1)H_1 + (\chi_1 \varphi_0 H_1)\chi_0 H_0, \tag{12.15}$$

which is the equivalent of (10.23) here in case (ii). The same notation employed there is deliberately also used here. The Paley-Wiener conditions are satisfied since $|H_\eta(i\rho)|_{\mathcal{B}(X)} \to 0$ in case (i), and $\varphi_0(i\rho) = \hat{b}_0(i\rho) \to 0$ in case (ii), as $|\rho| \to \infty$, provided R has been chosen large enough. Then in case (i), G_Λ is given by $\tilde{G}_\Lambda = \chi_0 H$, and $\tilde{G}_0 = \chi_0 H_0 + \varphi_0 H_1$ in case (ii). The remaining properties of G_Λ follow easily from the properties of H_η resp. H_1. Finally, to prove r_μ-admissibility of $\mathcal{H}_\Lambda(X)$, observe that G_Λ is integrable in $\mathcal{B}(X)$, and

$$\begin{aligned}
A\tilde{r}_\mu \tilde{G}_\Lambda &= (1 + \mu\hat{a})^{-1}\chi_0\hat{a}AH = (1 - \mu\tilde{r}_\mu)\chi_0(H - 1/i\rho) \\
&= (1 - \mu\tilde{r}_\mu)\tilde{G}_\Lambda - (1 - \mu\tilde{r}_\mu)\chi_0/i\rho = \tilde{R}_\mu,
\end{aligned}$$

for some integrable function $R_\mu(t)$. $\square$

Observe that $G_\Lambda(t)$ satisfies

$$G_\Lambda x = A(a * G_\Lambda x) + e_0 x - 1 * \varphi_0 x, \quad x \in X, \tag{12.16}$$

by uniqueness of the Fourier transform and closedness of A. $G_\Lambda(t)$ evidently commutes with A, and (12.16) implies

$$\frac{d}{dt}(G_\Lambda x - h_0 x - Aa_0 * G_\Lambda x) = da_1 * AG_\Lambda x - \varphi_0 x, \quad x \in D(A). \tag{12.17}$$

Combining Theorem 12.2 with the perturbation result Theorem 12.1 now leads to sufficient conditions for admissibility of equations with main part. We leave the formulation of such a result to the reader.

12.3 Admissibility in Hilbert Spaces
If X and Y are both Hilbert spaces one may use Parseval's theorem to obtain admissibility of homogeneous spaces. The simplest result in this direction is certainly the following which concerns admissibility of $L_\Lambda^2(X)$.

Proposition 12.1 *Suppose X, Y are Hilbert spaces, $\Lambda \subset \mathbb{R}$ is closed, and $b \in \mathcal{W}(\mathbb{R})$. Then $L_\Lambda^2(X)$ is b-admissible if $\Lambda \cap \Lambda_0 = \emptyset$ and*

$$|H(i\rho)|_{\mathcal{B}(X)} + |\tilde{b}(\rho)||H(i\rho)|_{\mathcal{B}(X,Y)} \leq M, \quad \rho \in \Lambda, \tag{12.18}$$

for some constant $M \geq 1$.

This is an easy consequence of the representation

$$\tilde{u}(\rho) = H(i\rho)\tilde{f}(\rho), \quad \rho \in \Lambda,$$

of the solution of (12.1) for $f \in L_\Lambda^2(X)$ with $\sigma(f)$ compact, and Parseval's theorem; cp. Proposition 11.6. The details are left to the reader.

If one imposes mild regularity restrictions on $A(t)$ and the function $b(t)$, and assumes existence of an exponentially bounded b-regular resolvent $S(t)$, the full converse of Proposition 11.5 can be obtained. The resulting theorem contains as a special case Gearhart's characterization of the spectrum of C_0-semigroups in Hilbert spaces.

Theorem 12.3 *Suppose X, Y are Hilbert spaces, $\Lambda \subset \mathbb{R}$ is closed, and let $b \in L^1(\mathbb{R}_+)$ be 1-regular and such that $\hat{b}(0) \neq 0$. Assume that (6.1) admits a b-regular pseudo-resolvent, exponentially bounded, and in addition*

$$|\lambda \hat{A}'(\lambda)|_{\mathcal{B}(Y,X)} \leq C|\hat{b}(\lambda)|, \qquad Re\ \lambda > 0,\ |\lambda| \geq R, \tag{12.19}$$

for some constants $R, C > 0$. Then $\Lambda(X)$ is b-admissible if and only if $\Lambda \cap \Lambda_0 = \emptyset$ and (12.18) holds. In this case, for each $x \in X$ there is $h_x \in L^1(X) \cap L^2(X)$ such that $b * h_x \in L^1(Y) \cap L^2(Y)$ and $\widetilde{h_x}(\rho) = H(i\rho)x$ for all $\rho \in \Lambda$. Moreover, if Λ_0 is compact, then there is a Λ-kernel $G_\Lambda : \mathbb{R} \to \mathcal{B}(X)$, strongly continuous for $t \neq 0$, such that $G_\Lambda(\cdot)x \in L^1(\mathbb{R}) \cap L^\infty(\mathbb{R})$, $G_\Lambda(0+)x - G_\Lambda(0-)x = x$, $\lim_{|t| \to \infty} G_\Lambda(t)x = 0$, and $\tilde{G}_\Lambda(\rho)x = H(i\rho)x$ on Λ, for each $x \in X$.

Proof: The necessity part follows from Proposition 11.5. To prove the sufficiency of (12.18) we may assume, as in the proof of Theorem 12.1, $\Lambda \cap \bar{B}_N(0) = \emptyset$, where $N > 0$ can be arbitrarily large.
(a) First we show that (12.18) holds also on $\Lambda_\varepsilon = \Lambda + \bar{B}_\varepsilon(0)$ with M replaced by $4M$, provided $\varepsilon > 0$ is chosen small enough. For this purpose consider the identity

$$
\begin{aligned}
I - \hat{A}(i\rho) &= [I + i\rho_0(\hat{A}(i\rho_0) - \hat{A}(i\rho))H(i\rho_0)](I - \hat{A}(i\rho_0)) \\
&= [I + K(\rho, \rho_0)](I - \hat{A}(i\rho_0)), \quad \rho_0 \in \Lambda,\ |\rho - \rho_0| \leq \varepsilon;
\end{aligned}
$$

we show $|K(\rho, \rho_0)| \leq 1/2$ whenever $\rho_0 \in \Lambda$, $|\rho - \rho_0| \leq \varepsilon$, provided $\varepsilon > 0$ is sufficiently small. This then implies $\Lambda_0 \cap \Lambda_\varepsilon = \emptyset$ and (12.18) with M replaced by $4M$.

From 1-regularity of $b(t)$ we obtain the estimate

$$|\hat{b}(i\rho)| \leq |\hat{b}(i\rho_0)| + \int_{\rho_0}^{\rho} |\hat{b}'(i\tau)|d\tau \leq |\hat{b}(i\rho_0)| + \frac{c}{|\rho_0| - \varepsilon} \int_{\rho_0 - \varepsilon}^{\rho_0 + \varepsilon} |\hat{b}(i\tau)|d\tau,$$

hence

$$\int_{\rho_0 - \varepsilon}^{\rho_0 + \varepsilon} |\hat{b}(is)|ds \leq 2\varepsilon(1 - \frac{2c\varepsilon}{|\rho_0| - \varepsilon})^{-1}|\hat{b}(i\rho_0)| \leq 4\varepsilon|\hat{b}(i\rho_0)|,$$

if $\rho_0 \in \Lambda$ and $\varepsilon \leq N/(1 + 4c)$. This yields with (12.19)

$$
\begin{aligned}
|K(\rho, \rho_0)|_{\mathcal{B}(X)} &\leq |\rho_0| \int_{\rho_0}^{\rho} |d\hat{A}(is)|_{\mathcal{B}(Y,X)} \cdot |H(i\rho_0)|_{\mathcal{B}(X,Y)} \\
&\leq \frac{CM|\rho_0|}{|\rho_0| - \varepsilon} \int_{\rho_0 - \varepsilon}^{\rho_0 + \varepsilon} |\hat{b}(is)|ds \leq 8CM\varepsilon,
\end{aligned}
$$

for each $\rho_0 \in \Lambda$, $|\rho - \rho_0| \leq \varepsilon$, provided $\varepsilon > 0$ is small enough; recall $|\rho_0| \geq N$.
(b) By assumption, (6.1) admits a resolvent $S(t)$ such that

$$|S(t)|_{B(X)} + |(b * S)(t)|_{B(X,Y)} \leq M_0 e^{(\omega-1)t}, \quad t \geq 0,$$

for some constants $M_0, \omega > 0$. But then the function $u(t) = e^{-\omega t} S(t) x e_0(t)$ belongs
to $L^2(X)$ for each $x \in X$, and similarly $v(t) = e^{-\omega t}(b*S)(t) x e_0(t)$ belongs to $L^2(Y)$
for each $x \in X$. By Parseval's theorem therefore $\tilde{u}(\rho) = \hat{S}(\omega + i\rho)x = H(\omega + i\rho)x$
belongs to $L^2(X)$ as well, and similarly $\tilde{v}(\rho) = \hat{b}(\omega + i\rho)H(\omega + i\rho)x$ is in $L^2(Y)$
for every $x \in X$.
(c) Choose a function $\varphi \in C^\infty(\mathbb{R})$ such that $0 \leq \varphi \leq 1$, $|\varphi'|_\infty < \infty$, $\varphi \equiv 1$ on
$\Lambda_{\varepsilon/3}$, $\varphi \equiv 0$ on $\mathbb{R}\backslash\Lambda_{2\varepsilon/3}$, and consider the identity

$$H(i\rho)x = (1 + \frac{\omega}{i\rho})H(\omega + i\rho)x + H(i\rho)(\omega + i\rho)[\hat{A}(i\rho) - \hat{A}(\omega + i\rho)]H(\omega + i\rho)x,$$

which after multiplication with $\varphi(\rho)$ yields

$$\varphi(\rho)H(i\rho)x = (1 + \frac{\omega}{i\rho})\varphi(\rho)\tilde{u}(\rho) + \varphi(\rho)H(i\rho)\frac{(\omega + i\rho)}{\hat{b}(\omega + i\rho)}[\hat{A}(i\rho) - \hat{A}(\omega + i\rho)]\tilde{v}(\rho).$$

$$(12.20)$$

Since $\tilde{u} \in L^2(X)$, $\tilde{v} \in L^2(Y)$ and $0 \notin \text{supp }\varphi$ there follows $\varphi(\rho)H(i\rho)x \in L^2(X)$ and
similarly $\varphi(\rho)\hat{b}(i\rho)H(i\rho)x \in L^2(Y)$, provided $\hat{b}(i\rho)/\hat{b}(\omega + i\rho)$ and $(\omega + i\rho)[\hat{A}(i\rho) -$
$\hat{A}(\omega + i\rho)]/\hat{b}(\omega + i\rho)$ are bounded for $\rho \in \Lambda_\varepsilon$. To prove this we employ again
1-regularity of $b(t)$ and (12.19). The identity

$$\hat{b}(i\rho + s) = \hat{b}(i\rho + \omega) - \int_s^\omega \hat{b}'(i\rho + \tau)d\tau$$

yields with 1-regularity

$$|\hat{b}(i\rho + s)| \leq |\hat{b}(i\rho + \omega)| + \frac{c}{|\rho|}\int_0^\omega |\hat{b}(i\rho + \tau)|d\tau, \quad 0 \leq s \leq \omega,$$

hence

$$\int_0^\omega |\hat{b}(i\rho + \tau)|d\tau \leq \frac{N\omega}{N - c\omega}|\hat{b}(i\rho + \omega)|, \quad |\rho| > N, \qquad (12.21)$$

provided $N > c\omega$; in particular $\hat{b}(i\rho)/\hat{b}(\omega + i\rho)$ is bounded on Λ_ε. On the other
hand, (12.19) and (12.21) yield

$$|\hat{A}(i\rho) - \hat{A}(\omega + i\rho)|_{B(Y,X)} \leq \int_0^\omega |\hat{A}'(s + i\rho)|ds \leq \frac{C}{|\rho|}\int_0^\omega |\hat{b}(s + i\rho)|ds$$

$$\leq \frac{C}{|\rho|} \cdot \frac{N\omega}{N - c\omega}|\hat{b}(\omega + i\rho)|, \quad \rho \in \Lambda_\varepsilon,$$

hence $(\omega + i\rho)[\hat{A}(i\rho) - \hat{A}(\omega + i\rho)]/\hat{b}(\omega + i\rho)$ is bounded on Λ_ε as well. Thus, by
Parseval's theorem, for each $x \in X$ there is $h_x \in L^2(X)$ such that $b * h_x \in L^2(Y)$
and $\widetilde{h_x}(\rho) = \varphi(\rho)H(i\rho)x$, in particular $\widetilde{h_x}(\rho) = H(i\rho)x$ for $\rho \in \Lambda_\varepsilon$.

(d) Next we show $h_x \in L^1(X)$ and $b * h_x \in L^1(Y)$ for each $x \in X$. For this observe first that in virtue of (12.19), $\hat{A}(i\rho)$ is not only Lipschitz on Λ_ε but also strongly differentiable a.e., since X and Y are Hilbert spaces, which enjoy the Radon-Nikodym property. Differentation of $\widetilde{h_x}$ therefore yields

$$
\begin{aligned}
\widetilde{h_x}'(\rho) &= \varphi'(\rho)H(i\rho)x + i\varphi(\rho)H'(i\rho)x \\
&= \varphi'(\rho)H(i\rho)x - \varphi(\rho)H(i\rho)x/\rho + i\varphi(\rho)H(i\rho)i\rho\hat{A}'(i\rho)H(i\rho)x;
\end{aligned}
$$

since $0 \notin \operatorname{supp} \varphi$, φ' is bounded and $\operatorname{supp} \varphi' \subset \Lambda_\varepsilon$, by (12.19) we see that $\widetilde{h_x}' \in L^2(X)$, hence $th_x \in L^2(X)$ by Parseval's theorem. Similarly, employing 1-regularity of $b(t)$ once more we obtain also $t \cdot b * h_x \in L^2(Y)$. But then

$$
|h_x|_1 = \int_{-1}^{1} |h_x(t)|dt + \int_{|t| \geq 1} |h_x(t)t|dt/|t| \leq \sqrt{2}(|h_x|_2 + |th_x|_2) < \infty,
$$

i.e. $h_x \in L^1(X)$; similarly $b * h_x \in L^1(Y)$ for each $x \in X$.

(e) Next we apply a duality argument to obtain $h_x^* \in L^1(X) \cap L^2(X)$ such that $\widetilde{h_x^*}(-\rho) = \varphi(\rho)H(i\rho)^*x$ for each $x \in X$. For this purpose observe that $S^*(t)x$ is weakly continuous and strongly measurable on $\mathbb{R}_+$; the latter follows again from the Radon-Nikodym property. Therefore $H(\omega + i\rho)^*x$ belongs to $L^2(X)$ for each $x \in X$, and by boundedness of $(\omega + i\rho)(\hat{A}(i\rho) - \hat{A}(\omega + i\rho))/\hat{b}(i\rho)$, as above we obtain also $\varphi(\rho)H(i\rho)^*x \in L^2(X)$, for each $x \in X$. Also by duality we get $(\varphi(\rho)H(i\rho)^*x)' \in L^2(X)$, hence as in (c) and (d) there is $h_x^* \in L^1(X) \cap L^2(X)$ such that $\widetilde{h_x^*}(-\rho) = \varphi(\rho)H(i\rho)^*x$, $\rho \in \mathbb{R}$, $x \in X$.

By similar arguments, for each $y \in Y$ there is $g_y^* \in L^1(X) \cap L^2(X)$ such that $\widetilde{g_y^*}(-\rho) = \varphi(\rho)(\hat{b}(i\rho)H(i\rho))^*y$.

(f) The mappings $x \mapsto h_x$, $x \mapsto h_x^*$ from X to $L^1(X)$ are linear and closed; hence by the closed graph theorem, there is a constant $C > 0$ such that

$$
|h_x|_1 + |h_x^*|_1 \leq C|x|_X, \quad x \in X.
$$

Similarly,

$$
|g_y^*|_1 \leq C|y|_Y, \ y \in Y, \ \text{and} \ |b * h_x|_{L^1(Y)} \leq C|x|_X, \ x \in X.
$$

To prove boundedness of the solution operator G from $\Lambda(X)$ to $C_{ub}(X)$ and of $b * G$ from $\Lambda(X)$ to $C_{ub}(Y)$ it is enough to verify the relations

$$
(x, Gf(t))_X = \int_{-\infty}^{\infty} (h_x^*(t - \tau), f(\tau))_X d\tau, \quad t \in \mathbb{R}, \tag{12.22}
$$

and

$$
(y, b * Gf(t))_Y = \int_{-\infty}^{\infty} (g_y^*(t - \tau), f(\tau))_X d\tau, \quad t \in \mathbb{R}, \tag{12.23}
$$

for each $f \in \Lambda(X)$ such that $\sigma(f)$ is compact. In fact, (12.22) and (12.23) yield the estimates

$$
|Gf|_{L^\infty(X)} \leq C|f|_{L^\infty(X)} \ \text{and} \ |b * Gf|_{L^\infty(Y)} \leq C|f|_{L^\infty(X)},
$$

for each $f \in \Lambda(X)$ with compact spectrum, which by Proposition 11.5 is sufficient for b-admissibility of $\Lambda(X)$.

So let such f be given, choose another cutoff function $\tilde{\varphi}_0 \in C_0^\infty(\mathbb{R})$, which is real and symmetric, such that $\tilde{\varphi}_0 \equiv 1$ on a neighborhood of $\sigma(f)$, and let $G_0 \in W^{\infty,1}(\mathbb{R}; \mathcal{B}(X,Y))$ denote the kernel from Lemma 11.1 such that $\tilde{G}_0(\rho) = \tilde{\varphi}(\rho)H(i\rho)\tilde{\varphi}_0(\rho)$, $\rho \in \mathbb{R}$. Then with $\varphi_0 * f = f$ and $\widetilde{G_0^*}(\rho)x = \tilde{G}_0(-\rho)^*x = \tilde{\varphi}_0(\rho)\tilde{h}_x^*(\rho)$, i.e. $G_0^*x = \varphi_0 * h_x^*$, we obtain

$$
\begin{aligned}
(x, Gf(t))_X &= (x, (G_0 * f)(t))_X = \int_{-\infty}^{\infty} (G_0^*(t-\tau)x, f(\tau))_X \, d\tau \\
&= \int_{-\infty}^{\infty} (h_x^*(t-\tau), (\varphi_0 * f)(\tau))_X \, d\tau = \int_{-\infty}^{\infty} (h_x^*(t-\tau), f(\tau))_X \, d\tau,
\end{aligned}
$$

i.e. (12.22) holds. Similarly, one also obtains (12.23).

(g) Finally, suppose Λ_0 is compact; w.o.l.g. we then may choose $\varphi(\rho) \equiv 1$ for $|\rho|$ sufficiently large, say $|\rho| \geq N_0$. Then $1 - \varphi \in C_0^\infty(\mathbb{R})$, $\varphi(0) = 0$, hence $1 - \varphi = \tilde{\psi}$ for some $\psi \in W^{\infty,1}(\mathbb{R})$ with $\int_{-\infty}^{\infty} \psi(t)dt = 1$. Then the decomposition of the first term in (12.20),

$$
\varphi(\rho)(1 + \frac{\omega}{i\rho})H(\omega + i\rho)x = H(\omega + i\rho)x - \tilde{\psi}(\rho)H(\omega + i\rho)x + \omega\frac{\varphi(\rho)}{i\rho}H(\omega + i\rho)x,
$$

shows that this term is the Fourier transform of a bounded function $w_1(t)$ which is continuous for $t \neq 0$, satisfies the jump relation $\lim_{t\to 0+} w_1(t) - \lim_{t\to 0-} w_1(t) = w_1(0+) - w_1(0-) = x$ as well as $\lim_{|t|\to\infty} w_1(t) = 0$. On the other hand, the second term in (12.20) can be written in the form

$$
H(\omega + i\rho)\frac{(\omega + i\rho)}{\hat{b}(i\rho)}[\hat{A}(i\rho) - \hat{A}(\omega + i\rho)]\varphi(\rho)\hat{b}(i\rho)H(i\rho)x = H(\omega + i\rho)\tilde{w}_2(\rho),
$$

where $w_2 \in L^2(X)$, hence is the Fourier transform of $Se^{-\omega \cdot}e_0 * w_2$, which belongs to $C_0(\mathbb{R}; X)$. The proof is complete. $\square$

From the proof of Theorem 12.3 it follows that in case $A(t)$ decomposes as $A(t) = B(t) + C(t)$, where $B \in L_{loc}^1(\mathbb{R}_+; \mathcal{B}(Y,X))$ and $C \in L_{loc}^1(\mathbb{R}_+; \mathcal{B}(X))$, only $B(t)$ need be subject to (12.19). For $C(t)$ the weaker condition

$$
|\lambda\hat{C}'(\lambda)| \leq M, \quad \text{Re } \lambda > 0, \ |\lambda| \geq R, \tag{12.24}
$$

is sufficient. Observe that (12.24) is implied by

$$
\int_0^\infty t|dC(t)|_{\mathcal{B}(X)} < \infty,
$$

whenever $C \in BV_{loc}((0,\infty);\mathcal{B}(X))$. This remark will turn out to be quite useful in Section 13.

By an interpolation argument the admissibility assertion in Theorem 12.3 can be extended to $L^p_\Lambda(X)$, for all $1 \leq p < \infty$.

Corollary 12.1 *Let the assumptions of Theorem 12.3 be satisfied and let (12.18) be fullfilled. Then $L^p_\Lambda(X)$ is b-admissible, $1 \leq p < \infty$.*

Proof: For a simple function $f = \sum_{i=1}^n x_i \chi_i$, where $x_i \in X$ and χ_i are characteristic functions of disjoint measurable sets of finite measure, we define

$$Gf = \sum_{i=1}^n h_{x_i} * \chi_i.$$

We then have obviously the estimate $|Gf|_1 \leq C|f|_1$; cp. the first step in (f) of the proof of Theorem 12.4. On the other hand, as there we also have $|Gf|_\infty \leq C|f|_\infty$, hence by the vector-valued Riesz-Thorin interpolation theorem (see e.g. Bergh and Löfström [22]) $G \in \mathcal{B}(L^p(X))$ for each $1 \leq p < \infty$. Similarly, $b * G \in \mathcal{B}(L^p(X), L^p(Y))$, and therefore $L^p_\Lambda(X)$ is b-admissible for every $p \in [1,\infty)$. □

Since for hyperbolic problems Λ_0 need not be compact in general, it remains open whether in the situation of Theorem 12.3 a Λ-kernel G_Λ always exists. Even if so, it seems to be very difficult to obtain pointwise estimates of $G_\Lambda(t)$, in particular to prove the integrability of G_Λ. For this reason it also seems to be quite unclear whether Corollary 12.1 can be extended to homogeneous spaces other than $L^p_\Lambda(X)$, e.g. it is not known whether $BM^0_\Lambda(X)$ is b-admissible, in this situation.

Let us conclude this section with an illustrative application of Theorem 12.3 to obtain the characterization of the spectrum of a C_0-semigroup in a Hilbert space, i.e. Gearharts theorem mentioned above; see Section 12.7 for further comments.

Example 12.1 Let A generate a C_0-semigroup e^{At} in the Hilbert space X. Choose $Y = X_A$, $b(t) = e^{-t}e_0(t)$, $\Lambda = (2\pi/\tau)\mathbb{Z}$. With $A(t) \equiv A$ the assumptions of Theorem 12.3 are easily seen to be true; the resolvent $S(t)$ is of course the C_0-semigroup e^{At}. A moment of reflection shows also that $\Lambda(X)$ is b-admissible if and only if the evolution equation

$$\dot{u} = Au + f \tag{12.25}$$

admits a mild τ-periodic solution u, for each $f \in C_{ub}(X)$ τ-periodic. Via the variation of parameters formula the latter is equivalent to $1 \in \rho(e^{A\tau})$. Theorem 12.3 characterizes this property by

$$\{2\pi n i/\tau : n \in \mathbb{Z}\} \subset \rho(A), \; \sup\{|(2\pi n i/\tau - A)^{-1}| : n \in \mathbb{Z}\} < \infty, \tag{12.26}$$

which is precisely Gearhart's condition.

12.4 Λ-kernels for Parabolic Problems

Consider again (12.1) assuming the decomposition (12.2). In Section 11.3 we have seen that for compact $\Lambda \subset \mathbb{R}$, the spectral condition $\Lambda \cap \Lambda_0 = \emptyset$ is sufficient for b-admissibility of $\mathcal{H}_\Lambda(X)$, for any homogeneous space $\mathcal{H}(X)$ and any $b \in \mathcal{W}(\mathbb{R})$. For problems of parabolic type this result can be extended to noncompact Λ, however, b will no longer be arbitrary then. We begin with the case when (7.1) admits an analytic resolvent.

Theorem 12.4 *Suppose (7.1) admits a weak analytic resolvent $S(t)$ of type (ω_0, θ_0), and let (AN4) with $a = b \in \mathcal{W}(\mathbb{R}_+)$ be satisfied; cp. Corollary 7.2. Let $\mathcal{H}(X)$ be homogeneous, and assume $\Lambda \subset \mathbb{R}$ is closed and such that $\Lambda \cap \Lambda_0 = \emptyset$. Then $\mathcal{H}_\Lambda(X)$ is b-admissible, and there is a Λ-kernel $G_\Lambda \in L^1(\mathbb{R}; \mathcal{B}(X))$, $b * G_\Lambda \in L^1(\mathbb{R}; \mathcal{B}(X,Y))$ for $\mathcal{H}_\Lambda(X)$. Moreover, $G_0(t) = G_\Lambda(t) - S(t)e_0(t)$ admits extension to an entire function with values in $\mathcal{B}(X,Y)$ of exponential growth, and $|G_0(t)|_{\mathcal{B}(X,Y)} \leq C_\varepsilon \max(1, e^{(\omega_0 + \varepsilon)t})$, $t \in \mathbb{R}$. There is $\varphi_0 \in C_0^\infty(\mathbb{R})$ such that*

$$\dot{G}_0 - \varphi_0 = A_0 * \dot{G}_0 + dA_1 * G_0 = \dot{G}_0 * A_0 + G_0 * dA_1 \quad on \ \mathbb{R}. \quad (12.27)$$

Proof: Corollaries 7.1 and 7.2 imply that the family $H(\lambda) = \hat{S}(\lambda)$ is holomorphic on $\Sigma(\omega_0, \theta_0 + \pi/2)$ with values in $\mathcal{B}(X,Y)$ and for each $\omega > \omega_0$, $\theta < \theta_0$ there is a constant $C = C(\omega, \theta)$ such that

$$|H(\lambda)|_{\mathcal{B}(X)} + |\hat{b}(\lambda)H(\lambda)|_{\mathcal{B}(X,Y)} \leq C/|\lambda - \omega|, \quad \lambda \in \Sigma(\omega, \theta + \pi/2). \quad (12.28)$$

Therefore, Λ_0 is compact, $\Lambda_0 \subset [-N_0, N_0]$, where $N_0 = \omega \cot \theta_0$, and so $d(\Lambda_0, \Lambda) > 0$. As in the proof of Theorem 12.1 we may assume that $\Lambda \cap B_{N_0+3}(0) = \emptyset$. Choose a cut-off function $\psi(i\rho)$ of class C^∞ such that $\psi(i\rho) = 1$ for $|\rho| \geq N_0 + 2$, $\psi(i\rho) = 0$ for $|\rho| \leq N_0 + 1$, and $0 \leq \psi \leq 1$. Then $\psi(i\rho) - 1$ is of class $C_0^\infty(\mathbb{R})$, hence $\psi(i\rho) - 1 = \tilde{\varphi}_0(\rho)$, for some $\varphi_0 \in C^\infty(\mathbb{R}) \cap L^1(\mathbb{R})$.

Set $\psi(\lambda) = 1$ for Re $\lambda \neq 0$, and choose a smooth curve $\Gamma_+ \subset \Sigma(\omega_0, \theta_0 + \pi/2) \cap \{\lambda \in \mathbb{C} : \text{Re } \lambda \leq \omega\}$ running from $i(N_0 + 3)$ to $-i(N_0 + 3)$, $\omega > \omega_0$, let $\Gamma_0 = \Gamma_+ \cup [-i(N_0 + 3), i(N_0 + 3)]$, and define

$$G_0(z) = \frac{1}{2\pi i} \int_{\Gamma_0} \psi(\lambda)H(\lambda)e^{\lambda z}d\lambda, \quad z \in \mathbb{C}. \quad (12.29)$$

Then $G_0 : \mathbb{C} \to \mathcal{B}(X,Y)$ is entire, $|G_0(z)|_{\mathcal{B}(X,Y)} \leq C_\omega e^{\omega|z|}$ for some $C_\omega > 0$, and $|G^{(n)}(t)|_{\mathcal{B}(X,Y)} \leq C_n$ for $t \leq 0$, $n \in \mathbb{N}_0$. A direct computation shows that

$$\dot{G}_0 - A_0 * \dot{G}_0 - dA_1 * G_0 = \dot{G}_0 - \dot{G}_0 * A_0 - G_0 * dA_1$$

$$= \frac{1}{2\pi i} \int_{\Gamma_0} \psi(\lambda)e^{\lambda t}d\lambda = \frac{1}{2\pi} \int_{-\infty}^{\infty} (\psi(i\rho) - 1)e^{i\rho t} = \varphi_0(t), \quad t \in \mathbb{R},$$

i.e. (12.27) holds. Define $G_\Lambda(t)$ according to

$$G_\Lambda(t) = G_0(t) + S(t)e_0(t), \quad t \in \mathbb{R}; \quad (12.30)$$

then by boundedness of $S(t)$, $G_\Lambda(t)$ is bounded in $\mathcal{B}(X)$, uniformly on each interval of the form $(-\infty, t_0]$. Cauchy's theorem and (12.28) show that we have the representation

$$G_\Lambda(t) = \frac{1}{2\pi} \lim_{N \to \infty} \int_{-N}^{N} \psi(i\rho) H(i\rho) e^{i\rho t} d\rho, \quad t \neq 0. \tag{12.31}$$

To prove integrability of G_Λ we integrate twice by parts. Since ψ vanishes on $[-N_0 - 1, N_0 + 1]$, the function $\Phi(i\rho) = \psi(i\rho) H(i\rho)$ is of class C^∞ on $\mathbb{R}$, and by Cauchy's theorem we have

$$|\Phi^{(j)}(\rho)| \leq M_j / (1 + |\rho|)^{j+1}, \quad \text{for all } \rho \in \mathbb{R}, \; j \in \mathbb{N}_0. \tag{12.32}$$

Integrating twice by parts we obtain

$$G_\Lambda(t) = -\frac{1}{2\pi t^2} \int_{-\infty}^{\infty} \Phi''(\rho) e^{i\rho t} d\rho, \quad t \in \mathbb{R},$$

hence $G_\Lambda \in L^1(\mathbb{R}; \mathcal{B}(X))$. Replacing $H(\lambda)$ by $\hat{b}(\lambda) H(\lambda)$ in these arguments, we also obtain $b * G_\Lambda \in L^1(\mathbb{R}; \mathcal{B}(X, Y))$.

Finally, we show that the kernel G_Λ represents the solution operator G by (11.14); b-admissibility of any $\mathcal{H}_\Lambda(X)$ is then an easy consequence of the integrability properties of G_Λ and of Proposition 11.6. From the first resolvent equation $S = I + A * S$ and b-regularity of $S(t)$ there follows by (12.27) an identity for $R = b * G_\Lambda$ in $\mathcal{B}(X)$,

$$\dot{R} = A_0 * \dot{R} + dA_1 * R + b * \varphi_0 + b.$$

Hence $u = R * f$ is a solution of (12.1) with inhomogeneity $b * f + \varphi_0 * b * f$, for each $f \in \mathcal{H}_\Lambda(X)$. However, since $\tilde{\varphi}_0(\rho) = \psi(i\rho) - 1 = 0$ for $\rho \in \sigma(f)$, we have $\sigma(\varphi_0 * f) \subset \sigma(\varphi_0) \cap \sigma(f) = \emptyset$, hence $\varphi_0 * f = 0$. Thus $u = R * f = G_\Lambda * b * f$ is a strong solution of (12.1) with inhomogeneity $b * f$, and so (11.14) is valid. $\quad \square$

The main property of $H(i\rho)$ which was used in the proof of Theorem 12.4 is the estimate (12.32). This decay property is also present in the case of weakly parabolic equations with weakly 2-regular kernels satisfying (NP4); see Section 7.2. Somewhat more generally we have

Corollary 12.2 *Suppose $b \in L^1(\mathbb{R}_+)$ is 2-regular, $\hat{b}(0) \neq 0$, and let the following conditions hold.*
(i) For each $\mathrm{Re}\, \lambda \geq 0$, $|\lambda| \geq R$, $I - A \in \mathcal{B}(Y, X)$ is bijective and $H(\lambda) = (I - \hat{A}(\lambda))^{-1}/\lambda$ satisfies for some $R > 0$

$$|H(\lambda)|_{\mathcal{B}(X)} + |\hat{b}(\lambda) H(\lambda)|_{\mathcal{B}(X,Y)} \leq \frac{M}{|\lambda|}, \quad \text{for all } \mathrm{Re}\, \lambda \geq 0, \; |\lambda| \geq R.$$

(ii) There is a constant $C > 0$ such that

$$|\lambda \hat{A}'(\lambda)|_{\mathcal{B}(Y,X)} + |\lambda^2 \hat{A}''(\lambda)|_{\mathcal{B}(Y,X)} \leq |\hat{b}(\lambda)|, \quad \text{for all } \mathrm{Re}\, \lambda \geq 0, \; |\lambda| \geq R.$$

Let $\mathcal{H}(X)$ be homogeneous, and assume $\Lambda \subset \mathbb{R}$ is closed and such that $\Lambda \cap \Lambda_0 = \emptyset$.
Then the assertions of Theorem 12.4 are valid.

For the proof observe only that $H(\lambda)$ satisfies by (i) and (ii) the crucial estimate
(12.32) and that after an exponential shift of size $\omega \geq R$ the assumptions of
Corollary 7.4 are satisfied, hence there is a b-regular weak resolvent $S(t)$. Therefore
the proof of Theorem 12.4 carries over.

12.5 Maximal Regularity on the Line

As one might expect, for parabolic equations there are also maximal regularity
results on the line. We again employ the notation introduced in Section 7.4, with
$J = \mathbb{R}$; for $\alpha \in (0,1)$, $p \in [1,\infty)$, $q \in [1,\infty]$ we let

$$B_p^{\alpha,q}(X) = (L^p(X), W^{1,p}(X))_{\alpha,q},$$

denote the real interpolation space between $L^p(X)$ and $W^{1,p}(X)$ of order α and
power q, and similarly for $p = \infty$

$$B_\infty^{\alpha,q}(X) = (C_{ub}(X), C_{ub}^1(X))_{\alpha,q};$$

see Section 7.4 for the definition of the norms. These spaces are easily seen to be
homogeneous.

Theorem 12.5 *Let the assumptions of Theorem 12.4 or Corollary 12.2 be fulfilled.*
Then for $\alpha \in (0,1)$, $p,q \in [1,\infty]$, the spaces $B_{p,\Lambda}^{\alpha,q}(X)$ are spaces of maximal
regularity for (12.1), i.e. if $f \in B_p^{\alpha,q}(X)$, $\sigma(f) \subset \Lambda$, then the solution $u = Gf = G_\Lambda * f$ of (12.1) satisfies $u, \dot{u} \in B_p^{\alpha,q}(X)$ and $b * u$, $b * \dot{u} \in B_p^{\alpha,q}(Y)$.

Proof: (a) Suppose the assumptions of Theorem 12.4 or Corollary 12.2 are fulfilled,
and let $G_\Lambda(t)$ denote the corresponding Λ-kernel for the closed set $\Lambda \subset \mathbb{R}$; w.l.o.g.
we may assume $\Lambda = \{\rho \in \mathbb{R} : |\rho| \geq R\}$, for $R > 0$ sufficiently large. Also, with
some fixed $\omega > R + 1$ we let $S_\omega(t) = e^{-\omega t} e_0(t) S(t)$, $t \in \mathbb{R}$, where $S(t)$ denotes the
weak resolvent, which exists as was observed before. Then the results of Section
7.1 and 7.2 show that S_ω satisfies

$$|S_\omega(t)|_{\mathcal{B}(X)} + |t\dot{S}_\omega(t)|_{\mathcal{B}(X)} \leq Ne^{-t}, \quad t > 0,$$

$$|\dot{S}_\omega(t) - \dot{S}_\omega(s)|_{\mathcal{B}(X)} \leq N\frac{t-s}{ts}(1 + \log\frac{t}{t-s}), \quad t > s > 0; \qquad (12.33)$$

cp. (7.29). With $b_\omega(t) = e^{-\omega t} b(t)$, $T_\omega = b_\omega * S_\omega$ is also subject to (12.33), where
$\mathcal{B}(X)$ can be replaced by $\mathcal{B}(X,Y)$.
(b) Define $G_1(t)$ by

$$G_1(t) = G_\Lambda(t) - S_\omega(t), \quad t \in \mathbb{R}; \qquad (12.34)$$

then (12.33) and uniform integrability of $G_\Lambda(t)$ yield $G_1 \in L^1(\mathcal{B}(X))$. We show
that $G_1 \in W^{1,1}(\mathcal{B}(X))$. For this purpose consider $i\rho\widehat{G_1}(\rho)$ and decompose as

follows

$$
\begin{aligned}
i\rho\widetilde{G_1}(\rho) &= \psi(i\rho)i\rho H(i\rho) - i\rho H(\omega + i\rho) \\
&= \psi(i\rho)(i\rho H(i\rho) - (\omega + i\rho)H(\omega + i\rho)) + \omega\psi(i\rho)H(\omega + i\rho) \\
&+ (\psi(i\rho) - 1)i\rho H(\omega + i\rho) \\
&= \Phi_2(\rho) + \omega(1 + \tilde{\varphi}_0(\rho))\tilde{S}_\omega(\rho) + \tilde{\varphi}_0(\rho)\tilde{S}_\omega(\rho),
\end{aligned}
\tag{12.35}
$$

where ψ and φ_0 are as in the proof of Theorem 12.4. If the assumptions of Corollary 12.2 are satisfied we may apply the arguments following (12.21) in the proof of Theorem 12.3 to the result that

$$
\Phi_2(\rho) = \psi(i\rho)(I - \hat{A}(i\rho))^{-1}(\hat{A}(i\rho) - \hat{A}(\omega + i\rho))(I - \hat{A}(\omega + i\rho))^{-1}
$$

is twice differentiable and subject to

$$
|\Phi_2(\rho)| + (1 + |\rho|)(|\Phi_2'(\rho)| + |\Phi_2''(\rho)|) \le M_0(1 + |\rho|)^{-1}, \quad \rho \in \mathbb{R}.
\tag{12.36}
$$

In the situation of Theorem 12.4, Cauchy type estimates lead to (12.36) as well. Integrating by parts, (12.35) gives

$$
\begin{aligned}
G_2(t) &:= \lim_{N \to \infty} \frac{1}{2\pi} \int_{-N}^{N} \Phi_2(\rho)e^{i\rho t} d\rho \\
&= \frac{-1}{2\pi i t} \int_{-\infty}^{\infty} \Phi_2'(\rho)(e^{i\rho t} - 1)d\rho = \frac{-1}{2\pi t^2} \int_{-\infty}^{\infty} \Phi_2''(\rho)e^{i\rho t}d\rho,
\end{aligned}
$$

where the last integrals are absolutely convergent; observe that $\int_{-\infty}^{\infty} \Phi_2'(\rho)d\rho = 0$ by (12.36). Thus $G_2(t)$ is well-defined for $t \ne 0$, and direct estimates yield

$$
|G_2(t)| \le M_1 \min\{1/t^2, 1 + \log(1 + 1/|t|)\}, \quad t \ne 0,
$$

hence $G_2 \in L^1(\mathcal{B}(X))$. But since the last two terms of (12.35) are the Fourier transforms of $\omega(S_\omega + \varphi_0 * S_\omega)$ and $\dot{\varphi}_0 * S_\omega$, respectively, we have obtained $\dot{G}_1 \in L^1(\mathcal{B}(X))$ by uniqueness of the Fourier transform. Similarly one also proves $b*G_1 \in L^1(\mathcal{B}(X,Y))$.

(c) Now let $f \in B_p^{\alpha,q}(X)$, $\sigma(f) \subset \Lambda$, where $\alpha \in (0,1)$, $p, q \in [1, \infty]$, and let $u = Gf = G_\Lambda * f = G_1 * f + S_\omega * f$ denote the solution of (12.1). It is then not difficult to show by a limiting argument that u is differentiable a.e. and

$$
\dot{u}(t) = \int_{-\infty}^{\infty} \dot{G}_1(\tau)f(t - \tau)d\tau + \int_0^{\infty} \dot{S}_\omega(\tau)(f(t - \tau) - f(t))d\tau, \quad \text{for a.a. } t \in \mathbb{R},
\tag{12.37}
$$

holds. Since $\dot{G}_1 \in L^1(\mathcal{B}(X))$ the first term leaves every homogeneous $\mathcal{H}(X)$ invariant, in particular the spaces $B_p^{\alpha,q}(X)$. On the other hand, by means of (12.33) a minor modification of the proof of Theorem 7.5 shows that the last term has this property too. Therefore $u, \dot{u} \in B_p^{\alpha,q}(X)$. Since T_ω is subject to (12.33) with $\mathcal{B}(X)$ replaced by $\mathcal{B}(X,Y)$ we obtain $b * u, b * \dot{u} \in B_p^{\alpha,q}(Y)$. The proof is complete. $\square$

12.6 Perturbations: Parabolic Case

Via the usual fixed point argument, Theorem 12.5 can be extended to the perturbed problem

$$\dot{u} - A_0 * \dot{u} - dA_1 * u = B_0 * \dot{u} + dB_1 * u + b * (C_0 * \dot{u} + dC_1 * u) + f \quad (12.38)$$

where $B_0 \in L^1(\mathbb{R}_+; \mathcal{B}(X))$, $B_1 \in BV(\mathbb{R}_+; \mathcal{B}(X))$, $C_0 \in L^1(\mathbb{R}_+; \mathcal{B}(Y, X))$, and $C_1 \in BV(\mathbb{R}_+; \mathcal{B}(Y, X))$.

Corollary 12.3 *Let the assumptions of Theorem 12.4 or Corollary 12.2 be satisfied, where Λ_0 is replaced by the spectrum Λ_1 of the perturbed problem (12.38). Then for each $\alpha \in (0, 1)$, $p, q \in [1, \infty]$, the spaces $B_{p,\Lambda}^{\alpha,q}(X)$ are spaces of maximal regularity for (12.38).*

Proof: W.o.l.g. $\Lambda = \{\rho \in \mathbb{R} : |\rho| \geq R\}$, where R is sufficiently large. Let $B_0^N(t)$ denote the Fejer approximation of B_0, i.e.

$$B_0^N(t) = \frac{1}{2\pi} \int_{-N}^{N} (1 - \frac{|\rho|}{N}) \widetilde{B_0}(\rho) e^{i\rho t} d\rho, \quad t \in \mathbb{R};$$

$\dot{B}_1^N$, C_0^N, $\dot{C}_1^N$ are defined similarly. Then by choosing N large enough

$$|B_0^N - B_0|_{L^1(\mathcal{B}(X))} + |e^{-t} * (dB_1 - \dot{B}_1^N)|_{L^1(\mathcal{B}(X))}$$
$$+ |C_0^N - C_0|_{L^1(\mathcal{B}(Y,X))} + |e^{-t} * (dC_1 - \dot{C}_1^N)|_{L^1(\mathcal{B}(Y,X))} \leq \eta(N),$$

where $\eta(N) \to 0$ as $N \to \infty$. Since supp $\widetilde{B_0^N} \subset \overline{B_N}(0)$ we may replace B_0 by $B_0 - B_0^N$ in (12.38) and similarly C_0 by $C_0 - C_0^N$, in case $\sigma(f) \cap B_{N+1}(0) = \emptyset$. On the other hand, we may write

$$dB_1 * u = (dB_1 - \dot{B}_1^N) * u = e^{-t} * (dB_1 - \dot{B}_1^N) * (u + \dot{u}),$$

and similarly the term $dC_1 * b * u$ is treated. These considerations show that (12.38) can be transformed into the problem

$$\dot{u} - A_0 * \dot{u} - dA_1 * u = B_2 * \dot{u} + B_3 * u + b * (C_2 * \dot{u} + C_3 * u) + f, \quad (12.39)$$

where the L^1-norms of B_j and C_j are as small as we please, provided $\sigma(f) \cap B_{N+1}(0) = \emptyset$ for a sufficiently large N. This means that we have the fixed point problem

$$u = Gf + G(B_2 * \dot{u} + B_3 * u) + G(C_2 * b * \dot{u} + C_3 * b * u),$$

which by the contraction mapping principle in virtue of Theorem 12.5 is uniquely solvable in each of the spaces $B_{p,\Lambda}^{\alpha,q}(X)$, provided R is large enough. □

To construct a Λ-kernel G_Λ for (12.38) consider the convolution equation

$$G_\Lambda = G_\Lambda^0 + \frac{d}{dt} G_\Lambda^0 * (B_2 * G_\Lambda + C_2 * b * G_\Lambda) + G_\Lambda^0 * (B_3 * G_\Lambda + C_3 * b * G_\Lambda); \quad (12.40)$$

since (12.39) is equivalent to (12.38) for $\Lambda = \{\rho \in \mathbb{R} : |\rho| \geq R\}$, R sufficiently large. Here G_Λ^0 denotes the Λ-kernel for (12.1) according to Theorem 12.4 or Corollary 12.2. We are now looking for a solution $G_\Lambda \in L^1(\mathcal{B}(X))$ with $b*G_\Lambda \in L^1(\mathcal{B}(X,Y))$. Note that the Paley-Wiener lemma Theorem 0.6 cannot be applied here, since there is no chance to prove that the involved kernel is L^1, without imposing extra assumptions at least.

However, estimates (12.33) for $S_\omega(t)$ and the estimate for G_2 derived in (b) of the proof of Theorem 12.5 show

$$G_\Lambda^0 \in \bigcap_{p\in(1,\infty)} B_p^{1/p,\infty}(\mathcal{B}(X)) \cap \bigcap_{\alpha<1} B_1^{\alpha,\infty}(\mathcal{B}(X));$$

similarly we also have

$$b * G_\Lambda^0 \in \bigcap_{p\in(1,\infty)} B_p^{1/p,\infty}(\mathcal{B}(X,Y)) \cap \bigcap_{\alpha<1} B_1^{\alpha,\infty}(\mathcal{B}(X,Y)).$$

Therefore by Theorem 12.5 and the contraction mapping principle, there is a unique solution $G_\Lambda \in B_1^{\alpha,\infty}(\mathcal{B}(X)) \cap B_p^{1/p,\infty}(\mathcal{B}(X))$, with $b*G_\Lambda \in B_1^{\alpha,\infty}(\mathcal{B}(X,Y)) \cap B_p^{1/p,\infty}(\mathcal{B}(X,Y))$, whenever $\alpha < 1$ and $p \in (1,\infty)$ are fixed. This is the Λ-kernel for (12.38).

Corollary 12.4 *Let the assumptions of Theorem 12.4 or Corollary 12.2 be satisfied, where Λ_0 is replaced by the spectrum Λ_1 of the perturbed problem (12.38). Then each space $\mathcal{H}_\Lambda(X)$ is b-admissable for (12.38), and there is a Λ-kernel $G_\Lambda \in L^1(\mathcal{B}(X))$ such that $b * G_\Lambda \in L^1(\mathcal{B}(X,Y))$.*

Actually one can show even the decomposition

$$G_\Lambda(t) = S_\omega(t) + G_\omega(t), \quad t \in \mathbb{R},$$

where $G_\omega \in C_0(\mathcal{B}(X))$ and $b * G_\omega \in C_0(\mathcal{B}(X,Y))$. This can be achieved by using the L^∞-bounds derived in Section 7.5 and 7.6, but we do not want to repeat this procedure here.

We conclude this section with a remark on the bounded case where $Y = X$, which of course contains the case $\dim X < \infty$. In this case it is enough to consider (12.38) with $A_0 = A_1 = C_0 = C_1 = 0$ and $b = \delta_0 \in \mathcal{W}(\mathbb{R}_+)$, i.e. the equation

$$\dot{u} - B_0 * \dot{u} = dB_1 * u + f, \tag{12.41}$$

where $B_0 \in L^1(\mathbb{R}_+;\mathcal{B}(X))$ and $B_1 \in BV(\mathbb{R}_+;X)$. Corollary 12.4 yields for this case the following result.

Corollary 12.5 *Let $\mathcal{H}(X)$ be homogeneous and $\Lambda \subset \mathbb{R}$ closed such that $\Lambda \cap \Lambda_2 = \emptyset$, where Λ_2 denotes the spectrum of (12.41). Then $\mathcal{H}_\Lambda(X)$ is 0-admissable and there is a Λ-kernel $G_\Lambda \in L^1(\mathcal{B}(X))$.*

Better results on properties of G_Λ for equation (12.41) can be proved like $G_\Lambda = e^{-t}e_0(t) + G_0$, where $G_0 \in C_0^1(\mathcal{B}(X)) \cap W^{1,1}(\mathcal{B}(X))$. However, the bounded case is not the subject of this book, and so we conclude here.

12.7 Comments

a) Theorem 12.3 extends Theorem 3 of Prüss [275]; see also Prüss [271]. There the case $A(t) = A_0 + B(t)$ was studied, where A_0 generates a C_0-semigroup in the Hilbert space X, $Y = X_{A_0}$, and $B \in W^{1,1}_{loc}(\mathbb{R}_+; \mathcal{B}(Y,X))$, $B(0) = 0$, $\dot{B} \in BV(\mathbb{R}_+; \mathcal{B}(Y,X)) \cap L^1(\mathbb{R}_+; \mathcal{B}(Y,X))$, and $\dot{B}$ as well as $d\dot{B}$ have first moments. However, (12.19) is already implied by $\int_0^\infty [|\dot{B}|_{\mathcal{B}(Y,X)} dt + t|d\dot{B}|_{\mathcal{B}(Y,X)}] < \infty$. Observe that it remains an open question whether a Λ-kernel exists if Λ_0 is non-compact. It is also unclear whether Corollary 12.1 holds for any homogeneous space even in case Λ_0 is compact. The reason for this is that the Λ-kernel if it exists at all is not known to be integrable rather than strongly integrable.

b) Theorem 12.4 and Corollary 12.2 are extensions of results of Da Prato and Lunardi [73] who consider equations with main part A, a generator of an analytic semigroup (however, in this paper it is not assumed that $\mathcal{D}(A)$ is dense in X). In this situation Theorem 12.5 is proved there for $p = q = \infty$, $\Lambda = \mathbb{R}$ and $\Lambda = 2\pi\mathbb{Z}/\tau$, i.e. the τ-periodic case. A simple resonance case is treated as well, where $\Lambda = (2\pi/\tau)\mathbb{Z}$, $\Lambda \cap \Lambda_0 = (2\pi/\tau)\{k_1, \ldots, k_n\}$, and each $q_j = 2\pi k_j/\tau$ is a simple pole of $H(\lambda)$. Corollary 12.4 extends Theorem 2 of Prüss [275].

c) It has been known for a long time that the spectral mapping theorem $\sigma(e^{At}) = e^{\sigma(A)t}$ is in general false for C_0-semigroups, the spectrum of e^{At} may be strictly larger than $e^{\sigma(A)t}$; see Hille and Phillips [180], Chapter XVI. Zabczyk [348] was the first to give a counterexample even in the Hilbert space case. The spectral mapping theorem, however, is true in a general Banach space if the semigroup is eventually norm-continuous, i.e. if e^{At} is $\mathcal{B}(X)$-continuous for $t > t_0 \geq 0$. For C_0-semigroups of contractions in a Hilbert space, Gearhart [127] was the first who obtained a complete characterization of $\sigma(e^{At})$ in terms of the spectrum of A and boundedness properties of $(\lambda - A)^{-1}$; see Example 12.1. His result was later extended to general C_0-semigroups in Hilbert spaces independently by Herbst [179], Howland [184], and Prüss [272]; the simplest proof is now contained in Greiner [136], see also Greiner and Schwarz [137]. Gearhart's theorem does not extend to Banach spaces, as a counterexample due to Greiner, Voigt and Wolff [138] even for positive semigroups shows. We refer to the textbooks of semigroup theory mentioned in the Introduction for further information.

13 Further Applications and Complements

The first three subsections are devoted to applications of the results of Sections 10, 11, and 12 to some of the problems introduced in Sections 5 and 9. These include the hyperbolic viscoelastic Timoshenko beam, heat conduction in isotropic materials with memory, and boundary value problems for electrodynamics with memory.

In the remaining part of this section, discussions of several problems are presented which are strongly related to the contents of this book but out of its scope due to space considerations. One interesting subject which so far has only been studied in the scalar case is the ergodic theory for evolutionary integral equations which is strongly connected with the Tauberian theory of the vector-valued Laplace transform. By means of the variation of parameters formula and fixed point theorems, semilinear equations can be handled by the usual perturbation method. An outline of the many different semigroup approaches is given in Section 13.6. The section is concluded with the subordination principle, which allows for a considerable extension in the first order case, by admitting A to be nonlinear and multivalued but m-accretive.

13.1 Viscoelastic Timoshenko Beams

Consider the viscoelastic Timoshenko beam model introduced in Section 9.1.

$$\begin{aligned}
\ddot{w} &= da_1 * (w_{xx} + \phi_x) + f_s, \\
\ddot{\phi} &= de_1 * \phi_{xx} - \gamma da_1 * (w_x + \phi) + f_b,
\end{aligned} \tag{13.1}$$

where we have assumed for simplicity that the density $\rho_0 = 1$, the shear correction coefficient $\kappa = 1$, and the length of the beam $l = 1$. We want to consider (13.1) on the line, for the case when one end is clamped and the other is free, i.e. w.r.t. the boundary conditions

$$w(t, 0) = \phi(t, 0) = 0, \quad \phi_x(t, 1) = w_x(t, 1) + \phi(t, 1) = 0. \tag{13.2}$$

Shear and tensile moduli $a_1(t)$ and $e_1(t)$ will be assumed to be those of a rigid solid, which means

$$\begin{aligned}
& a_1(t), \ e_1(t) \text{ are nonnegative, nonincreasing, convex;} \\
& a_1(0+), \ e_1(0+) < \infty; \quad e_\infty = e_1(\infty), \ a_\infty = a_1(\infty) > 0.
\end{aligned} \tag{13.3}$$

Thus (13.1) is of hyperbolic type and almost synchronous, since

$$\frac{\widehat{de_1}(\lambda)}{\widehat{da_1}(\lambda)} = \frac{e_1(0+) + \widehat{\dot{e}_1}(\lambda)}{a_1(0+) + \widehat{\dot{a}_1}(\lambda)} = \widehat{dk}(\lambda), \quad \lambda > 0,$$

for some $k \in BV_{loc}(\mathbb{R}_+)$ with $k(0+) = \lim_{\lambda \to \infty} \widehat{dk}(\lambda) = e_1(0+)/a_1(0+) > 0$. Let $X = L^2(0, 1) \times L^2(0, 1)$ and

$$Y = \{(w, \phi) \in W^{2,2}([0, 1]) \times W^{2,2}([0, 1]) : w(0) = \phi(0) = \phi'(1) = w'(1) + \phi(1) = 0\}$$

be equipped with their natural norms, and define $A_1(t) \in \mathcal{B}(Y, X)$ by means of

$$A_1(t) = \begin{pmatrix} a_1(t)\partial_x^2, & a_1(t)\partial_x \\ -\gamma a_1(t)\partial_x, & e_1(t)\partial_x^2 - \gamma a_1(t) \end{pmatrix}, \quad t \in \mathbb{R}_+.$$

Then with $u = (w, \phi)^T$, (13.1) can be rewritten as

$$\ddot{u} = dA_1 * u + f, \quad t \in \mathbb{R}, \tag{13.4}$$

a second order problem. Apparently, (13.4) is not of the form (11.3), however, convolving (13.4) with $c(t) = e^{-t}e_0(t)$, e_0 the Heaviside function, it becomes

$$\dot{u} = c * \dot{u} + c * dA_1 * u + g, \quad t \in \mathbb{R}, \tag{13.5}$$

where $g = c * f$, which is of the form (11.3). Since $c \in \mathcal{W}(\mathbb{R}_+)$ it is easily seen that (13.4) and (13.5) are equivalent.

We want to apply Theorem 12.3 to obtain admissibility of $\Lambda(X)$, as well as Corollary 12.1 for $L_\Lambda^p(X)$, $1 \leq p < \infty$, where $\Lambda \subset \mathbb{R}$ closed is such that $\Lambda \cap \Lambda_0 = \emptyset$. For this purpose recall first from Section 9.1 that there is a b-regular exponentially bounded resolvent $S(t)$ for the local versions of (13.4) and (13.5), where e.g. $b(t) = tc(t)$ is a possible choice. Obviously $b \in L^1(\mathbb{R}_+)$ is 1-regular and such that $\hat{b}(0) \neq 0$.

To verify (12.19) for (13.5), observe that we have to show

$$|\lambda[\hat{c}\hat{A}_1]'(\lambda)|_{\mathcal{B}(Y,X)} \leq C|\hat{b}(\lambda)|, \quad \text{Re } \lambda > 0, \ |\lambda| \geq R,$$

and

$$|\lambda\hat{c}'(\lambda)| \leq M, \quad \text{Re } \lambda > 0, \ |\lambda| \geq R,$$

by the remark following the proof of Theorem 12.3. The second of these inequalities is obvious, while the first follows from

$$|\lambda^2 \hat{a}_1'(\lambda)| + |\lambda^2 \hat{e}_1'(\lambda)| \leq M_0,$$

as an easy calculation shows; but the latter is a consequence of (13.3), since e.g.

$$|\lambda^2 \hat{a}_1'(\lambda)| = |\widehat{td\dot{a}_1}(\lambda) + 2\hat{\dot{a}}_1(\lambda)| \leq \int_0^\infty td\dot{a}_1(t) - 2\int_0^\infty \dot{a}_1(t)dt \leq 3a_1(0+),$$

by convexity of $a_1(t)$.

Thus the assumptions of Theorem 12.3 are satisfied and it remains to compute Λ_0 and to show boundedness of $H(i\rho)$ and $\hat{b}(i\rho)H(i\rho)$ in $\mathcal{B}(X)$ resp. in $\mathcal{B}(X, Y)$. This means that we have to study the invertibility properties of $\lambda^2 - \widehat{dA}_1(\lambda)$ on the imaginary axis, i.e. for $\lambda = i\rho$, $\rho \in \mathbb{R}$. So let $u = (w, \phi)^\tau$, $f = (g, \psi)^\tau$ and consider the equation $(\rho^2 + \widehat{dA}(i\rho))u = f$, i.e.

$$\begin{aligned}
\rho^2 w &+ \widehat{da}_1(i\rho)w'' + \widehat{da}_1(i\rho)\phi' = g, \\
\rho^2 \phi &+ \widehat{de}_1(i\rho)\phi'' - \gamma\widehat{da}_1(i\rho)(w' + \phi) = \psi, \\
w(0) &= \phi(0) = 0, \ \phi'(1) = w'(1) + \phi(1) = 0.
\end{aligned} \tag{13.6}$$

Taking the inner product in $L^2(0,1)$ of the first equation with w, integrating by parts and employing the boundary conditions we obtain the identity

$$\rho^2 |w|^2 = \widehat{da_1}(i\rho)[|w'|^2 + (\phi, w')] + (g, w); \tag{13.7}$$

similarly, the second equation yields after multiplication with ϕ

$$\rho^2 |\phi|^2 = \widehat{de_1}(i\rho)|\phi'|^2 + \gamma\widehat{da_1}(i\rho)[(w', \phi) + |\phi|^2] + (\psi, \phi). \tag{13.8}$$

Here we used the notation $(\cdot, \cdot)$ for the inner product and $|\cdot|$ for the norm in $L^2(0,1)$. Multiplying (13.7) with γ and adding the result to (13.8) leads to

$$\rho^2 (\gamma |w|^2 + |\phi|^2) = \widehat{de_1}(i\rho)|\phi'|^2 + \gamma\widehat{da_1}(i\rho)|w' + \phi|^2 + \gamma(g, w) + (\psi, \phi), \tag{13.9}$$

and taking real and imaginary parts in this identity we finally obtain

$$\begin{aligned}
\rho^2 (\gamma |w|^2 + |\phi|^2) &= \operatorname{Re} \widehat{de_1}(i\rho)|\phi'|^2 + \gamma \operatorname{Re} \widehat{da_1}(i\rho)|w' + \phi|^2 \\
&\quad + \gamma \operatorname{Re} (g, w) + \operatorname{Re} (\psi, \phi),
\end{aligned} \tag{13.10}$$

and

$$0 = \operatorname{Im} \widehat{de_1}(i\rho)|\phi'|^2 + \gamma \operatorname{Im} \widehat{da_1}(i\rho)|w' + \phi|^2 + \gamma \operatorname{Im} (g, w) + \operatorname{Im} (\psi, \phi). \tag{13.11}$$

The properties (13.3) of e_1 and a_1 imply

$$\operatorname{Re} \widehat{de_1}(i\rho) \geq e_\infty > 0 \quad , \quad \operatorname{Re} \widehat{da_1}(i\rho) \geq a_\infty > 0.$$
$$\rho \operatorname{Im} \widehat{de_1}(i\rho) \geq 0 \quad , \quad \rho \operatorname{Im} \widehat{da_1}(i\rho) \geq 0,$$
$$\lim_{|\rho| \to \infty} \widehat{de_1}(i\rho) = e_1(0+) \quad , \quad \lim_{|\rho| \to \infty} \widehat{da_1}(i\rho) = a_1(0+),$$

as is easily verified. Moreover, for $\rho > 0$, $\operatorname{Im} \widehat{de_1}(i\rho) = 0$ if and only if $e_1(t)$ is piecewise linear, with nodes only at $t \in (2\pi/\rho)\mathbb{N}$. If $e_1(t) \neq e_\infty$, then there is a smallest such ρ, say $\rho_e > 0$, and then $\operatorname{Im} \widehat{de_1}(i\rho) = 0$ if and only if $\rho \in \rho_e\mathbb{Z}$. The same is true of course for a_1, with a smallest $\rho_a > 0$, in case $a_1(t) \neq a_\infty$.

To compute Λ_0 observe that either $\rho^2 + \widehat{dA}(i\rho)$ is invertible in $\mathcal{B}(X, Y)$ or 0 is an eigenvalue of $\rho^2 + \widehat{dA}(i\rho)$; the latter is equivalent to $\rho \in \Lambda_0$. Let $g = \psi = 0$ and assume $\operatorname{Im} \widehat{de_1}(i\rho) \neq 0$; then (13.11) implies $\phi' \equiv 0$, hence $\phi \equiv 0$ by (13.6), and then $\widehat{da_1}(i\rho)(w' + \phi) = 0$, hence $\rho^2 w = 0$, by (13.6) again. Therefore $\operatorname{Im} \widehat{de_1}(i\rho) \neq 0$ implies $\rho \notin \Lambda_0$. Similarly, if $g = \psi = 0$ and $\operatorname{Im} \widehat{da_1}(i\rho) \neq 0$, (13.11) yields $w' + \phi \equiv 0$, and then by (13.6) again $\rho^2 w \equiv 0$, i.e. $w \equiv 0$ and $\phi \equiv 0$, i.e. $\rho \notin \Lambda_0$.

Thus $\rho \in \Lambda_0$ implies $\operatorname{Im} \widehat{de_1}(i\rho) = \operatorname{Im} \widehat{da_1}(i\rho) = 0$, i.e. since $0 \notin \Lambda$, $e_1(t)$ and $a_1(t)$ must be of the special form explained above. In particular, if e_1 or a_1 is log-convex, or $-\dot{e}_1$ or $-\dot{a}_1$ is convex, then $\Lambda_0 = \emptyset$.

Now let $\Lambda \subset \mathbb{R}$ be closed such that $\operatorname{Im} \widehat{de}_1(i\rho) \neq 0$, $\operatorname{Im} \widehat{da}_1(i\rho) \neq 0$ for all $\rho \in \Lambda$ and

$$\sup\left\{\frac{\operatorname{Re} \widehat{de}_1(i\rho)}{\rho \operatorname{Im} \widehat{de}_1(i\rho)}, \frac{\operatorname{Re} \widehat{da}_1(i\rho)}{\rho \operatorname{Im} \widehat{da}_1(i\rho)} : \rho \in \Lambda\right\} = N < \infty. \tag{13.12}$$

Then (13.10) and (13.11) yield for $|\rho| \geq 1$

$$\begin{aligned}
\rho^2(\gamma|w|^2 + |\phi|^2) &\leq \gamma \operatorname{Re}(g,w) + \operatorname{Re}(\psi,\phi) - N\rho(\gamma \operatorname{Im}(g,w) + \operatorname{Im}(\psi,\phi)) \\
&\leq C_1|\rho|(\gamma|g||w| + |\psi||\phi|),
\end{aligned}$$

hence

$$\rho^2(\gamma|w|^2 + |\phi|^2) \leq C_1^2(\gamma|g|^2 + |\psi|^2), \quad \rho \in \Lambda, \ |\rho| \geq 1. \tag{13.13}$$

With (13.10) this implies by $\operatorname{Re} \widehat{de}_1(i\rho) \geq e_\infty > 0$, $\operatorname{Re} \widehat{da}_1(i\rho) \geq a_\infty > 0$,

$$e_\infty|\phi'|^2 + \gamma a_\infty|w' + \phi|^2 \leq C_2^2(\gamma|g|^2 + |\psi|^2), \tag{13.14}$$

and (13.6) gives

$$|w''|^2 + |\phi''|^2 \leq C_3^2(\gamma|g|^2 + |\psi|^2)\rho^2, \quad \rho \in \Lambda, \ |\rho| \geq 1. \tag{13.15}$$

Estimates (13.13) and (13.15) yield

$$|\rho(\rho^2 + \widehat{dA}_1(i\rho))^{-1}|_{\mathcal{B}(X)} + \frac{1}{|\rho|}|(\rho^2 + \widehat{dA}_1(i\rho))^{-1}|_{\mathcal{B}(X,Y)} \leq C_4, \quad \rho \in \Lambda, \ |\rho| \geq 1,$$

which implies

$$|H(i\rho)|_{\mathcal{B}(X)} + |\hat{b}(i\rho)||H(i\rho)|_{\mathcal{B}(X,Y)} \leq C_5, \quad \rho \in \Lambda, \tag{13.16}$$

for (13.5). Invoking Theorem 12.3 then yields b-admissibility of $\Lambda(X)$, and Corollary 12.1 that of $L_\Lambda^p(X)$. We may now apply Theorem 11.2 or Corollary 11.2 and 11.3 to deduce results on periodic or almost periodic solutions of (13.1); the details of this, however, are left to the reader.

Concluding we refer to the recent paper Desch and Grimmer [85] where a detailed analysis of the spectrum of the viscoelastic Timoshenko beam has been performed in case the relaxation moduli are completely monotonic. Under certain assumptions on the behaviour of their Laplace transforms for large frequencies it is shown that the spectrum of the viscoelastic Timoshenko beam asymptotically decomposes into the spectra of the two decoupled viscoelastic wave equations arising from (13.1) by neglecting the lower order terms $da * \phi_x$ and $\gamma da * (w_x + \phi)$.

13.2 Heat Conduction in Materials with Memory

Consider the problem of heat conduction in materials with memory described in Section 5.3, where the rigid body Ω is assumed to be homogeneous, isotropic and bounded. The dynamic equations on the line are then given by

$$\begin{aligned}
dm * \dot{\theta} &= dc * \Delta\theta + f, \quad t \in \mathbb{R}, \ x \in \Omega, \\
\theta &= 0, \quad t \in \mathbb{R}, \ x \in \Gamma_b, \\
\frac{\partial\theta}{\partial n} &= 0, \quad t \in \mathbb{R}, \ x \in \Gamma_f,
\end{aligned} \tag{13.17}$$

where Γ_b, $\Gamma_f \subset \partial\Omega$ are closed, $\Gamma_b \cap \Gamma_f = \emptyset$, and $\Gamma_b \cup \Gamma_f = \partial\Omega$. Concerning the heat capacity function $m(t)$ we require

$$m(t) = m_0 + \int_0^t m_1(\tau)d\tau, \quad t \in \mathbb{R}, \tag{13.18}$$

where $m_0 > 0$ and $m_1 \in L^1(\mathbb{R}_+)$; cp. the discussion in Section 5.3. We are here interested in the parabolic case, hence the heat conduction function $c(t)$ is supposed to be of the form

$$c(t) = c_0 + \int_0^t c_1(\tau)d\tau, \quad t \in \mathbb{R}, \tag{13.19}$$

where $c_0 > 0$ and $c_1 \in L^1(\mathbb{R})$. By a proper rescaling of the time variable we may then assume w.l.o.g. $m_0 = c_0 = 1$.

Let $X_q = L^q(\Omega)$, $1 \le q < \infty$, $X_0 = C_{\Gamma_b}(\bar{\Omega}) = \{\theta \in C(\bar{\Omega}) : \theta|_{\Gamma_b} = 0\}$ be equipped with their natural norms and define A_q by means of

$$(A_q\theta)(x) = \Delta\theta(x), \quad x \in \Omega,$$

with domains

$$\mathcal{D}(A_q) = \left\{\theta \in W^{2,q}(\bar{\Omega}) : \theta|_{\Gamma_b} = 0, \frac{\partial\theta}{\partial n}\Big|_{\Gamma_f} = 0\right\}, \quad 1 < q < \infty,$$

$$\mathcal{D}(A_0) = \{\theta \in \cap_{q>1}\mathcal{D}(A_q) : A_0\theta \in C_{\Gamma_b}(\bar{\Omega})\},$$

and in X_1 define A_1 as the closure of e.g. A_2. Let $\Omega \subset \mathbb{R}^N$ be bounded and $\partial\Omega$ smooth; then it is well known that A_q generates a C_0-semigroup which is analytic in $\mathbb{C}_+$ and bounded on each sector $\Sigma(0,\phi)$, $\phi < \pi/2$. In the Hilbert space case $q = 2$, the operator A_2 is even selfadjoint and negative semidefinite. The spectrum $\sigma(A_q)$ is discrete and independent of q, it consists only of simple eigenvalues, $(\mu - A_q)^{-1}$ is compact in X_q, for each $\mu \in \rho(A_q)$. Let

$$\sigma(A) = \{-\mu_j\}_{j=0}^\infty, \quad 0 \le \mu_0 < \mu_1 < \dots, \tag{13.20}$$

and observe that $\mu_0 > 0$, unless $\Gamma_b = \emptyset$, i.e. for the pure Neumann problem.

With these notations and with $b(t) = e^{-t}e_0(t)$, (13.17) can be rewritten as

$$\dot{v} - A_q v = -m_1 * \dot{v} + b * (c_1 * A_q\dot{v} + c_1 * A_q v) + f, \tag{13.21}$$

where $v(t) = \theta(t,\cdot)$. This problem is of the perturbation form (12.38), with $A_0 \equiv 0$, $A_1 \equiv A_q$, $B_0 \equiv -m_1$, $B_1 \equiv 0$, $C_0 \equiv c_1 A_q$, $C_1 \equiv e_0 * c_1 A_q$. With $Y_q = X_{A_q}$, therefore all the assumptions of Corollaries 12.3 and 12.4 are satisfied, and it remains to compute the spectra of (13.21), Λ_0 and Σ_0.

For this purpose define the functions

$$\chi_j(\lambda) = \lambda(1 + \hat{m}_1(\lambda)) + (1 + \hat{c}_1(\lambda))\mu_j, \quad j \in \mathbb{N}_0,$$
$$\chi_\infty(\lambda) = 1 + \hat{c}_1(\lambda), \quad \text{Re } \lambda \ge 0; \tag{13.22}$$

since $\lambda \in \Sigma_0$ iff $\lambda(1 + \hat{m}_1(\lambda)) - (1 + \hat{c}_1(\lambda))A_q$ is not invertible in $\mathcal{B}(Y_q, X_q)$, we see that

$$\Sigma_0 = \{\lambda \in \overline{\mathbb{C}}_+ : \chi_j(\lambda) = 0 \text{ for some } j \in \mathbb{N}_0 \cup \{\infty\}\} \tag{13.23}$$

and

$$\Lambda_0 = \{\rho \in \mathbb{R} : \chi_j(i\rho) = 0 \text{ for some } j \in \mathbb{N}_0 \cup \{\infty\}\}. \tag{13.24}$$

Since $m_1, c_1 \in L^1(\mathbb{R})_+$ by assumption, we have $\hat{c}_1(\lambda), \hat{m}_1(\lambda) \to 0$ as $|\lambda| \to \infty$, $\lambda \in \overline{\mathbb{C}}_+$, hence Σ_0 as well as Λ_0 are compact, and $H_q(\lambda)$ satisfies

$$|\lambda H_q(\lambda)|_{\mathcal{B}(X_q)} + |H_q(\lambda)|_{\mathcal{B}(X_q, Y_q)} \le M, \quad \lambda \in \overline{\mathbb{C}}_+, \ |\lambda| \ge R, \tag{13.25}$$

where R is sufficiently large. Furthermore, $\Sigma_0 \cap \mathbb{C}_+$ is at most countable, Λ_0 has Lebesgue measure 0, and $\lambda_0 \in \Sigma_0$, $\mathrm{Re}\,\lambda_0 > 0$, is a cluster point of Σ_0 iff $\chi_\infty(\lambda_0) = 0$.

Next we observe that $\chi_\infty(\lambda) \ne 0$ on $\overline{\mathbb{C}}_+$ if either $|c_1| < 1$ or c_1 is of positive type; cp. Section 5.3. If this is the case and if $m_1(t)$ is nonnegative, nonincreasing, then also $\chi_j(\lambda) \ne 0$ on $\overline{\mathbb{C}}_+$ for $j \in \mathbb{N}$, and $\chi_0(\lambda) \ne 0$ on $\overline{\mathbb{C}}_+ \setminus \{0\}$. Thus except for the pure Neumann case $\Gamma_b = \emptyset$, in the physically interesting situation we have $\Sigma_0 = \emptyset$, hence also $\Lambda_0 = \emptyset$. On the other hand, if $\Gamma_b = \emptyset$ then $\mu_0 = 0$, and so $\Lambda_0 = \Sigma_0 = \{0\}$.

In the latter situation we have $X_q = \mathcal{N}(A_q) \oplus \mathcal{R}(A_q)$, hence (13.21) splits into an infinite dimensional problem on $\mathcal{R}(A_q)$ with $\Sigma_0 = \emptyset$, and into a one-dimensional equation for $\varphi(t) = \int_\Omega v(t, x)dx$, the mean temperature, which reads

$$\dot{\varphi} + m_1 * \dot{\varphi} = \psi, \quad t \in \mathbb{R}, \tag{13.26}$$

where $\psi(t) = \int_\Omega r(t, x)dx$.

Returning to the general case, the results of Section 11 and 12 yield e.g. the following properties of (13.21), i.e. (13.17).

(a) If $\Lambda \cap \Lambda_0 = \emptyset$, then the spaces $B_{p,\Lambda}^{\alpha,r}(X_q)$ are spaces of maximal regularity for (13.21), where $\alpha \in (0, 1)$, $p, r \in [1, \infty]$ (Corollary 12.3).

(b) If $\Lambda \cap \Lambda_0 = \emptyset$, $b(t) = e^{-t}e_0(t)$, $\mathcal{H}(X_q)$ homogeneous, then $\mathcal{H}_\Lambda(X_q)$ is b-admissable for (13.21), and there is a Λ-kernel $G_\Lambda \in L^1(\mathcal{B}(X_q))$ such that $b * G_\Lambda \in L^1(\mathcal{B}(X_q, Y_q))$ (Corollary 12.4).

(c) If $\Lambda \cap \Lambda_0 = \emptyset$, $f \in W_\Lambda^+(X)$, then $v = G_\Lambda * f \in W_\Lambda^+(X)$, $\sigma(v) = \sigma(f)$, $\exp(v) = \exp(f)$, and

$$\alpha(\rho; v) = H(i\rho)\alpha(\rho; f) \quad \text{for all } \rho \in \exp(f).$$

The same assertions hold for $AAP_\Lambda^+(X)$ instead of $W_\Lambda^+(X)$. If $0 \in \Lambda$, $f \in C_l^+(X)$ and $\sigma(f) \subset \Lambda$ imply $v = G_\Lambda * f \in C_l^+(X)$, $\sigma(v) = \sigma(f)$, and

$$v(\infty) = H(0)f(\infty)$$

(Corollary 11.3 and Theorem 11.2).

(d) There is a b-regular resolvent $S(t)$ for the local version of (13.21), and it has the maximal regularity property of type $B_p^{\alpha,r}(J; X_q)$, and in case $1 < p, q < \infty$ also of type $L^p(J; X_q)$ (Theorems 7.6, 7.5, and 8.7).

(e) If $\Lambda_0 = \emptyset$, then $S(t) = S_0(t) + G(t)$, (13.21) and its local version are asymptotically equivalent w.r.t. bounded solutions (Theorems 11.3 and 11.4).

(f) If $\Sigma_0 = \emptyset$, then S is uniformly integrable, and in case $1 < q < \infty$, the local version of (13.21) is strongly L^p-stable on the halfline (Theorems 11.3 and 8.8).

13.3 Electrodynamics with Memory

Let $\Omega \subset \mathbb{R}^3$ be domain with smooth boundary $\partial\Omega$, which forms a rigid linear isotropic and homogeneous medium. Then Maxwell's equation are (cp. Section 9.5)

$$\mathcal{B}_t + \operatorname{curl} \mathcal{E} = \mathcal{K}_e, \quad \operatorname{div} \mathcal{B} = \rho_m$$
$$\mathcal{D}_t - \operatorname{curl} \mathcal{H} + \mathcal{J} = \mathcal{J}_e, \quad \operatorname{div} \mathcal{D} = \rho_e \tag{13.27}$$

for $t \in \mathbb{R}$, $x \in \Omega$; here $\mathcal{K}_e$, $\mathcal{J}_e$ denote the impressed magnetic and electric currents ($\mathcal{K}_e = 0$ in reality), and ρ_e, ρ_m the free electric resp. magnetic charges ($\rho_m = 0$ in reality). As a boundary condition we impose

$$n \times \mathcal{E} = 0, \quad t \in \mathbb{R}, \ x \in \partial\Omega, \tag{13.28}$$

where $n(x)$ denotes the outer normal of Ω at $x \in \partial\Omega$. Other boundary conditions are of course also of interest, like $n \times \mathcal{H} = 0$, but the latter can be treated similar to the approach given below. According to Section 9.5 the constitutive laws are

$$\mathcal{B} = d\mu * \mathcal{H}, \quad \mathcal{D} = d\varepsilon * \mathcal{E}, \quad \mathcal{J} = d\sigma * \mathcal{E}, \tag{13.29}$$

where the material functions $\mu(t)$, $\varepsilon(t)$, $\sigma(t)$ are of the form

$$\mu(t) = \mu_0 + \int_0^t \mu_1(\tau)d\tau, \quad \varepsilon(t) = \varepsilon_0 + \int_0^t \varepsilon_1(\tau)d\tau,$$

$$\sigma(t) = \sigma_0 + \int_0^t \sigma_1(\tau)d\tau, \quad t > 0, \tag{13.30}$$

with $\mu_1, \varepsilon_1, \sigma_1 \in L^1(\mathbb{R}_+)$ of positive type, $\sigma_0 \geq 0$, and $\mu_0 > 0$, $\varepsilon_0 > 0$ the fundamental constant of the vacuum, $c_0 = (\varepsilon_0\mu_0)^{-1/2}$ the speed of light in vacuum.

In rewriting this equation in the standard form of Section 11, we proceed slightly different from Sections 9.5 or 9.6. Since $\mu_0, \varepsilon_0 > 0$, $\operatorname{Re} \hat{\mu}_1(\lambda)$, $\operatorname{Re} \hat{\varepsilon}_1(\lambda) \geq 0$ on $\mathbb{C}_+$, and $\mu_1, \varepsilon_1 \in L^1(\mathbb{R}_+)$, by the Wiener-Levy theorem, there are functions $\eta^+, \eta^-, \nu^+, \nu^- \in BV(\mathbb{R}_+)$ such that

$$\widehat{d\eta}^\pm(\lambda) = [\widehat{d\varepsilon}(\lambda)]^{\pm 1/2}, \quad \widehat{d\nu}^\pm(\lambda) = [\widehat{d\mu}(\lambda)]^{\pm 1/2}, \quad \operatorname{Re} \lambda \geq 0; \tag{13.31}$$

observe that $\eta^\pm(0+) = \varepsilon_0^{\pm 1/2}$, $\nu^\pm(0) = \mu_0^{\pm 1/2}$, $\eta^\pm, \nu^\pm \in W_{loc}^{1,1}(\mathbb{R}_+)$ and $\eta_1^\pm := \dot{\eta}^\pm$, $\nu_1^\pm := \dot{\nu}^\pm \in L^1(\mathbb{R}_+)$. The transformations

$$v = d\eta^+ * \mathcal{E} \quad , \quad w = d\nu^+ * \mathcal{H} \tag{13.32}$$

then induce isomorphisms on any homogeous space, and lead to a particularly nice form of the dynamic equations.

In fact, let $X = L^2(\Omega; \mathbb{R}^3) \times L^2(\Omega; \mathbb{R}^3)$, $u = (v, w)^T$, and define a linear operator A_0 in X by means of

$$A_0 u = A_0 \begin{pmatrix} v \\ w \end{pmatrix} = \begin{pmatrix} 0 & \mathrm{curl} \\ -\mathrm{curl} & 0 \end{pmatrix} \begin{pmatrix} v \\ w \end{pmatrix}, \quad u \in \mathcal{D}(A), \tag{13.33}$$

with domain

$$\mathcal{D}(A_0) = \{ u = \begin{pmatrix} v \\ w \end{pmatrix} : v, w \in W^{1,2}(\Omega; \mathbb{R}^3), \ n \times v|_{\partial\Omega} = 0 \};$$

and let $Y = X_A$, where $A = \overline{A}_0$. It is well-known that A is skewadjoint, hence generates a C_0-group of unitary operators in X. The direct sum decomposition

$$X = \overline{\mathcal{R}(A)} \oplus \mathcal{N}(A)$$

which is also orthogonal, holds, and $\sigma(A) \subset i\mathbb{R}$. If Ω is an exterior domain then $\sigma_p(A) = \{0\}$, and $\sigma_c(A) = i\mathbb{R} \setminus \{0\}$. For a bounded domain Ω, the range of A is closed and $A|_{\overline{\mathcal{R}(A)}}^{-1}$ is compact, hence $\sigma(A) = \sigma_p(A)$ and $\sigma_p(A) \setminus \{0\}$ consists of eigenvalues $\pm i\xi_n$, $0 < \xi_n \to \infty$, as $n \to \infty$, with corresponding eigenvectors forming an orthonormal base of $\mathcal{R}(A)$. For a proof of these facts see e.g. the monograph Leis [208], Chap. 8.

Ignoring the equations $\mathrm{div}\, \mathcal{B} = \rho_m$ and $\mathrm{div}\, \mathcal{D} = \rho_e$ (these can be considered as the definitions of ρ_m, ρ_e), we obtain by simple manipulations the following reformulation of (13.27), (13.28) and (13.29).

$$\dot{u} = da * Au + dB * u + f, \quad t \in \mathbb{R}, \tag{13.34}$$

where $da = d\eta^- * d\nu^-$, $f = (d\eta^- * \mathcal{J}_e, d\nu^- * \mathcal{K}_e)^T$, and the operator-valued kernel $B \in BV(\mathbb{R}_+; \mathcal{B}(X) \cap \mathcal{B}(Y))$ is given by

$$B(t) = \begin{pmatrix} -d\eta^- * d\eta^- * \sigma(t) & 0 \\ 0 & 0 \end{pmatrix}. \tag{13.35}$$

Along with the problem (13.34) on the line we also consider the corresponding problem on the halfline

$$\dot{z} = da * Az + dB * z + f, \quad t \geq 0, \ z(0) = z_0. \tag{13.36}$$

We are now in position to apply the theory developed in this book to the study of (13.34) and (13.36); observe that these problems are of hyperbolic type.

There are different lines of approach to the problem in question; we use here that based on subordination and perturbation, which yields the strongest result. However, we have to assume that μ, ε are bounded Bernstein functions.

(a) As in Section 9.5, $a(t)$ is completely monotonic by Lemma 4.3. The characteristic numbers κ, α, ω for $a(t)$ are

$$\kappa = 1/c_0, \quad \omega = 0, \quad \alpha = \frac{1}{2}(\mu_1(0+)\sqrt{\frac{\varepsilon_0}{\mu_0}} + \varepsilon_1(0+)\sqrt{\frac{\mu_0}{\varepsilon_0}}), \tag{13.37}$$

and for the case $\sigma \equiv 0$, by Theorem 4.2 and 4.4, the resolvent $S_0(t)$ for (13.36) with $\sigma \equiv 0$ is

$$S_0(t) = e^{c_0 At} e^{-\alpha c_0 t} + S_1(t), \quad t \geq 0, \tag{13.38}$$

where e^{At} denotes the C_0-group generated by A and $S_1 \in C((0, \infty); \mathcal{B}(X))$ is bounded, S_1 is also $\mathcal{B}(X)$-continuous at $t = 0$ and $S_1(0) = 0$ in case $\alpha < \infty$. Observe that $S_0(t)$ is $\mathcal{B}(X)$-continuous on $\mathbb{R}_+$ iff $\alpha = \infty$, i.e. if $\mu_1(0+) = \infty$ or $\varepsilon_1(0+) = \infty$.

The general case is now a consequence of Theorem 6.1 in the form of Corollary 6.2, to the result that (13.36) admits an exponentially bounded e^{-t}-regular resolvent $S(t)$.

(b) Next we may apply Theorem 12.2, assuming again $\sigma = 0$ first. With $b(t) \equiv 1$, $\omega_b = \beta = 0$, this result yields a Λ-kernel $G_\Lambda(t)$ for (13.34) whenever $\alpha > 0$ and $\Lambda \cap \Lambda_0 = \emptyset$. $G_\Lambda(t)$ is of the form

$$G_\Lambda(t) = e^{c_0 At} e^{-\alpha c_0 t} e_0(t) + G_0(t), \quad t \in \mathbb{R}, \tag{13.39}$$

with $G_0 \in L^1(\mathbb{R}_+; \mathcal{B}(X)) \cap L^\infty(\mathbb{R}; \mathcal{B}(X))$ $\mathcal{B}(X)$-continuous on $\mathbb{R}$, $G_\Lambda(t) \to 0$ in $\mathcal{B}(X)$ as $|t| \to \infty$, and the jump relation

$$\lim_{t \to 0+} G_\Lambda(t) - \lim_{t \to 0-} G_\Lambda(t) = I,$$

holds in the strong sense. Moreover, each homogeneous space $\mathcal{H}_\Lambda(X)$ is e^{-t}-admissable if $\Lambda \cap \Lambda_0 = \emptyset$. Observe that Λ_0 is compact anyway, and that G_Λ is integrable, not just strongly integrable, and even uniformly integrable in case $\alpha = \infty$.

Via perturbation results these properties remain valid for $\sigma \neq 0$; this is clear for $\sigma_0 = 0$, by Theorem 12.1, but also true in case $\sigma_0 > 0$. For the latter one has to modify A by incorporating the damping term $-\sigma_0 \sqrt{\varepsilon_0/\mu_0}$ in the left upper corner of A defined by (13.33). e^{At} is then no longer unitary, but is still a bounded C_0-semigroup, which respects the decomposition $X = \overline{\mathcal{R}(A)} \oplus \mathcal{N}(A)$.

(c) It remains to compute the spectra Λ_0 and Σ_0 of (13.34) and (13.36). For this purpose observe that the equation

$$(\lambda - \widehat{da}(\lambda)A - \widehat{dB}(\lambda))u = f$$

with $u = (v, w)^T$ and $f = (g, h)^T$ reduces to the boundary value problem

$$\begin{aligned} \varphi(\lambda)v - \operatorname{curl}\ w &= \widehat{dk}(\lambda)g \quad \text{in } \Omega, \\ n \times v &= 0 \quad \text{on } \partial\Omega, \\ \psi(\lambda)w + \operatorname{curl}\ v &= \widehat{dk}(\lambda)h \quad \text{in } \Omega, \end{aligned} \tag{13.40}$$

where $\widehat{dk} = 1/\widehat{da} = \sqrt{\widehat{d\varepsilon}\widehat{d\mu}}$, $\psi = \lambda\widehat{dk}$, and $\varphi = \psi + \widehat{dk}\widehat{d\sigma}/\widehat{d\varepsilon}$. Taking the inner product with v in the first equation of (13.40), with w in the third, addition and an integration by parts yield

$$\varphi|v|_2^2 + \overline{\psi}|w|_2^2 = \widehat{dk}(g, v) + \overline{\widehat{dk}}(h, w). \tag{13.41}$$

Since $k \in BV(\mathbb{R}_+)$, this identity implies an estimate of the form

$$|v|_2^2 + |w|_2^2 \le C(\lambda)^2 (|g|_2^2 + |h|_2^2)$$

for all $v, w \in L^2(\Omega; \mathbb{R}^3)$, provided $\varphi\psi \notin (-\infty, 0]$. Therefore we obtain

$$\Sigma_0 \subset \{\lambda \in \overline{\mathbb{C}}_+ : \varphi(\lambda)\psi(\lambda) \in (-\infty, 0]\}.$$

However, the definitions of φ and ψ imply

$$\varphi\psi = (\lambda\widehat{d\mu})(\lambda\widehat{d\varepsilon} + \widehat{d\sigma});$$

since ε, μ are Bernstein functions we have Re $(\lambda\widehat{d\mu}) > 0$ and Re $(\lambda\widehat{d\varepsilon}) > 0$ for Re $\lambda > 0$, and so the assumption that σ_1 is of positive type shows $\Sigma_0 = i\Lambda_0$, i.e. $\varphi\psi \in (-\infty, 0]$ can only happen for $\lambda \in i\mathbb{R}$. But then for $\lambda = i\rho \ne 0$ we must have

$$-\rho \, \mathrm{Im} \, \widehat{\mu_1}(i\rho) = 0, \quad -\rho \, \mathrm{Im} \, \widehat{\varepsilon_1}(i\rho) = 0, \quad \sigma_0 = 0, \quad \mathrm{Re} \, \widehat{\sigma_1}(i\rho) = 0;$$

since μ_1, ε_1 are completely monotonic this is only possible in case $\mu_1 \equiv \varepsilon_1 \equiv 0$, which contradicts the assumption $\alpha > 0$ made above. Thus we conclude $\Sigma_0 = \Lambda_0 \subset \{0\}$.

(d) Let us consider the point $\lambda = 0$ separately. If the domain Ω is not bounded then $\mathcal{R}(A)$ will in general not be closed, hence there is no hope for $\Lambda_0 = \emptyset$, without changing the topology, at least. On the other hand, if Ω is bounded then $\mathcal{R}(A)$ is closed; hence we may split the problem into one in $\mathcal{R}(A)$ and one in $\mathcal{N}(A)$, since $B(t)$ respects the orthogonal decomposition $X = \mathcal{R}(A) \oplus \mathcal{N}(A)$.

In $\mathcal{N}(A)$ we obtain an ordinary system of Volterra equations given by

$$\dot{v} + d\eta^- * d\eta^- * d\sigma * v = g, \quad \dot{w} = h; \tag{13.42}$$

if $\sigma(\infty) > 0$ the first of these equations can be solved in any homogeneous function space $\mathcal{H}(\mathbb{C}^3)$, while the second (both if $\sigma(\infty) = 0$) depends on the invariance of $\mathcal{H}(\mathbb{C}^3)$ w.r.t. integration.

The problem restricted to $\mathcal{R}(A)$ has the property $\Lambda_0 = \Sigma_0 = \emptyset$; recall we are assuming Ω bounded, here. In fact, since $A|_{\mathcal{R}(A)}^{-1}$ is compact, either $(A + B(\infty))|_{\mathcal{R}(A)}$ is invertible, i.e. $0 \notin \Lambda_0$, or 0 is an eigenvalue of $(A + B(\infty))|_{\mathcal{R}(A)}$. To exclude the latter assume $u \in \mathcal{R}(A)$ and $Au + B(\infty)u = 0$; with $u = (v, w)^T$ this implies

$$\mathrm{curl} \, w = \gamma v, \quad \mathrm{curl} \, v = 0, \quad n \times v|_{\partial\Omega} = 0,$$

where $\gamma = \sigma(\infty)\sqrt{\mu(\infty)/\varepsilon(\infty)}$. If $\gamma = 0$ then obviously $u \in \mathcal{N}(A)$, hence $u = 0$. If $\gamma > 0$ then

$$0 = (\, \mathrm{curl} \, v, w) = (v, \, \mathrm{curl} \, w) = \gamma|v|^2,$$

i.e. $v = 0$, and therefore $\mathrm{curl} \, w = 0$ which means $u \in \mathcal{N}(A)$ and so again $u = 0$. This shows $0 \notin \Lambda_0$, hence $\Lambda_0 = \Sigma_0 = \emptyset$, when (13.34) and (13.36) are restricted to $\mathcal{R}(A)$.

(e) To conclude the discussion, let Ω be bounded (hence $\mathcal{R}(A)$ closed) and simply connected, $\alpha > 0$ (which excludes $\varepsilon(t) \equiv \varepsilon_0$, $\mu(t) \equiv \mu_0$). Then the resolvent

$S(t)$ is integrable on $\mathcal{R}(A)$, even uniformly if $\alpha = -\infty$, every homogeneous space $\mathcal{H}(\mathcal{R}(A))$ is e^{-t}-admissible for (13.34), and (13.34) and (13.36) are asymptotically equivalent, by Theorems 11.3 and 11.4. If in addition $\sigma(\infty) > 0$, the first equation in (13.42) admits also an integrable resolvent. Since there are no magnetic currents in reality, we actually have $h = 0$, and so the resolvent $S(t)$ is even integrable on $X_0 = \{u \in X : \operatorname{div} w = 0, w \cdot n|_{\partial\Omega} = 0\}$ the space of physical interest.

Thus the presence of a memory in at least one of the material functions $\mu(t)$ or $\varepsilon(t)$ stabilizes the high frequencies; while for low frequencies a stationary resistance $1/\sigma(\infty) < \infty$ does this job no matter how large ε_1, μ_1 and σ are.

13.4 Ergodic Theory

A question very much related to the subject of integrability of resolvents $S(t)$ for

$$u(t) = f(t) + \int_0^t a(t-\tau)Au(\tau)d\tau, \quad t \geq 0, \tag{13.43}$$

or

$$u(t) = f(t) + \int_0^t A(t-\tau)u(\tau)d\tau, \quad t \geq 0, \tag{13.44}$$

is that of existence of $P = \lim_{t\to\infty} S(t)$ in various senses, and what the nature of this limit will be. More precisely, the resolvent $S(t)$ and (13.43) resp. (13.44) are called *uniformly* resp. *strongly* resp. *weakly ergodic* if $P = \lim_{t\to\infty} S(t)$ exists in the uniform resp. strong resp. weak operator topology. This type of ergodicity will be denoted by (j, E)-*ergodicity*, where j denotes any of the symbols u, s, w representing uniformly, strongly, weakly, respectively. If instead only the Cesaro-means $t^{-1}\int_0^t S(\tau)d\tau$ converge to P in one of these operator topologies, then $S(t)$ and (13.43) resp. (13.44) will be called *uniformly* resp. *strongly* resp. *weakly Cesaro ergodic*, and this type will be denoted by (j, C)-*ergodicity*. If merely the Abel-means $\lambda H(\lambda) = \int_0^\infty [\lambda e^{-t\lambda}]S(t)dt$ converge as $\lambda \to 0$ then the term *Abel-ergodicity* or more precisely (j, A)-*ergodicity* will be used. Observe that the limit P is unique and the following implication scheme is valid

$$
\begin{array}{ccccc}
(u, E) & \Rightarrow & (s, E) & \Rightarrow & (w, E) \\
\Downarrow & & \Downarrow & & \Downarrow \\
(u, C) & \Rightarrow & (s, C) & \Rightarrow & (w, C) \\
\Downarrow & & \Downarrow & & \Downarrow \\
(u, A) & \Rightarrow & (s, A) & \Rightarrow & (w, A).
\end{array}
$$

Here the horizontal implications are obvious, while the vertical arrows are implied by so-called *Abelian theorems*. In general, none of these implications can be reversed; a result asserting the converse of a vertical arrow is called a *Tauberian theorem*, in honour of A. Tauber who was the first to use this type of argument in the case of numerical power series; see Tauber [323]. The recent paper by Arendt and Prüss [14] contains an account of the Tauberian theory for the vector-valued Laplace transform; see also Hille and Phillips [180], Chapter XVIII, Arendt and

Batty [11], [12], and Batty [20]. For the classical Tauberian theory we refer to the monographs of Doetsch [100] and Widder [339], [340].

Let us briefly discuss ergodic theory in the semigroup case $a(t) \equiv 1$; see Hille and Phillips [180]. If $\omega_0(A) \leq 0$ (the type of e^{At}) then (j, A)-ergodicity can be completely characterized. e^{At} is (w, A)-ergodic iff it is (s, A)-ergodic iff $\lambda H(\lambda)$ is bounded as $\lambda \to 0+$, and $\mathcal{N}(A)^{\perp} \cap \mathcal{N}(A^*) = \{0\}$; e^{At} is (u, A)-ergodic iff it is (w, A)-ergodic and $\mathcal{R}(A)$ is closed. Note that $\mathcal{N}(A)^{\perp} \cap \mathcal{N}(A^*) = \{0\}$ is automatically satisfied in case X is sun-reflexive w.r.t. e^{At}, in particular if X itself is reflexive, in virtue of de Pagter's characterization of sun-reflexivity; see de Pagter [80]. The Abelian ergodic limit P if it exists has a very simple structure, namely it is the projection onto $\mathcal{N}(A)$ along $\mathcal{R}(A)$. By means of vector-valued real Tauberian theorems, e^{At} is (s, C)-ergodic if it is (w, A)-ergodic and e^{At} is bounded, and it is (u, C)-ergodic if it is (u, A)-ergodic and e^{At} is bounded and eventually $\mathcal{B}(X)$-continuous. Furthermore, (j, A)-ergodicity and boundedness of $|e^{At}| + t|Ae^{At}|$ imply (j, E)-ergodicity, since in this case e^{At} is feebly oscillating; note that bounded analytic semigroups are covered by this result. Another very interesting theorem was recently discovered independently by Arendt and Batty [11] and Lyubich and Phong [229]; it states that (s, A)-ergodicity, boundedness of e^{At}, $\sigma(A)$ countable, and $\sigma_p(A^*) \cap i\mathbb{R} \subset \{0\}$ imply (s, E)-ergodicity. This result has been used e.g. by Wyler [341] to obtain asymptotic stability of several classes of dissipative wave equations. To conclude this discussion, observe that integrability of e^{At} is equivalent to $\omega_0(A) < 0$, in which case e^{At} is (u, E)-ergodic with ergodic limit $P = 0$. In the Hilbert space case, by Gearhart's theorem the same result is valid for strongly integrable semigroups; cp. Section 12.7.

Returning to evolutionary integral equations, let us first observe that in any of the integrability results for the resolvent proved in Sections 10 and 12, $S(t) \to 0$ in $\mathcal{B}(X)$, except for Section 10.4 where the convergence still takes place in the strong operator topology. But although, contrary to the semigroup case, integrability of $S(t)$ can in general not be expected to imply (s, E)-ergodicity, the situation is much more interesting if the resolvent is not integrable. To understand this, let us consider linear isotropic incompressible viscoelasticity once more. If as in Section 5.6, A denotes the Stokes operator in $L_0^2(\Omega; \mathbb{R}^3)$ and $a(t)$ is a creep function with $\log a_1(t)$ convex, then Corollary 10.3 states that $S(t)$ is integrable iff $a(t) \not\equiv a_\infty t$ and $0 \in \rho(A)$. The latter is the case if the underlying domain $\Omega \subset \mathbb{R}^3$ is bounded, and it is also true for some types of unbounded domains, however, not for all. In Arendt and Prüss [14] it has been shown that one always has $S(t) \to 0$ as $t \to \infty$ in the strong sense, provided $a(t) \not\equiv a_\infty t$, i.e. the material is not purely elastic. This result as well as others follow from the main result of the paper just quoted, which applies to (13.43).

Theorem 13.1 *Suppose (13.43) admits a resolvent $S(t)$, bounded on $\mathbb{R}_+$, let $a(t)$ be of subexponential growth and such that $\hat{a}(\lambda)$ admits continuous extension in $\mathbb{C} \cup \{\infty\}$ to $\overline{\mathbb{C}}_+$; assume that $P = \lim_{\lambda \to 0+} \lambda H(\lambda)$ exists in the strong sense, and define*

$$\rho(a, A) = \{i\eta \in i\mathbb{R} : \quad \text{there is } \varepsilon > 0 \text{ such that } [(1 - \hat{a}(\lambda)A)^{-1} - P]/\lambda \quad (13.45)$$

admits a strongly continuous extension to $\mathbb{C}_+ \cup i[\eta - \varepsilon, \eta + \varepsilon]\}$.

In addition suppose the following conditions are satisfied.

(E1)   *The singular set* $iE := i\mathbb{R} \setminus \rho(a; A)$ *is countable, and*

$$0 \neq \mu \in E, \ \hat{a}(i\mu) \neq 0, \infty \quad \Rightarrow \quad \overline{\mathcal{R}(1 - \hat{a}(i\mu)A)} = X;$$

$$0 \neq \mu \in E, \ \hat{a}(i\mu) = \infty \quad \Rightarrow \quad \mathcal{N}(A)^\perp \cap \mathcal{N}(A^*) = \{0\}.$$

(E2)   *For all* $\mu \in E$ *there exists a constant* $C(\mu) > 0$ *such that for all* $x \in \mathcal{D}(A)$

$$|\int_0^t e^{-i\mu s}(a * S(s) - \hat{a}(i\mu)S(s)Ax)ds| \leq C(\mu)|x|_A \quad \textit{if } \hat{a}(i\mu) \in \mathbb{C}, \textit{ and}$$

$$|\int_0^t e^{-i\mu s}S(s)Ax ds| \leq C(\mu)|x|_A \quad \textit{if } \hat{a}(i\mu) = \infty.$$

(E3)   *There exist* $\tau \geq 0$, $M \geq 0$ *such that* $|\dot{S}(t)x| \leq M|x|_A$, *for all* $x \in \mathcal{D}(A)$, $t \geq \tau$.

   Then $\lim_{t \to \infty} S(t)x = Px$ *for all* $x \in X$, *where* $P = (1 - \hat{a}(0)A)^{-1}$ *if* $\hat{a}(0) \in \mathbb{C}$, *and* P *is the projection onto* $\mathcal{N}(A)$ *along* $\overline{\mathcal{R}(A)}$ *if* $\hat{a}(0) = \infty$.

This result contains as a special case the convergence theorem due to Arendt and Batty, and Lyubich and Phong which was mentioned above; take $a(t) \equiv 1$. Although the conditions seem to be difficult, a check can be made in many specific instances, including the application to viscoelasticity discussed above; see Arendt and Prüss [14].

   As an illustrative example, suppose that A is an m-dissipative operator in a Hilbert space X such that $\rho(A) \supset i\mathbb{R}$, and consider $a(t) = \cos t$; then Theorem 13.1 implies $S(t) \to I$ strongly as $t \to \infty$.

   In the paper just quoted one also finds complete characterizations of Abel ergodicity of $S(t)$ for equations of scalar type. Together with boundedness of $S(t)$ on $\mathbb{R}_+$ these imply Cesaro ergodicity, i.e. yield the mean ergodic theorem for resolvents. A different mean ergodic theorem has been recently obtained in Lizama [217]. There are yet no such studies for nonscalar problems (13.44), nor are there results on asymptotic almost periodicity of resolvents like the theorem of Jacobs, Glicksberg and Deleuuw for C_0-semigroups; cp. Section 11.4.

13.5 Semilinear Equations

For about three decades semilinear abstract Cauchy problems of the form

$$\dot{u}(t) = Au(t) + F(t, u(t)), \ u(0) = u_0, \quad t \geq 0, \tag{13.46}$$

in a Banach space X have been studied in many different settings. Here A denotes the generator of a C_0-semigroup $T(t)$ in X and F a nonlinear function which is considered as a perturbation of the main term Au. We refer to the monographs Friedman [122], Goldstein [133], Haraux [172], Henry [178], Martin [235], Pazy

[267], Reed [287], Tanabe [319], and Webb [338] for details of the approaches explained briefly below, as well as for the original literature.

The basic idea for this type of approach is to consider the integral equation

$$u(t) = T(t)u_0 + \int_0^t T(t-s)F(s, u(s))ds, \quad t \geq 0, \tag{13.47}$$

which follows from the variation of parameters formula for the linear Cauchy problem $\dot{u} = Au + f$, $u(0) = u_0$. In this formulation the unbounded operator A does only appear in terms of the semigroup $T(t)$, which makes e.g. the application of fixed point theorems for the solution of the integral equation possible.

A typical argument uses the contraction mapping principle in the space $Z = C(J; X)$, where $J = [0, \tau]$. Suppose the nonlinearity $F : J \times X \to X$ is continuous and globally Lipschitz, i.e.

$$|F(t, u) - F(t, v)| \leq L|u - v|, \quad \text{for all } u, v \in X, \ t \in J;$$

then the map $K : Z \to Z$ defined by the right hand side of (13.47) is well-defined and easily seen to be a contraction w.r.t the norm

$$||u|| = \sup\{|u(t)|e^{-\omega t} : t \in J\}, \quad u \in Z,$$

provided ω is chosen large enough. Consequently, (13.47) has a unique continuous solution $u(t)$, for every $u_0 \in X$, and the map $u_0 \mapsto u(\cdot)$ is Lipschitz from X to Z. Solutions of the integral equation (13.47) are called mild solutions of (13.46). This argument has been refined in many directions, e.g. localizing the Lipschitz condition, or replacing it by compactness assumptions involving measures of non-compactness and applying fixed point theorems for set contractions, or by allowing A to depend on t. This way also a qualitative theory for mild solutions of (13.46) can be developed which to a large extent parallels that of ordinary differential equations in finite dimensions.

If (13.46) is parabolic in the sense that the semigroup $T(t)$ generated by A is analytic (and w.o.l.g. bounded on $\mathbb{R}_+$), the assumptions on F can be relaxed considerably, due to the regularizing effect of the convolution with $T(t)$. In this situation it is enough to assume that $F : \mathbb{R}_+ \times X_\alpha \to X$ is continuous and e.g. Lipschitz in its second argument, where $X_\alpha = D((-A)^\alpha)$ equipped with its graph norm, and $\alpha \in [0, 1)$.

In the parabolic case one can go a step further, utilizing the available maximal regularity results. In fact, by means of maximal regularity of type C^β, the choice $\alpha = 1$ is possible, if in addition F is Hölder continuous of degree $\beta > 0$ w.r.t. its first argument, the compatibility condition $Au_0 + F(0, u_0) \in D_A(\beta, \infty)$ is satisfied, and $F_u(0, u_0) = 0$ holds for the Fréchet-derivative F_u of F w.r.t. u. The solutions obtained this way will even be strong solutions and enjoy the extra regularity $u \in C^\beta(J; X_A)$, $\dot{u} \in C^\beta(J; X)$.

Being only based on the variation of parameters formula, this approach can almost directly be generalized to semilinear evolutionary integral equations of the form

$$\dot{u}(t) = \int_0^t dA(\tau)u(t-\tau) + V(u)(t), \ u(0) = u_0, \quad t \geq 0, \tag{13.48}$$

where $A \in BV_{loc}(\mathbb{R}_+; \mathcal{B}(Y,X))$ is such that the corresponding linear problem $u = A * u + f$ admits a weak resolvent $S(t)$ in X, and $V : C(\mathbb{R}_+; X) \to X$ is a Volterra operator on the halfline in the sense that $V(u)(t)$ depends only on the values $\{u(s) : s \in [0,t]\}$, and satisfies appropriate smoothness or compactness conditions. The integral equation corresponding to (13.48) is of course

$$u(t) = S(t)u_0 + \int_0^t S(t-\tau)V(u)(\tau)ds, \quad t \geq 0. \tag{13.49}$$

Examples for possible nonlinearities V which fit into this framework are e.g. Volterra operators like

$$V(u)(t) = F(t, u(t), \int_0^t k(t,s,u(s))ds).$$

Depending on the regularity properties of the resolvent $S(t)$ for the linear part of (13.48) and those of V, the solvability behaviour of (13.48) can be studied as indicated above for the pure abstract semilinear Cauchy problem (13.46).

Similarly, the equation on the line

$$\dot{u}(t) = \int_0^\infty dA(\tau)u(t-\tau) + V(u)(t), \quad t \in \mathbb{R}, \tag{13.50}$$

can be treated, where now $A \in BV(\mathbb{R}_+; \mathcal{B}(Y,X))$, the corresponding linear equation on the line $\dot{u} = dA * u + f$ admits a $\mathbb{R}$-kernel in the sense of Section 11.5, and V is a nonlinear Volterra operator on the line, i.e. $V(u)(t)$ depends only on the values $\{u(s) : s \leq t\}$, the history of u at time t. The equivalent integral equation is then given by

$$u(t) = \int_{-\infty}^t G(t-\tau)V(u)(\tau)ds, \quad t \in \mathbb{R}. \tag{13.51}$$

By means of this reformulation, based on the theory of reolvents presented in this book, also a qualitative theory for (13.48) and (13.50) can be developed, in particular stability results, saddle point properties, Hopf bifurcation, etc. can be obtained as in the case of (13.46).

There is a already a number of papers available where these ideas have been carried out in various settings, hyperbolic as well as parabolic. Mostly the cases $A(t) \equiv A$, i.e. the resolvent is a C_0-semigroup, and $A(t) \equiv tA$, i.e. $S(t)$ is a cosine family have been treated so far; see e.g. Aizicovici [4], Da Prato and Lunardi [72], Engler [109], [111], Heard [174], [175], [176], Heard and Rankin [177], Lunardi [224], Lunardi and Sinestrari [228], [227], Milota [247], [248], Milota and Petzeltová [250], [249], Petzeltová [268], Sinestrari [306], Tesei [324], Travis and Webb [327], Webb [336], [337], Yamada [344], Yamada and Niikura [345], Yoshida [346]. For the general case, however, much work is left to be done.

13.6 Semigroup Approaches

In this subsection some of the meanwhile numerous different semigroup approaches to evolutionary integral equations will be presented and discussed. Since the work of Hale [160] on functional differential equations in finite dimensions these methods have become quite popular and have been employed by many people; see Gripenberg, London and Staffans [156] for the finite-dimensional case. For evolutionary integral equations they go back to Miller [243], Grimmer and Miller [142], [143], Miller and Wheeler [245], and have been further developed and extended by Chen and Grimmer [40], [41], Desch and Grimmer [82], [83], Desch, Grimmer and Schappacher [86], Desch and Schappacher [92], [93], Di Blasio, Kunisch and Sinestrari [96], [97], Grimmer [139], Grimmer and Schappacher [147], Grimmer and Zeman [148], Kunisch and Schappacher [202], Jeong [186], Nakagiri [256], Navarro [257], Sinestrari [307], Slemrod [308], [309], Staffans [315], and Tanabe [322]; see also Renardy, Hrusa, and Nohel [289], Chapter V. This list does not claim any completeness.

a) To illustrate the basic ideas, let us consider an equation with main part of the form

$$\dot{u}(t) = Au(t) + \int_0^t B(t-s)u(s)ds + f(t), \quad u(0) = x, \quad t \geq 0, \qquad (13.52)$$

in a Banach space X, where A is a closed linear densely defined operator in X, and $B \in L^1_{loc}(\mathbb{R}_+; \mathcal{B}(X_A, X))$; the data should satisfy $x \in X$ and $f \in L^1_{loc}(\mathbb{R}_+; X)$ for the moment. We want to rewrite (13.52) as an abstract Cauchy problem of the form

$$\dot{z}(t) = \mathcal{A}z(t), \quad z(0) = z_0, \quad t \geq 0, \qquad (13.53)$$

in an appropriate Banach space Z. There are many different ways how this can be done. The so-called *forcing function approach* proceeds as follows. Choose a function space $\mathcal{F}(\mathbb{R}_+; X)$ where the forcing functions f one is interested in should belong to, e.g. $\mathcal{F} = C_{ub}$, or $\mathcal{F} = L^p$, etc., and let $Z = X \times \mathcal{F}(\mathbb{R}_+; X)$. For a given strong solution $u(t)$ of (13.52) define

$$z(t) = (u(t), g(t)), \quad g(t)(r) = f(t+r) + \int_0^t B(t+r-s)u(s)ds, \quad t, r \geq 0; \quad (13.54)$$

in particular, $z(0) = z_0 = (u(0), g(0)) = (x, f)$. It is then straight forward to compute that, formally, $z(t)$ satisfies (13.53), with $\mathcal{A}$ defined by

$$\mathcal{A} = \begin{pmatrix} A & \delta_0 \\ B(\cdot) & D \end{pmatrix}, \qquad (13.55)$$

where $\delta_0 f = f(0)$, $B(\cdot)x(r) = B(r)x$, and $Df(r) = f'(r)$. The natural domain of definition for $\mathcal{A}$ is of course

$$\mathcal{D}(\mathcal{A}) = \mathcal{D}(A) \times \mathcal{D}(D),$$

as soon as $B(\cdot)x \in \mathcal{F}(\mathbb{R}_+; X)$ for each $x \in \mathcal{D}(A)$. For the solutions $u(t)$ of (13.52) we have the representation via the variation of parameters formula

$$u(t) = S(t)x + \int_0^t S(t - \tau)f(\tau)d\tau, \quad t \geq 0,$$

where $S(t)$ denotes the resolvent of (13.52). On the other hand, the solution of (13.53) is given by

$$z(t) = e^{\mathcal{A}t}z_0 \quad t \geq 0;$$

this leads to the following formal representation of $e^{\mathcal{A}t}$ in terms of $S(t)$ and $B(t)$.

$$e^{\mathcal{A}t} = \begin{pmatrix} S(t) & [S*](t) \\ \int_0^t B(t - \tau + \cdot)S(\tau)d\tau & T_t + [R(\cdot)*](t) \end{pmatrix}, \quad t \geq 0. \tag{13.56}$$

Here $(T_t f)(r) = f(t+r)$ denotes the translation semigroup, $[S*](t)f = (S*f)(t)$ is a funny notation for the convolution, and $R(r)(t) = \int_0^t B(r + t - \tau)S(\tau)d\tau$. Thus the first component of $z(t) = e^{\mathcal{A}t}z_0$ is nothing else than the variation of parameters formula.

Since $S(t)$ appears in the left upper corner of $e^{\mathcal{A}t}$, well-posedness of (13.53) implies existence of the resolvent $S(t)$ for (13.52), no matter which function space $\mathcal{F}$ has been chosen. Therefore whenever it can be proved that $\mathcal{A}$ generates a C_0-semigroup in Z then well-posedness of (13.52) follows, a very appealing aspect of this approach. The converse of this, however, is in general not true since it is a priori not clear that the entries in the second row of $e^{\mathcal{A}t}$ are well-defined; note that $B(t)$ is only defined on X_A. For the same reason $\mathcal{A}$ need not even be closable, at least without further assumptions, which is clearly a drawback of this method.

In general, it also seems to be very difficult to apply directly the Hille-Yosida theorem on generation of C_0-semigroups, well-posedness of (13.53) is usually obtained via perturbation methods. A typical result requires A to be a generator; then the operator $\mathcal{A}_0$ defined by the diagonal part of $\mathcal{A}$ is again a generator - at least if the translation semigroup is strongly continuous in $\mathcal{F}$. Therefore it is natural to consider the off-diagonal part of $\mathcal{A}$ as a perturbation $\mathcal{B}$ of $\mathcal{A}_0$, and to apply one of the many perturbation results for semigroups. This works e.g. in case $\mathcal{F}(\mathbb{R}_+; X)) \subset C(\mathbb{R}_+; X)$ and $B(\cdot)x \in \mathcal{D}(D)$ for all $x \in \mathcal{D}(A)$.

Another possibility is to consider the case $B \equiv 0$ first; the representation (13.56) shows that then $\mathcal{A}$ is a generator in Z whenever A is a generator in X, the translation semigroup is strongly continuous in $\mathcal{F}(\mathbb{R}_+; X)$, and this space is continuously embedded into $L^1_{loc}(\mathbb{R}_+; X)$. A perturbation result due to Desch and Schappacher [92] then implies that (13.53) is well-posed if $B(\cdot)x \in \widetilde{\mathcal{D}}(D)$, for each $x \in \mathcal{D}(A)$; here $\widetilde{\mathcal{D}}(D)$ denotes the generalized domain of D.

Note, however, that the semigroups obtained via these methods will never be analytic nor compact, this has nothing to do with the corresponding properties of $S(t)$. Therefore maximal regularity results for (13.52) cannot be proved this way. The same remark applies to stability questions; even if $S(t)$ is integrable, the type of $e^{\mathcal{A}t}$ cannot be negative unless the type of $S(t)$ already is; this implies that it is not possible to obtain sharp integrability results for the resolvent like those obtained in Section 10 by means of this approach.

b) Another semigroup formulation which has been developed, in particular for the parabolic case is the *history function approach*. To explain it in some detail, consider the problem

$$\dot{u}(t) = Au(t) + \int_{-\infty}^{t} B(t-s)u(s)ds, \quad t \geq 0, \tag{13.57}$$

$$u(0) = x, \ u(t) = \phi(t), \quad t < 0,$$

where X, A, and B are as before. Fix a function space $\mathcal{F}(\mathbb{R}_-; X)$ serving as the space where the history functions ϕ belong to, and let $Z = X \times \mathcal{F}(\mathbb{R}_-; X)$; the initial value for the Cauchy problem will be $z_0 = (x, \phi)$. If $u(t)$, $t \in \mathbb{R}$, denotes the solutions of (13.57) then $z(t) = (u(t), u(t + \cdot)) \in Z$ should be the solution of (13.53), where $\mathcal{A}$ still has to be determined. For the infinite memory problem (13.57) the variation of parameters formula becomes

$$u(t) = S(t)x + \int_0^t S(t-r) \int_{-\infty}^0 B(r-\tau)\phi(\tau)d\tau dr, \quad t \geq 0;$$

therefore a formal computation shows that $\mathcal{A}$ is given by

$$\mathcal{A} = \begin{pmatrix} A & < B(\cdot), \cdot > \\ 0 & D \end{pmatrix}, \tag{13.58}$$

where $< B(\cdot), \phi > = \int_0^\infty B(s)\phi(-s)ds$. Here, however, the domain of $\mathcal{A}$ is restricted by the requirement $\phi(0) = x$.

Proceeding as before, consider the case $B \equiv 0$ first, and let $\mathcal{A}_0$ denote the corresponding operator in Z; if A is a generator in X and $\mathcal{F}$ is such that the translation semigroup is strongly continuous then $\mathcal{A}_0$ is a generator on $\overline{\mathcal{D}(\mathcal{A}_0)}$, where

$$\mathcal{D}(\mathcal{A}_0) = \{z = (x, \phi) \in \mathcal{D}(A) \times \mathcal{D}(D) : \phi(0) = x\}.$$

In fact, the action of the semigroup is given by

$$e^{\mathcal{A}_0 t}(x, \phi) = (e^{At}x, \begin{cases} e^{A(t+s)}x & \text{for } t+s > 0 \\ \phi(t+s) & \text{for } t+s < 0 \end{cases}). \tag{13.59}$$

But now there is a problem with the perturbation term $< B(\cdot), \phi >$; namely this term is not defined at all unless $B(s)$ is bounded in X, the trivial case. One might think that a way out would be to consider history functions ϕ with values in X_A only; but then the problem with $B \equiv 0$ is not well-posed, as the representation (13.59) shows.

One possibilty to overcome these difficulties is to consider kernels $B(t)$ of the form

$$B(t) = AL(t) + K(t), \ t > 0, \quad \text{with } L(t), K(t) \in \mathcal{B}(X).$$

Then the perturbation term coming from K leads to a bounded perturbation of A_0 -at least under appropriate regularity assumptions- hence causes no problems; so we may assume $K \equiv 0$ for simplicity. But then $\mathcal{A}$ is of the form $\mathcal{A} = \mathcal{A}_0(\mathcal{I}+\mathcal{L})$, where $\mathcal{I}$ denotes the identity in Z, and

$$\mathcal{L}(x,\phi) = (\int_0^\infty L(s)\phi(-s)ds, 0), \quad (x,\phi) \in Z.$$

It is easy to check that $\mathcal{I}+\mathcal{L}$ is invertible, therefore $\mathcal{A}$ is a generator iff $(\mathcal{I}+\mathcal{L})\mathcal{A}_0$ is such; but

$$\mathcal{L}\mathcal{A}_0(x,\phi) = (\int_0^\infty L(s)\phi'(-s)ds, 0) = (L(0)x + \int_0^\infty \dot{L}(s)\phi(-s)ds, 0),$$

for all $(x,\phi) \in \mathcal{D}(\mathcal{A}_0)$, shows that $\mathcal{L}\mathcal{A}_0$ is a bounded perturbation, provided $\dot{L}$ satisfies appropriate regularity assumptions. The domain of $\mathcal{A}$ will then be

$$\mathcal{D}(\mathcal{A}) = \{z = (x,\phi) \in Z : x + \int_0^\infty L(s)\phi(-s)ds \in \mathcal{D}(A), \ \phi \in \mathcal{D}(D), \ \phi(0) = x.\}$$

Observe that strong solutions of (13.53) will only be mild solutions of (13.52), in general, since the first component of $z(t)$ does not necessarily belong to $\mathcal{D}(A)$.

c) Since many people still believe that it is useful to have a semigroup formulation of evolutionary integral equations let us show by means of results proved in this book, how semigroups can be constructed. For this purpose consider the infinite memory problem

$$\dot{u}(t) = Au(t) + \int_{-\infty}^t dB(t-s)u(s)ds, \quad t \geq 0, \tag{13.60}$$

$$u(0) = x, \ u(t) = \phi(t), \quad t < 0,$$

where A generates an analytic C_0-semigroup in X, and $B \in BV(\mathbb{R}_+; \mathcal{B}(X_A, X))$ with $B(0) = B(0+) = 0$. As a phase space we choose e.g.

$$Z = \{(x,\phi) \in X_A \times C^\alpha(\mathbb{R}_+; X_A) : y(x,\phi) \in D_A(\alpha,\infty), \phi(0) = x\},$$

where

$$y(x,\phi) = Ax + \int_0^\infty dB(s)\phi(-s)ds.$$

Then for this situation the maximal regularity result Theorem 7.4 with $S_0(t) = e^{At}$ shows that the solution $u(t)$ of (13.60) belongs to $C^\alpha(-\infty, T]; X_A)$, whenever $(x,\phi) \in Z$. For this one has to observe that $u(t)$ can be written as

$$u = x + 1 * Sy + S * f, \quad \text{for some } f \in C_0^\alpha(\mathbb{R}_+; X),$$

and $1 * Sy \in C_0^\alpha(\mathbb{R}_+; X_A)$ by the compatibility condition. It is not difficult to see that the latter, which may be written as $\dot{u}(t) \in D_A(\alpha,\infty)$, is also preserved, and therefore the map $(x,\phi) \mapsto (u(t), u(t+\cdot))$ is a strongly continuous semigroup in Z.

The maximal regularity results from Sections 7 and 8 yield many other choices of phase spaces Z for which the semigroup can be constructed.

d) Besides the semigroup approaches discussed above there are many other possibilities, especially for concrete partial differential-integral equations like those arising in viscoelasticity, for example. This is of course due to the additional structure such problems enjoy, and which to a large part is abandoned when formulating them abstractly. For semigroups constructed in this spirit the interested reader is referred to the papers of Slemrod [308], [309], Navarro [257], and to the more recent ones of Desch and Grimmer [83], Desch and Miller [90], Leugering [212], and Staffans [315].

13.7 Nonlinear Equations with Accretive Operators
Since the early 1970s Volterra equations of scalar type have been under consideration also in the nonlinear case. More precisely, equations of the form

$$u(t) + \int_0^t a(t-s)Au(s)ds \ni f(t), \quad t \geq 0, \tag{13.61}$$

have been studied in general Banach spaces, where $a \in L^1_{loc}(\mathbb{R}_+)$, $f \in L^1_{loc}(\mathbb{R}_+; X)$, and A is a possibly nonlinear multivalued m-accretive operator in X.

Recall that an operator $A : X \to 2^X$ is called accretive if

$$|x_1 - x_2| \leq |x_1 - x_2 + \mu(y_1 - y_2)|, \quad \text{for all } y_j \in Ax_j, \mu > 0, \tag{13.62}$$

is satisfied; A is called m-accretive if it is accretive and in addition $\mathcal{R}(I + \mu A) = X$ holds for some $\mu > 0$. The latter means that for each $y \in X$ there is $x \in \mathcal{D}(A) = \{x \in X : Ax \neq \emptyset\}$ such that $\mu^{-1}(y - x) \in Ax$. For the theory of accretive and m-accretive operators and nonlinear semigroup theory we refer e.g. to the monograph Deimling [81]; note that 'm-accretive' is called 'hyperaccretive' there.

A function $u \in L^1_{loc}(\mathbb{R}_+; X)$ is called a *strong solution* of (13.61) if $u(t) \in \mathcal{D}(A)$ for a.a. $t \geq 0$, and there is a function $w \in L^1_{loc}(\mathbb{R}_+; X)$ with $w(t) \in Au(t)$ a.e., and $u + a * w = f$ a.e. on $\mathbb{R}_+$. $u \in L^1_{loc}(\mathbb{R}_+; X)$ is called a *mild solution* of (13.61) if there are functions $f_n \to f$ in $L^1_{loc}(\mathbb{R}_+; X)$, and strong solutions u_n of (13.61) with f replaced by f_n, such that $u_n \to u$ in $L^1_{loc}(\mathbb{R}_+; X)$.

Beginning with the papers McCamy [230], Barbu [17], and Londen [218], (13.61) for nonlinear A has been studied in many different settings under various assumptions by means of different techniques. The main emphasis has been on existence of solutions and well-posedness questions, and on the asymptotic behaviour of the solutions. In the papers Barbu [17] and Londen [218] it is assumed that X is a Hilbert space, and that $A = \partial\phi$ is the subdifferential of a lower semi-continuous (l.s.c.) proper convex functional; see e.g. Deimling [81] for the definition and properties of such functionals. The nonlinear operator A was then replaced by its Yosida approximation A_μ defined by

$$A_\mu = (I - J_\mu)/\mu, \quad J_\mu = (I + A)^{-1}, \quad \mu > 0, \tag{13.63}$$

which is globally Lipschitz; hence there are unique solutions u_μ of the approximating equations

$$u_\mu(t) + \int_0^t a(t-s)A_\mu u_\mu(s)ds \ni f(t), \quad t \geq 0. \tag{13.64}$$

For $0 \neq a \in BV(\mathbb{R}_+)$ 1-monotone and of positive type, Barbu obtained bounds on u_μ and $A_\mu u_\mu$ which allowed for passing to the limit $\mu \to 0+$, thus obtaining strong solutions, and some further asymptotic properties in case $a(\infty) > 0$. Instead, Londen assumed $a, \dot{a} \in BV_{loc}(\mathbb{R}_+)$ and $a(0) > 0$ to obtain essentially the same conclusion; Gripenberg [150] extended these results to general maximal monotone operators A in a Hilbert space X; see also Londen [219]. For a further development of the asymptotic theory of (13.61) in this setting we refer to Clément, McCamy, and Nohel [48].

An essential step forward was made in the paper by Crandall and Nohel [62]. They consider the perturbed nonlinear Cauchy problem

$$\dot{u} + Au \ni G(u), \quad u(0) = x, \quad t \in J = [0, T], \tag{13.65}$$

where A is m-accretive in the Banach space X and $G : C(J; \overline{\mathcal{D}(A)}) \to L^1(J; X)$ is a functional operator subject to the Lipschitz estimate

$$|G(u) - G(v)|_{L^1((0,t);X)} \leq \int_0^t \gamma(s)|u - v|_{L^\infty((0,s);X)}ds, \quad t \in J.$$

They show the existence, uniqueness and continuous dependence of (continuous) *integral solutions* for (13.65), which are even strong solutions if $x \in \mathcal{D}(A)$, X is reflexive, and $\mathcal{R}(G) \subset BV(J; X)$. Their method is basically a perturbation argument, it relies on the Crandall-Liggett theorem of nonlinear semigroup theory (see e.g. Deimling [81]) and on the contraction mapping principle. Observe that (13.61) can be reduced to (13.65) by inverting the convolution with a, provided there are $k_0 > 0$, $k_1 \in BV(J)$ such that $k_0 a + k_1 * a = 1$, which means that $a, \dot{a} \in BV(J)$, $a(0+) > 0$, i.e. Londen's condition. Even more, the approximate solutions u_μ converge uniformly in X to u as $\mu \to 0+$. Observe that all results mentioned so far require $0 < a(0+) < \infty$.

Clément and Nohel [50] discovered that kernels $a(t)$ which later on have been called completely positive play an important role also in the nonlinear case. Specifically they proved the positivity (w.r.t. a cone P) of mild solutions of (13.61) which are obtained as weak limits of the u_μ, whenever $(I + \mu A)^{-1}P \subset P$ for each $\mu > 0$, $a(t)$ is a completely positive function, and f is of the form $f = x + a * g$, with $x \in P$ and $g(t) \in P$ for a.a. t. In Clément [43] the important inequality

$$|u_1(t) - u_2(t)| \leq |x_1 - x_2| + \int_0^t a(t-s)|g_1(s) - g_2(s)|ds, \quad t > 0, \tag{13.66}$$

is proved for mild solutions u_j of (13.61) with f replaced by $f_j = x_j + a * g_j$, which are limits of the approximate solutions $u_{\mu,j}$. This estimate is the basis on the results on asymptotic behaviour of mild solutions obtained in Clément [43], and Clément and Nohel [51]. The asymptotic theory was carried on by Aizicovici [5], Baillon and Clément [15], N. Hirano [181], N. Kato, K. Kobayasi, and I. Miyadera [192], and N. Kato [190], [191]. See also Clément and Mitidieri [49] for other qualitative properties of (13.61) like invariance, monotonicity, and comparison principles, whenever a is completely positive and A m-accretive.

Let $a(t)$ be completely positive and $k(t)$ the creep function associated with a, as in Section 4. If f is of the form $f = x + a * g$, $x \in X$ and $g \in L^1_{loc}(\mathbb{R}_+; X)$, then (13.61) is equivalent to

$$\frac{d}{dt}[k_0 u(t) + \int_0^t k_1(t-s)u(s)ds] + k_\infty u(t) + Au(t) \ni (k_1(t) + k_\infty)x + g(t),$$

$$k_0 u(0) + (k_1 * u)(0) = k_0 x, \quad t \in \mathbb{R}_+. \tag{13.67}$$

Since the linear Volterra operator L defined by

$$(Lu)(t) = \frac{d}{dt}[k_0 u(t) + \int_0^t k_1(t-s)u(s)ds] + k_\infty u(t), \quad t \in \mathbb{R}_+,$$

is m-accretive in $L^p(\mathbb{R}_+; X)$ (see also Clément and Mitidieri [49] and Clément and Prüss [54]), the idea in Gripenberg [153] was to approximate L by means of its Yosida approximation, rather than A. This has the advantage that much more structural information is available for L_μ than for A_μ, since L is linear, m-accretive, and the semigroups generated by $-L$ are positive in any space $L^p(J; X)$ (w.r.t. the standard cone generated by a cone P in X), where $J = [0, T]$ or $J = \mathbb{R}_+$. The paper Gripenberg [153] contains the most comprehensive treatment of (13.67), hence also of (13.61). He does not need to assume $k_0 > 0$ nor is $k_1(0+) < \infty$ required, in particular $a(0+) = \infty$ is admitted. By means of the inequality (13.66) the perturbation argument involving the contraction mapping principle allows also nonlinear functional right hand sides $G(u)(t)$ instead of $g(t)$ in (13.67). Thus these results extend those of Crandall and Nohel [62] considerably.

Another recent approach taken in Egberts [108] (see also Clément and Egberts [46]) follows the route taken in Section 8 for the linear case, i.e. uses the sum approach writing (13.67) in abstract form as $Lu + Au = h$. This approach yields better results if X is a Hilbert space and either $A = \partial\phi$ and $k(t)$ is a creep function, or if A is m-accretive and $k_0 = 0$, $|\arg k_1(\lambda)| \leq \theta \leq \pi/2$ on $\mathbb{C}_+$, i.e. if $a(t)$ is θ-sectorial; cp. Section 8.

There are many papers that consider equations which generalize (13.61) in one way or another, and it would be too much here to list all references. We only mention the following.

In the papers of Barbu and Malik [19], Crandall, Londen, and Nohel [61], Engler [110], Londen and Nohel [220], Staffans [312], the problem

$$\dot{u}(t) + Bu(t) + \int_0^t a(t-s)Au(s)ds \ni f(t), \quad u(0) = x, \quad t \geq 0, \tag{13.68}$$

is studied in a Hilbert space X, where A and B are both subdifferentials of proper l.s.c. convex functionals which are related by an angle condition and subject to some compactness assumptions.

Perturbation results for (13.61) of various kinds have been obtained in Aizicovici [3], [5], [6], Aizicovici, Londen, and Reich [7], Gripenberg [151], [149], Kiffe [195], Mitidieri [251], Mitidieri and Tosques [252].

The kernel $a(t)$ is allowed to take values in $\mathcal{B}(X)$, X a Hilbert space, in the papers McCamy [230], McCamy and Mizel [231], McCamy and Wong [232], and Staffans [312], [313].

For references concerning concrete partial integro-differential problems of the forms (13.61) or (13.68) with A and B nonlinear partial differential operators and X being an appropriate function space we refer to the monograph Renardy, Hrusa, and Nohel [289].

Bibliography

[1] P. Acquistapace. Existence and maximal time regularity for linear parabolic integrodifferential equations. *J. Int. Equations*, 10:5–43, 1985.

[2] S. Adali. Existence and asymptotic stability of solutions of an abstract integrodifferential equation with applications to viscoelasticity. *SIAM J. Math. Anal.*, 9:185–206, 1978.

[3] S. Aizicovici. On an abstract Volterra equation. In S.O. Londen and O.J. Staffans, editors, *Volterra Equations*, pages 1–8, Berlin, 1979. Springer Verlag.

[4] S. Aizicovici. On a semilinear Volterra integrodifferential equation. *Israel J. Math.*, 36:273–284, 1980.

[5] S. Aizicovici. On the asymptotic behaviour of solutions of Volterra equations in Hilbert space. *Nonl. Anal.: M.T.A.*, 7:271–278, 1983.

[6] S. Aizicovivi. Asymptotic properties of solutions of time-dependent Volterra integral equations. *J. Math. Anal. Appl.*, 131:421–440, 1988.

[7] S. Aizicovivi, S.O. Londen, and S. Reich. Asymptotic behavior of solutions to a class of nonlinear Volterra equations. *Diff. Int. Eqns*, 3:813–825, 1990.

[8] H. Amann. Dual semigroups and second order linear elliptic boundary value problems. *Israel J. Math.*, 45:225–254, 1983.

[9] L. Amerio and G. Prouse. *Almost-Periodic Functions and Functional Equations*. van Nostrand Reinhold Co., New York, 1971.

[10] W. Arendt. Vector Laplace transforms and Cauchy problems. *Israel J. Math.*, 59:327–352, 1987.

[11] W. Arendt and C.J.K. Batty. Tauberian theorems and stability of one-parameter semigroups. *Trans. Amer. Math. Soc.*, 306:837–852, 1988.

[12] W. Arendt and C.J.K. Batty. A complex Tauberian theorem and mean ergodic semigroups. To appear, 1992.

[13] W. Arendt and H. Kellerman. Integrated solutions of Volterra integrodifferential equations and applications. In G. Da Prato and M. Iannelli, editors, *Volterra Integrodifferential Equations in Banach Spaces and Applications*, pages 21–51, Harlow, Essex, 1989. Longman Sci. Tech.

[14] W. Arendt and J. Prüss. Vector-valued Tauberian theorems and asymptotic behavior of linear Volterra equations. *SIAM J. Math. Anal.*, 23:412–448, 1992.

[15] J.B. Baillon and Ph. Clément. Ergodic theorems for non-linear Volterra equations in Hilbert space. *Nonlin. Anal.: TMA*, 5:789–801, 1981.

[16] J.B. Baillon and Ph. Clément. Examples of unbounded imaginary powers of operators. *J. Funct. Anal.*, 100:419–434, 1991.

[17] V. Barbu. Nonlinear Volterra equations in a Hilbert space. *SIAM J. Math. Anal.*, 6:728–741, 1975.

[18] V. Barbu, G. Da Prato, and M. Iannelli. Stability results for some integrodifferential equations of hyperbolic type in Hilbert space. *J. Int. Equations*, 2:93–110, 1980.

[19] V. Barbu and M.A. Malik. Semilinear integrodifferential equations in Hilbert space. *J. Math. Anal. Appl.*, 67:452–475, 1979.

[20] C.J.K. Batty. Tauberian theorems for the Laplace-Stieltjes transform. *Trans. Amer. Math. Soc.*, 322:783–804, 1990.

[21] C. Berg and G. Forst. *Potential Theory of Locally Compact Groups*, volume 87 of *Ergebn. Math. Grenzgeb.* Springer Verlag, Berlin, 1975.

[22] J. Bergh and J. Löfström. *Interpolation Spaces. An Introduction*, volume 223 of *Grundl. Math. Wiss.* Springer Verlag, Berlin, 1976.

[23] R.B. Bird, R.C. Armstrong, and O. Hassager. *Dynamics of Polymeric Liquids, Vol. 1: Fluid Mechanics*. John Wiley and Sons, New York, 1977.

[24] D.R. Bland. *The Theory of Linear Viscoelasticity*, volume 10 of *Pure and Applied Mathematics*. Pergamon Press, New York, 1960.

[25] F. Bloom. *Ill-posed Problems for Integrodifferential Equations in Mechanics and Electromagnetic Theory*, volume 3 of *Studies in Applied Mathematics*. SIAM, Philadelphia, 1981.

[26] S. Bochner. *Harmonic Analysis and the Theory of Probability*. Univ. of California Press, Berkley, California, 1955.

[27] H. Bohr. *Fastperiodische Funktionen*. Springer Verlag, Berlin, 1932.

[28] L. Boltzmann. Zur Theorie der elastischen Nachwirkung. *Ann. Phys. Chem.*, 7:624–654, 1876.

[29] W. Borchers and H. Sohr. On the semigroup of the Stokes operator in exterior domains. *Math. Z.*, 196:415–425, 1987.

[30] R. Bouc, G. Geymonat, M. Jean, and B. Nayroles. Solution périodique du problème quasi-statique d'un solide viscoélastique à coefficients périodique. *J. Mécanique*, 14:609–637, 1975.

[31] J. Bourgain. Some remarks on Banach spaces in which martingale difference sequences are unconditional. *Ark. Mat.*, 22:163–168, 1983.

[32] J. Bourgain. Vector-valued singular integrals and the $H^1 - BMO$ duality. In D. Burkholder, editor, *Probability Theory and Harmonic Analysis*, pages 1–19, New York, 1986. Marcel Dekker.

[33] K. Boyadzhiev and R. deLaubenfels. H^∞-functional calculus for perturbations of generators of holomorphic semigroups. *Housten J. Math.*, 17:131–147, 1991.

[34] K. Boyadzhiev and R. deLaubenfels. Semigroups and resolvents of bounded variation, imaginary powers and H^∞ functional calculus. *Semigroup Forum*, 45:372–384, 1992.

[35] D.L. Burkholder. Martingales and Fourier analysis in Banach spaces. In G. Letta and M. Pratelli, editors, *Probability and Analysis*, volume 1206 of *Lect. Notes Math.*, pages 61–108, Berlin, 1986. Springer Verlag.

[36] P.L. Butzer and H. Berens. *Semi-Groups of Operators and Approximation*, volume 145 of *Grundlehren math. Wiss.* Springer Verlag, Berlin, 1967.

[37] T. Carleman. *L'intégrale de Fourier et questions qui s'y rattachent*. Almqvist and Wiksell, Uppsala, 1944.

[38] R.W. Carr and K.B. Hannsgen. A nonhomogeneous integrodifferential equation in Hilbert space. *SIAM J. Math. Anal.*, 10:961–984, 1979.

[39] R.W. Carr and K.B. Hannsgen. Resolvent formulas for a Volterra equation in Hilbert space. *SIAM J. Math. Anal.*, 13:453–483, 1982.

[40] G. Chen and R. Grimmer. Semigroups and integral equations. *J. Int. Equations*, 2:133–154, 1980.

[41] G. Chen and R. Grimmer. Integral equations as evolution equations. *J. Diff. Equations*, 45:53–74, 1982.

[42] R.M. Christensen. *Theory of Viscoelasticity*. Acad. Press, New York, second edition, 1982.

[43] Ph. Clément. On abstract Volterra equations with kernels having a positive resolvent. *Israel J. Math.*, 36:193–200, 1980.

[44] Ph. Clément. On L^p-L^q coerciveness for a class of integrodifferential equations on the line. Winter School of Mathematics, Voronezh, 1991.

[45] Ph. Clément. L^p-L^q coerciveness and applications to nonlinear integrodifferential equations. In *Proc. Int. Symp. Nonlinear Problems in Engineering and Science*, Beijing, 1992.

[46] Ph. Clément and P. Egberts. On the sum of maximal monotone operators. *Diff. Integral Eqns.*, 3:1127–1138, 1990.

[47] Ph. Clément, H.J.A.M. Heijmans, S. Angenent, C.J. van Duijn, and B. de Pagter. *One-Parameter semigroups*, volume 5 of *CWI Monographs*. North-Holland, Amsterdam, 1987.

[48] Ph. Clément, R.C. MacCamy, and J.A. Nohel. Asymptotic properties of solutions of nonlinear abstract Volterra equations. *J. Int. Equations*, 3:185–216, 1981.

[49] Ph. Clément and E. Mitidieri. Qualitative properties of solutions of Volterra equations in Banach spaces. *Israel J. Math.*, 64:1–24, 1988.

[50] Ph. Clément and J.A. Nohel. Abstract linear and nonlinear Volterra equations preserving positivity. *SIAM J. Math. Anal.*, 10:365–388, 1979.

[51] Ph. Clément and J.A. Nohel. Asymptotic behavior of solutions of nonlinear Volterra equations with completely positive kernels. *SIAM J. Math. Anal.*, 12:514–535, 1981.

[52] Ph. Clément and G. Da Prato. Existence and regularity results for an integral equation with infinite delay in a Banach space. *Int. Eqns. Operator Theory*, 11:480–500, 1988.

[53] Ph. Clément and J. Prüss. On second order differential equations in Hilbert space. *Boll. U.M.I.*, 3-B:623–638, 1989.

[54] Ph. Clément and J. Prüss. Completely positive measures and Feller semigroups. *Math. Ann.*, 287:73–105, 1990.

[55] Ph. Clément and J. Prüss. Global existence for a semilinear parabolic Volterra equation. *Math. Z.*, 209:17–26, 1992.

[56] R.R. Coifmann and G. Weiss. *Transference Methods in Analysis*, volume 31 of *Conf. Board Math. Sci., Reg. Conf. Series Math.* Amer. Math. Soc., Providence, Rhode Island, 1977.

[57] B.D. Coleman. Thermodynamics of materials with memory. *Arch. Rat. Mech. Anal.*, 17:1–46, 1964.

[58] B.D. Coleman and M.E. Gurtin. Equipresence and constitutive equations for rigid heat conductors. *Z. Angew. Math. Phys.*, 18:199–208, 1967.

[59] J.B. Conway. *Functions of One Complex Variable*, volume 11 of *Graduate Texts in Mathematics*. Springer Verlag, Berlin, second edition, 1978.

[60] C. Corduneanu. *Integral Equations and Stability of Feedback Systems*, volume 104 of *Mathematics in Science and Engineering*. Acad. Press, 1973.

[61] M.G. Crandall, S.O. Londen, and J.A. Nohel. An abstract nonlinear Volterra integrodifferential equation. *J. Math. Anal. Appl.*, 64:701–735, 1978.

[62] M.G. Crandall and J.A. Nohel. An abstract functional differential equation and a related nonlinear Volterra equation. *Israel J. Math.*, 29:313–328, 1978.

[63] G. Da Prato and E. Giusti. Una caratterizzazione dei generatori di funzioni coseno astratte. *Boll. U.M.I.*, 3:1–6, 1967.

[64] G. Da Prato and P. Grisvard. Sommes d'opérateurs linéaires et équations différentielles opérationelles. *J. Math. Pures Appl.*, 54:305–387, 1975.

[65] G. Da Prato and M. Iannelli. Linear abstract integrodifferential equations of hyperbolic type in Hilbert spaces. *Rend. Sem. Mat. Padova*, 62:191–206, 1980.

[66] G. Da Prato and M. Iannelli. Linear integrodifferential equations in Banach spaces. *Rend. Sem. Mat. Padova*, 62:207–219, 1980.

[67] G. Da Prato and M. Iannelli. Distribution resolvents for Volterra equations in Banach space. *J. Int. Equations*, 6:93–103, 1984.

[68] G. Da Prato and M. Iannelli. Existence and regularity for a class of integrodifferential equations of parabolic type. *J. Math. Anal. Appl.*, 112:36–55, 1985.

[69] G. Da Prato, M. Iannelli, and E. Sinestrari. Temporal regularity for a class of integrodifferential equations in Banach spaces. *Boll. U.M.I.*, 2:171–185, 1983.

[70] G. Da Prato, M. Iannelli, and E. Sinestrari. Regularity of solutions of a class of linear integrodifferential equations in Banach spaces. *J. Int. Equations*, 8:27–40, 1985.

[71] G. Da Prato and A. Lunardi. Periodic solutions of linear integrodifferential equations with infinite delay in Banach spaces. In A. Favini and E. Obrecht, editors, *Differential Equations in Banach Spaces*, pages 49–60, Berlin, 1986. Springer Verlag.

[72] G. Da Prato and A. Lunardi. Hopf bifurcation for nonlinear integrodifferential equations in Banach spaces with infinite delay. *Indiana Univ. Math. J.*, 36, 1987.

[73] G. Da Prato and A. Lunardi. Solvability on the real line of a class of linear Volterra integrodifferential equations of parabolic type. *Ann. Mat. Pura Appl.*, 55:67–118, 1988.

[74] G. Da Prato and A. Lunardi. Stabilizability of integrodifferential parabolic equations. *J. Int. Eqns. Appl.*, 2:281–304, 1990.

[75] C.M. Dafermos. An abstract Volterra equation with applications to linear viscoelasticity. *J. Diff. Equations*, 7:554–569, 1970.

[76] C.M. Dafermos. Asymptotic stability in viscoelasticity. *Arch. Rat. Mech. Anal.*, 37:297–308, 1970.

[77] E.B. Davies. *One-Parameter Semigroups*, volume 15 of *London Math. Soc. Monographs*. Acad. Press, New York, 1980.

[78] P.L. Davis. On the hyperbolicity of the equations of linear theory of heat conduction for materials with memory. *SIAM J. Math. Anal.*, 30:75–80, 1976.

[79] P.L. Davis. On the linear theory of heat conduction for materials with memory. *SIAM J. Math. Anal.*, 9:49–53, 1978.

[80] B. de Pagter. A characterization of sun-reflexivity. *Math. Ann.*, 283:511–518, 1989.

[81] K. Deimling. *Nonlinear Functional Analysis*. Springer Verlag, Berlin, 1985.

[82] W. Desch and R. Grimmer. Initial-boundary value problems for integrodifferential equations. *J. Int. Equations*, 10:73–97, 1985.

[83] W. Desch and R. Grimmer. Singular relaxation moduli and smoothing in three-dimensional viscoelasticity. *Trans. Amer. Math. Soc.*, 314:381–404, 1989.

[84] W. Desch and R. Grimmer. Smoothing properties of linear Volterra integrodifferential equations. *SIAM J. Math. Anal.*, 20:116–132, 1989.

[85] W. Desch and R. Grimmer. The spectrum of a viscoelastic Timoshenko beam. Preprint, 1992.

[86] W. Desch, R. Grimmer, and W. Schappacher. Well-posedness and wave propagation for a class of integrodifferential equations in Banach space. *J. Diff. Equations*, 74:391–411, 1988.

[87] W. Desch, K.B. Hannsgen, Y. Renardy, and R.L. Wheeler. Boundary stabilization of an Euler-Bernoulli beam with viscoelastic damping. In *Decision and Control*, pages 1792–1795. IEEE, 1987.

[88] W. Desch, K.B. Hannsgen, and R.L. Wheeler. Feedback stabiliztion of a viscoelastic rod. *Adv. Comp. Control*, pages 331–337, 1989.

[89] W. Desch and R.K. Miller. Exponential stabilization of Volterra integrodifferential equations in Hilbert space. *J. Diff. Equations*, 70:366–389, 1987.

[90] W. Desch and R.K. Miller. Exponential stabilization of Volterra integral equations with singular kernels. *J. Integral Eqns. Appl.*, 1:397–433, 1988.

[91] W. Desch and J. Prüss. Counterexamples for abstract linear Volterra equations. *J. Integral Eqns. Appl.*, 1993. to appear.

[92] W. Desch and W. Schappacher. On relatively bounded perturbations of linear C_0-semigroups. *Ann. Sc. Norm. Sup. Pisa*, 11:327–341, 1984.

[93] W. Desch and W. Schappacher. A semigroup approach to integrodifferential equations in Banach spaces. *J. Int. Equations*, 10:95–110, 1985.

[94] W. Desch and R.L. Wheeler. Destabilization due to delay in one-dimensional feedback. In *Control and Estimation of Distributed Parameter Systems*, Basel, 1989. Birkhäuser Verlag.

[95] G. Di Blasio. Nonautonomous integrodifferential equations in L^p-spaces. *J. Integral Eqns.*, 10:111–122, 1985.

[96] G. Di Blasio, K. Kunisch, and E. Sinestrari. L^2-regularity for parabolic partial integrodifferential equations with delay in the highest order derivatives. *J. Math. Anal. Appl.*, 102:38–57, 1984.

[97] G. Di Blasio, K. Kunisch, and E. Sinestrari. Stability for abstract linear functional differential equations. *Israel J. Math.*, 50:231–263, 1985.

[98] J. Diestel. *Geometry of Banach Spaces*, volume 485 of *Lect. Notes Math.* Springer Verlag, Berlin, 1975.

[99] J. Diestel and J.J. Uhl, Jr. *Vector Measures*, volume 15 of *Mathematical Surveys*. Amer. Math. Soc., Providence, Rhode Island, 1977.

[100] G. Doetsch. *Handbuch der Laplace-Transformation*. Birkhäuser Verlag, Basel, 1971.

[101] G. Dore and A. Venni. On the closedness of the sum of two closed operators. *Math. Z.*, 196:189–201, 1987.

[102] G. Dore and A. Venni. Some results about complex powers of closed operators. *J. Math. Anal. Appl.*, 149:124–136, 1990.

[103] J.N. Dunford and J.T. Schwartz. *Linear Operators*, volume 7 of *Pure Appl. Math.* Interscience Inc., New York, 1957.

[104] X. T. Duong. H_∞-functional calculus for second order elliptic partial differential operators on L^p-spaces. In I. Doust, B. Jefferies, C. Li, and A. McIntosh, editors, *Operators in Analysis*, pages 91–102, Sydney, 1989. Australian National University.

[105] X. T. Duong. H_∞-functional calculus of elliptic operators with C^∞-coefficients on L^p-spaces of smooth domains. *J. Austral. Math. Soc.*, 48:113–123, 1990.

[106] P.L. Duren. *Theory of H^p Spaces*, volume 38 of *Pure Appl. Math.* Acad. Press, New York, 1970.

[107] W.F. Eberlein. Abstract ergodic theorems and weak almost periodicity. *Trans. Amer. Math. Soc.*, 67:217–240, 1949.

[108] P. Egberts. *On the sum of accretive operators*. PhD thesis, Tech. Univ. Delft, 1992.

[109] H. Engler. Functional differential equations in Banach spaces: Growth and decay of solutions. *J. Reine Angewandte Math.*, 322:53–73, 1981.

[110] H. Engler. A version of the chain rule and integrodifferential equations in Hilbert spaces. *SIAM J. Math. Anal.*, 13:801–810, 1982.

[111] H. Engler. Stabilization of solutions for a class of parabolic integro-differential equations. *Nonl. Anal.: T.M.A.*, 8:1337–1371, 1984.

[112] H. Engler. Regularity properties of solutions of linear integrodifferential equations with singular kernels. *J. Integral Eqns. Appl.*, 2:403–420, 1990.

[113] H.O. Fattorini. *The Cauchy Problem*, volume 18 of *Encyclopedia of Mathematics*. Addison Wesley, London, 1983.

[114] H.O. Fattorini. *Second Order Linear Differential Equations in Banach Spaces*, volume 108 of *Mathematics Studies*. North Holland, Amsterdam, 1985.

[115] W. Feller. *An Introduction to Probability Theory and Its Applications, Vol. 1 and Vol. 2*. John Wiley and Sons, New York, second edition, 1970.

[116] J.M. Finn and L.T. Wheeler. Wave propagation aspects of the generalized theory of heat conduction. *Z. Angew. Math. Phys.*, 23:927–940, 1972.

[117] W. Flügge. *Viscoelasticity*. Springer Verlag, Berlin, second edition, 1975.

[118] C. Foias and B. Sz.-Nagy. *Harmonic Analysis of Operators on Hilbert Space*. North Holland, Amsterdam, 1970.

[119] M. Fréchet. Les fonctions asymptotiquement presque-périodiques. *Rev. Sci.*, 79:341–354, 1941.

[120] A. Friedman. On integral equations of Volterra type. *J. Anal. Math.*, 11:381–413, 1963.

[121] A. Friedman. Monotonicity of solutions of Volterra integral equations in Banach space. *Trans. Amer. Math. Soc.*, 198:129–148, 1969.

[122] A. Friedman. *Partial Differential Equations*. Holt, Rinehart and Winston, New York, 1969.

[123] A. Friedman and M. Shinbrot. Volterra integral equations in Banach spaces. *Trans. Amer. Math. Soc.*, 126:131–179, 1967.

[124] Y. Fujita. Integrodifferential equations which interpolates the heat equation and the wave equation. *Osaka J. Math.*, 27:309–321, 1990.

[125] D. Fujiwara. Fractional powers of second order elliptic operators. *J. Math. Soc. Japan*, 21:481–522, 1969.

[126] D. Fujiwara and H. Morimoto. An L_r-theorem of the Helmholtz decomposition of vector fields. *J. Fac. Sci. Univ. Tokyo*, 24:685–700, 1977.

[127] L. Gearhart. Spectral theory for contraction semigroups on Hilbert spaces. *Trans. Amer. Math. Soc.*, 236:385–394, 1978.

[128] I.M. Gel'fand, D.A. Raikov, and G.E. Shilov. *Commutative Normed Rings*. Chelsea, New York, 1964.

[129] Y. Giga. Analyticity of the semigroup generated by the Stokes operator in L_r-spaces. *Math. Z.*, 178:297–329, 1981.

[130] Y. Giga. Domains of fractional powers of the Stokes operator in L_r-spaces. *Arch. Rat. Mech. Anal.*, 89:251–265, 1985.

[131] Y. Giga and H. Sohr. On the Stokes operator in exterior domains. *J. Fac. Sci. Univ. Tokyo*, 36:103–130, 1989.

[132] Y. Giga and H. Sohr. Abstract L^p estimates for the Cauchy problem with applications to the Navier-Stokes equation in exterior domains. *J. Funct. Anal.*, 102:72–94, 1991.

[133] J. Goldstein. *Semigroups of Linear Operators and Applications*. Oxford Math. Monographs. Clarendon Press, Oxford, 1985.

[134] H. Grabmüller. *Über die Lösbarkeit einer Integrodifferentialgleichung aus der Theorie der Wärmeleitung*, volume 10, pages 117–137. Bibliographisches Institut, Mannheim, 1973.

[135] I.S. Grant and W.R. Phillips. *Electromagnetism*. John Wiley and Sons, New York, 1975.

[136] G. Greiner. A short proof of Gearhart's theorem. *Semesterberichte Funktionalanalysis, Univ. Tübingen*, 16:89–92, 1989.

[137] G. Greiner and M. Schwarz. Weak spectral mapping theorems for functional differential equations. *J. Diff. Equations*, 94:205–216, 1989.

[138] G. Greiner, J. Voigt, and M. Wolff. On the spectral bound of the generator of a semigroup of positive operators. *J. Operator Theory*, 5:245–256, 1981.

[139] R. Grimmer. Resolvent operators for integral equations in a Banach space. *Trans. Amer. Math. Soc.*, 273:333–349, 1982.

[140] R. Grimmer and F. Kappel. Series expansions for resolvents of Volterra integrodifferential equations in Banach spaces. *SIAM J. Math. Anal.*, 15:595–604, 1984.

[141] R. Grimmer, R. Lenczewski, and W. Schappacher. Well-posedness of hyperbolic equations with delay in the boundary conditions. In Ph. Clément, S. Invernizzi, E. Mitidieri, and I. Vrabie, editors, *Trends in Semigroup Theory and Applications*, pages 215–228. Marcel Dekker, 1989.

[142] R. Grimmer and R.K. Miller. Existence, uniqueness and continuity for integral equations in a Banach space. *J. Math. Anal. Appl.*, 57:429–447, 1977.

[143] R. Grimmer and R.K. Miller. Well-posedness of Volterra integral equations in Hilbert space. *J. Int. Equations*, 1:201–216, 1979.

[144] R. Grimmer and A.J. Pritchard. Analytic resolvent operators for integral equations in Banach spaces. *J. Diff. Equations*, 50:234–259, 1983.

[145] R. Grimmer and J. Prüss. On linear Volterra equations in Banach spaces. *J. Comp. Appl. Math.*, 11:189–205, 1985.

[146] R. Grimmer and W. Schappacher. Weak solutions of integrodifferential equations and applications. *J. Int. Equations*, 6:205–229, 1984.

[147] R. Grimmer and W. Schappacher. Integrodifferential equations in Banach space with infinite memory. In G. Da Prato and M. Iannelli, editors, *Volterra Integrodifferential Equations in Banach Spaces and Applications*, pages 167–176. Longman Sci. Tech., 1989.

[148] R. Grimmer and M. Zeman. Nonlinear Volterra equations in a Banach space. *Israel J. Math.*, 42:162–176, 1982.

[149] G. Gripenberg. An abstract nonlinear Volterra equation. *Israel J. Math.*, 34:198–212, 1978.

[150] G. Gripenberg. An existence result for nonlinear Volterra integral equations in a Hilbert space. *SIAM J. Math. Anal.*, 9:793–805, 1978.

[151] G. Gripenberg. On some integral and integro-differential equations in a Hilbert space. *Ann. Mat. Pura Appl.*, 118:181–198, 1978.

[152] G. Gripenberg. On Volterra equations of the first kind. *Int. Eqns. Operator Theory*, 3:473–488, 1980.

[153] G. Gripenberg. Volterra integro-differential equations with accretive nonlinearity. *J. Diff. Equations*, 60:57–79, 1985.

[154] G. Gripenberg. Asymptotic behaviour of resolvents of abstract Volterra equations. *J. Math. Anal. Appl.*, 122:427–438, 1987.

[155] G. Gripenberg. Stability of Volterra equations with measure kernels in Banach spaces. *J. Math. Anal. Appl.*, 1992. Preprint.

[156] G. Gripenberg, S.O. Londen, and O.J. Staffans. *Volterra Integral and Functional Equations*. Cambridge Univ. Press, Cambridge, 1990.

[157] M.E. Gurtin and A.C. Pipkin. A general theory of heat conduction with finite wave speed. *Arch. Rat. Mech. Anal.*, 31:113–126, 1968.

[158] M.E. Gurtin. On the thermodynamics of materials with memory. *Arch. Rat. Mech. Anal.*, 28:40–50, 1968.

[159] M.E. Gurtin. *An Introduction to Continuum Mechanics*. Acad. Press, New York, 1981.

[160] J. Hale. *Functional Differential Equations*, volume 3 of *Appl. Math. Sci.* Springer Verlag, Berlin, 1971.

[161] K.B. Hannsgen. Indirect Abelian theorems and a linear Volterra equation. *Trans. Amer. Math. Soc.*, 142:539–555, 1969.

[162] K.B. Hannsgen. A linear Volterra equation in Hilbert space. *SIAM J. Math. Anal.*, 5:927–940, 1974.

[163] K.B. Hannsgen. A Volterra equation in Hilbert space. *SIAM J. Math. Anal.*, 5:412–416, 1974.

[164] K.B. Hannsgen. Uniform L^1-behavior for an integrodifferential equation with parameter. *SIAM J. Math. Anal.*, 8:626–639, 1977.

[165] K.B. Hannsgen. A linear integrodifferential equation for viscoelastic rods and plates. *Quarterly Appl. Math.*, pages 75–83, 1983.

[166] K.B. Hannsgen, Y. Renardy, and R.L. Wheeler. Effectiveness and robustness with respect to time delays of boundary feedback stabilization in one-dimensional viscoelsticity. *SIAM J. Control and Optimization*, 26:1200–1234, 1988.

[167] K.B. Hannsgen and R.L. Wheeler. Time delays and boundary feedback stabilization in one-dimensional viscoelasticity. Proceedings of Vorau, Austria conference, July 1986.

[168] K.B. Hannsgen and R.L. Wheeler. Behavior of the solution of a Volterra equation as a parameter tends to infinity. *J. Int. Equations*, 7:229–237, 1984.

[169] K.B. Hannsgen and R.L. Wheeler. Uniform L^1 behavior in classes of integrodifferential equations with completely monotonic kernels. *SIAM J. Math. Anal.*, 15:579–594, 1984.

[170] K.B. Hannsgen and R.L. Wheeler. Existence and decay estimates for boundary feedback stabilization of torsional vibrations in a viscoelastic rod. In G. Da Prato and M. Iannelli, editors, *Volterra Integrodifferential Equations in Banach Spaces and Applications*, pages 177–183, Harlow, Essex, 1989. Longman Sci. Tech.

[171] K.B. Hannsgen and R.L. Wheeler. Viscoelastic and boundary feedback damping: Precise energy decay rates when creep modes are dominant. *J. Integral Eqns. Appl.*, 2:495–527, 1990.

[172] A. Haraux. *Nonlinear Evolution Equations*, volume 841 of *Lecture Notes in Mathematics*. Springer Verlag, Berlin, 1981.

[173] K. Hashimoto and S. Oharu. Gelfand integrals and generalized derivatives of vector measures. *Hiroshima Math. J.*, 13:301–326, 1983.

[174] M.L. Heard. An abstract semilinear hyperbolic Volterra integrodifferential equation. *J. Math. Anal. Appl.*, 80:175–202, 1981.

[175] M.L. Heard. An abstract parabolic Volterra integrodifferential equation. *SIAM J. Math. Anal.*, 13:81–105, 1982.

[176] M.L. Heard. A class of hyperbolic Volterra integrodifferential equation. *Nonl. Anal.: T.M.A.*, 8:79–93, 1984.

[177] M.L. Heard and S.M. Rankin. Weak solutions for a class of parabolic Volterra integrodiffferential equation. In G. Da Prato and M. Iannelli, editors, *Volterra Integrodifferential Equations in Banach Spaces and Applications*, pages 184–220, Harlow, Essex, 1989. Longman Sci. Tech.

[178] D. Henry. *Geometric Theory of Semilinear Parabolic Equations*, volume 840 of *Lect. Notes Math.* Springer Verlag, Berlin, 1981.

[179] I. Herbst. The spectrum of Hilbert space semigroups. *J. Operator Th.*, 10:87–94, 1983.

[180] E. Hille and R.S. Phillips. *Functional Analysis and Semi-groups*, volume 31 of *Amer. Math. Soc. Colloq. Publ.* Amer. Math. Soc., Providence, Rhode Island, 1957.

[181] N. Hirano. Asymptotic behavior of solutions of nonlinear Volterra equations. *J. Diff. Equations*, 47:163–179, 1983.

[182] F. Hirsch. Familles resolvantes, generateurs, cogenerateurs. *Ann. Inst. Fourier*, 22:89–210, 1972.

[183] J. Hopkinson. The residual charge of a Leyden jar. *Phil. Trans. Royal Soc. London*, 167:599–626, 1877.

[184] J.S. Howland. On a theorem of Gearhart. *Int. Eqns. Operator Th.*, 7:138–142, 1984.

[185] W.J. Hrusa and M. Renardy. On wave propagation in linear viscoelasticity. *Quarterly Appl. Math.*, 43:237–254, 1985.

[186] J.M. Jeong. Spectral properties of the operator associated with a retarded functional differential equations in Hilbert space. *Proc. Japan Acad.*, 65:98–101, 1989.

[187] J.M. Jeong. Retarded functional differential equations with L^1-valued control, 1992. Preprint.

[188] J.M. Jeong. Stabilizability of retarded functional differential equations in Hilbert space, 1992. Preprint.

[189] G.S. Jordan, O.J. Staffans, and R.L. Wheeler. Local analyticity on weighted L^1-spaces and applications to stability problems for Volterra equations. *Trans. Amer. Math. Soc.*, 274:749–782, 1982.

[190] N. Kato. On the asymptotic behavior of solutions of nonlinear Volterra equations. *J. Math. Anal. Appl.*, 120:647–654, 1986.

[191] N. Kato. Unbounded behavior and convergence of solutions of nonlinear Volterra equations in Banach spaces. *Nonl. Anal.: T.M.A.*, 12:1193–1201, 1988.

[192] N. Kato, K. Kobayasi, and I. Miyadera. On the asymptotic behavior of solutions of evolution equations associated with nonlinear Volterra equations. *Nonl. Anal.: T.M.A.*, 9:419–430, 1985.

[193] T. Kato. *Perturbation Theory for Linear Operators*, volume 132 of *Grundlehren Math. Wiss.* Springer Verlag, Berlin, second edition, 1976.

[194] Y. Katznelson. *An Introduction to Harmonic Analysis.* Dover Publ. Inc., New York, second edition, 1976.

[195] T. Kiffe. A perturbation of an abstract Volterra equation. *SIAM J. Math. Anal.*, 11:1036–1046, 1980.

[196] T. Kiffe and M. Stecher. Properties and applications of the resolvent operator to a Volterra integral equation in Hilbert space. *SIAM J. Math. Anal.*, 11:82–91, 1980.

[197] J.U. Kim and Y. Renardy. Boundary control of the Timoshenko beam. *SIAM J. Control Opt.*, 25:1417–1429, 1987.

[198] J.F.C. Kingman. *Regenerative Phenomena.* John Wiley and Sons, London, 1972.

[199] H. Komatsu. Fractional powers of operators. *Pacific J. Math.*, 1:285–346, 1966.

[200] S.G. Krein. *Linear Differential Equations in Banach Space*, volume 29 of *Transl. Math. Monographs.* Amer. Math. Soc., Providence, Rhode Island, 1972.

[201] U. Krengel. *Ergodic Theorems*, volume 6 of *de Gruyter Studies in Mathematics.* de Gruyter, Berlin, 1985.

[202] K. Kunisch and W. Schappacher. Necessary conditions for partial differential equations with delay to generate C_0 semigroups. *J. Diff. Equations*, 50:49–79, 1983.

[203] J.E. Lagnese. *Boundary Stabilization of Thin Plates.* SIAM Studies in Appl. Math. SIAM, Philadelphia, 1989.

[204] J.E. Lagnese and J.L. Lions. *Modelling Analysis and Control of Thin Plates*, volume 6 of *Recherches Math. Appl.* Springer Verlag, Berlin, 1989.

[205] L.D. Landau and E.M. Lifschitz. *Lehrbuch der Theoretischen Physik, Bd.II: Klassische Feldtheorie.* Akademie Verlag, fifth edition, 1971.

[206] L.D. Landau and E.M. Lifschitz. *Lehrbuch der Theoretischen Physik, Bd.VIII: Elektrodynamik der Kontinua.* Akademie Verlag, second edition, 1971.

[207] L.D. Landau and E.M. Lifschitz. *Lehrbuch der Theoretischen Physik, Bd.VII: Elastizitätstheorie.* Akademie Verlag, fifth edition, 1983.

[208] R. Leis. *Initial Boundary Value Problems in Mathematical Physics.* Teubner Verlag, Stuttgart, 1986.

[209] G. Leugering. On boundary feedback stabilization of a viscoelastic membrane. *Dyn. Stability Syst.*, 4:71–79, 1989.

[210] G. Leugering. On the reachability problem of a viscoelasic beam during a slewing maneuver. *Int. Series Numerical Math.*, 91:249–261, 1989.

[211] G. Leugering. On boundary feedback stabilization of a viscoelastic beam. *Proc. Royal Soc. Edinburgh*, 114:57–69, 1990.

[212] G. Leugering. A decomposition method for integro-partial differential equations and applications. *J. Math. Pure Appl.*, 1992. To appear.

[213] J.J. Levin. Resolvents and bounds for linear and nonlinear Volterra equations. *Trans. Amer. Math. Soc.*, 228:207–222, 1977.

[214] B. M. Levitan and V.V. Zhikov. *Almost Periodic Functions and Differential Equations.* Cambridge Univ. Press, Cambridge, 1982.

[215] J.L. Lions. *Équations différentielles-opérationnelles.* Springer Verlag, Berlin, 1961.

[216] C. Lizama. On an extension of the Trotter-Kato theorem for resolvent families of operators. *J. Integral Eqns. Appl.*, 2:269–280, 1990.

[217] C. Lizama. A mean ergodic theorem for resolvent operators. *Semigroup Forum*, 1992.

[218] S.O. Londen. On an integral equation in a Hilbert space. *SIAM J. Math. Anal.*, 8:950–970, 1977.

[219] S.O. Londen. An existence result on a Volterra equation in a Banach space. *Trans. Amer. Math. Soc.*, 235:285–304, 1978.

[220] S.O. Londen and J.A. Nohel. A nonlinear Volterra integrodifferential equation occuring in heat flow. *J. Int. Equations*, 6:11–50, 1984.

[221] A. Lorenzi. An inverse problem in the theory of materials with memory ii. In Ph. Clément, S. Invernizzi, E. Mitidieri, and I. Vrabie, editors, *Semigroup Theory and Applications*, volume 116 of *Lect. Notes Pure Appl. Math.*, pages 261–290, New York, 1989. Marcel Dekker Inc.

[222] A. Lorenzi and E. Sinestrari. An inverse problem in the theory of materials with memory. *Nonlin. Anal.: T.M.A.*, 13, 1989.

[223] A. Lorenzi and E. Sinestrari. Stability results for a partial integrodifferential inverse problem. In G. Da Prato and M. Iannelli, editors, *Volterra Integrodifferential Equations in Banach Spaces and Applications*, pages 271–294, Harlow, Essex, 1989. Longman Sci. Tech.

[224] A. Lunardi. Laplace transform methods in integrodifferential equations. *J. Int. Equation*, 10:185–211, 1985.

[225] A. Lunardi. Regular solutions for time dependent abstract integrodifferential equations with singular kernel. *J. Math. Anal. Appl.*, 130:1–21, 1988.

[226] A. Lunardi. On the linear heat equation with fading memory. *SIAM J. Math. Anal.*, 21:1213–1224, 1990.

[227] A. Lunardi and E. Sinestrari. Existence in the large and stability for nonlinear Volterra equations. *J. Int. Eqns.*, 10:213–239, 1985.

[228] A. Lunardi and E. Sinestrari. Fully nonlinear integrodifferential equations in general Banach space. *Math. Z.*, 190:225–248, 1985.

[229] Y.I. Lyubich and V.Q. Phong. Asymptotic stability of linear differential equations in Banach spaces. *Studia Math.*, 88:37–42, 1988.

[230] R.C. Mac Camy. Nonlinear Volterra equations on a Hilbert space. *J. Diff. Equations*, 16:373–393, 1974.

[231] R.C. Mac Camy and V.J. Mizel. Nonlinear vectorvalued hereditary equations on the line. In S.O. Londen and O.J. Staffans, editors, *Volterra Equations*, pages 206–219, Berlin, 1979. Springer Verlag.

[232] R.C. Mac Camy and J.S.W. Wong. Stability theorems for some functional equations. *Amer. Math. Soc.*, 164:1–37, 1972.

[233] F. Mainardi, editor. *Wave propagation in viscoelastic media*, volume 52 of *Research Notes Math.*, London, 1982. Pitman.

[234] J.E. Marsden and T.J.R. Hughes. *Mathematical Foundations of Elasticity*. Prentice-Hall, Englewood Cliffs, New Jersey, 1983.

[235] R.H. Martin. *Nonlinear Operators and Differential Equations in Banach Spaces*. John Wiley and Sons, New York, 1976.

[236] J.C. Maxwell. On the dynamical theory of gases. *Philos. Trans. Royal Soc. London*, 157:49–88, 1867.

[237] T.R. McConnell. On Fourier multiplier transformations of Banach-valued functions. *Trans. Amer. Math. Soc.*, 285:739–757, 1984.

[238] A. McIntosh. Operators which have an H^∞ functional calculus. In *Miniconference on Operator Theory and PDE*, volume 14 of *Proc. Center Math. Analysis*, pages 210–231, Canberra, 1986. Australian Nat. Univ.

[239] A. McIntosh and A. Yagi. Operators of type ω without a bounded H^∞ functional calculus. In *Miniconference on Operators in Analysis*, volume 24 of *Proc. Center Math. Analysis*, Canberra, 1989. Australian Nat. Univ.

[240] J. Meixner. On the linear theory of heat conduction. *Arch. Rat. Mech. Anal.*, 39:108–130, 1971.

[241] R.K. Miller. On Volterra integral equations with nonnegative integrable resolvents. *J. Math. Anal. Appl.*, 22:319–340, 1968.

[242] R.K. Miller. *Nonlinear Volterra Integral Equations*. Benjamin Inc., Menlo Park, California, 1971.

[243] R.K. Miller. Volterra integral equations in a Banach space. *Funkcial. Ekvac.*, 18:163–194, 1975.

[244] R.K. Miller. An integrodifferential equation for rigid heat conductors with memory. *J. Math. Anal. Appl.*, 66:313–332, 1978.

[245] R.K. Miller and R.L. Wheeler. Asymptotic behavior for a linear Volterra integral equation in Hilbert space. *J. Diff. Equations*, 23:270–284, 1977.

[246] R.K. Miller and R.L. Wheeler. Well-posedness and stability of linear Volterra integrodifferential equations in abstract spaces. *Funkcial. Ekvac.*, 21:279–305, 1978.

[247] J. Milota. Stability and saddle-point property for a linear autonomous functional parabolic equation. *Comm. Math. Univ. Carolinae*, 27:87–101, 1986.

[248] J. Milota. Asymptotic behaviour of parabolic equations with infinite delay. In G. Da Prato and M. Iannelli, editors, *Volterra Integrodifferential Equations in Banach Spaces and Applications*, pages 295–305, Harlow, Essex, 1989. Longman Sci. Tech.

[249] J. Milota and H. Petzeltová. Continuous dependence for semilinear parabolic functional equations without uniqueness. *Časopis Pěst. Mat.*, 110:394–402, 1985.

[250] J. Milota and H. Petzeltová. An existence theorem for semilinear functional parabolic equations. *Časopis Pěst. Mat.*, 110:274–288, 1985.

[251] E. Mitidieri. Asymptotic behavior of the solutions of a class of functional differential equations: remarks on a related Volterra equation. *J. Math. Anal. Appl.*, 1989.

[252] E. Mitidieri and M. Tosques. Volterra integral equations associated with a class of nonlinear operators in Hilbert spaces. *Ann. Fac. Sci. Toulouse Math.*, 8:131–158, 1987.

[253] T. Miyakawa. On nonstationary solutions of the Navier-Stokes equations in exterior domains. *Hiroshima Math. J.*, 12:115–140, 1982.

[254] R.R. Nachlinger and L. Wheeler. A uniqueness theorem for rigid heat conductors with memory. *Quart. Applied Math.*, 31:267–273, 1973.

[255] R. Nagel. *One-parameter Semigroups of Positive Operators*, volume 1184 of *Lecture Notes in Mathematics*. Springer Verlag, Berlin, 1986.

[256] S. Nakagiri. Structural properties of functional differential equations in Banach spaces. *Osaka J. Math.*, 25:353–398, 1989.

[257] C. Navarro. Asymptotic stability in linear thermovisco-elasticity. *J. Math. Anal. Appl.*, 65:399–431, 1978.

[258] J.A. Nohel and D.F. Shea. Frequency domain methods for Volterra equations. *Advances Math.*, 22:278–304, 1976.

[259] R.D. Noren. A linear Volterra integrodifferential equation for viscoelastic rods and plates. *Quarterly Appl. Math.*, 45:503–514, 1987.

[260] R.D. Noren. Uniform L^1-behavior of the solution of a Volterra equation with parameter. *SIAM J. Math. Anal.*, 19:270–286, 1988.

[261] J.W. Nunziato. On heat conduction in materials with memory. *Quarterly Appl. Math.*, 29:187–204, 1971.

[262] J.W. Nunziato. A note on uniqueness in the linear theory of heat conduction with finite wave speeds. *SIAM J. Appl. Math.*, 25:1–4, 1973.

[263] E. Obrecht. Evolution operators for higher order abstract parabolic equations. *Czechoslavak Math. J.*, 36:210–222, 1984.

[264] E. Obrecht. The Cauchy problem for time-dependent abstract parabolic equations of higher order. *J. Math. Anal. Appl.*, 125:508–530, 1987.

[265] V.P. Orlov. On the stability of the zero solution of one one-dimensional mathematical model of a viscoelastisity. Preprint, 1990.

[266] R. Paley and N. Wiener. *Fourier Transforms in the Complex Domain*, volume 19 of *Amer. Math. Soc. Colloq. Publ.* Amer. Math. Soc, Providence, Rhode Island, 1934.

[267] A. Pazy. *Semigroups of Linear Operators and Applications to Partial Differential Equations*. Springer Verlag, Berlin, 1983.

[268] H. Petzeltová. The Hopf bifurcation theorem for parabolic equations with infinite delay. Preprint, 1992.

[269] A.C. Pipkin. *Lectures on Viscoelasticity Theory*, volume 7 of *Appl. Math. Sci.* Springer Verlag, Berlin, 1972.

[270] J. Prüss. On resolvent operators for linear integrodifferential equations of Volterra type. *J. Int. Equations*, 5:211–236, 1983.

[271] J. Prüss. Lineare Volterra Gleichungen in Banach-Räumen. Habilitationsschrift, Universität-GH Paderborn, 1984.

[272] J. Prüss. On the spectrum of C_0-semigroups. *Trans. Amer. Math. Soc.*, 284:847–857, 1984.

[273] J. Prüss. On linear Volterra equations of parabolic type in Banach spaces. *Trans. Amer. Math. Soc.*, 301:691–721, 1987.

[274] J. Prüss. Positivity and regularity of hyperbolic Volterra equations in Banach spaces. *Math. Ann.*, 279:317–344, 1987.

[275] J. Prüss. Bounded solutions of Volterra equations. *SIAM J. Math. Anal.*, 19:133–149, 1988.

[276] J. Prüss. Linear hyperbolic Volterra equations of scalar type. In Ph. Clément, S. Invernizzi, E. Mitidieri, and I.Vrabie, editors, *Semigroup Theory and Applications*, pages 367–384, New York, 1989. Marcel Dekker Inc.

[277] J. Prüss. Regularity and integrability of resolvents of linear Volterra equations. In M. Iannelli G.Da Prato, editor, *Volterra Integrodifferential Equations in Banach Spaces and Applications*, pages 339–367, Harlow, Essex, 1989. Longman Sci. Tech.

[278] J. Prüss. Maximal regularity of linear vector valued parabolic Volterra equations. *J. Integral Eqns. Appl.*, 3:63–83, 1991.

[279] J. Prüss. Quasilinear parabolic Volterra equations in spaces of integrable functions. In B. de Pagter Ph. Clément, E. Mitidieri, editor, *Semigroup Theory and Evolution Equations*, volume 135 of *Lect. Notes Pure Appl. Math.*, pages 401–420, New York, 1991. Marcel Dekker Inc.

[280] J. Prüss. Solvability behaviour of linear evolutionary integral equations on the line. In Ph. Clément, G. Lumer, and B. de Pagter, editors, *Semigroup Theory and Evolution Equations*, Lect. Notes Pure Appl. Math., New York, 1993. Marcel Dekker Inc. To appear.

[281] J. Prüss. Stability of linear evolutionary systems in viscoelasticity. In A. Favini, E. Obrecht, and A. Venni, editors, *Differential Equations in Banach Spaces*, New York, 1993. To appear.

[282] J. Prüss and W. Ruess. Weak almost periodicity of convolutions. *J. Integral Eqns. Appl.* To appear, 1993.

[283] J. Prüss and H. Sohr. On operators with bounded imaginary powers in Banach spaces. *Math. Z.*, 203:429–452, 1990.

[284] J. Prüss and H. Sohr. Boundedness of imaginary powers of second-order elliptic differential operators in L^p. *Hiroshima Math. J.* To appear, 1993.

[285] A. Pugliese. Some questions on the integrodifferential equation $u' = AK * u + BM * u$. In A. Favini, E. Obrecht, and A. Venni, editors, *Differential Equations in Banach Spaces*, pages 227–242, Berlin, 1986. Springer Verlag.

[286] A. Pugliese. Hille-Yosida-type theorems for integrodifferential equations. In Ph. Clément, S. Invernizzi, E. Mitidieri, and I.Vrabie, editors, *Semigroup Theory and Applications*, pages 341–353, New York, 1989. Marcel Dekker Inc.

[287] M. Reed. *Abstract Nonlinear Wave Equations*, volume 507 of *Lect. Notes Math.* Springer Verlag, Berlin, 1976.

[288] M. Renardy. Some remarks on the propagation and nonpropagation of discontinuities in linearly viscoelastic liquids. *Rheol. Acta*, 21:251–254, 1982.

[289] M. Renardy, W.J. Hrusa, and J.A. Nohel. *Mathematical Problems in Viscoelasticity*, volume 35 of *Pitman Monographs Pure Appl. Math.* Longman Sci. Tech., Harlow, Essex, 1988.

[290] G. Reuter. Über eine Volterrasche Integralgleichung mit totalmonotonem Kern. *Arch. Math.*, 7:59–66, 1956.

[291] A. Rhandi. Positive perturbations of linear Volterra equations and sine functions of operators. *J. Integral Eqns. Appl.*, 4:409–420, 1992.

[292] J. Rosenblatt, W.M. Ruess, and D. Sentilles. On the critical part of a weakly almost periodic function. *Houston J. Math.*, 17:237–249, 1991.

[293] W.M. Ruess. Almost periodicity properties of solutions to the nonlinear Cauchy problem in Banach spaces. In Ph. Clément, E. Mitidieri, and B. de Pagter, editors, *Semigroup Theory and Evolution Equations*, New York, 1991. Marcel Dekker Inc.

[294] W.M. Ruess and W.H. Summers. Asymptotic almost periodicity and motions of semigroups of operators. *Lin. Algebra Appls.*, 84:335–351, 1986.

[295] W.M. Ruess and W.H. Summers. Compactness in spaces of vector valued continuous functions and asymptotic almost periodicity. *Math. Nachr.*, 135:7–33, 1988.

[296] W.M. Ruess and W.H. Summers. Integration of asymptotically almost periodic functions and weak asymptotic almost periodicity. *Diss. Math.*, 279, 1989.

[297] W.M. Ruess and W.H. Summers. Weakly almost periodic semigroups of operators. *Pacific J. Math.*, 143:175–193, 1990.

[298] W.M. Ruess and W.H. Summers. Ergodic theorems for semigroups of operators. *Proc. Amer. Math. Soc.*, 114:423–432, 1992.

[299] R. Seeley. Norms and domains of the complex powers A_B^z. *Amer. J. Math.*, 93:299–309, 1971.

[300] D. Sforza. On the Hille-Yosida theorem for integrodifferential equations in Banach spaces. *Boll. U.N.I.*, 5:169–173, 1987.

[301] D. Sforza. Regularity results for an integrodifferential equation of convolution type in Hilbert space. In G. Da Prato and M. Iannelli, editors, *Volterra Integrodifferential Equations in Banach Spaces and Applications*, pages 368–386, Harlow, Essex, 1989. Longman Sci. Tech.

[302] D.F. Shea and S. Wainger. Variants of the Wiener-Levy theorem, with applications to problems for some Volterra integral equations. *Amer. J. Math.*, 97:312–343, 1975.

[303] R.E. Showalter. *Hilbert Space Methods for Partial Differential Equations*. Pitman, London, 1977.

[304] G. G. Simader and H. Sohr. *A New Approach to the Helmholtz Decomposition in L_q-Spaces for Bounded and Exterior Domains*, volume 2 of *Advances in Mathematics for the Applied Sciences*. World Scientific, 1992.

[305] K. Simonyi. *Theoretische Elektrotechnik*, volume 20 of *Hochschulbücher für Physik*. Deutscher Verlag der Wissenschaften, Berlin, seventh edition, 1979.

[306] E. Sinestrari. Continuous interpolation spaces and spatial regularity in nonlinear Volterra integro-differential equations. *J. Int. Equations*, 5:287–308, 1983.

[307] E. Sinestrari. On a class of retarded partial differential equations. *Math. Z.*, 186:223–246, 1984.

[308] M. Slemrod. A hereditary partial differential equation with applications in the theory of simple fluids. *Arch. Rat. Mech. Anal.*, 62:303–321, 1976.

[309] M. Slemrod. An energy stability method for simple fluids. *Arch. Rat. Mech. Anal.*, 68:1–18, 1978.

[310] P.E. Sobolevskii. Fractional powers of coercively positive sums of operators. *Soviet Math. Dokl.*, 16:1638–1641, 1975.

[311] V.A. Solonnikov. Estimates for solutions of nonstationary Navier-Stokes equations. *J. Soviet Math.*, 8:467–523, 1977.

[312] O.J. Staffans. An asymptotic problem for a positive definite operator-valued Volterra kernel. *SIAM J. Math. Anal.*, 9:855–866, 1978.

[313] O.J. Staffans. Some energy estimates for a nondifferentiated Volterra equation. *J. Diff. Equations*, 32:285–293, 1979.

[314] O.J. Staffans. Periodic L^2-solutions of an integrodifferential equation in a Hilbert space. *Proc. Amer. Math. Soc.*, 1992. To appear.

[315] O.J. Staffans. Well-posedness and stabilizability of a viscoelastic equation in energy space. *Research Report, Helsinki University*, A312, 1992.

[316] E. Stein. *Singular Integrals and Differentiability Properties of Functions*. Princeton Univ. Press, Princeton, 1970.

[317] H.B. Stewart. Generation of analytic semigroups by strongly elliptic operators. *Trans. Amer. Math. Soc.*, 192:141–162, 1974.

[318] H.B. Stewart. Generation of analytic semigroups by strongly elliptic operators under general boundary conditions. *Trans. Amer. Math. Soc.*, 259:299–310, 1980.

[319] H. Tanabe. *Equations of Evolution*, volume 6 of *Monographs and Studies in Mathematics*. Pitman, London, 1979.

[320] H. Tanabe. Linear Volterra integral equations of parabolic type. *Hokkaido Math. J.*, 12:265–275, 1983.

[321] H. Tanabe. Remarks on linear Volterra integral equations of parabolic type. *Osaka J. Math.*, 22:519–531, 1985.

[322] H. Tanabe. Fundamental solutions for linear retarded functional differential equations in Banach space. *Funkcial. Ekavac.*, 35:149–177, 1992.

[323] A. Tauber. Ein Satz aus der Theorie der unendlichen Reihen. *Monatshefte Math. Phys.*, 8:273–277, 1897.

[324] A. Tesei. Stability properties for partial Volterra integrodifferential equations. *Ann. Mat. Pura Appl.*, 126:103–115, 1980.

[325] S.P. Timoshenko and S. Woinowsky-Krieger. *Theory of Plates and Shells*. McGraw-Hill, New York, second edition, 1959.

[326] R. Touplin and R. Rivlin. Linear functional electromagnetic constitutive relations and plane waves in a hemihedral isotropic material. *Arch. Rat. Mech. Anal.*, 6:188–197, 1960.

[327] C.C. Travis and G.F. Webb. An abstract second order semilinear Volterra integrodifferential equation. *SIAM J. Math. Anal.*, 10:412–424, 1979.

[328] H. Triebel. *Interpolation Theory, Function Spaces, Differential Operators*. North Holland, Amsterdam, 1978.

[329] K. Tsuruta. Regularity of solutions to an abstract inhomogeneous linear integrodifferential equation. *Math. Japonica*, 26:65–76, 1981.

[330] K. Tsuruta. Bounded linear operators satisfying second-order integrodiffferential equations in a Banach space. *J. Integral Eqns.*, 6:231–268, 1984.

[331] V. Volterra. Sulle equazioni dell'elettrodinamica. *Rend. Accad. Lincei*, 18:203–211, 1909.

[332] V. Volterra. Sulle equazioni integro differenziali della teoria dell'elasticità. *Rend. Accad. Lincei*, 18:295–301, 1909.

[333] V. Volterra. *Lecons sur les équations intégrales et les équations intégro-differentielles*. Gauthier Villars, Paris, 1913.

[334] W. von Wahl. *The Equations of Navier-Stokes and Abstract Parabolic Equations*, volume 8 of *Aspects of Mathematics*. Vieweg Verlag, Braunschweig, 1985.

[335] M. Watanabe. Cosine families of operators and applications. In E. Obrecht A. Favini, editor, *Differential Equations in Banach Spaces*, pages 278–292, Berlin, 1986. Springer Verlag.

[336] G.F. Webb. An abstract semilinear Volterra integrodifferential equation. *Proc. Amer. Math. Soc.*, 69:255–260, 1978.

[337] G.F. Webb. Abstract Volterra integrodifferential equations and a class of reaction-diffusion equations. In S.O. Londen and O.J. Staffans, editors, *Volterra Equations*, pages 295–303, Berlin, 1979. Springer Verlag.

[338] G.F. Webb. *Theory of Nonlinear Age-Dependent Population Dynamics*, volume 89 of *Monogr. Pure Appl. Math.* Marcel Dekker Inc., New York, 1985.

[339] D.V. Widder. *The Laplace Transform.* Princeton Univ. Press, Princeton, 1941.

[340] D.V. Widder. *An Introduction to Transform Theory.* Acad. Press, New York, 1971.

[341] A. Wyler. *Über die Stabilität von dissipativen Wellengleichungen.* PhD thesis, Univ. Zürich, 1992.

[342] A. Yagi. Coincidence entre des espaces d'interpolation et des domaines de puissances fractionnaires d'operateurs. *C.R. Acad. Sci. Paris*, 299:173–176, 1984.

[343] A. Yagi. Applications of the purely imaginary powers of operators in Hilbert spaces. *J. Funct. Anal.*, 73:216–231, 1987.

[344] Y. Yamada. Asymptotic stability for some systems of semilinear Volterra diffusion equations. *J. Diff. Equations*, 52:295–326, 1984.

[345] Y. Yamada and Y. Niikura. Bifurcation of periodic solutions for nonlinear parabolic equations with infinite delays. *Funkc. Ekvac.*, 29:309–333, 1986.

[346] K. Yoshida. The Hopf bifurcation and its stability for semilinear diffusion equations with time delay arising in ecology. *Hiroshima Math. J.*, 12:321–348, 1982.

[347] K. Yosida. *Functional Analysis*, volume 123 of *Grundlehren Math. Wiss.* Springer Verlag, Berlin, third edition, 1971.

[348] J. Zabczyk. A note on C_0-semigroups. *Bull. Acad. Polon. Sci.*, 23:895–898, 1975.

[349] F. Zimmermann. On vector-valued Fourier multiplier theorems. *Studia Math.*, 93:201–222, 1989.

Index